Brain Edema VIII

Proceedings of the Eighth International Symposium
Bern, June 17–20, 1990

Edited by

H.-J. Reulen, A. Baethmann, J. Fenstermacher,
A. Marmarou, M. Spatz

Acta Neurochirurgica
Supplementum 51

Springer-Verlag Wien New York

Professor Dr. Hans-J. Reulen
Department of Neurosurgery, Bern, Switzerland

Professor Dr. Alexander Baethmann
Institut für chirurgische Forschung, Ludwig-Maximilians-Universität, Klinikum Grosshadern, München, Federal Republic of Germany

Professor Dr. Joseph Fenstermacher
Department of Neurological Surgery, State University of New York at Stony Brook, New York, U.S.A.

Professor Dr. Anthony Marmarou
Division of Neurosurgery, Medical College of Virginia, Richmond, Virginia, U.S.A.

Professor Dr. Maria Spatz
Laboratory of Neuropathology and Neuroanatomical Sciences, National Institutes of Health, Bethesda, Maryland, U.S.A.

Proceedings of the Eighth International Symposium on Brain Edema

Honorary Member: I. Klatzo

Organizers: H.-J. Reulen, A. Baethmann, A. Marmarou

Advisory Board: J. Fenstermacher, G. Go, J. T. Hoff, K. A. Hossmann, U. Ito, D. Long, A. R. Shaknovitch, M. Spatz

With 203 Figures

ISSN 0065-1419
ISBN 978-3-7091-9117-0   ISBN 978-3-7091-9115-6 (eBook)
DOI 10.1007/978-3-7091-9115-6

Preface

25 years have passed since a small group met for the First International Symposia on Brain Edema in Vienna. Subsequent Symposia were held in Mainz, Montreal, Berlin, Groningen, Tokyo and Baltimore. During this time we have witnessed a virtual explosion of the number of publications in this field and our basic and clinical understanding of this disease process has increased tremendously. Our meetings have always been a landmark to take stock of our experience so far and to provide perspectives toward future developments. In addition, it always was a good opportunity to renew old friendship and to make new friends.

This volume is a compilation of papers presented at the Eighth International Symposium on Brain Edema held on June 17–20, 1990 in Bern, Switzerland. During this Symposium 158 papers were presented as oral or poster presentations. This considerable number of papers was chosen from more than 230 abstracts that were received. The organizers wish to thank the Advisory Committee for the work done in paper selection and focus on the Symposium. Appreciation is also given to all persons, who have contributed to the success of this meeting, the Chairmen, the participants and last but not least all the staff who worked behind the scene.

If it is to be of value, the written record of the meeting, "the Proceedings" should be quickly available and relatively inexpensive. In order to come close to such an ideal, the authors, the editors and the publishers have to follow clear and strict rules. The manuscripts must be delivered before deadline and must be strictly prepared according to the "Guidelines". This means that any delay as for instance by missing summaries, inappropriate English etc., must be avoided. With the present proceedings, the eighth volume in the series, the authors and editors have made such an effort. Regarding this, the editors extend their compliments to the authors who enabled us to meet the deadline. We also wish to acknowledge our gratitude to Springer-Verlag, Vienna, for their technical aid and for their prompt publication of this volume.

As in the past, brain edema resulting from acute insults, such as disturbances of the cerebral perfusion, or head injury remains a central issue of research. Under these circumstances, the supportive role of biochemically active mediators substances in opening of the blood brain barrier, spread of edema, or in cell swelling continues to elicit great scientific as well as clinical interest. Respective experimental and clinical efforts may eventually produce methods of their specific inhibition as a particularly effective form of therapeutic intervention as compared to what is currently available. It can be hoped thereby to interfere not only with the development of brain edema, but also with the secondary destruction of viable brain parenchyma.

This volume provides a comprehensive survey of the state-of-art of this topic as of June 1990. For the reader's convenience the papers are structured according to the various disease processes which are associated with edema: tumor, trauma, ischemia, hypertension, infection, hydrocephalus etc. We do hope that you enjoy the articles and that they provide impetus and insight for further work.

Prof. Dr. Dr. h. c. K.-J. Zülch, our Honorary Member, one of the pioneers in the study of brain edema, has deceased on 2nd December 1988. It was decided to dedicate this volume to him.

The Ninth International Symposium on Brain Edema will be held in Japan in 1993.

Hans-J. Reulen
Axel Baethmann
Joseph Fenstermacher
Antony Marmarou
Maria Spatz

October 1990

Contents

Pathophysiology III (Formation, Resolution)

Pathophysiology IV (Treatment)

Tumour and Brain Oedema (I)

Tumour and Brain Oedema (II)

Ischaemia and Brain Oedema I (Global Ischaemia)

Ischemia and Brain Oedema II (Focal Ischemia)

Ischaemia and Brain Oedema III

Trauma and Brain Oedema (I)

Trauma and Brain Oedema (II)

Various Disease Processes and Brain Oedema I (Hypertension, Hyperammonemia, Hydrocephalus)

Various Disease Processes and Brain Oedema II (Pseudotumour, Irradiation, Infection, Stress)

Brain Oedema, Mass Effects, CBF

Listed in Current Contents

Acta Neurochirurgica, Suppl. 51, 1–2 (1990)

Prof. Dr. Dr. h.c. Klaus Joachim Zülch

In Memoriam

I. Klatzo

Laboratory of Neuropathology and Neuroanatomical Sciences, National Institute of Neurological Health, Bethesda, Maryland, U.S.A.

Human progress is related to continuous discovery of new horizons by countless scientific workers who contribute their share of effort, though each differing in abilities and style. Some workers are brilliant "breakthrough" types, explosively opening new vistas of science, whereas others, more unassuming, through their consistent, high quality work provide firm launching pads for other explorers to surge forward. I believe, Klaus Joachim Zülch belonged to this latter category.

He was born on April 11th, 1910 in Allenstein, East Prussia and, after graduation from the humanistic type of gymnasium, he studied medicine in the traditional German way by getting exposure from various universities. In his case, he started in Marburg, going then through Rostock, Vienna and Heidelberg, to graduate in medicine from university of Berlin in 1935. The following year he wrote his doctor of medicine thesis on "Primary cerebellar atrophy" in Breslau, under the supervision of the brilliant German neurosurgeon Otfried Foerster, who became responsible for Zülch's persistent interest in two main neurosurgical conditions, namely, brain tumours and brain oedema. To study these problems he acquired solid foundations in neuropathology from Prof. Schaltenbrandt in Würzburg, where Zülch also started his longlasting association with the famous neurosurgeon Wilhelm Tönnis. In 1937 Tönnis asked Zülch to develope a laboratory of experimental neurpathology concerning brain tumours at the Kaiser Wilhelm Institute for Brain Research in Berlin-Buch. This institute headed by Oskar and Cecille Vogt was the world center of brain research and a Mecca for neuroscientists. From Berlin-Buch Zülch had access to probably the richest neuropathological tumour collec-tion at that time at the Charité Hospital in Berlin, which provided the foundation for his later classification of brain tumours. With this persistent interest in neurosurgery, Zülch turned his attention to the study of mechanisms of increased intracranial pressure, a common complication of brain tumours. I believe, Zülch's strong interest and future contributions to the problem of brain oedema started at that time, as he observed so often oedema being responsible for elevation of intracranial pressure in brain tumours.

The war, with its horrors of cranial injuries, increased his recognition that brain oedema represents the most important neurosurgical complication. He was drafted at the beginning of the war and served mainly in Russia, although having occasional opportunities to study head and peripheral nerve injuries at the special military hospital in Hamburg. After the war, Zülch went through several, mainly clinical positions serving as neurologist in Bochum, Hamburg and as visiting professor in Rio de Janeiro, until he finally acquired an appointment at the Max-Planck Institute in Cologne in 1952, becoming its director in 1959.

His aims were high, and were centered on establishing the Max-Planck Insitute in Cologne as one of the preeminent centers for neurological research. To achieve this objective he provided balanced support to both clinical and basic research departments, expecting them to interact and stimulate each other. In Cologne his interests increasingly became directed towards cerebrovascular disorders and he gained reputation as one of the leading clinicians in this field, occasionally being called as consultant in cases involving some very prominent people. Although educated with firm roots in classical, clinical neurology and neuropathology, he was very adaptable and receptive to all new approaches and developments, as well as new ideas, particularly to those deriving from basic research. With Zülch's outstanding ability to attract and select bright young people, the Cologne Institute became a leading research center in cerebrovascular disorders. His steady guidance and foresight accounted for a smooth, well planned transfer of the Intitute's leadership when the time for Zülch's retirement arrived in 1978. To his credit goes the right choice of Professors Hossmann and Heiss, who, since then, have further enhanced recognition of Cologne as one of the outstanding centers for cerebrovascular research.

Concerning Zülch's contributions to the problem of brain oedema, they were many and significant. He started by accepting Reichard's classification of brain oedema into Hirnschwellung (Brain Swelling) and Hirnödem (Brain Oedema), which was prevailing at that time in Germany being based primarily on gross observations and measurements of cranium and brain volumes. He came to realize, however, that Reichard's criteria frequently were difficult to apply and to interpret. He shifted then his interest to studies of oedema on the microscopic level, and this resulted in two comprehensive publications[1, 2], which contained the most detailed and fine microscopic descriptions of oedematous changes in the human brain. Introducing Masson's trichrome stain in his studies, Zülch clearly demonstrated extravasation of serum proteins from leaky vessels and thus he was able to recognize with this approach several types of oedema. He associated pictures of swollen myelinated fibres and oligodendroglia as equivalent to Reichard's Hirnschwellung, whereas in Hirnödem he recognized two forms of microscopic substrates: 1) with slight serum exudation and 2) with heavy protein extravasation, seen mainly in the white matter. In his later papers Zülch provided descriptions and interpretations of dynamics in brain oedemas associated with tumours, inflammations, circulatory disorders, trauma, intoxications, etc., emphasizing their intra- and extracellular character, particularly the preference of the latter for the white matter[3]. This differentiation between intra- and extracellular fluid inspired the later classification of oedema into the cytotoxic and vasogenic types[4].

As a person, Zülch to non-Germans appeared first as a slightly intimidating classical type of German professor, with his Mensuren-scarred face giving an almost militaristic impression. But one had only to meet him a little closer to recognize his warm and friendly nature. He was uncompromisingly honest and straight forward in matters concerning his work and relationship with people. He was totally devoted to his family and to the Cologne Institute, which to him were most important. He believed in universal values of sciences and his friends included pupils and associates from many countries. I believe, his favoured countries were Brazil and the United States. In Brazil he spent time as Visiting Professor in Rio de Janeiro (1951); with the U.S., his ties dated to his ancestors who established a foothold in Texas, the small town near Houston still bearing the name of North Zulch.

Zülch's work and accomplishments in the neurological field brought him many honors and awards. He received an honorary doctorate from the University of Mainz, was named Honorary President of World Congress of Neurology in Hamburg (1985), was awarded the order of Cruzeiro do Sul (1966, Brazil) and of the Rising Sun (1982, Japan), and many other distinctions. Receiving high honors left Zülch, however, unchanged in his modesty and outgoing simplicity in the relationships with his friends and associates. He left many who miss him very much.

References

1. Zülch KJ (1943) Hirnödem and Hirnschwellung. Virchows Arch Path Anat 310: 1–58
2. Zülch KJ (1953) Hirschwellung and Hirnödem. Deutsch Nervenheilkunde 170: 179–208
3. Zülch KJ (1977) Edema in cerebrovascular disease. Bibl Anat (Basel) 16: 238–241
4. Klatzo I (1967) Presidential address: Neuropathological aspects of brain edema. J Neuropathol Exp Neurol 26: 1–13

Correspondence: Dr. I. Klatzo, National Institutes of Health, 9000 Rockville Pike, Building 36, Room 4D04, Bethesda, MD 20892, U.S.A.

Acta Neurochirurgica, Suppl. 51, 3–6 (1990)
© by Springer-Verlag 1990

Pathophysiology I (Cell Volume Regulation, Methods)

Effects of Lactacidosis on Volume and Viability of Glial Cells

F. Staub, A. Baethmann, J. Peters, and O. Kempski

Institute for Surgical Research, Klinikum Großhadern, Ludwig-Maximilians-University, München, Federal Republic of Germany

Summary

The effect of lactacidosis was analyzed *in vitro* by employment of C 6 glioma cells and astrocytes from primary culture. The cells were suspended in an incubation chamber under continuous control of pH, temperature and pO_2. Cell swelling and viability were quantified by flow cytometry using propidium iodide for staining of dead cells. After a control period, the pH of the suspension medium was titrated to levels between pH 6.8 down to 4.2 by addition of isotonic lactic acid. Acidification below pH 6.8 led to an immediate swelling of C 6 glioma cells as well as of astrocytes. The degree of cell swelling was related to the decrease in pH and the duration of exposure. For instance, lactacidosis of 60 min at pH 6.2 resulted in an increase of glial volume to $124.5 \pm 4.6\%$, while pH 4.2 in an increase to $190.9 \pm 8.4\%$. Cell viability remained unchanged down to pH 6.2. At pH 5.6 and below viability decreased in relation to the severity of acidosis. When sulfuric acid was used, the extent of cell swelling at pH 5.6 was only 50% of what was found by addition of lactic acid, whereas cell viability was not differently affected. The results demonstrate a specific efficacy of lactic acid to induce glial swelling, which might be due to a cellular accumulation of the compound.

Introduction

Cerebral ischaemia, seizures, and severe head injuries among others are associated with an enhanced stimulation of anaerobic metabolism and, thereby, development of intra- and extracellular lactacidosis[1, 10, 15]. Lactacidosis is thought to be an important factor which enhances brain damage under these conditions, *e.g.* leading to the formation of cytotoxic brain oedema and, eventually irreversible death of glial cells and neurons[16]. Lactic acid formed in the intracellular compartment may leak from necrotic or severely damaged cells of an ischaemic or traumatic focus, resulting in an extracellular acidosis in the lesion proper and surrounding perifocal brain tissue. It has been shown that in cerebral ischaemia the extracellular pH may fall below pH 6.0 with accumulation of lactic acid to a level of $20–30\,\mu mol/g$[11].

Because of the complexity of brain tissue and the fact that *in vivo* many parameters are affected simultaneously, assessment of molecular mechanisms underlying cell damage is rather difficult. *In vitro* conditions in contrast permit a higher level of control allowing to study the significance of a single pathophysiological factor in isolation. Following that concept, the effect of lactacidosis on swelling and damage of brain cells was investigated under defined conditions *in vitro*. C 6 glioma cells, which have been successfully employed in many studies on glial function, were used as model cells[13]. Many results obtained with this cell line in our laboratory were confirmed in experiments using astrocytes from primary culture[8]. Currently, flow cytometric studies on glial cells were conducted to assess changes of the cell volume in association with cell viability using cell staining with the fluorescent dye propidium iodide as a measure of cell death.

Materials and Methods

The experimental model has been previously described in detail[7, 9]. In brief, C 6 glioma cells originate from a chemically induced glial tumour in rats[2]. These cells have similar properties as glial cells isolated from brain tissue[13]. Glial cells from primary culture were obtained from 3-day-old DB 9 rats according to a modified method of Frangakis and Kimelberg[4]. Both cell types were cultivated as monolayers in Petri dishes using Dulbecco's modified minimal essential medium (DMEM) with 25 mM bicarbonate. The medium was supplemented with 10% fetal calf serum (FCS) and 100 IU/ml penicillin G and 50 µg/ml streptomycin. The cells were grown in a humidified atmosphere of 5% CO_2 and 95% room air at 37 °C. For the experiments only confluent cultures were used. The cells were harvested with 0.05% trypsin-0.02% EDTA in phosphate-buffered saline and washed twice thereafter. After resuspension in serum-free medium the glial cells were transferred into a plexiglas incubation chamber supplied with electrodes to control pH, temperature, and pO_2. A gas-permeable silicon rubber tube in the chamber provided the single cell suspension with a mixture of O_2, CO_2, and N_2. Sedimentation of the cells was prevented by a magnetic stirrer.

The cell volume of glial cells was determined by flow cytometry using an advanced Coulter system with hydrodynamic focusing[5]. Flow cytometry was also employed for measurement of cell viability by uptake of the fluorescent dye propidium iodide (final concentration: 40 µg/ml) for staining of dead cells[14]. Excitation was induced through a 500 nm short pass filter. The maximum emission of propidium iodide at 630 nm was measured by using a 580 nm long pass filter. A window integration system was employed for discrimination of propidium-stained (dead) from non-fluorescing (viable) cells[6].

The experiments were performed after a 45 minute control period used for measurements of normal cell volume, viability, and medium osmolality. The medium buffered by 25 mM bicarbonate was titrated then from pH 7.4 (control) to 6.8, 6.2, 5.6, 5.0, 4.6, or 4.2 by addition of isotonic lactic acid. The pCO_2 in the medium was increased to 80–100 mmHg in order to avoid evaporation of CO_2 through the membrane oxygenator. The cell volume and viability were monitored for 60 minutes during lactacidosis. Parallel experiments were performed using sulfuric acid to assess the specificity of changes caused by lactic acid. In further experiments, cell swelling and viability were studied in sodium-free medium where Na^+-ions were exchanged against choline and bicarbonate against 10 mM HEPES.

Results and Discussion

Acidification of the suspension medium below pH 6.8 led to an immediate volume increase of the C 6 glioma cells. The degree of cell swelling was related to the decrease in extracellular pH and the duration of exposure. At pH 6.2, cell volume increased within 1 min to 111.6 ± 1.3% (mean ± SEM) of control (p < 0.001), at pH 5.0 to 120.7 ± 3.1%, and at pH 4.2 to 131.7 ± 1.9% (p < 0.001). Exposure for 30 min led to a further enhancement of cell swelling. At pH 6.2, cell volume reached 123.4 ± 1.3%, at pH 5.0 it was 145.6 ± 5.0%, and at pH 4.2 even 162.8 ± 4.0%. Cell volume remained apparently constant then for the next half hour of lactacidosis, except for the experiments performed at pH 4.2, where cell swelling continued to 190.9 ± 8.5% of control.

At normal pH under control conditions 89.1 ± 1.1% of the cells were found viable. Cell viability of the C 6 glioma cells was maintained also after decreasing the pH to 6.8, or 6.2. Likewise, the viability of the cells at pH 5.6 was unaffected during the first half hour, but decreased significantly then to 73.4 ± 2.2% after 60 min (p < 0.01). Lowering of the

pH to 5.0, or 4.6 decreased the number of viable cells to 47.8 ± 32.3%, or 40.3 ± 6.0%, respectively. At pH 4.2 only 21.1 ± 10.4% of C 6 cells survived 1 hour of lactacidosis.

When isotonic sulfuric acid was used to lower the pH, the extent of cell swelling at pH 5.6 was only about half of what was found at the same level of acidosis induced by lactic acid. Viability of the C 6 glioma cells, however, was not differently affected in the experiments with lactic or sulfuric acid. The acidosis induced swelling of C 6 glioma cells could be nearly completely inhibited when the experiments were conducted under omission of the Na^+ ions in the medium which were replaced by choline and of bicarbonate replaced by HEPES. Under those conditions, the largest volume increase seen at pH 5.6 was only 104.2 ± 1.5% of control. On the other hand, acidosis of this severity was found to significantly reduce the cell viability as compared to the acidosis experiments conducted in the presence of a normal Na^+-concentration in the medium. The results obtained with C 6 glioma cells could be confirmed in experiments with glial cells obtained from primary culture.

The current studies demonstrate a specific significance of lactic acid in the swelling process. Absence of cell swelling in Na^+-free medium indicates that an influx of Na^+ and Cl^- ion resulting in an increase of osmotic solutes in the cell is a major mechanism of the acidosis induced glial swelling. These findings confirm the previously described concept of events involved in cell swelling upon extracellular acidosis[8, 9]. Accordingly, H^+-ions which are accumulating in the extracellular compartment are buffered by bicarbonate resulting in the formation of CO_2 and water (Fig. 1). CO_2 diffuses into the intracellular compartment resulting in intracellular acidification. This process is catalyzed by carbonic anhydrase which is present in glial cells. H_2CO_3 thereby formed dissociates immediately into H^+ and bicarbonate ions, which are subsequently exchanged against extracellular Na^+ and Cl^- ions by activation of Na^+/H^+- and Cl^-/HCO_3^- antiporters. A futile cycle may develop, where the H^+ ions transported into the extracellular space sustain the above processes with further formation of CO_2, its intracellular diffusion, and intracellular acidification. The resulting net influx of Na^+ and Cl^- ions into the cells increases the intracellular osmotically active solute concentration as the ultimate basis of cell swelling (Fig. 1). The activation of the Na^+/H^+-antiporter by intracellular acidosis can be considered as an attempt to maintain, or reestablish, respectively the normal intracel-

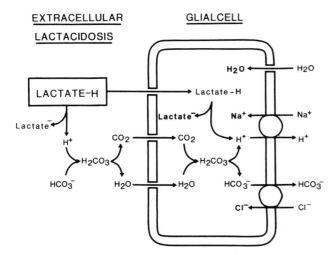

Fig. 1. Mechanisms of glial swelling from lactacidosis. Addition of lactic acid into the extracellular space causes buffering of the H^+-ions by bicarbonate and, consequently, formation of CO_2. Carbon-dioxide diffuses into the glial cells resulting in acidification of the intracellular compartment by formation of H_2CO_3, which is cata-lyzed by carboanhydrase. The H^+- and HCO_3^--ions thereby gen-erated inside the cell are exchanged against extracellular Na^+ and Cl^- ions by activation of the Na^+/H^+- and HCO_3^-/Cl^--anti-porters. Consequently, net amounts of Na^+ and Cl^- are accumu-lating in the cell as a final mechanism of cell swelling, whereas CO_2, H^+ and HCO_3^- ions are recycled into and out of the cells. In addition, entrance of lactic acid in non-polar associated form con-tributes to the glial swelling by two different mechanisms. One is dissociation in the cell and exchange of the H^+ ions against Na^+, the other accumulation of lactate anions, which cannot leave the cell as charged molecules

lular pH. This, however, is brought about at the expense of normal cell volume. Further swelling mechanisms can be considered in the case of lactacidosis.

As shown, the extent of cell swelling induced by lactacidosis was significantly larger than by addition of sulfuric acid at a given pH level. The specificity of the volume increase induced by lactic acid may be ex-plained by a cellular uptake of the acid in non-disso-ciated form. Inside the cell, lactic acid dissociates into H^+ and lactate anions further raising the osmotic load (Fig. 1). The lactic acid anion is in equilibrium with the protonated form. The latter can freely penetrate the cell membrane due to its lipophilicity as compared to the lactate anion[3]. The lactate anion, however, formed after entry of protonated lactic acid can as a polar compound not leave the cell. Both, accumulation of lactate anions together with the sustained operation of the Na^+/H^+-antiporter increase the osmotic solutes in the cell leading to uptake of water and, thus, swelling.

Our data confirm the capability of glial cells to resist even long periods of acidosis without a loss of viability.

Maintenance of a normal cell viability even under in-creased levels of acidosis in the medium may be at-tributed to the efficient control of the intracellular pH at or close to the normal level under these conditions. This conclusion is supported by the significant decline in viability of the C6 glioma cells suspended in acid medium under omission of Na^+ ions. Then, a major mechanism regulating the intracellular pH, the ex-change of extracellular Na^+ ions against H^+ ions by the antiporter was non-functional.

The viability of the glial cells decreased in relation to the level of acidosis and duration of exposure, once the pH fell below a threshold of 5.6. Such low pH values are found *in vivo* only in severe cases of cerebral ischaemia in combination with hyperglycaemia[10]. The decline in cell viability beginning at pH 5.6 to 5.0 is in good agreement with morphological studies of glial cells obtained from primary culture[12] or with respective findings *in vivo*[11]. The present findings on the lactaci-dosis induced cell swelling underline a specific role of lactic acid in the development of cell swelling and cell damage under conditions of an extracellular acidosis in the brain. Consequently, control of the brain tissue pH under respective pathological conditions should be expected to salvage brain tissue liable to perish, such as the penumbra zone around an ischaemic focus.

Acknowledgements

The excellent technical and secretarial assistance of Helga Kleylein, Dennise Linke, and Barbara Stöpfel is gratefully acknowledged. The current project was supported by grants provided by the Deutsche Forschungsgemeinschaft: Ba 452/6–7.

References

1. Andersen BJ, Unterberg AW, Clarke GD, Marmarou A (1988) Effect of posttraumatic hypoventilation on cerebral energy me-tabolism. J Neurosurg 68: 601–607
2. Benda P, Lightbody J, Sato G, Levine L, Sweet W (1968) Dif-ferentiated rat glial cell strain in tissue culture. Science 161: 370–371
3. Boron WF (1983) Transport of H^+ and of ionic weak acids and bases. J Membrane Biol 72: 1–16
4. Frangakis MV, Kimelberg HK (1984) Dissociation of neonatal rat brain by dispase for preparation of primary astrocyte cul-tures. Neurochem Res 9: 1689–1698
5. Kachel V, Glossner E, Kordwig G, Ruhenstroth-Bauer G (1977) Fluvo-Metricell, a combined cell volume and cell fluorescence analyzer. J Histochem Cytochem 25: 804–812
6. Kachel V (1986) Interactive multi-window integration of two-parameter flow cytometric data fields. Cytometry 7: 89–92
7. Kempski O, Chaussy L, Groß U, Zimmer M, Baethmann A (1983) Volume regulation and metabolism of suspended C6 glioma cells: An *in vitro* model to study cytotoxic brain oedema. Brain Res 279: 217–228

8. Kempski O, Staub F, Jansen M, Schödel F, Baethmann A
    (1988 a) Glial swelling during extracellular acidosis *in vitro*.
    Stroke 19: 385–392
 9. Kempski O, Staub F, v Rosen F, Zimmer M, Neu A, Baethmann
    A (1988 b) Molecular mechanisms of glial swelling *in vitro*. Neu-
    rochem Pathol 9: 109–125
10. Kraig RP, Pulsinelli WA, Plum F (1985) Heterogeneous distri-
    bution of hydrogen and bicarbonate ions during complete brain
    ischaemia. In: Kogure K, Hossmann KA, Siesjö BK, Welch FA
    (eds) Progress in brain research, vol 63. Elsevier, Amsterdam,
    pp 155–166
11. Kraig RP, Petito CK, Plum F, Pulsinelli WA (1987) Hydrogen
    ions kill brain at concentrations reached in ischaemia. J Cereb
    Blood Flow Metab 7: 379–386
12. Norenberg MD, Mozes LW, Gregorios JB, Norenberg L-OB
    (1987) Effects of lactic acid on astrocytes in primary culture. J
    Neuropathol Exp Neurol 46: 154–166
13. Pfeiffer SE, Betschart B, Cook J, Mancini P, Morris R (1977)
    Glial cell lines. In: Fedoroff S, Hertz L (eds) Cell, tissue, and
    organ cultures in neurobiology. Academic Press, New York,
    pp 287–346
14. Rothe G, Valet G (1988) Phagocytosis, intracellular pH, and
    cell volume in the multifunctional analysis of granulocytes by
    flow cytometry. Cytometry 9: 316–324
15. Siesjö BK, v Hanwehr R, Nergelius G, Nevander G, Ingvar M
    (1985) Extra- and intracellular pH in the brain during seizures
    and in the recovery period following the arrest of seizure activity.
    J Cereb Blood Flow Metab 5: 47–57
16. Siesjö BK (1988) Acidosis and ischaemic brain damage. Neu-
    rochem Pathol 9: 31–88

Correspondence: Dr. F. Staub, Institute for Surgical Research, Klinikum Großhadern, Ludwig-Maximilians-University, Marchion-inistrasse 15, D-8000 München 70, Federal Republic of Germany.

Acta Neurochirurgica, Suppl. 51, 7–10 (1990)

Mechanisms Underlying Glutamate-induced Swelling of Astrocytes in Primary Culture

P. H. Chan and **L. Chu**

CNS Injury and Edema Center, Department of Neurological Surgery and Neurology, School of Medicine, University of California, San Francisco, California, U.S.A.

Summary

The effects of glutamate and its agonists and antagonists on the swelling of primary astrocytes were studied. Glutamate, aspartate, homocysteate, and quisqualate caused a significant increase in astrocytic swelling as measured by 3-0-methyl-$[^{14}C]$-glucose uptake and by phase-contrast microscopic observations. N-methyl-D-aspartate and kainate were not effective, nor were the competitive NMDA receptor antagonists. Ketamine and MK-801, the non-competitive NMDA receptor antagonists protected the cultured astrocytes against glutamate-induced swelling. Moreover, Glu-induced astrocytic swelling was significantly reduced when sodium-ions were depleted from the culture medium. The sodium ion dependence of Glu-induced astrocytic swelling is rather specific, since depletion of Ca^{2+} or Mg^{2+} had no effect. Our data suggest that Na^+-dependent, Glu-mediated cell swelling occurs in primary cell cultures of astrocytes.

Introduction

Glutamate has been implicated in the pathophysiology of neuronal cell death in hypoxic-ischaemic brain injury (Rothman 1984, Choi 1987) and in epileptic brain damage and other neurodegenerative disorders (Rothman and Olney 1986). We have demonstrated previously that exogenous Glu and its dicarboxylic acid analogs cause cell swelling and decrease the extracellular [inulin]space in brain slices (Chan et al. 1979). Recently a great deal of effort has been devoted to the mechanisms of neuronal cell death involved with excitotoxins (Rothman and Olney 1986, Choi 1987). It has been suggested that NMDA receptors are involved in the Glu-induced neuronal cell death, since blockage of NMDA receptors with antagonists provides protection against Glu neurotoxicity in both, in vitro neuronal cell cultures (Choi et al. 1988) and ischaemic and hypoglycaemic cell death in hippocampal neurons in vivo (Simon et al. 1984, Wieloch 1985). However, the role of Glu in glial swelling is not clear. Besides an extremely high capacity of astrocytes to take up extracellular Glu (Yu et al. 1986), Glu and several of its agonists could depolarize the cell membrane of astrocytic cultures (Bowman and Kimelberg 1984). These studies have, thus, raised the question of whether Glu receptor-mediated excitotoxic damage may occur in astrocytes. The present studies aim to elucidate the mechanism of Glu-induced swelling of astrocytes in primary cultures.

Materials and Methods

Primary cultures of cerebral cortical astrocytes were prepared from newborn Sprague-Dawley rats as described previously (Yu et al. 1986, 1989). The cultures were used for experiments between ages of 4 to 5 weeks old. At this stage, more than 95% of the cells are astrocytes and more than 98% are viable cell as judged by glial fibrillary acidic protein antibody staining and trypan blue exclusion, respectively. The intracellular water space of intact astrocytes in culture was measured by the method of Kletzien et al. (1975). The membrane integrity and cell viability were determined by measuring lactate dehydrogenase (LDH, EC 1.1.1.27) activity in culture medium. The cell viability was studied using trypan blue staining. The cells were observed directly under the phase-contrast microscope and photos were taken.

Results

The time course of Glu effects on astrocytic swelling were studied and are shown in Fig. 1. Glu at 1.0 mM concentration caused a transient increase in intracellular water space (IWS) at 1 and 4 hr; it returned to normal at 24 hr. The morphological studies confirmed the transient biochemical changes in IWS, where swelling of astrocytes induced by Glu was observed (Fig. 1). As observed under phase-contrast microscopy, the swelling of astrocytes was characterized by a gradual

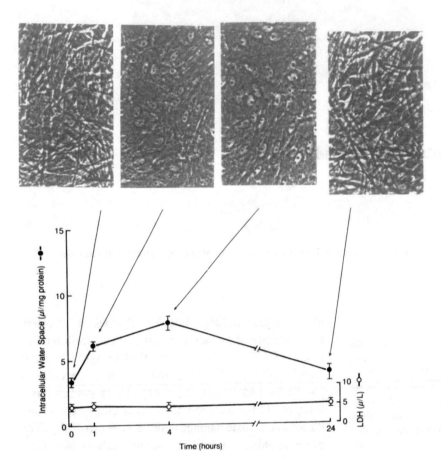

Fig. 1. A time-course study. Upper panel: Morphological appearance of astrocytic swelling observed under phase-contrast microscopy. Bar equals 40 μM. Lower panel: Intracellular water space of astrocytes and LDH levels in cultured medium. Glu concentration: 1 mM

decrease in appearance of astrocytic cell bodies and processes and the obvious and swollen nuclei at 1 hr following the addition of Glu. This swelling persisted for up to 8 hr, and all volume gradually returned to normal at 24 hr (Fig. 1).

The level of LDH released in the incubation medium remained normal throughout the time course studied, indicating the integrity of the cellular membranes and cell viability despite swelling. Furthermore, there was no observable trypan blue staining of the cells, again suggesting that the cells were viable despite the appearance of swelling. Incubation with Glu, Asp, HCA, and Quis at 1 mM caused an increase in IWS by 227%, 221%, 293%, and 209%, at 4 hr, respectively. Glu receptor agonists KA and NMDA and quinolinate were not effective in inducing astrocytic swelling (Fig. 2). The NMDA receptor agonists D-APV and D-APH and the Glu nonspecific receptor antagonist Kynu were not able to induce astrocytic swelling. Furthermore, preincubation with D-APV and D-APH and with Kynu 1 hr prior to the addition of Glu failed to inhibit astrocytic swelling measured at 4 hr. However, MK-801 [(+)-5-methyl-10,11-dihydro-5 H-dibenzo[a,d] cyclohepten-5,10-imine maleate], and ketamine, both noncompetitive antagonists of NMDA receptors, when prein-

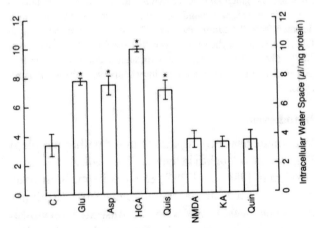

Fig. 2. Effects of glutamate agonists on the swelling of cultured astrocytes; *p < 0.01, using analysis of variance. Mean ± S.E. Agonists concentration: 1 mM

cubated with astrocytes at 1.0 mM concentration completely inhibited the Glu-induced IWS (Chan and Chu 1989, Chan et al. 1990). MK-801 or ketamine alone was not effective.

The ionic dependence of swelling of astrocytes induced by Glu was further investigated. Depletion of Mg^{2+} or Ca^{2+} in the incubation medium did not offer any protection against swelling nor did it exacerbate the swelling. However, Na^+ ions, when depleted from the incubation medium, significantly reduced the Glu-induced swelling.

Discussion

The present studies have demonstrated that the intracellular water space (measured by the uptake of 3-MG of cultured astrocytes) was increased by the exposure to Glu. The induction of astrocytic swelling appeared to be dose-dependent, since lower concentrations (e.g., 0.1 mM) were not effective. Our data have shown nevertheless that Glu at 0.5 mM exhibited a swelling-inducing effect on cultured astrocytes, although to a lesser degree. Using a similar astrocytic culture derived from newborn rats, Choi et al. (1987) reported a lack of Glu gliotoxicity when the cultured astrocytes were exposed briefly (for 5 min) to 0.5 mM Glu and the morphology was studied under phase-contrast microscopy 24 hr later. A lack of Glu effects on astrocytic swelling at this time point would coincide with the transient and reversible action of Glu on intracellular water space observed in our studies.

Among the Glu agonists, Asp, HCA, and Quis caused astrocytic swelling whereas KA, NMDA, and Quin were ineffective. The reason for the divergent action of Glu agonists is not clear at present. One possibility is that depolarization of the membrane by Glu would open up ion channels, which leads to an uptake of water and astrocytic swelling (Kimelberg and Ransom 1986). However, KA was equally effective in inducing membrane depolarization whereas it was ineffective in causing astrocytic swelling. Also, Quis induced astrocytic swelling but only slightly affected membrane depolarization, suggesting that the mechanism of membrane depolarization can not totally account for the divergent transient excitotoxic action of Glu agonists on astrocytic swelling.

The evidence for an exclusion of the involvement of the NMDA receptor in Glu-induced astrocytic swelling was complicated by the fact that the noncompetitive NMDA receptor antagonists, MK-801 and ketamine, completely abolished the Glu-induced swelling in cultured astrocytes. MK-801 and ketamine exert protection of ischaemic neurons by binding to the NMDA-associated cation channels. However, whether this NMDA-receptor-prefering noncompetitive antagonist exerts its glioprotective effect through its binding to the ionic channels is not known at present, since experimental data are lacking that MK-801 does bind to glial membranes and exerts glial protective effects.

An alternative explanation is that MK-801, and ketamine, block the ion channels independent of NMDA receptors (Chan and Chu 1989, Chan et al. 1990). This explanation is more likely since astrocyte swelling usually occurs by an increased Na^+ influx followed by the influx of osmotically obligated water. Thus the blockage of Glu-induced Na^+ influx by MK-801 and by ketamine would reduce the astrocyte swelling. Evidence provided from recent studies that Glu opens Na^+ channels and increases Na^+ uptake in cultured astrocytes is in line with this suggestion (Kimelberg 1987, Sontheimer et al. 1988). Our recent studies also indicate that sodium ions, rather than Ca^{2+} or Mg^{2+}, play an important role in Glu-induced swelling, since the depletion of sodium ions (replaced by choline chloride) in an isoosmotic medium completely inhibited the Glu-induced swelling in astrocytes.

Acknowledgements

We wish to thank Professor Robert A. Fishman for his comments. This work was supported by NIH grants NS-14543 and NS-25372. We thank Marilyn Stubblebine for editorial assistance.

References

1. Bowman CL, Kimelberg HK (1984) Excitatory amino acid directly depolarizes rat brain astrocytes in primary culture. Nature 311: 656–659
2. Chan PH, Fishman RA, Lee JL, Candelise L (1979) Effects of excitatory neurotransmitter amino acids on swelling of rat brain cortical slices. J Neurochem 33: 1309–1315
3. Chan PH, Chu L (1989) Ketamine protects cultured astrocytes from glutamate-induced swelling. Brain Res 487: 380–383
4. Chan PH, Chu L, Chen S (1990) Effects of MK-801 on glutamate-induced swelling of astrocytes in primary cell culture. J Neurosci Res 25: 87–93
5. Choi DW (1987) Ionic dependence of glutamate neurotoxicity. J Neurosci 7: 369–379
6. Choi DW, Koh J-Y, Peters S (1988) Pharmacology of glutamate neurotoxicity in cortical cell culture: Attenuation by NMDA antagonists. J Neurosci 8: 185–196
7. Kimelberg HK (1987) Anisotonic media and glutamate-induced ion transport and volume responses in primary astrocyte cultures. J Physiol (Paris) 82: 294–303
8. Kimelberg HK, Ransom BR (1986) Physiological and pathological aspects of astrocytic swelling. In: Federoff S, Vernadakis A (eds) Astrocytes, vol 3. Academic Press, Orlando, FL, pp 129–166
9. Kletzien RF, Pariza MW, Becker JE, Patter VR (1975) A method using 3-0-methyl-D-glucose and phloretin for the determination of intracellular water space of cells in monolayer culture. Anal Biochem 68: 537–544
10. Rothman S (1984) Synaptic release of excitatory amino acid neurotransmitter mediates anoxic neuronal death. J Neurosci 4: 1884–1891
11. Rothman SM, Olney JW (1986) Glutamate and pathophysiology of hypoxic-ischaemic brain damage. Ann Neurol 19: 105–111

12. Simon RP, Swan JH, Griffiths T, Meldrum BS (1984) Blockade of N-methyl-D-aspartate receptors may protect against ischaemic damage in the brain. Science 226: 850–852

13. Sontheimer H, Kettenmann H, Backus KH, Schachner M (1988) Glutamate opens Na^+/K^+ channels in cultured astrocytes. Glia 1: 328–336

14. Wieloch T (1985) Hypoglycaemia-induced neuronal damage prevented by an N-methyl-D-aspartate antagonist. Science 230: 681–683

15. Yu ACH, Chan PH, Fishman RA (1986) Effects of arachidonic acid on glutamate and gamma-aminobutyric acid uptake in primary cultures of rat cerebral cortical astrocytes and neurons. J Neurochem 47: 1181–1189

16. Yu ACH, Gregory GA, Chan PH (1989) Hypoxia-induced dysfunctions and injury of astrocytes in primary cell cultures. J Cereb Blood Flow Metab 9: 20–28

Correspondence: Dr. P. H. Chan, CNS Injury and Edema Center, Department of Neurological Surgery and Neurology, Box 0114, School of Medicine, University of California, San Francisco, CA 94143, U.S.A.

Acta Neurochirurgica, Suppl. 51, 11–13 (1990)

The Mechanical Filtration Coefficient (L_p) of the Cell Membrane of Cultured Glioma Cells (C_6)

M. Tomita, F. Gotoh, N. Tanahashi, M. Kobari, T. Shinohara, Y. Terayama, T. Yamawaki, B. Mihara, K. Ohta, and A. Kaneko[1]

Department of Neurology, School of Medicine, Keio University, Tokyo and [1] National Institute for Physiological Sciences, Okazaki, Japan

Summary

The mechanical filtration coefficient (L_p) of the membrane of cultured glioma cells was determined from the rate of swelling of the cells. The swelling was induced by exposing the cells to distilled water. Assuming that cells swell in a symmetrical manner, L_p was 2.2×10^{-8} cm/s·mmHg, or 1.7×10^{-4} μm/s·cmH$_2$O when calculated from changes of the cell diameter.

Introduction

Glial swelling is considered to play an essential role in cytotoxic brain oedema. We have attempted to estimate the hydraulic conductivity of the cell membrane of cultured glioma cells (C_6) in terms of the mechanical filtration coefficient (L_p).

Theory

According to the theory of nonequilibrium thermodynamics (Katchalsky and Curran 1967), the volume flow (J_v) across a membrane is expressed as

$$J_v = L_p(\triangle P - \Sigma \sigma \triangle \Pi) . \tag{1}$$

For fluids consisting of small ions, such as Na$^+$, K$^+$, and Cl$^-$, intracellular proteins (pr), and water under the condition of $\triangle P = 0$, Eq. 1 develops to

$$J_v = L_p(\sigma_{pr} \triangle \Pi_{pr} - \sigma_{ion} \triangle \Pi_{ion}) . \tag{2}$$

Since the protein concentration in the extracellular fluid is negligibly small, and $\sigma_{pr} = 1$, this equation is practically

$$J_v = L_p\{\Pi_{pr} - (\sigma_{ion}(\Pi(e)_{ion} - \Pi(i)_{ion})\} \tag{3}$$

where (e) denotes "extracellular" and (i) "intracellular". If $J_v = 0$, Leaf's double Donnan equilibrium (Leaf 1956) applies in which the high intracellular colloid osmotic pressure is counterbalanced by the osmotic pressure of small ions extruded by the membrane. In a previous communication, we reported that the intracellular colloid osmotic pressure of brain cells using our electronic osmometer was $\Pi_{pr} = 213.3$ mmHg (Tomita *et al.* 1988). Under the experimental conditions described below, $\Pi(e)_{ion}$ is the ionic composition (mEq/l) of the medium (Na : 155, K : 5, Cl : 118 or 278 mEq/l *in toto*), while $\Pi(i)_{ion}$ is that of the intracellular fluid, which was reported to be Na : 17, K : 156, Cl : 4 or 177 mEq/l *in toto* (Tomita *et al.* 1988). Since 1 mol of solute in solvent exerts 22.4 atm of pressure (or 22.4×760 mmHg), we have

$$J_v = 0 = L_p\{(213.3 - \sigma_{ion}(0.278 - 0.177) \times 22.4 \times 760\} . \tag{4}$$

Therefore, $\sigma_{ion} = 0.124$.

When the cells are exposed to a large osmotic gradient by replacing the DMEM with distilled water, the volume flow J_v will be

$$J_v = L_p\{213.3 + 0.124(0.177 \times 22.4 \times 760)\} . \tag{5}$$

$$\text{or } J_v = L_p \times 586.9 . \tag{6}$$

The *a priori* assumptions in the following experiments are as follows; 1. the cell membrane is so flaccid that it provides no restriction during the process of cellular swelling; 2. fluids facing the cell membrane are uniform, consisting of proteins, small ions and water; 3. the average reflection coefficients of the membrane for the small ions (σ_{ion}) are the same and constant as a whole.

The cell swelling or volume flow into the cell can be quantitatively measured in terms of (dV/dt)/A or practically as $\triangle V/A$, where $\triangle V$ is the volume increase of the cell per unit time and A the surface area of the cell. When the diameter change of the cell ($\triangle D$) is measured, $\triangle V/A$ can be calculated as

$$\Delta A = (1/6)\pi\{(D + \triangle D)^3 - D^3\}/\pi D^2 = \triangle D/2 . \qquad (7)$$

It is evident that $J_v = \triangle V/A$. Therefore, we can obtain the value of L_p if we have $\triangle D$ as a measure of cell volume change.

$$L_p = (\triangle D/2)/586.9 \, [\text{cm/s} \cdot \text{mmHg}] . \qquad (8)$$

Method

We used cultured glioma C_6 cells (provided via the following route: Nierenberg Inst., NIH – Prof. Haruhiro Higashida, Department of Biophysics, Neuro-information Research Institute, Kanazawa University School of Medicine – Prof. Yukuta Nagata, Department of Physiology, Fujita-Gakuen University School of Medicine). The cells were cultured in flasks containing 10% Dulbecco's modified Eagle's medium plus 10% Gibco newborn calf serum (DMEM) at 37 °C in 5% CO_2 and room air. Immediately before the experiment, the cells were harvested and placed in a culture dish mounted on the stage of an inverted microscope with phase-contrast optics (Nikon, TMD), so that a few cells could be seen in the high-power field of an inverted microscope. The cells were videotaped at 30 frames/s while being superperfused with DMEM in a system constructed by one of the authors (A.K.). A superfusing solution was introduced into the dish forming as an observation chamber through a fine plastic tube from a reservoir placed at a higher position and the flow was stopped by applying a clamp to the tube. First, the response time or delay of the superfusion system was examined. While the optical density of the solution in the observation chamber was continuously recorded with a small silicon photodiode attached on top of the observation chamber, a diluted carbon black solution was introduced into the inlet of the tube of the superfusing system. From the response curve to the step change, the delay was calculated to be 20 s based on the equation

$$MTT = \bar{t}_{ap} + \int_{0}^{\infty} (1\text{-}H(t))dt . \qquad (9)$$

where H(dt) is the cumulative function obtained by making the response curve dimensionless.

After a control period with superfusion of the cells with DMEM, the connecting tube was switched to a reservoir of distilled water. The changes in the cell volume were continuously observed and videotaped before and after exposure to distilled water. By playing back the videotape, $\triangle V/A$ was measured during cell swelling.

Results

When the superfusing fluid was switched from DMEM to distilled water to expose the cells to a large osmotic gradient, the cells began to swell. Figure 1 demonstrates the process of cell swelling (control and at 50 s, 70 s, and 90 s). The cells showed rapid swelling: not only the cell body but also the nucleus and intracellular small organelles were swollen. The cells finally exploded, releasing intracellular fluid together with small organ-

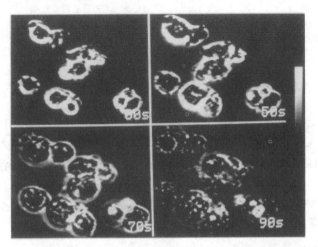

Fig. 1. Swelling of cultured glioma cells (C_6). The cells were suspended in DMEM (00 s); 30 s after exposure to distilled water (50 s); 50 s in distilled water (70 s); and 70 s in distilled water (90 s) (there was a delay time of 20 s; see text for further explanation). It must be noted that swelling is apparent not only in the cell body, but also in all intracellular organelles

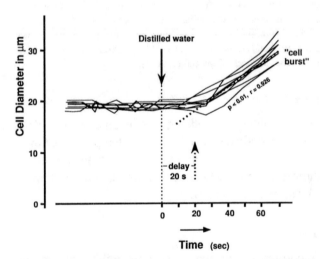

Fig. 2. Cell diameter change plotted against time in sec. The solid arrow indicates the time at which the tube for the perfusion fluid was connected to the reservoir of distilled water from DMEM. Note that the diameter change was linear with time when the starting point was adjusted for the delay by the dotted arrow

elles. Figure 2 shows the diameter change of the cells plotted against time in sec. It should be noted that the diameter increase was linear with time ($p < 0.01$, $r = 0.926$) when adjusted to the delay of 20 s. The constant increase in diameter continued until the cell exploded. The value of the mean diameter change ($\triangle D/2$) was $(0.13 \pm 0.16) \times 10^{-4}$ cm/s obtained from equation 8:

$$L_p = (\triangle D/2)/586.9 \qquad (10)$$
$$= 2.2 \times 10^{-8} \, [\text{cm/s} \cdot \text{mmHg}] .$$

Discussion

The value for the mechanical filtration coefficient (L_p) obtained from the present experiment was 2.2×10^{-8} cm/s · mmHg. The units refer to the volume of water in cm^3 that will enter, in 1 s, $1 cm^2$ of cell surface exposed to a concentration difference equivalent to 1 mmHg. For the sake of comparison, since a wide variety of units have been used for L_p in the literature, the present value may be expressed as 1.7×10^{-4} µm/s · cmH$_2$O, or 1.65×10^{-11} cm^3/dyne · s. This figure agrees almost with, but is slightly larger than the 0.92×10^{-11} cm^3/dyne · s for human red blood cells given by Sidel and Solomon (1957). Their original value was 0.23 cm^4/osm,s which was converted to the above units by Katchalsky and Curran (1967). The figure is also larger than a value of 0.15–1.5×10^{-4} µm/s · cmH$_2$O estimated for the single capillary of frog mesenterium (Mitchel 1980). This may reflect the fact that the capillary wall consists of two endothelial layers of luminal and antiluminal membranes.

The linear increment of cell diameter shown in Fig. 2 suggests that the cell membrane is by no means rigid but rather extremely flaccid. From the unchanged membrane characteristics during the process of swelling, we speculate that the membrane is folded or wrinkled under physiological conditions. As the cell expands, the membrane tends to straighten and at the limit, it would become torn. Another unexpected finding in the present experiment was that there appears to be no concentration gradient in the swelling cell, since the intracellular organelles were swelling at the same rate as the cell body.

References

Katchalsky A, Curran PF (1967) Nonequilibrium thermodynamics in biophysics. Harvard University Press, Cambridge

Mitchel CC (1980) Filtration coefficients and osmotic reflexion coefficients of the cells of single frog mesenteric capillaries. J Physiol 309: 341–355

Sidel VW, Solomon AK (1957) Entrance of water into human red cells under a osmotic pressure gradient. J Gen Physiol 41: 243–257

Tomita M, Gotoh F, Kobari M (1988) Colloid osmotic pressure of cat brain homogenate separated from autogenous CSF by a copper ferrocyanide membrane. Brain Res 474: 165–173

Tomita M, Gotoh F, Tanahashi N, Kobari M, Terayama Y, Mihara B, Ohta K (1987) Thermodynamic energy for maintaining brain volume and preventing swelling. J Cereb Blood Flow Metabol 7 [Suppl 1]: S 122

Correspondence: M. Tomita, M.D., Department of Neurology, School of Medicine, Keio University, Tokyo, Japan.

Acta Neurochirurgica, Suppl. 51, 14–16 (1990)

Swelling of Astroglia *in vitro* and the Effect of Arginine Vasopressin and Atrial Natriuretic Peptide

M. R. Del Bigio and **S. Fedoroff**

Department of Anatomy, The University of Saskatchewan, Saskatoon, Canada

Summary

Numerous neuropathological conditions are accompanied by astroglial swelling which may contribute to secondary neuronal damage. From 4–27 days *in vitro*, swelling of astroglial cells was induced by hypo-osmolar solutions and high K^+ solutions and the cells were examined by microscopy and flow cytometry. Regulatory volume decrease was observed beginning after approximately 30 minutes and was accompanied by increased membrane activity. Arginine vasopressin caused gradual persistent astroglial swelling. Atrial natriuretic peptide caused retraction of the astroglial cell processes but did not significantly alter the swelling induced by hypo-osmolar or elevated K^+ solutions. The membrane mechanisms responsible for the control of astroglial cell volume did not appear to be specific for mature astroglia, rather they are present at all developmental stages.

Introduction

Brain oedema is believed to be one of the major causes of secondary neuronal injury and astroglial swelling plays an important role in the pathophysiology of oedema. Astroglial cells in culture can be induced to swell by changes in the extracellular environment including hypo-osmolarity, elevation of K^+, acidosis, or hypoxia[6]. Arginine vasopressin (AVP) and atrial natriuretic peptide (ANP) have been implicated in regulation of brain water, AVP tending to elevate water content and ANP tending to decrease it[4]. Astrocytes in culture respond to and have specific receptors for both of these hormones[2, 3]. We have studied the effect of these two hormones on astroglial swelling at various developmental stages *in vitro*.

Materials and Methods

The neopallia of newborn Swiss mice were dissociated and the cells were plated on plastic dishes in modified Eagle's Minimum Essential Medium containing 5% adult horse serum and cultured at 37 °C in a humid atmosphere of 5% CO_2 in air. For experiments the growth medium was replaced by a control solution containing the same concentrations of salts and glucose. This was then replaced by a hypo-osmolar (200 mOsm) solution or one containing 15 or 60 mM K^+[8]. AVP or rat ANP were added to the control solution at 10^{-5} or 10^{-6} M concentrations.

After 4–27 days *in vitro* the cells were viewed during and after these changes using a phase contrast optical microscope housed in an incubating chamber and equipped with a video camera and recorder. After 60–90 minutes the experiment was terminated and the cultured cells were fixed in methanol for glial fibrillary acidic protein (GFAP) immunohistochemical study or in 1% glutaraldehyde and 0.5% osmium tetroxide in phosphate buffer for electron microscopy.

Flow cytometric analysis was performed with a Coulter EPICS V flow cytometer using incident laser light at 488 nm. Forward angle light scatter (FALS) information was collected. Trypsin (0.025%) was used to release the astroglial cells into a suspension which was divided into three aliquots. Sequential cytometric data was obtained for parallel control, 60 mM K^+, or 190 mOsm solutions over 60 minutes. In some experiments AVP or ANP (10^{-5} M) were added to the solutions. The results were tested by paired T-tests.

Results

The cultured brain cells developed in monolayer colonies, and the majority were GFAP immunoreactive indicating an astroglial lineage. The central cells of the colonies are the immature dividing glioblasts and these mature peripherally into stellate or senescent polygonal astroglia[5]. The cells were observed under the phase contrast microscope to be fairly stable in the control solution with only minimal ruffling of their membranes. Scanning and transmission electron microscopy revealed relatively flat surfaces with evenly distributed small microvilli.

The hypo-osmolar medium caused the cells to swell. Within 1 minute their membranes were smooth and bulging and the cytoplasm was lucent. After 15–30

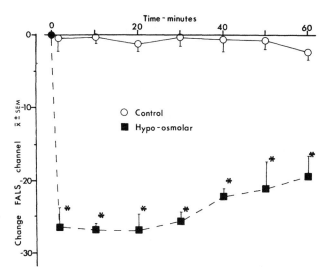

Fig. 1. Graph depicting the change in forward angle light scatter (FALS) by suspended astroglia following the exposure to a hypo-osmolar medium. As compared to the cells in the control solution, the cells swelled rapidly and at 30 minutes began to show a regulatory volume decrease. *, p < 0.01

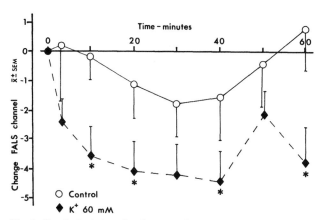

Fig. 2. Graph depicting the change in light scattering by suspended astroglial cells following exposure to a high K^+ concentration environment. Compared with the hypo-osmolar environment (Fig. 1), the cells swelled more gradually and to a lesser degree. The regulatory volume decrease was later and more subtle. *, p < 0.05 compared with parallel controls

minutes the cells began to retract to their previous shape. The application of 15 or 60 mM K^+ caused mild lateral spreading of the cell membranes, especially the fine peripheral processes, within 5 minutes. This was followed by retraction of the same membranes after 30 minutes. During this period there was an increase in the number of long microvilli and in the ruffling of the cell edges. AVP caused ruffling and subtle spreading of the cell membranes noticeable after 20–30 minutes. ANP caused increased ruffling and retraction of the

cell processes. These changes were similar regardless of the developmental stage of the astroglial cells.

When swelling was induced, suspended astroglia changed the pattern of light scatter measured by flow cytometry. Within 1 minute in the hypo-osmolar environment, FALS dropped precipitously then began to increase slowly after 20–30 minutes (Fig. 1). Following the addition of 60 mM K^+, there was a significant but more gradual and less pronounced change in FALS (Fig. 2). AVP 10^{-5} M caused a significant decrease in FALS of -8.36 ± 2.37 channels at sixty minutes. The addition of ANP 10^{-5} M caused no significant change in light scatter. Nor did the preincubation of the cells in ANP significantly modify the response to either the hypo-osmolar or the K^+ solutions.

Discussion

The swelling of astroglial cells has been studied by others through the application of hypo-osmolar solutions to mimic extracellular oedema or elevated K^+ solutions to mimic the release of potassium by damaged neurons. A comparison of these experimental conditions indicates that the induced changes differ both in magnitude and in morphology. When suspended cells swell due to osmotic stress, the scattering of laser light measured by flow cytometry tends to decrease, perhaps due to changes in the contour of the membrane[7] or due to dilution of the cytoplasmic contents. The subsequent regulatory volume decrease is associated with large microvilli and increased membrane motility, perhaps reflecting the molecular pump activity required to adjust the internal environment of the cells.

Consistent with the results of previous authors[4], AVP caused astroglial swelling. ANP caused retraction of astroglial processes when the cells were in monolayers. However, in suspension this hormone failed to modify the change in volume of the cells due to changes in the K^+ or osmolality. This may reflect trypsin-induced damage sustained by the receptors or the observation that ANP receptors on astrocytes are more concerned with regulation of ANP access to neurons than with ion fluxes across astroglial membranes[1].

In conclusion, the membrane processes that occur during the swelling of astroglia *in vitro* do not appear to be specific for the mature cells, rather they occur at all developmental stages. Care must be taken when comparing different models of cell swelling. Various hormones may play a role in astroglial volume regulation and further work is needed in this regard.

Acknowledgements

Financial support for these studies was provided by grants from the Medical Research Council of Canada (MT-4235) and the Saskatoon Cancer Foundation Grant for Flow Cytometry. We are grateful to Iris Cheng for operating the flow cytometry equipment.

References

1. Beaumont K, Tan PK (1990) Effects of atrial and brain natriuretic peptides upon cyclic GMP levels, potassium transport, and receptor binding in rat astrocytes. J Neurosci Res 25: 256–262
2. Bender AS, Hertz L (1987) Inhibition of [^{3}H]diazepam binding in primary cultures of astrocytes by atrial natriuretic peptide and by a cyclic GMP analog. Brain Res 436: 189–192
3. Cholewinski AJ, Wilkin GP (1988) Peptide receptors on astrocytes: evidence for regional heterogeneity. Biochem Soc Trans 16: 429–432
4. Doczi T, Joo F, Szerdahelyi P, Bodosi M (1988) Regulation of brain water and electrolyte contents: the opposite actions of central vasopressin and atrial natriuretic factor (ANF). Acta Neurochir (Wien) [Suppl] 43: 186–188
5. Fedoroff S, Neal J, Opas M, Kalnins VI (1984) Astrocyte cell lineage. III. The morphology of differentiating mouse astrocytes in colony culture. J Neurocytol 13: 1–20
6. Kimelberg HK, Ransom BR (1986) Physiological and pathological aspects of astrocytic swelling. In: Fedoroff S, Vernadakis A (eds) Astrocytes, vol 3. Academic Press, Orlando FL, pp 129–166
7. McGann LE, Walterson ML, Hogg LM (1988) Light scattering and cell volumes in osmotically stressed and frozen-thawed cells. Cytometry 9: 33–38
8. Walz W (1987) Swelling and potassium uptake in cultured astrocytes. Can J Physiol Pharmacol 65: 1051–1057

Correspondence: M. R. Del Bigio, M.D., Ph.D., Division of Neuropathology, The Hospital for Sick Children, 555 University Avenue, Toronto, ON, M5G 1X8 Canada.

Acta Neurochirurgica, Suppl. 51, 17–18 (1990)
© by Springer-Verlag 1990

Brain Interstitial Composition During Acute Hyponatremia

J. A. Lundbæk[1], T. Tønnesen[1], H. Laursen[2], and A. J. Hansen[3]

[1] Institute of General Physiology, The Panum Institute, University of Copenhagen, Denmark, [2] Institute of Neuropathology, The Teilum Building, University of Copenhagen, Denmark, [3] Novo-Nordisk A/S, CNS-Division, Sydmarken 5, 2860 Søborg, Denmark

Summary

Brain cortical water spaces were determined in rat brain following an intraperitoneal infusion of distilled water (15% of body weight). Total water content was determined by gravimetric methods while the size of the interstitial space and the interstitial ion concentrations were measured by ion-selective microelectrodes. Ion content of brain gray matter and plasma were determined by standard methods. During 2 h the plasma osmolarity changed from 301 to 251 mosmol/l while cortical water content increased from 4.04 ml/g dry weight (gdw) to 4.42 ml/gdw. It was observed that the size of the interstitial space remained stable during the two hours of osmotic challenge. Although the brain showed a volume regulatory response by an outward transport of ions, it was not sufficient since there was an increase of brain volume of 10%. The accumulated water was distributed in the interstitial and intracellular compartments according to their individual sizes. It is concluded that cells in brain gray matter are not able to regulate their volume in cytotoxic brain oedema.

Introduction

The brain responds to an acute hypoosmolar challenge by swelling. The accumulation of water is less than anticipated from the behaviour of a perfect osmometer since the brain regulates its volume by a loss of ions. The mechanisms by which the brain reacts to osmotic stresses and how the individual water spaces are regulated are still largely unknown. A number of investigations have been carried out on acute hyponatremia[2, 3] but they have not characterized brain interstitial ions nor measured the size of the interstitial and intracellular spaces. The purpose of the present investigation was to determine the size of the intra- and the extracellular spaces as well as their ionic content in brain gray matter during an acute lowering of plasma osmolarity.

Materials and Methods

Pentobarbital anaesthetized male Wistar rats weighing 350 g were used. Arterial blood pressure and arterial blood gases were monitored and the body temperature controlled. The head was secured in a headholder and a small burr hole (2 mm in diameter) was made in the parietal bone. After the dura was pierced ion-selective microelectrodes was positioned 500–700 µm below the brain surface.

The plasma osmolarity was lowered by infusion of water (37 °C) in the peritoneal cavity (initially 10% of body weight and subsequently 5% after 40 min)[2, 3]. Ion concentrations and osmolarity in plasma was measured at selected moments. After 2 hours the rat was decapitated, the brain quickly removed and cortical gray matter isolated and weighed. The dry weight was determined after 48 h at 105 °C, and the ion content determined by standard methods.

Ion selective microelectrodes. Ion concentrations in brain interstitial space was measured by doubled-barrelled ion selective microelectrodes (K^+, Cl^-, Na^+).

The size of the interstitial space was determined by iontophoresis of tetramethylammonium (TMA^+) from a micropipette. The concentration of TMA at a given distance (50–150 µm) was measured by a microelectrode and fitted to the equation:

$$C(r,t) = Q\lambda^2/4\pi D\alpha r \; erfc \; (r\lambda/2\sqrt{(Dt)})$$

in which α is the size of the distribution volume of TMA corresponding to the interstitial space. For further information see[4]. This technique makes it possible to determine the space size at frequent intervals and at discretes depth in the tissue.

Results

Plasma

The plasma osmolarity declined steadily from *301 mosmol/l* after ingestion of water and reached a constant level after 60 min og *251 mosmol/l*. The decrease can be accounted for by the lowering of plasma (Na) (137 to 107 mmol/l) and (Cl) (105 to 85 mmol/l).

Brain Cortex

The percentage of water in brain gray matter increased during the two hours from 80.17% ± .39 to 81.54% ± .36 corresponding to an increase of water content and in volume of ca. 10% (4.04 ml/g dw to

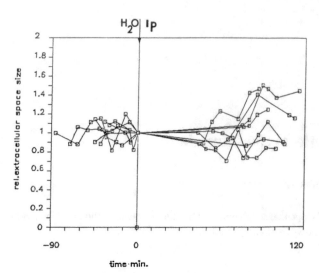

Fig. 1. The size of the interstitial space of rat brain cortex determined by a microelectrode technique as described in methods. The values are relative to the size determined in the period before hyponatremia was induced. No apparent change occurs during the 120 min period of hyposmolar challenge

4.42 ml/g dw). The swelling was accompanied by a loss of principle ions (μmol/g dw). Na^+ from 238 ± 17 to 197 ± 14, Cl: 164 ± 5 to 125 ± 8 and K: 558 ± 21 to 574 ± 24 or a total of about 70 μosmol/g dw.

The interstitial K^+ concentration remained unchanged while (Na^+) decreased (141 to 124 mmol/l) and Cl decreased (127 to 106 mmol/l) during the two hours.

Interstitial Space Size (α)

Figure 1 shows the time course of α in brain cortex. Although there is a relatively large scatter there was no evidence of a significant change during hypoosmolar condition. During control conditions α = 0.18 ml/ml and the value after two h was 0.19.

Discussion

The present study measured the interstitial space during acute hyponatremia. The study suggests that the accumulation of water, which amounted 10% of brain volume, was distributed in the interstitial and the in-tracellular compartments according to their individual volumes.

The amount of oedema was less than expected if the brain behaved like a perfect osmometer. Hence the brain should have increased its water content by 301/251, *i.e.* 20%. The lesser swelling is represents a regulatory response which involves loss of electrolytes from both the intercellular and the interstitial space.

The 65% decrease of chloride content in the cellular compartment (not shown) needs an explanation. It is tempting to suggest that the glial compartment is responding stronger because glial have higher chloride content and they seem able to adapt following swelling in hypotonic media[1].

Exposure of the brain to hypernatremia has shown a distinct volume control of brain cortical cells[5]. Using the same technique as presented here it was shown that brain matter shrinkage preferentially was located to the interstitial space while brain cellular volume remained constant. The present study, however, suggests, that oedema fluid in hyponatremia is located in both compartments without preservation of cell volume. The swelling of the tissue is only possible at the expense of brain ventricles since we observed a 40% reduction in their volume during the hypoosmolar treatment.

References

1. Kempski O, Chaussy L, Gross U, Zimmer M, Baethmann A (1983) Volume regulation and metabolism of suspended C6 glioma cells: an *in vitro* model to study cytotoxic brain oedema. Brain Res 279: 217–228
2. Melton JE, Nattie EE (1983) Brain and CSF water and ions during dilutional and isoosmotic hyponatremia in the rat. Am J Physiol 244: R 724–R 732
3. Melton JE, Patlak CS, Pettigrew, Cserr HF (1987) Volume regulatory loss of Na, Cl, and K from rat brain during acute hyponatremia. Am J Physiol 252: F 661–F 669
4. Nicholson C, Cserr HF, DePasquale M, Patlak CS, Rice ME (1988) Selective control of cell water during acute hypernatremia in rat cerebral cortex. Soc Neurosci 14: 1037
5. Nicholson C, Phillips JM (1981) Ion diffusion modified by tortuosity and volume fraction in the extracellular microenvironment of the rat cerebellum. J Physiol 321: 225–257

Correspondence: J. A. Lundbæk, M.D., Institute of General Physiology, The Planum Institute, University of Copenhagen, Denmark.

Acta Neurochirurgica, Suppl. 51, 19–21 (1990)

Do the Skull and Dura Exert Influence on Brain Volume Regulation Following Hypo- and Hyperosmolar Fluid Treatment?

T. Dóczi, Á. Kuncz, and M. Bodosi

Albert Szent-Györgyi Medical University, Department of Neurosurgery, Szeged, Hungary

Summary

The present studies were performed to determine the response of the brain water and electrolytes to acute hypoosmolality and hyperosmolality in animals with intact skull and dura, in comparison with those subjected to extensive bilateral or unilateral craniectomy and dural opening. Four to 5 weeks following extensive unilateral or bilateral craniectomy and dural opening in rats, a 50 mosm/kg decrease in plasma osmolality was produced by systemic administration of distilled water ("water intoxication"), or a 28 mosm/kg increase in plasma osmolality was produced by systemic administration of either 1 M NaCl or 1 M mannitol in 0.34 M NaCl. Tissue water, Na, and K contents were determined after 120 minutes. Tissue water accumulation or water loss was proportional to the decrease or increase in plasma osmolality. However, the tissue water accumulation following "water intoxication" was less (40% of the predicted value) than that predicted for ideal osmotic behaviour. The brain tissue was also found to shrink less than predicted on the basis of ideal osmotic behaviour (40% of the predicted value after mannitol treatment, and 60% after NaCl administration). This non-ideal osmotic response of the brain tissue is consistent with the finding in other studies[1] and indicated a significant degree of volume regulation.

Water and electrolyte changes were not different in operated and non-operated animals, demonstrating no effects of extensive skull and dura defects on tissue volume regulation under hypo- and hyperosmolar conditions of a degree that may be encountered under clinical circumstances. The results support the view that the volume of brain tissue is controlled by an internal "osmometer" consisting of the capillary endothelium which is not affected by alterations in the brain's container. These observations indicate that the effect of osmotic therapy on the non-pathological tissue volume of craniectomy patients is not influenced by a skull or dura defect.

Introduction

Background and Work Which Has Led to the Project

The Monroe-Kellie doctrine and subsequent modifications assert that new volume introduced into the craniospinal system can only be accommodated by means of spatial compensation, i.e., a reciprocal reduction in the volume of one of the normal constituents of the system (mainly CSF and blood). A volume load can also be stored within the craniospinal system, thereby causing an equal increase in its total volume determined by the "elastic properties" of the system (mainly the dura mater).

Regulation of the volume of the brain tissue itself is accomplished mainly via the capillary endothelium by two simple mechanisms which limit water flow across the blood-brain barrier: the very low hydraulic conductivity of the brain capillaries, and the high osmotic activity of the major solutes[5, 6]. The influence of the limited intracranial space (i.e., intact skull and dura) upon regulation of the brain tissue volume (controlled mainly by the capillary endothelium) is rather controversial[2–4, 7].

The question is whether the constancy of the closed container influences the tissue volume-regulatory changes when oedema development is due to the increased driving forces across the intact barrier.

Material and Methods

The experiments were performed on albino Wistar rats anaesthetized with an intraperitoneal injection of pentobarbital sodium (50 mg/100 g). Bilateral extensive craniectomies were performed in 63 animals and unilateral craniectomy in 18 animals. The dura was incised crucially and reflected under the operating microscope. The integrity of the arachnoid was preserved.

In 8 animals the effect of craniectomy on the blood-brain barrier was assessed by measurement of Evans blue extravasation[7]. Four to 5 weeks after the operation, the animals were exposed to osmotic stress. The experimental protocols for the various experimental groups are shown in Table 1.

The rats were decapitated 120 minutes after the start of the treatment. Serum osmolality and electrolytes were determined. The hemispheres were investigated separately in all animals. The tissue water content and dry weight were established by weighing before and after drying for 40 hours at 110 °C.

For assessment of the osmotic volume regulation, changes in brain water following hypo- and hyperosmolar stress were estimated by assuming that the brain behaves as a perfect osmometer, i.e., assuming that there is no volume regulation. In this situation, the steady-state brain volume at the decreased or increased osmolality is given by the equation: $V'' = V'(C'/C'')$, where V' and V'' are the volumes of the brain water in ml/kg dry weight before and after osmotic treatment, respectively, and C' and C'' are the plasma osmolality values (mosm/kg) before and after treatment, respectively[1].

Results and Discussion

Bilateral craniectomy and opening of the dura mater 4 weeks prior to testing of the blood-brain barrier integrity did not cause albumin extravasation into the tissue.

The administration of distilled water (Groups B, F, and H) caused significant decreases in the serum osmolality and serum sodium concentration, but did not alter the serum potassium content to a significant extent. The sodium concentration and osmolality values for the individual groups B, F, and H were uniform, leading to a mean reduction in serum osmolality of about 50 mosm/kg.

Hyperosmolar treatment with NaCl caused significant increases in serum osmolality (mean increase: 28 mosm/kg) and serum sodium concentration in groups C and I; and mannitol administration increased the serum osmolality significantly (mean increase: 28 mosm/kg). The serum sodium and potassium concentrations did not change in groups D and J.

The hypoosmolar stress brought about a significant brain water accumulation in every group (B, F, and H) (Table 2). The increase in water content was uniform, amounting to 1.5 g of water in 100 g of wet brain tissue. The increase was less (40% of the predicted value), however, than that predicted for ideal osmotic behaviour. In Group F, the brain water contents were 79.7% and 79.8% in the craniectomized and non-operated hemispheres, respectively. The sodium and potassium contents did not differ either. In group E the water and tissue Na and K contents of the hemispheres were also uniform.

The production of extensive skull and dura defects did not alter the response of the brain tissue to hypoosmolar stress. The water accumulation was proportional to that found in animals with intact skull.

When the brain sodium and water contents were

Table 1. *Experimental Protocols*

	Groups	No. animals	Operation	Sytemic treatment
A.	control	10	none	none
B.	unoperated + hypoosmolar	10	none	distilled water to a volume of 10% of body weight; followed by a second water load 5% of body weight 40 min later
C.	unoperated salt hyperosmolar	10	none	1 M NaCl solution in a dose of 2 ml/100 g tissue weight
D.	unoperated + mannitol hyperosmolar	10	none	1 M mannitol in 0.34 M NaCl in a dose of 3 ml/100 g tissue weight
E.	unilaterally operated + untreated	8	unilateral craniectomy + dural opening	none
F.	unilaterally operated + hypoosmolar	10	unilateral craniectomy + dural opening	distilled water to a volume of 10% of body weight; followed by a second water load to 5% of body weight 40 min later
G.	bilaterally operated non-treated	10	bilateral craniectomy + dural opening	none
H.	bilaterally operated + hypoosmolar	10	bilateral craniectomy + dural opening	distilled water to a volume of 10% of body weight; followed by a second water load to 5% of body weight 40 min later
I.	bilaterally operated + salt hyperosmolar	10	bilateral craniectomy + dural opening	1 M NaCl solution in a dose of 2 ml/100 g tissue weight
J.	bilaterally operated + mannitol hyperosmolar	10	bilateral craniectomy + dural opening	1 M mannitol in 0.34 M NaCl in d dose of 3 ml/100 g tissue weight

Table 2. *Changes in Brain Water Content and Electrolytes*[a]

Groups		Brain water concentration: (g water/100 g wet brain = % water content)	Ion concentration (mmol/kg dry weight)	
			Sodium	Potassium
A.	control	78.3 ± 0.2	199 ± 4	416 ± 5
B.	unoperated + hypoosmolar	79.8 ± 0.4[b]	159 ± 3[b]	402 ± 8
C.	unoperated + salt hyperosmolar	77.1 ± 0.2[b]	221 ± 6[b]	429 ± 10
D.	unoperated + mannitol hyperosmolar	77.6 ± 0.5	210 ± 6[b]	443 ± 12[b]
E.	unilaterally operated + untreated	see text	see text	see text
F.	unilaterally operated + hypoosmolar	see text	see text	see text
G.	bilaterally operated + non-treated	78.1 ± 0.3	202 ± 9	412 ± 10
H.	bilaterally operated + hypoosmolar	79.7 ± 0.2[b]	164 ± 8[b]	398 ± 11
I.	bilaterally operated + salt hyperosmolar	77.3 ± 0.5[b]	219 ± 8[b]	429 ± 10
J.	bilaterally operated + mannitol hyperosmolar	77.1 ± 0.3[b]	211 ± 6[b]	433 ± 7[b]

[a] Values are means ± SD.
[b] Statistically different from control ($p < 0.05$).

assessed as a function of plasma osmolality, the reduction in volume was proportional to the increase in plasma osmolality in all hyperosmolar groups, indicating that there is no significant role of the closed skull and dura in osmotic brain volume regulation. In both hyperosmolar conditions, the decreases in brain volume were less than those predicted for ideal osmotic behaviour (40–60%).

References

1. Cserr HF, DePasquale M, Patlak CS (1987) Am J Physiol 253 (Renal Fluid Electrolyte Physiol 22): F 522–F 529
2. Fenstermacher JD (1984) In: Staub NC, Taylor AE (eds) Edema. Raven Press, New York, pp 383–404
3. Hatashita S, Hoff J (1987) J Neurosurg 67: 573–578
4. Hochwald GM *et al* (1972) Dev Med Child Neurol 14 [Suppl 27]: 65–69
5. Rapoport SI (1979) In Vasogenic brain oedema. In: Poppp AJ *et al* (eds) Neural trauma. Raven Press, New York, pp 51–62
6. Rössner W, Tempel K (1966) Med Pharmacol Exp 14: 169–182
7. Shapiro K, Takei F, Fried A, Kohn I (1985) J Neurosurg 63: 82–87

Correspondence: T. Dóczi, M.D., Albert Szent-Györgyi Medical University, Department of Neurosurgery, H-6701 Szeged, Pf.: 464, Hungary.

Acta Neurochirurgica, Suppl. 51, 22–24 (1990)

In vivo Measurement of Intra- and Extracellular Space of Brain Tissue by Electrical Impedance Method

S. Suga, S. Mitani, Y. Shimamoto, T. Kawase, S. Toya, K. Sakamoto[1], H. Kanai[1], M. Fukui[2], and N. Takeneka[3]

Department of Neurosurgery, School of Medicine, Keio University, Tokyo, Japan, [1] Faculty of Science and Technology, Sophia University, Bulgaria, [2] Faculty of Science and Engineering, Tokyo Denki University, Tokyo, Japan, [3] Department of Neurosurgery, Ashigkaga Red Cross Hospital, Tokyo, Japan

Summary

An impedance method was applied to evaluate transcellular fluid shifts in ischaemic brain oedema. The admittances (apparent electrical conductivities) of tissues were measured at varied frequencies based on a simple model of an electrical equivalent circuit for tissues, which consisted of Re (resistivity of extracellular fluid), Ri (resistivity of intracellular fluid) and Cm (capacitance of cell membrane). Calculated were Re, Ri, Rinf (resistivity of total fluid), Re/Ri and alpha (Cole-Cole distribution index) of brain tissue by Cole-Cole an arc of a circle. During ischaemia induced by cat middle cerebral artery occlusion (MCAO), the parameters were examined continuously.

After MCAO, cerebral blood flow (CBF) decreased to under 10 ml/100 g/min. Then Re and Re/Ri increased, but Ri decreased. These results indicated that fluid shift from extracellular (EC) to intracellular (IC) space occurred after ischaemic insult. Rinf showed no changes during ischaemia of 30 min, which demonstrated no changes of total fluid volume.

Using this impedance technique, fluid accumulation and shift may be examined by changes of Re, Ri, Re/Ri and Rinf in various types of brain oedema *in vivo*.

Introduction

Brain oedema is considered as fluid accumulation of the intracellular (IC) and/or extracellular (EC) space. As it is possible to examine changes of fluid accumulation and shift, it is very useful to evaluate this pathophysiological aspect in the formation of brain oedema. Various methods for measurement of fluid shifts between IC and EC space have been presented. Measurement of the electrical impedance is one of the method to evaluate brain oedema *in vivo*. But in most often previous studies, impedance was examined at a fixed frequency presenting only changes of the EC space. In this study, we demonstrate a new technique of impedance measurement at varied frequencies and

evaluate the possibility of assessment of Re and Ri separately. At the same time, we have calculated other parameters about fluid of tissues and discuss the efficiency of this method.

Theoretical Background

Biological tissues have electrical properties which are affected by the range of frequencies of a delivered current. Three dispersions (α, β and γ) of frequencies in tissues were presented by Schwan[4]. With the frequency of current in β dispersion, a biological tissue is represented as a simple electrical equivalent circuit, which consists of Re, Ri and Cm. In this electrical circuit, we

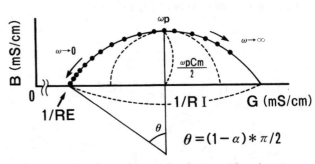

Fig. 1. Admittance locus of tissues. Conductance (G) and susceptance (B) of tissues were measured at varied frequencies. If cell volume is constant, the admittance locus is presented as a dotted semicircle. Considering that the volume of cells in the tissue varies, the admittance locus is presented as a solic arc of a circle (Cole-Cole an arc of a circle). 1/RE (resistivity of extracellular space) is equivalent with B as ω ($\omega = 2\pi *$ frequency) approaches 0. 1/Rinf (resistivity of total fluid space) is equivalent to B as ω approaches ∞. 1/RI (resistivity of intracellular space) is $\sim$ B as 1/RE taken from 1/Rinf. ω p is the angular frequency of the center of an arc of a circle. Cm is capacitance of the cell membrane

measured conductance (G) and suspectance (B) at varied frequencies and obtained thereby the admittance locus. Then Re, Ri, Cm and other parameters were calculated by utilizing the locus (Fig. 1).

Methods

1. Instruments

The admittance (apparent conductivity) of brain tissue *in vivo* was measured with the four electrodes method. The alternating current at varied frequencies was generated by a sine wave generator. For the measurements a frequency range of 0.4–1000 KHz was used (a range of β dispersion). The strength of current was kept at a constant level of 50 μA_{p-p}. The outer two electrodes were used to apply the alternating current, the inner two to measure the admittance. In this range admittance was measured at 15 chosen frequencies. A computer data management system was used to process the data according to a model described by Haeno *et al.*[1]. This makes possible calculation of parameters by means of a least-squares liner curve fitting combined with a computational iteration procedure.

2. Experimental Procedure

Adult cats were anaesthetized with ketamine (30 mg/kg, IM), for tracheostomy and catheterization of the femoral artery and vein. The cats were artificially ventilated and maintained with 1% α-chloralose and 10% urethane (5 ml/kg, IP). After the left middle cerebral artery (MCA) was exposed by transorbital approach, a left temporo-parietal craniotomy was performed. The probes for impedance and CBF measurements were placed at the ecto-sylvian gyrus above the dura. CBF was measured with a laser Doppler flowmeter.

Results

1. Control Values of Parameters

Before MCAO, measurements of parameters were made to obtain control values (n = 4). Re, Ri, Rinf, Re/Ri and α are presented in Table 1.

2. Changes of Parameters During Ischaemia

After measurement of control values, the MCA was occluded with a clip. Then, we measured CBF and admittance for 30 min. Once CBF was lower than 10 ml/100 g/min, the parameters began to change (n = 3). In these cases, Re increased and Ri decreased

Table 1. *Measured Control Values of Parameters* (n = 4)

Re (ohm · cm)	457.1 ± 54.5
Ri (ohm · cm)	1,528.4 ± 489.8
Rinf (ohm · cm)	348.9 ± 51.1
Re/Ri	0.314 ± 0.074
alpha	0.317 ± 0.069

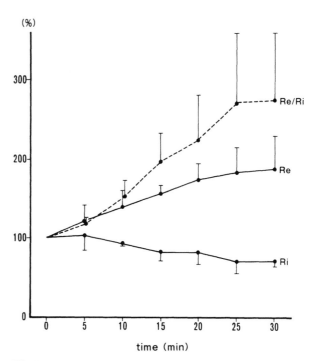

Fig. 2. Mean course of Ri, Re and Re/Ri during ischaemia of 30 min given as percentage of the control value (n = 3)

steadily (Fig. 2). But Re/Ri increased furthermore. Rinf and α showed no significant changes during ischaemia.

Discussion

Electrical impedance methods have been applied to evaluate brain oedema in various pathophysiological states. But impedance usually is examined at a fixed frequency in most studies and considered as Re which correlates with the EC volume. Recently, it was suggested that measurement of the electrical admittance at varied frequencies offers the possibility of evaluating Re and Ri[3]. In the previous experiments an increase of Re was regarded to indicate swelling of cells[5]. But the parameter corresponds directly to shrinkage of the EC space, not to swelling of cells itself. In fact, an increase of Re and Ri was observed during hemodialysis, with shrinkage of the EC and IC spaces simultaneously[1]. In this study, Re increased during ischaemia. This result is in agreement with previous studies[2]. Conversely, Ri decreased during ischaemia. These results certainly indicate shrinkage of the EC space and swelling of cells. As the increase of Re/Ri exceeded that of Re, changes of Re/Ri may indicate more clearly a fluid shift between the IC and EC space. Rinf is the resistivity of total fluid. Thus, changes of Rinf are considered to be correlated to changes of the total fluid space. In this study, Rinf did not change significantly during ischaemia. This corresponds with results described by Hossmann that the water content does not increase during ischaemia of 30 min[2].

In conclusion, a new impedance method is presented, which offers the possibility to assess Re, Ri, Re/Ri and Rinf for the analysis of fluid accumulation and shift in brain tissue.

References

1. Haeno M, Tagawa H, Sakamoto K, Kanai H (1985) Estimation and application of fluid distribution in legs by measurement of electrical impedances. Medical Electronics and Bioengineering 23: 354–359 (in Japanese)

2. Hossmann KA (1985) The pathophysiology of ischaemic brain swelling. In: Inaba Y, Klatzo I, Spatz M (eds) Brain oedema. Springer, Berlin Heidelberg New York, pp 367–384

3. Kanai H, Sakamoto K, Haeno M (1983) Electrical measurement of fluid distribution in human legs. J Microwave Power 18: 233–243

4. Schwan HP (1957) Electrical property of tissues and cells. Advances in Biol Med Physics 5: 147–207

5. Van Harreveld A, Ochs S (1956) Cerebral impedance changes after circulatory arrest. Am J Physiol 187: 180–192

Correspondence: Sadao Suga, M.D., Department of Neurosurgery, School of Medicine, Keio University, 35 Shinanomachi, Shinjuku-ku, Tokyo 160, Japan.

Acta Neurochirurgica, Suppl. 51, 25–27 (1990)

Intracellular pH Regulation of Normal Rat Brain: ^{31}P-MRS Study

F. Ohta, K. Moritake, T. Kagawa, M. Fukuda, and **A. Fukuma**

Department of Neurosurgery, Shimane Medical University, Izumo, Japan

Summary

Intracellular pH changes in rat brain tissue were investigated during low or high extracellular pH induced by acetazolamide or sodium bicarbonate, respectively. Intracellular pH was measured by ^{31}P-MRS in the brain of spontaneously breathing rats under intraperitoneal sodium pentobarbital anaesthesia. Extracellular pH was calculated from the results of blood gas analysis. After intravenous injection of sodium bicarbonate (280 mg/kg), the extracellular pH rose significantly ($p < 0.05$) from 7.47 ± 0.06 to 7.82 ± 0.15 (mean $\pm$ SE). After administration of acetazolamide (50 mg/kg), the extracellular pH dropped significantly ($p < 0.05$) from 7.45 ± 0.02 to 7.34 ± 0.03. Despite the changes in extracellular pH, the intracellular pH of rat brain did not change significantly under either condition. The following four factors are thought to contribute to the maintenance of intracellular pH in the normal brain: 1) production and consumption of H^+ by brain metabolism, 2) physicochemical buffering, 3) transmembrane transport of H^+ and its equivalent, 4) compensatory adaptation of circulatory factors. These mechanisms are not disturbed in the brain of rats that are breathing spontaneously, because the cerebral circulation and energy metabolism are preserved in the normal range.

Introduction

Cerebral intracellular pH is one of the most important physiologic parameters that changes significantly after ischaemia. A recent, ^{31}P-MRS study revealed that the intracellular pH of brain tissue became lower after ischaemic insult[2]. Numerous experimental studies have demonstrated that low intracellular pH of brain plays an important role in brain cell injury, namely it is not only the result but also the cause of acute ischaemic injury to neural tissue in acute brain ischaemia[3]. Few studies have, however, reported on the intracellular regulation of pH in the normal brain[11]. We have studied intracellular pH regulation in normal brain tissue of rats using ^{31}P-MRS, sodium bicarbonate and acetazolamide.

Materials and Methods

The femoral artery and vein of Sprague-Dawley rats weighing 400–570 g were cannulated under intraperitoneal pentobarbital sodium anaesthesia (40 mg/kg). Using an OTSUKA ELECTRONICS BEM 170/200 4.7 tesla NMR system and a round surface coil (20 mm diameter), the ^{31}P spectra were measured with a 2.0 sec interpulse delay and 240 acquisitions. Intracellular pH was calculated using the equation: $pH = 6.77 + \log\{F - 3.29)/(5.68 - F)\}$, $F = PCr : Pi$ chemical shift[6]. Extracellular pH was determined by the analysis of arterial blood gas. After baseline ^{31}P spectra were measured, sodium bicarbonate (280 mg/kg) or acetazolamide (50 mg/kg) was injected intravenously and ^{31}P spectra were obtained continuously. Blood gas analysis was performed 5 min before and after injection. Statistical analysis was performed by Student's unpaired t-test.

Results

After intravenous injection of sodium bicarbonate, the extracellular pH rose from 7.47 ± 0.06 to 7.82 ± 15 (mean $\pm$ SE). After intravenous injection of acetazolamide, the extracellular pH dropped from 7.45 ± 0.02 to 7.34 ± 0.03. These changes were statistically significant ($p < 0.05$). Intracellular pH obtained by ^{31}P spectra did not change (Figs. 1 and 2).

Discussion

Despite an increase or decrease in extracellular pH, the intracellular pH did not change significantly in brain tissue of normal rats. This experimental result suggests that the intracellular pH of the brain is strictly regulated under normal physiological conditions. Other studies indicate that four major mechanisms contribute to the maintenance of intracellular pH in the normal brain[8, 9, 10]: 1) production and consumption of H^+ by brain metabolism, 2) physicochemical buffering, 3) transmembrane transport of H^+ and its equivalent, 4) compensatory adaptation of circulatory factors. In aerobic brain cells whose oxygen is supplied by an intact cer-

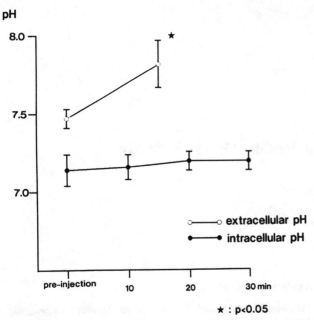

Fig. 1. Changes in intra- and extracellular pH after administration of sodium bicarbonate

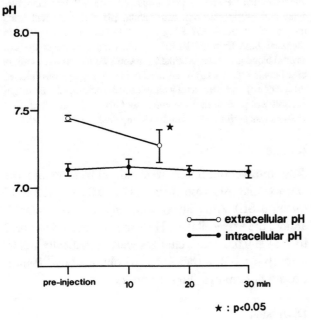

Fig. 2. Changes in intra- and extracellular pH after administration of acetazolamide

ible transfer of phosphate between phosphocreatine and ATP as follows:

$$PCr + ADP + H^+ <-> ATP + Cr \qquad (1)$$

This reaction is available only at physiological pH levels of 7.0–7.4, and results in production of ATP and consumes H^+ [5]. Although it is reported that PCr decreases remarkably 3 minutes after ischaemic insult[2]. PCr was maintained with an intact cerebral circulation in our experiments. Accordingly, the production and consumption mechanism of H^+ by cerebral metabolism was maintained in brain tissues of normal rats with an intact cerebral circulation. The other mechanism is physicochemical buffering. Physicochemical buffering consists of two buffer systems, bicarbonate and nonbicarbonate buffers. Siesjo *et al.*[9] indicated that regulation of intracellular pH due to the buffers is close to 40%. Physicochemical buffering is given as follows:

$$H^+ + HCO_3^- <-> H_2CO_3 <-> H_2O + CO_2 \qquad (2)$$

If the cerebral blood flow is preserved normally and the washout effect of CO_2 is maintained, the reaction in Equation (2) leads to generation of CO_2. Both H^+ produced by brain tissue and H^+ resulting from intravenous administration of acetazolamide are washed out via the microvessels by the bicarbonate buffering system as long as the cerebral circulation is preserved.

Transmembrane transport of H^+ and its equivalent depends on so-called acid-extrusion[4, 8, 9]. The acid extrusion is achieved by coupling the entry of Na^+ and HCO_3^- to the exit of Cl^- and possibly H^+. Na-H exchange is considered to be the main mechanism for acid extrusion and is ATP-dependent. After ischaemic insult, electrochemical disequilibrium of Na^+, K^+, and Ca^{2+} across the neural plasma membrane is disturbed due to energy failure, and the acid extrusion system is also impaired. Compensatory adaptation consists of factors such as the chemical control mechanism of the cerebral circulation (vasodilation by CO_2)[10]. These mechanisms minimize the decrease in intracellular pH after ischaemic insult. As long as cerebral blood flow is maintained normally and the wash-out mechanism of CO_2 by the cerebral circulation is not impaired in spontaneously breathing rats, these four mechanisms for intracellular pH regulation in brain tissues of rat are preserved.

Vorstrup *et al.* demonstrated that constant normal intracellular pH was maintained in brain during low extracellular pH induced by the intravenous administration of acetazolamide[11]. Their experiment was per-

ebral circulation, glucose is oxidized to carbon dioxide (CO_2) and water. Carbon dioxide is then washed out into the cerebral circulation by way of CO_2 penetration into normal cells[7]. In consequence, acidosis of normal brain cells does not develop. Under physiological conditions the brain has two mechanisms other than glycolysis and oxidative phosphorylation that contribute to maintenance of ATP level. One is by creatine phosphokinase. Creatine phosphokinase catalyzes a revers-

formed in six healthy human volunteers with spontaneous respiration. Their results are consistent with ours. Cohen *et al.*, on the other hand, showed that hypercapnia induced by mechanical ventilation in rats reduced the intracellular pH of brain tissues[1]. Their experiment was different from ours, since their rats were mechanically ventilated. Autoregulation of cerebral blood flow is thought to be impaired under severe respiratory acidosis induced by mechanical ventilation. As the wash-out effect of CO_2 by the cerebral circulation decreases, the pH regulation mechanism is preserved as mentioned above.

References

1. Cohen Y (1990) Stability of brain intracellular lactate and ^{31}P-metabolite levels at reduced intracellular pH during prolonged hypercapnia in rats. J Cereb Blood Flow Metab 10: 277–284
2. Germano IM (1989) Magnetic resonance imaging and ^{31}P magnetic resonance spectroscopy for evaluating focal cerebral ischaemia. J Neurosurg 70: 612–618
3. Goldman SA (1989) The effects of extracellular acidosis on neurons and glia *in vitro*. J Cereb Blood Flow Metab 9: 471–477
4. Hansen AJ (1981) Extracellular ion concentrations during spreading depression and ischaemia in the rat brain cortex. Acta Physiol Scand 113: 437–445
5. Hochachka P (1983) Protons and anaerobiosis. Science 219: 1391–1397
6. James TL (1984) *In vivo* nuclear magnetic resonance spectroscopy. In: Moss AA, Ring EJ, Higgins CB (eds) NMR, CT, and interventional radiology. Mosby, San Francisco, pp 235–244
7. Meyer JS (1961) CO_2 narcosis. An experimental study. Neurology 11: 524–537
8. Roos A (1981) Intracellular pH. Physiol Rev 61: 296–434
9. Siesjo BK (1978) Brain energy metabolism. John Wiley & Sons, Chichester
10. Tanaka K (1988) Cerebral circulation and metabolism in cerebral ischaemia: With particular reference to pH homeostasis. Shinkei Kenkyu No Shinpo 32: 235–247
11. Vorstrup S (1989) Neural pH regulation: Constant normal intracellular pH is maintained in brain during low extracellular pH induced by acetazolamide-^{31}P NMR study. J Cereb Blood Flow Metab 9: 417–421

Correspondence: Kouzo Moritake, M.D., Department of Neurosurgery, Shimane Medical University, Enya 89-1, Izumo 693, Japan.

Acta Neurochirurgica, Suppl. 51, 28–30 (1990)
© by Springer-Verlag 1990

Measurement of Canine Cerebral Oedema Using Time Domain Reflectometry

H. P. Kao[1], **E. R. Cardoso**[2], and **E. Shwedyk**[1]

[1] Department of Electrical Engineering, University of Manitoba, [2] Cerebral Hydrodynamics Laboratory,
Health Sciences Clinical Research Center, Winnipeg, Manitoba, Canada

Summary

The complex dielectric constant of canine cerebral white matter correlates with its water content. Time domain reflectometry was used to measure the real dielectric constant and loss factor of 27 oedematous brain samples at frequencies of 100, 250, 500, 750, and 1,000 MHz. The real dielectric constant increased with water content at each frequency. Correlation was least for 100 MHz (R = 0.76) and between 0.87 to 0.89 for the other frequencies. The loss factor decreased with frequency with a correlation coefficient from 0.81–0.84. The *in vitro* results indicate that use of time domain reflectometry can predict oedema at a sensitivity of approximately 1–2% change of tissue water content.

Introduction

Cerebral oedema remains a frequent cause of mortality and severe morbidity in medical practice. Though cerebral oedema can be easily identified and localized by modern imaging techniques, such as computerized tomography (CT scan) and magnetic resonance imaging (MRI), these techniques do not quantify or allow continuous measurement of cerebral oedema. Furthermore, MRI is not suitable for use with critically ill patients who benefit will the most from close oedema monitoring. Thus, the development of methods for continuous quantification of cerebral oedema should allow earlier neurosurgical intervention and prevention of severe secondary brain damage.

An investigation into the correlation between cerebral oedema and the dielectric properties of canine white matter (WM) is described in this paper. Time domain reflectometry (TDR) was used to measure the complex dielectric constant of white matter in which oedema had been induced osmotically.

Materials and Methods

Thirteen healthy adult mongrel dogs were sacrificed either through intravenous injection of KCl or a sodium pentobarbital (Nembutal) overdose. The cerebral hemispheres were removed immediately, with nine of them placed in a small thick plastic bag and stored at $-30\,°C$ for period of less than 1 week; the remaining four were immediately used.

Each brain was cut into 8 coronal slices. The left hemisphere slices were used as control, while slices from the right hemisphere were made oedematous by immersion in distilled water for periods of 1 to 6 hours. This was done with the four fresh brains and seven frozen brains (which were thawed at room temperature). Two of the frozen brains were kept normal.

The WM dielectric constant was measured with the TDR as

Table 1. *Determination of Minimum Sample Thickness*. The values in brackets are from [1]. All measurements taken at 24.2° C

Frequency (MHz)	Real dielectric constant					
	Accuracy test	Distance from beaker bottom (mm)				
		5	4	3	2	1
100.0	76.6 (78.0)	76.8	76.4	76.2	75.5	71.1
250.0	78.0 (77.6)	77.8	77.8	77.6	76.7	72.5
500.0	77.6 (77.3)	77.5	77.8	77.4	76.6	72.3
750.0	78.6 (77.2)	78.2	78.8	78.3	77.4	73.1
1,000.0	76.4 (77.1)	76.1	76.8	76.2	75.1	70.9

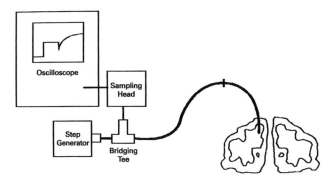

Fig. 1. Experimental setup to measure the white matter dielectric constant

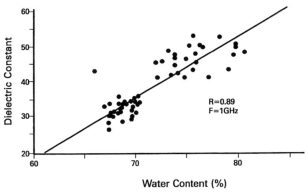

Fig. 2. Measured real dielectric constant of canine white matter

shown in Fig. 1. The real dielectric constant (RDC) and loss factor (LF) were determined at frequencies of 100, 250, 500, 750, and 1,000 MHz. A 0.141″ outside diameter, semi-rigid coaxial cable was used to collect first an incident waveform (used as a reference) with the probe in air. The probe tip was then inserted into WM to a depth of 1 mm and the reflected wave collected.

Immediately following the measurements, a WM sample was taken from directly beneath the cable tip and its water content measured using the specific gravity technique. Equilibrium depths of the samples were taken at 2 minutes after entry into the column. The graduated density column was calibrated by the equilibration depths of standard solutions K_2SO_4 of known specific gravity[3]. Only columns with linearity greater than 0.998 were used.

Only samples which had no grey matter within a 3 mm radius of the probe tip were used. This minimum sample volume had been previously determined by measuring the real dielectric constant of distilled water in a beaker with the probe tip at a varying distance from the beaker bottom. As seen from Table 1 at a distance of 2 mm or more the real dielectric constant is essentially constant, *i.e.*, the probe "sees" hemispheric tissue of approximately 2 mm in radius.

Results and Discussion

Twenty-nine normal samples were collected. The measured average specific gravity was 1.0435 ± 0.0015 SE corresponding to a water content of 68.71 ± 0.97 SE%. The specific gravity of the brain samples is in close agreement with that of 1.0437 reported for canine centrum semiovale[2].

Table 2. *Measured Average Real Dielectric Constant and Loss Factor for Normal Canine White Matter Samples*

Frequency (MHz)	Real dielectric constant	Loss factor
100.0	49.1 ± 3.2	53.9 ± 5.8
250.0	37.4 ± 2.5	26.8 ± 2.3
500.0	34.2 ± 2.3	15.7 ± 1.3
750.0	33.5 ± 2.3	11.3 ± 1.0
1,000.0	32.0 ± 2.1	9.2 ± 2.9

The average real dielectric constant and loss factor for the normal samples are presented in Table 2. The observed trends are in agreement with those reported by Schepps and Foster[4] for a temperature of 37 °C.

Twenty-seven oedematous samples were measured at temperatures between 20–22 °C. The RDC increased with water content at each frequency. The correlation was least for 100 MHz (R = 0.76) and was between 0.87 to 0.89 for the other frequencies. Figure 2 shows a typical plot. For these frequencies the slope was 1.5–1.7 dielectric units per percent water content. The LF decreased with water content, 0.81 < R < 0.84 with the largest change occurring at 100 MHz (29 units per percent water content).

The above results may be explained by modelling the increase in water content as a "dilution" of a mixture of water and proteins in the tissue slice. At the frequencies considered, cellular membranes have an extremely low impedance, and the tissue may be treated as a suspension of proteins in water[4].

Thus a linear mixture model

$$\varepsilon_{tissue} = \varepsilon_{normal} + \frac{\varepsilon_{water} - \varepsilon_{normal\ tissue}}{100 - (normal\ water\ content)}$$
$$\times\ (water\ content\ (\%) - (normal\ water\ content)$$

either the real dielectric constant or loss factor might be used to describe the dielectric tissue parameters as a function of water content. Using the linear equations fitted to the experimental data at each frequency, the linear mixture model has reasonable predictive power in the water content range of normal to maximum. It predicts the real dielectric constant at 100% water content to be within a maximum error of 10 dielectric units (12% error or less).

Conclusions

An *in vitro* study of the correlation of the dielectric properties, as measured by time domain reflectometry of canine white matter with osmotically induced oedema has been described. The real dielectric constants correlated well, $0.87 < R < 0.89$ for frequencies in the 250–1,000 MHz range. The *in vitro* results indicate that use of time domain reflectometry can predict oedema at a sensitivity of approximately 1–2% water content change. This needs to be verified by *in vivo* studies. The accuracy of the method is adequate for clinical purposes and thus offers promise as a continuous monitor of cerebral oedema.

Acknowledgement

Project supported by grants from Canadian Heart Stroke Foundation and Natural Science and Engineering Research Council (NSERC), Canada. H. P. Kao was supported by a NSERC scholarship.

References

1. Buckley F, Maryott AA (1958) Tables of dielectric dispersion data for pure liquids and dilute solutions. National Bureau of Standards. Circular 589
2. Inaba Y *et al* (1984) Evaluation of periventricular hypodensity in clinical and experimental hydrocephalus in Metrizamide computed tomography. In: Recent progress in the study and theory of brain oedema. Plenum Press, New York, pp 299–310
3. Marmarou A, Tawaka K, Shulman K (1982) An improved gravimetric measure of cerebral oedema. J Neurosurg 56: 246–253
4. Schepps JL, Foster KR (1980) The UHF and microwave dielectric properties of normal and tumour tissues: variations in dielectric properties with tissue water content. Phys Med Biol 25: 1149–1159

Correspondence: E. Shwedyk, M.D., Department of Electrical Engineering, University of Manitoba, Winnipeg, Manitoba, R3T-2N2, Canada.

Acta Neurochirurgica, Suppl. 51, 31–33 (1990)
© by Springer-Verlag 1990

A New Mapping Study of Superoxide Free Radicals, Vascular Permeability and Energy Metabolism in Central Nervous System

N. Hayashi, T. Tsubokawa, B. A. Green[1], B. D. Watson[1], and R. Prado[1]

Nihon University, Department of Neurological Surgery, Tokyo, Japan, [1] University of Miami, Department of Neurological Surgery, U.S.A.

Summary

To study the role of free radicals in various pathophysiologies in the central nervous system (CNS), we have introduced a new simultaneous mapping technique of superoxide free radicals, vascular permeability and energy metabolism. The detection of superoxide anion in the CNS is based on the 380 nm chemiluminescence of 2-methyl-6-phenyl-3,7-dihydroimidazo [1,2-alpyrazin-3-one(CLA-phenyl)] upon reaction with superoxide in the frozen tissue section. The results show topographical relationships between activation of superoxide free radicals, increased vascular permeability, and changes of hydrogen transport system in the energy metabolism (Fig. 2). Prolonged formation of superoxide free radicals in the white matter without vasogenic oedema and necrotic tissue are interesting as a mechanism of progression of secondary tissue damage in ischaemic insults.

Introduction

Superoxide anion, hydrogen peroxide and hydroxyl radicals are produced in biological systems by the single-electron reduction of oxygen[4]. Recent studies have been documented that infusion of xanthine oxidase/hypo-xanthine/ADF-Fe[3] in brain tissue produces severe brain injury and oedema[1, 6]. These highly reactive oxygen-derived free radicals are suspected to play an important role for the progress of oedema and cell membrane perturbation in ischaemic lesions[1, 4, 12], but are not confirmed to play a role *in vivo* following cerebral ischaemia. We have developed a new mapping method for assessment of superoxide anion free radicals, energy metabolism (ATP, NADH) and vascular permeability simultaneously, using frozen tissue sections. In this paper we present these techniques and results.

Materials and Methods

The investigations were carried out in various forms of ischaemia in the central nervous system (CNS), such as photochemical induced spinal corde ischaemia[9] in rats (n = 15), 15 minutes bilateral carotid ligation in gerbils (n = 20), and localized cerebral compression ischaemia in rats (n = 15). Photochemical injury in the spinal cord is produced by irradiation with an Argon laser (560 nm/CW/2.5 min) of the intact spinal column at the T 8 level after intravenous administration of rose Bengal (0.133 ml/100 g BW). The spinal cord and brain were frozen in situ by pouring liquid nitrogen into a funnel fitted to the spinal column and skull bone[8]. Tissue slices were prepared as 15 μ frozen sections by cryostat from a block of frozen tissues.

The mapping of superoxide free radicals. The assessment of superoxide anions in the CNS is based on the 380 nm chemiluminescence of CLA-phenyl upon reaction with superoxide in the tissue[7]. The frozen tissue section is mounted on millipore filter paper (CA 250/0 Schleicher & Schuell Com. W. Germany) soaked with saturated CLA-phenyl solution. The millipore filter paper is immediately mounted on an ASA-20,000 polaroid film in the dark room at room temperature. The superoxide-CLA-phenyl photochemical reaction in the millipore hole of the filter paper is visualized by 2 minutes exposure (Figs. 1 and 2). Other frozen tissue slices are prepared for the study of energy metabolism[3] and vascular permeability with extravasation of Evan's Blue.

Results

Normally, a low intensity of CLA-chemiluminescence from superoxide was observed as shown in Fig. 1. CLA-chemiluminescence was inhibited by ascorbic acid (0.25 mM), superoxide dismutase (10 units) and activated by Fe-Cl (0.01 mM) with oxygen saturation. The

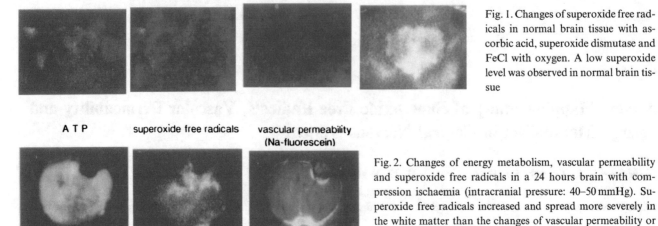

no medication   ascorbic ascid   superoxide dismutase   FeCl(0.01mM) + O_2
                     (0.25mM)   (10 units)

Fig. 1. Changes of superoxide free radicals in normal brain tissue with ascorbic acid, superoxide dismutase and FeCl with oxygen. A low superoxide level was observed in normal brain tissue

A T P   superoxide free radicals   vascular permeability
                                             (Na-fluorescein)

Fig. 2. Changes of energy metabolism, vascular permeability and superoxide free radicals in a 24 hours brain with compression ischaemia (intracranial pressure: 40–50 mmHg). Superoxide free radicals increased and spread more severely in the white matter than the changes of vascular permeability or the loss of ATP

relationships between the distribution of free radical generation, the changes of vascular permeability and energy metabolism in 24 hours with compressed cerebral ischaemia are presented in Fig. 2. Sequential changes of superoxide free radicals and energy metabolism were studied in the photochemical induced spinal cord ischaemia. These sequential studies have documented that severe activation of superoxide free radicals occurs before changes of ATP and the increase of vascular permeability. The activation of superoxide free radicals in the white matter without increasing the vascular permeability is interesting as a specific cause of formation of oedema in the white matter (Fig. 2). Prolonged free radical reactions over few days in the necrotic tissue have been observed in the photochemically induced spinal cord ischaemic lesion. The most intense generation of superoxide anions was unexpectedly always located in the center but not in the marginal area of the lesion.

Discussion

2-Methyl-6-phenyl-3,7-dihydroimidazole-[1,2-alprazin-3-one(CLA-phenyl)] produces at 380 nm a photochemical light response from superoxide free radical[7]. The photochemical reaction is extremely weak, and only detectable by supersensitive ASA 20 000 polaroid film. The quantitative photographical intensity of CLA-chemiluminescence upon reaction with superoxide is also affected by other factors, such as thickness of the frozen section, sensitivity of the polaroid film, exposure time, concentration of the CLA-phenyl solution. Further, artifacts play a role, such as contamination by extrachemiluminescence light during expo-

sure, movement of the millipore filter paper, and electrostatic reactions on the polaroid film from manipulation. Therefore, an exact 2 minutes exposure in a completely dark room is required for consistently good pictures. If the exposure time is prolonged over 3 minutes, the photochemical intensity assessment of the superoxide must be calibrated by formula 1.

$$PCU = K_1 \frac{\text{photographic intensity}}{\text{back grand intensity}} \times \frac{1}{\text{exposure times (min.)}} \quad (1)$$

We have no idea yet for conduction of a correct quantitative evaluation of superoxide free radicals by exposure of polaroid films in this stage. Only, the percent changes of regional superoxide free radicals in the lesion can be compared with that of normal tissue as calibration for a quantitative analysis.

As a source of superoxide anions in biological systems, oxidation of ubisemiquinone radicals in the respiratory chain[11], prostaglandin synthesis[5], conversion from available nucleotides (NADPH)[5], purine degradation with oxidation of xanthine oxidase, conversion of xanthine to urea, and activation of neutrophils have been reported[2]. A role of free radicals in ischaemia, trauma, inflammation and various kinds of injury has been considered as a causes of vasogenic oedema[1, 12] and neuronal cell membrane damage[10]. Especially, activation of superoxide anions by acidosis in the lesion is discussed as a mechanism of promoting secondary tissue damage. Under acid conditions, addition of protons converts the superoxide anion into perhydroxy and hydroxy radicals, which are very reactive toxic oxidizing agents[10]. This condition could be also analysed by topographical tissue pH mapping[3]. The new technique of simultaneous mapping of superoxide free radicals, vascular permeability, and energy metabolism opens up an opportunity for identifying regional effects of free radicals in the pathophysiology of CNS ischaemic injury.

References

1. Chan PH, Schmidley JW, Fishman RA, Longar SM (1984) Brain injury, edema, and vascular permeability induced by oxygen-derived free radicals. Neurology 34: 315–320
2. Fantone JC, Ward PA (1985) Polymorphonuclear leucocyte-mediated cell and tissue injury. Prog Pathol 16: 973–978
3. Kim SH, Handa H, Ishikawa M, Hirai O, Yoshida S, Imadaka K (1985) Brain tissue acidosis and changes of energy metabolism in mild incomplete ischemia-topographical study. J Cereb Blood Flow Metab 5: 432–438
4. Kellogg EW III, Fridovich I (1975) Superoxide, hydrogen peroxide, and singlet oxygen in lipid peroxidation by xanthine oxidase system. J Biol Chem 250: 8812–8817
5. Kukreja RC, Kontos HA, Hess MH, Ellis EF (1986) PGH synthase and lipoxygenase generate superoxide in the presence of NADH or NADPH. Circ Res 59: 612–619
6. McCord JM, Day DE Jr (1978) Superoxide-dependent production of hydroxy radical catalyzed by iron-EDTA complex. FEBS Lett 86: 139–142
7. Nakano M, Sugioka K, Ushijima Y, Goto T (1986) Chemiluminescence probe with cypridina luciferin analog, 2-methyl-6-phenyl-3,7-dihydroimidazo [1,2-alpyrazin-3-one] for estimating the ability of human granulocyte to generate superoxide. Anal Biochem 159: 363–369
8. Poten U, Ratcheson RA, Salford LG, Siesjo BK (1973) Optimal freezing conditions for cerebral metabolites in rats. J Neurochem 21: 1127–1138
9. Prado R, Dietrich DW, Watson BD, Ginsberg MD, Green BA (1987) Photochemically induced graded spinal cord infarction. J Neurosurg 67: 745–753
10. Siesjo Bo K, Bendek G, Koide T, Westerberg E, Wieloch T (1985) Influence of acidosis on lipid peroxidation in brain tissue *in vitro*. J Cereb Blood Flow Metab 5: 253–258
11. Turrens JF, Alexandre LA, Lehninger AL (1985) Ubisemiquinone is the electron donor for superoxide formatione by complex III of heart mitochondria. Arch Biochem Biophys 273: 408–414
12. Wei EP, Christman CW, Kontos HA, Povlishock JT (1985) Effects of oxygen radicals on cerebral arterioles. Am J Physiol 248: H 157–H 162

Correspondence: N. Hayashi, M.D., Nihon University, Department of Neurological Surgery, Oyaguchi Kamimachi 30, Itasbashi-ku, Tokyo 173, Japan.

Acta Neurochirurgica, Suppl. 51, 34–36 (1990)

Measurement of Brain Tissue Density Using Pycnometry

G. R. DiResta, J. Lee, N. Lau, F. Ali, J. H. Galicich, and **E. Arbit**

Neurosurgical Research Laboratory, Memorial Sloan Kettering Cancer Center, New York, N.Y., U.S.A.

Summary

A novel method to measure specific gravity (SG) of tissues, pycnometry (PYC), is described. This method utilizes a 2 ml glass pycnometer filled with distilled H_2O to determine the displacement volume of a tissue sample and an equation to compute SG from the sample's weight and the pycnometer's weight before and after adding the sample. The PYC method was validated using glass SG standards over the range 1.02–1.26, and against the column density gradient (DG) method using brain tissue from 250–300 g male rats. Factors which affect PYC accuracy, *i.e.* sample size, were also evaluated. Our results indicate that PYC SG values are highly correlated with the glass SG standards (slope = 1.0107, r = 0.9955, p < 0.001), and highly correlated with DG when ~ 0.120 ml tissue samples are used in the pycnometer. The DG method was preferable to the PYC method, however, when small tissue samples, *i.e.* 0.60 ml or less, were used.

Introduction

Measurement of a tissue's specific gravity (SG) or % water content are quantitative indices of the degree of oedema. The most straightforward method of measuring % water content is to determine the difference between wet and dry weights[6] of a sample. In spite of its simplicity, the method has two vulnerabilities which precludes rapid assessment of tissue water, *i.e.* its requirement for large samples to ensure reasonable accuracy, and a drying time that exceeds 24 hours[4]. SG measurements, performed using a density gradient (DG) column, provide an alternative index of oedema. The DG approach has been used by many investigators[1, 2, 3, 5]. It is simple to perform and shows reasonable reproducibility and precision with smaller samples in comparison with the dry/wet method. The accuracy of the method depends upon several factors such as the linearity of the density gradient, conditions of calibrating standards, temperature and humidity of dissection area, size of the sample, and the interval

between sample introduction into the column. The DG method uses large amounts of kerosene and bromobenzene, two volatile, flammable and toxic chemicals which may also affect the lipid content of the sample. The column preparation is a tedious process as well.

In this paper we introduce pycnometry, a simple method to measure tissue SG. We evaluated its accuracy against glass density standards, determine the optimal tissue specimen size using a uniform density material, and correlate it against the DG method with rat brain tissue samples.

Materials and Methods

Pycnometry

The pycnometer (PYC) method requires a micro-analytical balance and a small glass pycnometer. We used a 2 ml glass pycnometer (Fisher Inc.) with its center hole plugged to minimize evaporative losses and placed two small black alignment marks on the ground glass stopper and on the outside neck of the flask. These marks were aligned to each other to ensure that the fluid volume remained the same between measurements. The pycnometer was carefully cleaned prior to use and wiped dry using lint free paper. The fluid used in our studies was pure, distilled water.

The equation below was used to determine tissue density from the weight measurements.

$$d_s = s^* \, d_w/(m_{pf} - m_t + s)$$

where d_s = density of the sample (g/ml); d_w = density of water (g/ml); s = weight of sample (g); m_{pf} = weight of pycnometer and water (g); m_t = weight of pycnometer, water and sample (g).

Tissue pycnometry is implemented by the procedure below. All measurements must be made at the same temperature.

a. Fill the pycnometer with pure distilled water. Stopper, align pycnometer marks, dry the outside of the flask with lint-free wipes and weigh (m_{pf}). Tiny air bubbles can be removed by immersing the flask into an ultrasonic cleaner for several sec.

b. Weigh the tissue specimen (s).

c. Remove some of the water from within the pycnometer and transfer the tissue specimen into the flask. Re-fill the pycnometer with additional water, remove bubbles, align the marks, dry the outside of the flask with lint-free wipes then weigh (m_t).

d. Compute the SG using the equation above.

Density Gradient Column Method

The DG method presented by Marmarou *et al.*[2, 4] was implemented. Tissue samples weighing approximately 60 mg were taken from the parietal, frontal, occipital, medulla and cerebellum of fresh rat brain. The specimens were lowered into the column, at 5 min intervals, and allowed to descend. The depth of each sample was recorded after 2 minutes and related to its SG with the calibration plot.

Surgical Procedures

Sprague-Dawley male rats (250–300 g) were anaesthetized with the mixture of 5 mg/kg xylazine and 80 mg/kg ketamine, and then sacrificed by injection of saturated KCl. The rat brain was quickly removed from the cranium to avoid dehydration of the tissues. Tissue samples were taken from the frontal, parietal, occipital cortex and the cerebellum and medulla. Each sample was divided into two pieces, *i.e.* ∼ 0.120 ml for the PYC and ∼ 0.060 ml for the DG method.

Uniform Density Tissue Preparation

A rat brain was homogenized and mixed with 10 ml of warm, sterile 30% gelatine solution. The mixture was poured into a sterile petri dish, covered and allowed to set. It was then cut up into varying sized samples and used with both the DG and PYC method to determine the optimal sample volume for pycnometry.

Results

Minimum Sample Size for Pycnometry

Our data indicates that a sample volume of approximately 0.120 ml should be used to be confident of the accuracy of the measurement.

Accuracy of Pycnometer Method

Figure 1 displays the accuracy of pycnometry using glass, SG calibration beads (Techne, Inc.).

The average volume of these beads was ∼ 0.120 ml.

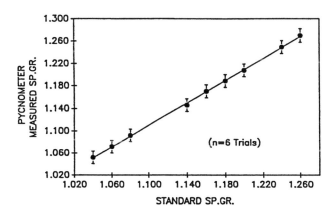

Fig. 1. Accuracy evaluation plot: measurement of specific gravity by pycnometry using glass, specific gravity calibration beads

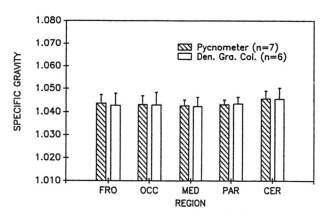

Fig. 2. Specific gravity measurements of rat brain tissue by pycnometry and the density gradient column. *FRO* frontal cortex; *OCC* occipital cortex; *MED* medulla; *CER* cerebellum; *PAR* parietal cortex

The slope of the linear regression line was 1.0107, r = 0.9955. No significant difference was observed between this slope and 1.0, the ideal slope (p < 0.001).

Correlation: Pycnometry against Density Gradient Column

The SG of tissue taken from five brain regions measured with pycnometry and the DG method are shown in Fig. 2; sample size was approximately 0.120 ml and 0.06 ml respectively.

There is no statistically significant difference between the values obtained by the two methods for any of the regions.

Discussion

Pycnometry is capable of measuring tissue SG accurately and quickly using specimens whose volume is approximately 0.120 ml or more. The major instrumental requirement is for an analytic balance with a resolution of at least 0.01 mg. Careful gravimetric technique, especially cleaning and handling of the PYC between weighings, is very important to performing the method. In contrast, the DG method has the capability of determining SG of smaller (< 0.12 ml) samples with a higher precision. The column mixture is, however, very toxic and a new gradient must be prepared after several measurements because of gradient disruption and cluttering of specimens within the column. All factors considered, pycnometry is a simple technique which can be used for rapid, measurements of tissue SG of most samples, whereas the DG method is preferred for particularly small samples.

References

1. Klatzo I, Chui E, Fujiwara K (1980) Resolution of vasogenic brain oedema. In: Cervos-Navaro J, Ferszt R (eds) Advances in neurology, Vol 28. Raven Press, New York
2. Marmarou A *et al* (1978) A simple technique for the measurement of cerebral oedema. J Neurosurg 49: 530–537
3. Marmarou A *et al* (1980) Biomechanics of brain oedema and effects on local cerebral blood flow. In: Cervos-Navarro J, Ferszt R (eds) Advances in neurology, Vol 28. Raven Press, New York
4. Marmarou A *et al* (1982) An improved gravimetric measure of cerebral oedema. J Neurosurg 56: 246–253
5. Shizuo Hatashita, Hoff JT (1988) Biomechanics of brain oedema in acute cerebral ischaemia in cats. Stroke 19, 1: 91–97
6. Steward-Wallace AM (1939) A biochemical study of cerebral tissue and of the changes in cerebral oedema. Brain 62: 426–438

Correspondence: G. R. DiResta, Nuclear Medicine Research Laboratory; Memorial Sloan Kettering Cancer Center; 425 E. 68th Street; New York, NY 10021, U.S.A.

Acta Neurochirurgica, Suppl. 51, 37–38 (1990)

Pathophysiology II (Methods, Mediators)

Experimental Studies for Use of Magnetic Resonance in Brain Water Measurements

P. P. Fatouros and A. Marmarou

Medical College of Virginia, Departments of Radiology and Neurosurgery, Richmond, Virginia, U.S.A.

Summary

The accurate description and quantification of altered brain water resulting from different pathologic conditions is of critical clinical importance. In this work we determined the influence of total water content, hydration fraction and magnetic field strength on observed proton relaxation rates by means of in vitro and in vivo model studies and developed a scheme for determining water content at any field strength. Equations relating T1 relaxation times and brain water content are derived. This allows a non-invasive measure of brain water to be determined in the clinical setting.

Introduction

The advent and widespread availability of MRI systems coupled with the high sensitivity of MRI to water content render this technoloy attractive for non-invasive measure of brain water [1, 2, 3]. Work in our laboratories has raised expectations that this approach may develop into a reliable method for characterization of brain edema. However, full and accurate utilization of this approach requires a detailed understanding of the state of accumulated water and the basic mechanisms underlying T1 and T2 relaxation. The objective of this research was to utilize both phantoms and experimental animal data to determine the correlation between the relaxation time and the amount of brain water present. With the latter measurements, the potential of this approach for accurate in vivo determinations of brain water content is explored and defined. The application of this method to humans and the correlation of brain water in patients is described in another report (see Marmarou et al., this volume).

Methods

Preparation of Gelatin Standards

Tissue mimicking gels of varying water content (68–90%) were prepared using appropriate proportions of doubly distilled water and commercial gelatin. The hot solutions were decanted into sample vials and sealed. Once cooled, samples were excised from each preparation for determination of the bound water fraction. The same set of six vials was used for all the field dependent relaxation time experiments.

Relaxation Time Measurements

Longitudinal relaxation time measurements for the gelatin solutions were carried out at 21 degrees C on four different magnetic resonance units operating at nominal proton Larmor frequencies of 5, 41, 63 and 100 MHz. The T1 determinations were made in two commercial imagers at 41 and 63 MHz (Siemens Magnetom and GE/Sigma, respectively).

Measurement of Bound Water Fraction

Bound water fraction of gelatin solutions and cat brain samples were performed with a Hart 7707 heat conduction scanning microcalorimeter consisting of four matched removable cells. The fraction of bound water was found from the ratio of the specific heats and melting of the sample and pure water. The hydration fraction k was found by dividing the mass of bound water by the mass of solute.

In-vivo Measurements of Relaxation Times

In this study, 5 adult cats anaesthetized with alpha chloralose and maintained normocapnic were used for determination of brain water content gravimetrically and from measurements of relaxation times. Brain tissue water was elevated using the infusion model of Marmarou[4]. Following the infusion of fluid, cats were imaged in the MRI unit. The inversion recovery method was used for T1 measurements. Following these measures, animals were sacrificed and brain tissue water determined at selected regions.

Results

The accuracy of the imaging protocol was verified by means of nickel-doped agarose gels having relaxation times simulating normal and oedematous brain white matter. The relaxation rates $(1/T_1)$ of all the gelatin solutions when plotted against their inverse total water content showed a linear behaviour as predicted by the fast two-state model. Similarly, linearity was obtained between the *in vivo* determined longitudinal relaxation times at 1 Tesla and their inverse water. The latter was measured by gravimetry. The equation describing this relationship for the cat white matter is

$$1/T_1 = -0.304 + 3.21 \, (1/W) \quad (r = 0.95)$$

Discussion

Investigations of the relationship between tissue water and relaxation times have been actively pursued for several years and are numerous. Several recent studies have underscored the dominant role of total water content in explaining the relaxation rates of oedematous brain tissue. A consistent picture emerges when this evidence is presented in the context of a fast proton exchange model. This model describes in a general sense tissue water as being either in a free bulk state or in an environment in which its motion if influenced by the presence of proteins and other macromolecules.

In the studies reported here, the gelatin solutions were selected to study the effect of total water and water changes because of the similarity in their relax-ation behaviour with brain tissue. The final equation evolving from our work relating T1 and water content equaled $1/T1 = (k/T1b)(1/W-1)$ where T1b reflects the entire frequency dependence of the measured T1 values, k the ratio bound water to solid tissue component and W the actual tissue water. The predicted linearity between $1/T1$ and $1/W$ was found to hold in both gelatin solutions and in experimental animal models of brain edema. Knowledge of this relationship along with precise measurements of $1/T1$ at a given field strength permit quantitative *in vivo* assessments of brain water content to be obtained with an uncertainty of less than 1%.

Acknowledgement

This work was supported in part by grants NS 19235 and NS 12587 from the National Institutes of Health. Additional facilities and support were provided by the Richard Roland Reynolds Neurosurgical Research Laboratories.

References

1. MacDonald HL, Bell BA, Smith MA, Kean DM, Tocher JL, Douglas RHB, Miller JD, Best JJK (1986) Br. J. Radiol 59: 355
2. Bell BA, Kean DM, MacDonald HL, Barnett GH, Douglas RHB, Smith MA, McGhee CNJ, Miller JD, Tocher JL, Best JJK (1987) Lancet i: 66
3. Kaneoke Y, Furuse M, Inao S, Saso K, Yoshida K, Motegi Y, Mizuno M, and Izawa A (1987) Magn. Reson. Imaging 5: 415
4. Marmarou A, Nakamura T, Tanaka K (1984) The kinetics of fluid movement through brain tissue. Seminars in Neurology, Vol 4, 439–444

Correspondence: P. Fatouros, Ph.D., Medical College of Virginia, Radiation Physics Division, MCV Station, Box 72, Richmond, VA 23298, U.S.A.

Acta Neurochirurgica, Suppl. 51, 39–42 (1990)
© by Springer-Verlag 1990

MR Studies of Brain Oedema in the Developing Animal

A. V. Lorenzo[1], R. V. Mulkern[2], S. T. S. Wong[2], V. M. Colucci[2], and F. A. Jolesz[2]

[1] Department of Neurosurgery, Children's Hospital and Brigham And Women's Hospital and [2] Department of Radiology, Brigham Women's Hospital, Boston, Massachusetts, U.S.A.

Summary

Assessment of perinatal brain oedema is complicated by normal changes in brain water that accompany the marked physiological, biochemical and morphological alterations occurring during this phase of development. Multiexponential analysis of transverse decay curves (TDCs), derived from 128 echo CPMG images, of white matter (WM) made oedematous by either exposure of animals to triethyltin (TET) or cryogenic cortical lesions revealed a second, slower decay component not apparent in controls. More significantly, an obvious difference was noted between the TET and cryogenic lesion fast decay components which might serve as a basis to differentiate non-invasively cytotoxic and vasogenic oedemas.

Introduction

Nuclear magnetic resonance (NMR) has been widely used to study the state of water in biological tissues. It has been used successfully to investigate myelin and myelin associated water compartments of brain, demyelination and experimental brain oedema[3, 4]. Much of this work has focused in the mature brain which prior to extensive aging has stable myelin and water compartments even though these compartments turn over constantly. In the newborn, these compartments turn over too but, in contrast to the adult, water and myelin content change dramatically, particularly during the perinatal period. MR studies of brain myelin during postnatal development have revealed reductions in the relaxation times T_1 and T_2 that may be attributable to a progressive decrease in the content or mobility of water protons[4].

The object of this study was to investigate changes in MR parameters observed during postnatal development in normal rabbits and in those made anoxic or exposed to triethyltin (TET) or cryogenic cortical lesions.

Materials and Methods

A total of 158 New Zealand White rabbits from 15 litters ranging in age from newborn to adult were used in this study. Animals were assigned to one of four groups. *Anoxic group* – Animals were made anoxic in a $45 \times 45 \times 36$ cm plexiglas chamber by gassing with 100% N_2 until their heart rates fell to 70% of normal. Some were killed immediately, others were returned to room air and allowed to recover for 5 or 240 min before being euthanized. *TET group* – Animals were injected daily for 5 days either with 1 mg/kg TET or saline. They were examined and weighed each day. *Cryogenic group* – Adult animals were anesthetized with ketamine plus xylazine. A trephine plug of parietal bone, which was 0.8 cm in diameter, was removed without damaging the underlying area. A brass cylinder (0.5 mm in diameter), cooled to −90 °C with liquid nitrogen, was placed on the dura for 3 sec or until the temperature rose to −50 °C. After closure of the wound the animals were killed immediately or at 24, 48 or 168 hours after cold injury. *Controls* – Littermates not exposed to the respective experimental lesions, but otherwise treated similarly, were used as controls.

Prior to sacrifice brains of selected animals in each group were imaged using a 6 cm 1.5 tesla imager. Images were acquired with a spin echo pulse sequence (TR = 2,000, TE = 80 ms) or a newly developed 128 echo CPMG sequence[5]. Biexponential analysis of TDCs, derived from the 128 echo CPMG images, were obtained from regions of interest. Subsequently, those less than 9 days of age were decapitated; older ones were euthanized with 50 mg/kg pentobarbital iv. The brains and other tissue samples were removed as quickly as possible, weighed and placed in a humidity chamber for dissection. Water and electrolyte content of selected brain areas and tissues was determined by the method of desiccation to constant weight and by flame photometry, respectively[4]. *In vitro* T_1 and T_2 relaxation times were determined using an IBM Minispec operating at a Larmor frequency of 10 MHz. T_1 was measured by an inversion recovery sequence, while T_2 was measured using a 10 spin echo Carr-Purcell-Meiboon-Gill (CPMG) sequence. Measurements were completed within 2–4 hours after death.

Results

During the first 30 postnatal days, water content decreased more rapidly in rostral then in caudal areas of

Table 1. *Water Content (%) of Selected Brain Areas and Muscle*

Group	Cortex	WM	Medulla	Muscle
Control				
Newborn [10]	88.10 ± 0.42	88.07 ± 0.43	86.48 ± 0.59	83.62 ± 0.77
Adult [7]	80.57 ± 0.15	68.81 ± 0.55	73.00 ± 0.25	75.22 ± 0.38
Anoxia				
Newborn* [6]	88.60 ± 0.45	88.60 ± 0.58	86.10 ± 0.60	83.59 ± 0.50
Adult** [5]	79.73 ± 0.54	69.01 ± 0.32	73.36 ± 0.60	75.53 ± 0.47
Triethyltin				
Newborn [5]	88.46 ± 0.23	92.64 ± 1.79	91.19 ± 1.14	83.22 ± 0.52
Adult [5]	80.46 ± 0.23	78.64 ± 1.29	76.76 ± 1.14	75.22 ± 0.52
Cryogenic lesion				
Newborn [0]				
Adult [5] (24 hr post)	82.72 ± 0.79	76.09 ± 1.47	73.34 ± 0.36	75.94 ± 0.94

Average exposure time to 100% N_2 was *16 ± 3 and **1.3 ± 0.3 min. Arterial blood pO_2 dropped from *87.5 ± 2.1 to 0.7 ± 0.6 mm Hg in 3–5 min and **90.3 ± 2.9 to 3.3 ± 0.4 mm Hg within 45 to 60 sec of onset of anoxia.

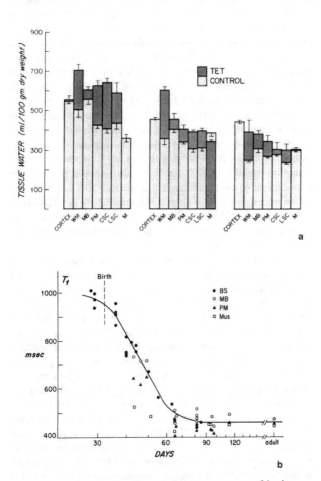

Fig. 1. (a) MR relaxation T_1 and (b) water content of brain areas and muscle (M) as a function of age

brain (Fig. 1). Thereafter, the decline was more gradual. T_1 and T_2 (not shown) declined in a parallel fashion from 1,000 to 500 msec postnatally and thereafter re-mained relatively unchanged. Neither brain water content, blood-brain barrier (bbb) permeability to ^{125}I-albumin or spectroscopic MR parameters were altered in newborn and adult rabbits following severe hypoxia (Table 1). Marked increases in water content occurred in all myelinated areas of brain and spinal cord, but not in cortex or muscle, following TET exposure (Table 1). In neonates these increases were most apparent in the spinal cord and caudal areas of brain while in older animals such increases were most apparent in the rostal subcortical WM (Fig. 1a). Parallel changes were observed in T_1 and T_2 (Fig. 1b) and in the relative signal intensity of T_2 weighed MR images from myelinated areas (Fig. 2). In contrast to the WM decay curves from controls, which decayed at an average rate of 12.6 ± 1.2 Hz and were essentially mon-oexponential those obtained from TET animals were biexponential. The decay components were of approx-imately equal size with the slowly relaxing component averaging 2.2 ± 0.2 Hz and faster component 9.1 ± 0.7 Hz. In cryogenic cortical lesioned animals in-creases in brain water were restricted to the ipsilateral cortex and to the ipsilateral and contralateral subcor-tical WM. Analysis of TDCs from the ipsi and con-tralateral WM revealed biexponential behavior with the slowly relaxing component of the contralateral side decaying at an average rate of 2.3 ± 0.2 Hz (relative size of the component 47.6 ± 10.7%) and the fast com-ponent at 6.6 ± 0.3 Hz (relative component size (52.4 ± 10.5%).

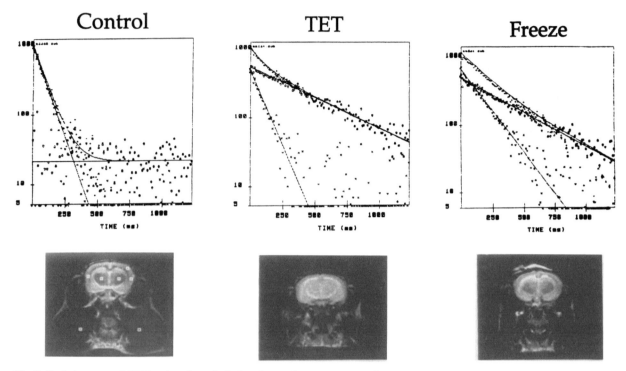

Fig. 2. Brain images and TDCs taken from 1 pixel regions (0.78 × 0.78 × 5 mm³) of subcortical WM from free separate animals. The images from TET and cryogenic treated animals display a relatively brighter subcortical WM than from controls. Note monoexponential decay rate of TDC for control WM vs the biexponential decay for TET and cryogenic treated animals

Discussion

Tissue dehydration is a normal accompaniment of growth but there occurs early in life a more precipitous loss of tissue water that is related more to weight gain than age[1]. Others have noted that the brain shares in this process[8] and the evidence presented in this study is consistent with this observation. The fact that in newborns brain water content is much greater than in older animals and that it decreases markedly during early postnatal life complicated the assessment of perinatal brain oedema.

Asphyxia is recognized as one of the major causes of perinatal brain oedema. However, under such conditions the brain becomes both hypoxic and ischemic. The suggestion that lack of oxygen alone will induce a cytotoxic edema in neonates is neither supported by our results or that of others[7].

TET induced oedema was limited to myelinated areas of the brain in both neonates and older animals. These changes were paralleled by prolongation of both T_1 and T_2 and an increase in the relative signal intensity of WM structures in the MR images. Interestingly, multiexponential analysis of the TDCs for WM, but not for cortex or muscle, revealed a second, slower component[2], equal in size to the faster component, not apparent in controls. A similar slow component was detected in oedematous WM of animals sustaining a cortical cryogenic lesion the decay rate of which was similar to that of the TET group. The fact that the fast component of the TDC for cryogenic induced oedema had a smaller relaxation rate than that for TET oedema appears paradoxical since the former has a higher protein content. However, relaxation rates, as analyzed within the framework of biexponential models, are complex and influenced by other factors including water content and equilibrium water exchange rates[6]. The results obtained imply that in oedematous WM water exchange rates are slower in the cryogenic than the TET induced lesion. Finally, it is concluded that the obvious and statically significant difference observed between decay rates of the respective fast components provides a basis for the non-invasive MRI differentiation of cytotoxic and vasogenic oedemas.

Acknowledgements

This work was supported by NIH grants NS 23093 and by MR Center Grant HD 18655.

References

1. Coulter DM, Avery ME (1980) Paradoxical reduction in tissue hydration with weight gain in neonatal rabbit pups. Pediatr Res 14: 1122–1126

2. Go GK, Edzes HT (1975) Water in brain oedema. Observations by the pulsed nuclear magnetic resonance technique. Arch Neurol 32: 462–465
3. Jolesz FA, Polack JF, Ruenzel PW *et al* (1984) Wallerian degeneration demonstrated by magnetic resonance: spectroscopic measurements on peripheral nerve. Radiology 152: 85–87
4. Lorenzo AV, Jolesz FA, Wallman JK *et al* (1989) Proton magnetic resonance studies of triethyltin-induced oedema during perinatal brain development in rabbits. J Neurosurg 70: 432–440
5. Mulkern RV, Bleier AR, Jakab P *et al* (1990) High field CPMG imaging sequences for transverse relaxation time studies. Magn Reson Med (in press)

6. Mulkern RV, Bleier AR, Adzamli IK *et al* (1989) Two-site exchange revisited: A new method for extracting exchange parameters in biological systems. Biophys J 55: 221–232
7. Norris JW, Pappius HM (1970) Cerebral water and electrolytes. Effect of asphyxia, hypoxia and hypercapnia. Arch Neurol 23: 248–258
8. Welch K (1980) The intracranial pressure in infants. J Neurosurg 52: 693–699

Correspondence: A. V. Lorenzo, M.D., Department of Neurosurgery, Children's Hospital, 300 Longwood Avenue, 02115 MA Boston, U.S.A.

Acta Neurochirurgica, Suppl. 51, 43–45 (1990)

Passage of DMP Across a Disrupted BBB in the Context of Antibody-mediated MR Imaging of Brain Metastases

J. W. M. Bulte[1], M. W. A. de Jonge[1], L. de Leij[1], T. H. The[1], R. L. Kamman[2], B. Blaauw[3], F. Zuiderveen[4], and K. G. Go[4]

Departments of [1] Clinical Immunology, [2] Magnetic Resonance, [3] Histology and Cell Biology and [4] Neurosurgery, University of Groningen, The Netherlands

Summary

To study the possible application of monoclonal antibody/dextran-magnetite conjugates in specific MR imaging of brain metastases, both components of these conjugates were tested for their ability to penetrate the endothelium during conditions of local blood-brain barrier (BBB) impairment.

The passage of dextran-magnetite particles (DMP) across a disrupted blood-brain barrier was studied in a freezing lesion model using electron microscopy (EM) and MR imaging. One hour after *i.v.* injection, focal accumulation of DMP in capillary endothelial cells within the freezing lesion was shown by EM. In parallel with this, MR imaging indicated a strong contrast enhancement in the lesion. EM observations showed that the particles were still present in the endothelial cells four and eight hours after injection.

The passage of an anti-small cell lung cancer (SCLC) monoclonal antibody across the endothelium of intracerebrally xenografted human SCLC was studied using immunohistological techniques. It was found that passage across endothelial cell occurred in the tumor within four hours after injection.

Introduction

At the present time magnetic resonance imaging (MRI) has the best spatial resolution for the visualization of cerebral lesions. To further improve the sensitivity of MRI in the depiction of these lesions, Gd-DTPA can be administered intravenously. Areas with abnormal blood perfusion as well as regions with an impaired BBB, both characteristic features of brain metastases, might thus be enhanced by the use of Gd-DTPA as contrast agent. However, the enhanced detection still lacks specificity and attempts to use Gd-DTPA labeled monoclonal antibodies (Mab) as an immunospecific MR contrast agent have not been successful[1]. This may be ascribed to the large difference between the minimal amount of Gd-DTPA which is required as label to induce adequate contrast and the maximum amount of label which is allowed to be attached to a Mab to conserve immunoreactivity.

As relaxivity (*i.e.*, the influence on MR parameters) is concerned, DMP is superior to Gd-DTPA. The coating of the magnetite particles with dextran enables the conjugation of Mab's[2], which may result in a strongly altered specific tumor signal as shown after *in vivo* targeting to subcutaneous xenografted neuroblastoma[3]. For a possible role of such DMP-Mab conjugates in the specific visualization of brain metastases, we have studied the passage of DMP across a disrupted BBB in the freezing lesion model using electron microscopy and MR imaging. In addition, using a rat model of intracerebrally xenografted human SCLC, we report here on the passage of the anti-SCLC Mab MOC-31 in this experimental brain metastasis model. A schematic overview is given in Fig. 1.

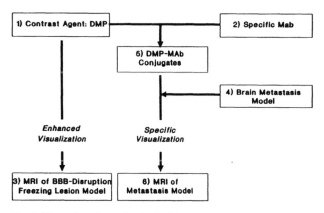

Fig. 1. Flow chart for technological development of MR imaging in the specific visualization of metastatic cerebral lesions

Material and Methods

Freezing Lesion Model and DMP Localization

The preparation of DMP was performed essentially as described[2]. Briefly, magnetite (Fe_3O_4) was precipitated by the addition of NH_4OH to a solution containing dextran ($Mw_{av} = 10\,kD$), $FeCl_2$ and $FeCl_3$. A freezing injury was induced in Wistar rats (weighing about 200 g) by placing a rod of dry ice ($-85\,°C$) to the dura of the left parietal cortex for 1 minute. Immediately following, 1 ml of DMP solution was injected *i.v.* in the tail vein at a dose of 450 µmol/kg. One hour after injection, MR imaging was conducted at 1.5 T (Philips Gyroscan S15, circular surface coil with ø = 8 cm) in a transverse plane using spin echo sequences with TR = 2,000 msec and TE = 50 msec. One (n = 3), four (n = 3) or eight (n = 3) hours after injection, the brains of the animals were fixed by cardiac perfusion of Karnofskyś fixative. After processing according to routine histological procedures (embedding in epon, staining with OsO_4, and uranyl acetate) ultrathin sections were made and examined with a transmission electron microscops.

Brain Metastasis Model and Mab Localization

GLC-28, a MOC-31 antigen-positive cell line established from a SCLC biopsy in our laboratory, was xenografted intracerebrally in 8 to 10 weeks old Wag/Rij nude rats. To this end, *in vitro* grown GLC-28 tumour cells were collected by centrifugation and suspended in tissue culture medium at a concentration of 5×10^6 or 2×10^7 cells/ml. Using a Hamilton microsyringe, 5 µl of this suspension was injected at a rate of 0.5 µl/min. For injection in the cerebral cortex, the needle was positioned from above the dura of the parietal cortex 4 mm deep in the bregma and 1 mm left to the sagittal fissure. For injection in the lateral ventricle, the needle was placed at the coronal suture 1.2 mm left of midline, and inserted to 6 mm depth. Immediately after injection, the perforation in the skull was sealed with methacrylate.

The mouse Mab MOC-31 (isotype IgG1), which recognizes a 40 kD membrane antigen present on human carcinoma, was purified by affinity chromatography using Protein A-Sepharose. Nude rats bearing GLC-28 xenografts were injected *i.p.* with 1 mg purified MOC-31 IgG in PBS and sacrificed at 4 or 24 hrs after injection. The *in vivo* localized MOC-31 was assessed by immunoperoxidase staining on cryostat sections made from the brain. To this end, the tissue sections were incubated with HRPO-conjugated rabbit anti-mouse Ig. Control incubations were performed by adding both MOC-31 and HRPO-conjugated rabbit anti-mouse Ig.

Results

Passage of DMP Across a Disrupted bbb

After induction of the freezing lesion and *i.v.* injection of DMP, the localization of the electron dense particles was studied by transmission electron microscopy. One hour after injection, accumulation of the particles in the cytoplasm of endothelial cells lining the brain capillaries in the freezing lesion was observed (see Fig. 2a). In addition to clusters of particles, individual particles with a diameter of about 5 nm (Fe-core) could be discriminated. Four as well as eight hours after injection, similar pictures were seen suggesting no further passage

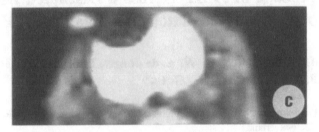

Fig. 2. Transmission electron micrographs (a, b; bar = 1 µm) and MR image (c) of DMP passage across a disrupted BBB 1 h (a, c) and 4 h (b) after injection

of DMP across the enothelial cell (see Fig. 2b). MR imaging showed that in the entire freezing lesion there was strong contrast enhancement induced by DMP (see Fig. 2c).

Intracerebrally Xenografted SCLC and in vivo Mab Binding

GLC-28 tumour cells were xenografted in the brain of nude rats. After a distinct time period, neurological disorders became apparent, which was taken at an indication of progressive tumour growth (see Table 1). When the tumour cells were injected into the cavity of the lateral ventricle, large tumours developed showing an abundant vascularization. A few tumour cells could also be detected in the cerebrospinal fluid of the cerebellum but not beyond the border of the ventricular spaces. When the tumour cells were injected into the cerebral cortex, small tumour noduli at the place of injection were formed. The development of blood vessels within these tumours was significantly less than within those growing in the ventricles. Occacionally, tumour cells could be detected in the ventricular spaces of the forebrain, whereas in some cases a thin layer of tumour cells was found on the dura of the parietal cortex just above the place of injection.

After *i.p.* administration of the Mab MOC-31, binding of the antibody to the tumour cells was found to occur within four hours after injection. This indicated that the antibody can penetrate the endothelium in the brain-localized tumour. However, the amount of *in vivo* localized Mab was low as compared to that found twenty-four hours after injection, since significantly more positive tumour cells could be detected. Positive staining was always observed around the nearby vicinity of blood vessels. Therefore, the rate of *in vivo* Mab localization between different tumours seems to vary depending on the amount of vascularization.

Discussion

In this study an experimental brain metastasis model is described which may be used for future antibody-mediated MR imaging. In this context, we have investigated the passage of both DMP and Mab across the endothelium. From these *in vivo* localization studies, it is concluded that DMP is an excellent contrast agent for the detection of cerebral lesions. It appeared that the particles could not pass the endothelium and that the particles cluster in the cytoplasm of disrupted endothelium. The size of these individual particles was about 5 nm (iron-core), comparable to that of an IgG

molecule. For endothelium of metastases in the brain, which tends to lack tight junctions, we have found transcapillary passage of Mab (IgG). Probably Mab-sized molecules such as DMP pass this endothelium by intercytoplasmic transport.

References

1. Eckelman WC, Tweedle M, Welch MJ (1988) NMR enhancement with Gd labeled antibodies. In: Srivastava SC (ed) Radiolabeled monoclonal antibodies for imaging and therapy. Plenum Press, New York, pp 571–579
2. Molday RS, MacKenzie D (1982) Immunospecific ferromagnetic iron-dextran reagents for the labelling and magnetic separation of cells. J Immunol Methods 52: 353–367
3. Renshaw PF, Owen CS, Evans AE, Leigh JS jr (1986) Immunospecific NMR contrast agents. Magn Reson Imaging 4: 351–357

Correspondence: K. G. Go, M.D., University Hospital Groningen, Department of Neurosurgery, Oostersingel 59, 9713 EZ Groningen, The Netherlands.

Acta Neurochirurgica, Suppl. 51, 46–48 (1990)

Continuous Monitoring of Blood-Brain Barrier Opening to Cr^{51}-EDTA by Microdialysis Following Probe Injury

O. Major[1], T. Shdanova[1], L. Duffek[2], and Z. Nagy[3]

National Institute of Neurosurgery, Budapest, [1] Institute of Physiology Acad. Sci. USSR, Leningrad, [2] Department of Nuclear Medicine, Semmelweis Medical University Budapest, [3] Department of Psychiatry of Semmelweis Medical University Budapest

Summary

Continuous detection of the permeability changes of the blood-brain (BB) interface became feasible with the recent development of the cerebral microdialysis technique[6].

Vasogenic brain oedema was induced by the insertion of the dialysis capillary into the cerebral cortex of rats, and blood-brain transfer was measured with Cr^{51}-EDTA as a blood-born tracer. The dialysis capillaries had an upper limit of 500 daltons (MW). Ringer solution was perfused through the dialysis tube. Samples were collected continuously. The permeability coefficient has been calculated from the radioactivity of the serum and of the washing fluid.

Introduction

The blood-brain transfer constant, (K_1) calculated from the parenchymal and blood tracer concentration, is a function of the effective volume fraction of the blood in which the substance resides, the cerebral blood flow (CBF), the permeability of the substance and the surface area of the blood-tissue interface. Blood-brain barrier (BBB) function in various experimental conditions can be evaluated by measuring K_1[2].

The cerebral microdialysis method is able to detect the presence of different compounds in the brain extracellular fluid[1, 5, 6, 7]. By combining this method with intravenous tracer administration, K_1 can be calculated continuously over a long period of time.

Materials and Methods

The experiments were carried out on 6 male Wistar rats (200–220 g). Following pentobarbital anaesthesia both femoral veins were catheterized. Through a parietal burr hole a dialysis probe (membrane length = 4 mm; 250 µm in wet diameter; permeability cut off = 5,000 Daltons MW) was implanted tangentially into the cerebral cortex. The probe as continuously perfused with Ringer solution by a microinfusion pump. The perfusion rate was 5 µl/min and dialysate was sampled in twelve 30 min fractions.

Twenty min after the implantation of the microdialysis tube, a bolus of Cr^{51}-EDTA was administered intravenously. Five min later the first blood sample (100 µl) was taken. Blood samples were obtained every 180 min thereafter. Before using the microdialysis system, all probes were calibrated with Cr^{51}-EDTA in Ringer solution. The radioactivity of serum and washing fluid samples was counted on a liquid scintillation counter (model: MK 300, GAMMA, Budapest).

Calculations

Following the approach of Ohno et al.[3], the unidirectional blood-to-brain transfer factor has been calculated as:

$$K_i = C_i(T) / \int_T^0 \cdot C_{p\,dt}$$

where C_i = parenchymal tracer concentration, C_p = blood concentration of tracer, and T = time of sampling. K_1 is related to the permeability-surface area product (PS) and CBF by $PS = CBF \ln (I-K_1/CBF)$.

The interstitial concentration can be estimated by a formula according to Benveniste et al.[1].

$$C_i = C_r \cdot [C_b (\lambda^2 \cdot t) / C_{rp} (t)] \cdot [\lambda^2/\alpha] \cdot K$$

where C_r = concentration of calibration media, C_b is the recovered tracer from the brain, and C_{rp} is the washed out tracer from calibration media during T time. Lambda (λ) is a tortuosity factor, α is the brain extracellular volume factor and factor K is expressed as: $[1 - (C_d/C_i) / [1 - (C_d' / C_r)]$. In the latter expression C_d is the concentration at the inner boundary of the dialysis membrane when placed in the brain, and C_d' is the same parameter when put into the calibration media [1].

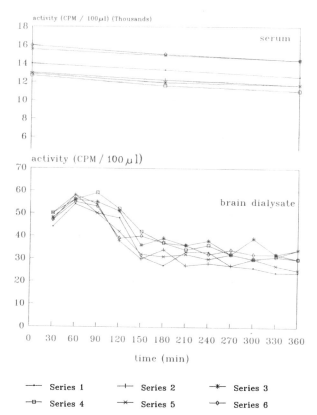

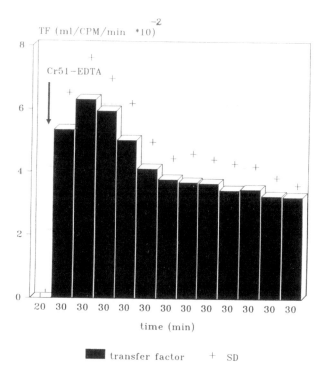

Fig. 2. Blood-brain transfer constant of ^{51}Cr-EDTA following brain injury. Bars indicate means; + indicates the SD

Fig. 1. Radioactivity in serum and brain dialysate after a bolus intravenous injection of [512]Cr-EDTA

In this experiment the flow rate was high, and factor K could be ignored. If a substance exchanges rapidly among different compartments of brain tissue, the λ^2/α ratio should be nearly 1.0. Therefore, with Ungerstedt's simplified formula[6], $C_i = C_r \cdot (C_b/C_{rp})$. The results are expressed as mean $\pm$ SD.

Results

A slight increase in radioactivity of the washing fluid was detected in the 2nd and 3rd 30 min samples (Fig. 1). Radioactivity decreased slowly in later samples of the washing solution. In the same period of time the activity of serum decreased steadily. There was a peak of the calculated transfer factor during the first two hours (Fig. 2). Extravasation of the radiotracer was detectable even 6 hours following probe "injury."

Discussion

The fundamental assumption of cerebral microdialysis is that the semipermeable membrane allows free diffusion of water and solutes between the cerebral interstitial space and washing fluid. That is, the concentration of a certain material in the dialysis perfusion fluid reflects its concentration in the interstitial space[5]. Cr[51]-EDTA does not readily penetrate the intact BBB. We could detect increased radioactivity in the washing fluid during the 6 hours of experiment. The radiotracer certainly crossed the BBB following the implantation of microdialysis probe. The implantation procedure is a "stab" injury that damages the BBB[4]. The dynamics of the opening is an important factor in the evaluation of all cerebral microdialysis experiments and may severely compromise the utility and interpretation of the resulting data.

Furthermore, this method allows continuous detection of fluid extravasation through the endothelium following BBB damage. The microdialysis method with the combination of the tracer technique could be a new approach to study the endothelial barrier in the brain.

References

1. Benveniste H, Hansen AJ, Ottosen NS (1989) Determination of Brain Interstitial Concentrations by Microdialysis. J Neurochem 52: 1741–1750
2. Fenstermacher J, Blasberg R, Patlak C (1981) Methods for quantifying the transport of drugs across brain barrier systems. Pharmacol Ther 14: 217–248
3. Ohno K, Pettigrew K, Rapoport S (1978) Lower limits of cerebrovascular permeability to non-electrolytes in the conscious rat. Am J Physiol 33: 635–645

4. Persson L, Hansson HA, Sourander P (1976) Extravasation, spread and cellular uptake of Evans blue-labelled albumin around a reproducible small stab-wound in the rat brain. Acta Neuropoath (Berlin) 34: 125–136
5. Phebus LA, Perry KW, Clemenc JA, Fuller RW (1986) Brain anoxia releases striatal dopoamine in rats. Life Sci 38: 2447–2453
6. Ungerstedt U, Herrera-Marchintz M, Jungnelius U, Stahle L, Tossman U, Zetterstrom T (1982) Dopamine synaptic mechanisms reflected in studies combining behavioural recordings and brain dialysis. In: Kotisaka M (ed) Advances in dopamine research. Pergamon Press, New York, pp 219–231

Correspondence: Z. Nagy, M.D., Department of Psychiatry, Semmelweis Medical University, Budapest, 1083, Balassa utca 6, Hungary.

Acta Neurochirurgica, Suppl. 51, 49–51 (1990)
© by Springer-Verlag 1990

Effects of Antineutrophil Serum (ANS) on Posttraumatic Brain Oedema in Rats

L. Schürer, U. Prügner, O. Kempski, K.-E. Arfors, and **A. Baethmann**

Institut of Surgical Research, Klinikum Großhadern, Ludwig-Maximilians-Universität München, München, Federal Republic of Germany

Summary

The role of white blood cells in acute central nervous system disorders, such as stroke or traumatic injury is poorly understood. In this experimental study the effect of neutropenia on posttraumatic brain oedema was investigated. Polyclonal antiserum against polymorphonuclear leukocytes was used to induce neutropenia. Control animals received normal serum or no treatment at all. Hemispheric swelling and brain water content after cryo-injury to the exposed left cerebral hemisphere was gravimetrically assessed, and by the wet weight-dry weight method. Administration of anti-neutrophil serum led to a significant reduction of circulating neutrophils. In untreated animals or in rats receiving normal serum, hemispheric swelling of the brain was $7.26 \pm 0.35\%$ or $8.52 \pm 0.26\%$, respectively. Neutropenia induced by anti-neutrophil serum resulted in a significant increase in hemispheric swelling to $11.44 \pm 0.84\%$ ($p < 0.001$). It may be noted, however, that the enhancement of brain swelling by neutropenia was not obvious when comparing the water contents of the affected hemispheres of animals subjected to cold injury with and without anti-neutrophil serum.

The enhancement of brain oedema by neutropenia suggests a protective function of neutrophils against extravasation in the presence of a broken blood-brain barrier.

Introduction

The most common forms of brain oedema in clinical practice are due to microvascular damage. This results in an increased cerebrovascular permeability and allows escape of serum constituents through the open blood-brain barrier into the brain parenchyma. Work of our group observing blood-brain barrier dynamics by using intravital microscopy combined with electron microscopy (EM) during cerebral exposure to arachidonic acid[4], suggested a role of white blood cells in the development of blood-brain barrier disruption. In animals with opening of the blood-brain barrier to macromolecules electron micrographs showed apparently activated polymorphonuclear leukocytes (PMNLs) adhering to, or penetrating through the vascular endothelium. In a recent abstract by Schoettle et al.[3], it was reported that accumulation of PMNLs in rat brain after mechanical trauma correlates well with the extent of cerebral oedema.

Based on these observations the following pathophysiological scenario might be suggested: White blood cells attracted by tissue injury with release of chemotaxins produce proteases and toxic oxygen radicals upon activation[5]. Thereby, these cells aquire a potential to destroy plasma-membranes of the vascular endothelium and to disrupt endothelial junctions. Opening of the blood-brain barrier with extravasation of oedema fluid might be the most important consequence. For example, neutrophil-dependent mediation of microvascular permeability has been demonstrated in the small intestine[1]. Therefore, we tried to answer in the present study, whether white blood cells have a role in traumatic brain oedema. For that purpose, neutrophils circulating in blood were depleted by a specific polyclonal antiserum against rat polymorphonuclear leukocytes (PMNLs).

Materials and Methods

Polyclonal antiserum against PMNLs of Wistar rats was raised in sheep according to the method of Sandler et al.[2]. The anti-serum was intraperitoneally injected into male Wistar rats 24, 12, and 1 hour before induction of a standard freezing injury ($-68\,°C$) of the left exposed cerebral hemisphere (KLATZO). Blood cell counts were made prior to injection of serum, before cold lesion, and at sacrifice. Only animals with less than 200 neutrophils per µl blood were included into the study. A second group of animals received normal sheep serum, which was processed as the anti-neutrophil serum. A group of control animals with cold injury did not receive any treatment. 24 hrs after cold injury the rats were sacrificed by bleeding.

Brain water content was determined by the wet-dry weight method and hemispheric brain swelling by gravimetrical assessment of the increase in weight over that of the contralateral hemisphere not exposed to trauma.

Results

Administration of anti-neutrophil serum resulted in a significant decrease (p < 0.001) of the circulating leukocytes. Counts of polymorphonuclear leukocytes of animals with normal sheep serum were not affected (Fig. 1). The columns to the left at 24 hrs prior to trauma (−24 h) show the neutrophil counts before injection of serum. Neutropenia from anti-neutrophil serum was pronounced 1 h before trauma and was still present 24 hrs after cold lesion. Lymphocyte and monocyte counts were not affected by administration of antiserum (data not shown). In untreated control animals, hemispheric swelling *i.e.* increase of hemispheric weight, was 7.26 ± 0.35%. Under treatment with normal sheep serum swelling was 8.52 ± 0.26%. Neutro-

penia induced by anti-neutrophil serum, however, resulted in a significant enhancement of hemispheric swelling to 11.44 ± 0.84% (Fig. 2). The increase in hemispheric swelling by antineutrophil serum was statistically significant at p < 0.001 when compared with the hemispheric swelling induced in experimental animals with normal neutrophil counts (no treatment, normal sheep serum). An enhancement of brain oedema by neutropenia could not be recognized, however, when comparing the water contents of the affected hemispheres of the three experimental groups (data not shown).

Discussion

In the present study, a marked increase of hemispheric swelling was found in animals made neutropenic by a specific antiserum against polymorphonuclear leukocytes. The currently found enhancement of extravasation by neutropenia is in contrast with results obtained from the small intestine subjected to ischaemia and reperfusion. There, antibodies against adhesion molecules (integrins), or systemic depletion of white cells by polyclonal antiserum significantly reduced the formation of postischaemic oedema[1]. Observations of Unterberg *et al.*[4] on leukocyte accumulation during cerebral superfusion with arachidonic acid causing opening of the blood-brain barrier might be interpreted as follows. It is conceivable that by superfusion of arachidonic acid chemotaxins were formed attracting white blood cells. These were found in electron micrographs to adhere at or penetrate through the vascular endothelium.

Observations on an accumulation of PMNLs 4–8 hrs after mechanical impact trauma to the brain[3] support our findings. The following pathophysiologic scenario may be developed. Polymorphonuclear leukocytes which become attracted to the brain by release of chemotaxins are attached to the sites of damage of the vascular endothelial surface. Thereby, together with blood platelets the leukocytes seal off capillary leaks by thrombus formation. Thus a lack of circulating neutrophils induced by antineutrophil serum may enhance extravasation of oedema fluid in the presence of a blood-brain barrier lesion. This was actually found in our current experiments.

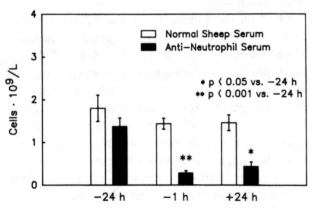

Fig. 1. Effect of anti-neutrophil serum on polymorphonuclear leukocyte count in peripheral blood of rats. The columns to the left, at 24 hrs prior to trauma (−24 h), represent control counts before injection of serum

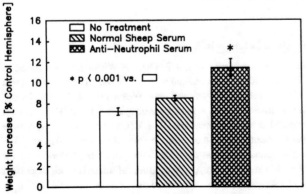

Fig. 2. Effect of normal sheep serum and anti-neutrophil serum in hemispheric brain swelling from a standardized cold injury

References

1. Hernandez LA, Grisham MB, Twohig B, Arfors K-E, Harlan JM, Granger DN (1987) Role of neutrophils in ischaemia-reperfusion-induced microvascular injury. Am J Physiol 253: H699–H703

2. Sandler H, Högstrop H, Lundberg C, Gerdin B (1987) Antiserum-induced neutropenia in the rat: characterization of a rabbit anti-rat neutrophil serum. Br J Exp Pathol 68: 71–80
3. Schoettle RJ, Kochanek PM, Nemoto EM, Barmada MA, Magaree MJ, Melick JA (1989) Granulocyte accumulation and oedema after cerebral trauma. Crit Care Med 17: S71
4. Unterberg A, Wahl M, Hammersen F, Baethmann A (1987) Permeability and vasomotor response of cerebral vessels during exposure to arachidonic acid. Acta Neuropathol (Berl) 73: 209–219
5. Weiss SJ (1989) Tissue destruction by neutrophilys. New Engl J Med 320: 365–375

Correspondence: Dr. L. Schürer, Institut for Surgical Research, Klinikum Großhadern, Ludwig-Maximilians-Universität München, Marchioninistrasse 15, D-8000 München 70, Federal Republic of Germany.

Acta Neurochirurgica, Suppl. 51, 52–54 (1990)

The Effect of Immunosuppression with Whole Body and Regional Irradiation on the Development of Cerebral Oedema in a Rat Model of Intracerebral Haemorrhage

P. J. Kane, P. Modha, R. D. Strachan, A. D. Mendelow, S. Cook, and I. R. Chambers[1]

Departments of Neurosurgery and [1] Medical Physics, University of Newcastle Upon Tyne, England

Summary

A lesion simulating intracerebral haemorrhage was produced in the right caudate nucleus of rats immunosuppressed with whole body or regional irradiation. Whole body irradiation produced significant leucopaenia and thrombocytopaenia and conferred protection against cerebral ischaemia and oedema when compared to nonirradiated control animals. Local radiation to the head or torso did not confer protection.

Introduction

Brain oedema secondary to spontaneous intracerebral haemorrhage is a major clinical problem and often proves refractory to treatment. This oedema is in part secondary to ischaemia[5]. Blood constituents also play a role in producing oedema. In animal models, injection of autologous blood into the basal ganglia has been shown to produce greater ischaemia[4] and oedema[11] than injection of an inert oil suggesting that they are not merely the result of mechanical compression of the brain by the haematoma.

Granulocytes and platelets have been shown to accumulate in ischaemic brain[3, 7]. Granulocytes are also implicated in the generation of ischaemic damage in heart, kidneys and lung. Romson demonstrated a 43% reduction in the size of myocardial infarction in leucopaenic dogs[9]. Our aim was to determine whether or not a global depletion of leucocytes and platelets using radiotherapy would reduce brain oedema in rats with an experimental intracerebral haematoma.

Methods

Seventy nine adult male Wistar rats (280–370 g) were studied. Intracerebral haemorrhage and ischaemic oedema were produced by inflation of a stereotaxically placed microballoon (Ingenor laboratories) in the right caudate nucleus[5]. Irradiation was achieved by placing the rat, including tail, into a perspex tube flanked by two X-ray generating tube lights. The rats received 5 Grey at 240 KvP voltage over approximately 15 minutes either as a whole body dose or, by appropriate shielding, to the head only or the body and tail only. Control animals were restrained in the tubes for an equivalent period. The investigators were blind to the treatment given until the experiments were completed. Statistical analysis between groups was performed using Students t-test.

The animals were divided into three groups:

Group 1: Whole body irradiation – 4 hour oedema. Surgery occurred 4 days after whole body irradiation (treated n = 7, control n = 5): After induction of anaesthesia in an atmosphere of 5% Halothane in nitrous oxide/oxygen a tracheostomy was performed and mechanical ventilation initiated. Inspired halothane was reduced to 1% to maintain anaesthesia. Normothermia was maintained using a homeothermic heating unit. The left femoral artery was cannulated to allow serial estimation of blood gases, glucose and haematocrit, constant monitoring of arterial blood pressure and blood sampling for cell count and differential. The left femoral vein was cannulated to allow fluid replacement at 2–3 ml/hr. Burrholes were made to allow insertion of the microballoon (2.8 mm lateral and 1.1 mm anterior to the bregma) and bilateral frontal and parietal (2.8 mm lateral and 3.0 mm anterior/posterior to the bregma) platinum cortical electrodes for cerebral blood flow (CBF) measurements using a hydrogen clearance technique. A microballoon mounted on the tip of a blunted 25 g needle was inserted to a depth of 5.5 mm into the right caudate nucleus and inflated to 50 microlitres with saline. It remained inflated for 240 minutes. CBF measurements were made at 5 and 30 minutes post inflation and every 30 minutes thereafter. At 240 minutes the rat was killed with an overdose of intravenous pentobarbitone and the brain removed rapidly for brain oedema measurements.

Group 2: Whole body irradiation – 72 hour oedema. Surgery occurred 4 days after whole body irradiation (treated n = 8, controls n = 7). Blood sample was taken prior to surgery for cell count and differential. Anaesthesia was induced by intraperitoneal injection of a mixture of Midazolam and Hypnorm. Using an aseptic technique a microballoon was inserted as for Group 1, inflated to 50 microlitres for 5 minutes, deflated and removed. The burrhole was sealed with bone wax, sprayed with topical antibiotic and the wound sutured. The rat was transfered to an incubator for recovery. At 72 hours the rat was anaesthetized in an atmosphere of 5% halothane, killed with an intracardiac injection of pentobarbitone and the brain rapidly removed for oedema measurements.

Group 3: Differential irradiation – 72 hour oedema. Surgery occurred 7 days post irradiation. Treatment groups included; Whole body irradiation n = 10; Head only irradiation n = 10; Body only irradiation n = 9; Controls n = 24. The animals were treated as for Group 2 except for the use of a microballoon of a different manufacturer (Balt laboratories).

Brain oedema measurements were made using the gravimetric technique described by Shigeno *et al.*[9]. Coronal slices 1 mm thick were taken either side of the point of entry of the microballoon on the right cortical surface. Under the operating microscope paired 1–2 mm^3 samples of tissue were taken from the caudate of both hemispheres. These were dropped into the gravimetric column and their descent after 1 minute recorded. The specific gravity of the sample was calculated using a linear regression analysis derived from calibration of the column with droplets of potassium sulphate of known specific gravity. All columns used had a correlation coefficient greater than 0.999.

Results

Group 1: Whole body irradiation – 4 hour oedema. No significant differences were found in the physiological variables between the two groups. The leucocyte and platelet counts were significantly reduced in the treated group (Table 1). CBF in the right frontal region was significantly reduced from baseline values in the control group at 60, 90, 120, 150 and 180 minutes post lesion (Fig. 1). In the irradiated group significant reduction in frontal flow was only observed at 210 minutes (Fig. 1). No significant changes in CBF were found in the left hemisphere. Brain oedema measurements showed a significant reduction in right caudate oedema in irradiated animals compared to nonirradiated controls (Table 1).

Table 1. *Haematological Indices and Right Caudate Specific Gravities for Group 1 and Group 2 Experiments*

	Leucocyte count ($\times 10^9$/l)	Platelet count ($\times 10^9$/l)	Right caudate Specific gravity
Group 1			
Control	5.8 ± 0.82	609 ± 79	1.0382 ± 0.0014
Irradiated	0.96 ± 0.12*	304 ± 22*	1.0417 ± 0.0008**
Group 2			
Control	7.5 ± 0.87	641 ± 47	1.0396 ± 0.0008
Irradiated	1.7 ± 0.33*	366 ± 20*	1.0425 ± 0.0006**

Values are Mean ± sem. * p < 0.05, ** p 0.01 (compared to control values).

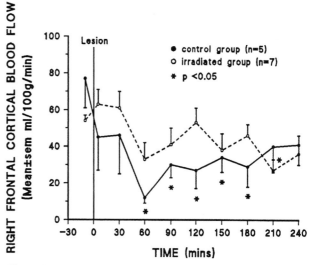

Fig. 1. Changes in right cortical blood flow with time following intracerebral haemorrhage in Group 1 experiments. * p < 0.05: Significant change from baseline flow

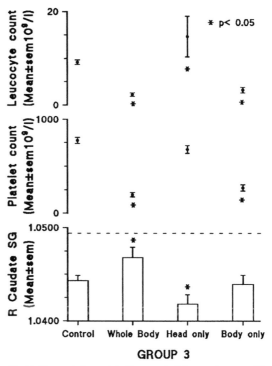

Fig. 2. Haematological indices and right caudate specific gravities for Group 3 experiments. *p < 0.05: Significant difference from control value. — — — Specific gravity of non lesioned left caudate in control group

Group 2: Whole body irradiation – 72 hour oedema. The leucocyte and platelet counts were significantly reduced in irradiated animals when compared to controls (Table 1). Irradiated animals showed significantly less right caudate oedema that nonirradiated controls (Table 1).

Group 3: Differential irradiation—72 hour oedema. (Fig. 2). "Whole body" and "Body only" irradiation significantly reduced platelet and leucocyte counts compared to controls, whereas "Head only" irradiation did not. The leucocyte/platelet counts in the "Body only" group were higher than the "Whole body" group reflecting the haemopoetic activity in skull marrow. "Whole body" irradiation resulted in significantly less oedema in the right caudate compared to control values whereas "Head only" irradiation made the oedema significantly worse. There was no difference in right caudate oedema between the "Body only" irradiated group and non irradiated control values.

Discussion

The involvement of leucocytes and platelets in cerebral ischaemia is recognized in experimental and clinical studies[1, 2, 3, 7, 8]. Our results suggest that depletion of leucocytes and platelets by irradiation with X-rays reduces the formation of brain oedema around an intracerebral haemorrhage. The blood flow in the overlying brain is also less deranged than in nonirradiated control animals. Irradiation of the brain alone does not confer protection against oedema formation. Recent studies by Dutka[1] and Grogaard[2] have suggested that granulocytopaenia protects against cerebral ischaemia. However, granulocyte depletion alone has failed to reduce oedema in our model[6]. The release of locally active substances by platelets and leucocytes in ischaemic brain may be the initiating step in the generation of cerebral oedema.

References

1. Dutka *et al* (1989) Influence of granulocytopaenia on canine cerebral ischaemia induced by air embolism. Stroke 20: 390–395
2. Grogaard *et al* (1989) Delayed hypoperfusion after incomplete forebrain ischaemia in the rat. The role of polymorphonuclear leucocytes. J Cereb Blood Flow Metab 9: 500–505
3. Hallenbeck JM *et al* ((1986) Polymorphonuclear leucocyte accumulation in brain regions with low blood flow during the early post ischaemic period. Stroke 17: 246–253
4. Jenkins A *et al* (1990) Experimental intracerebral haematoma: The role of blood constituents in early ischaemia. Br J Neurosurg 4: 45–52
5. Kingman TA *et al* (1988) Experimental intracerebral mass: Description of model, intracranial pressure changes and neuropathology. J Neuropath Exptl Neurol 47: 128–137
6. Modha P *et al* (1988) Experimental intracerebral haematoma: Ischaemic brain oedema in granulocytopaenic rats. Med Sci Res 16: 1031–1032
7. Obrenovitch TP *et al* (1985) Platelet accumulation in regions of low blood flow during the post ischaemic period. Stroke 16 (2): 224–234
8. Pozzilli C *et al* (1985) Imaging of leucocyte infiltration in human cerebral infarct. Stroke 16 (2): 251–255
9. Romson JL *et al* (1985) Reduction of the extent of ischaemic myocardial injury by neutrophil depletion in the dog. Circulation 67: 1016–1023
10. Shigeno T *et al* (1982) The determination of brain water content: microgravimetry versus dry weighing method. J Neurosurg 57: 99–107
11. Suzuki J *et al* Sequential changes in tissue surrounding ICH. In: Pia HW, Langmaid C, Zierski J (eds) Spontaneous intracerebral haematomas: Advances in diagnosis and therapy. Springer, Berlin Heidelberg New York, pp 121–128

Correspondence: A. D. Mendelow, Ph.D., F.R.C.S. (SN), Reader in Neurosurgery, Department of Neurosurgery, Newcastle General Hospital, Westgate Road, Newcastle Upon Tyne, NE4 6BE, England.

Acta Neurochirurgica, Suppl. 51, 55–57 (1990)

Effect of Stimulation of Leukocyte Chemotaxis by fMLP on White Blood Cell Behaviour in the Microcirculation of Rat Brain

S. Corvin, L. Schürer, C. Abels, O. Kempski, and A. Baethmann

Institute for Surgical Research, Klinikum Großhadern, Ludwig-Maximilians-Universität München, München, Federal Republic of Germany

Summary

The role of white blood cells in acute cerebral disorders such as ischemia or stroke is still unclear. Therefore, in the present study we investigated the effects of the leukotaxin n-formyl-methionyl-leucyl-phenylalanine (fMLP) on white blood cell endothelial-cell interactions in the rat brain surface microcirculation. An improved closed cranial window technique[5] was applied. Superfusion of fMLP in rising concentrations ($10^{-8} - 10^{-5}$M) was seen to induce rolling and adherence of leukocytes to the endothelium of small venules. Rolling was more effectively stimulated than firm attachment. fMLP-induced vasodilation was more pronounced in arterioles than in venules. In this study it has been shown that the hydrophilic fMLP is effectively stimulating neutrophil chemotaxis across the blood-brain barrier. Further, the closed cranial window preparation is useful to analyze quantitatively properties of activated leukocytes, which may be pertinent in injury to the blood-brain barrier and induction of microcirculatory disturbances.

Introduction

The role of white blood cells in acute cerebral disorders, such as ischemia, stroke or inflammation is not understood. Activated neutrophils might be involved, for example in damage of the cerebrovascular endothelium and opening of the blood-brain barrier[9]. Leukocytes are known to obstruct capillaries in heart muscle after reversible ischemia[1] and, thus, could be responsible for the no-reflow phenomenon, which inhibits nutritive flow and oxygen delivery. These results would be compatible with investigations of Ernst *et al.*[2], who found a reduced fluidity of white blood cells *in vitro* obtained from patients with stroke. This might be attributable either to an increased adhesiveness and/or a decreased deformability of leukocytes in the cerebral microcirculation. White blood cells adherence to the endothelial surface, however, is not only seen in ischemia[3], but also in other disorders of the brain. In subarachnoid hae-

morrhage, adhering leukocytes are assumed to cause inflammation of the vascular wall and, thus, to be involved in the pathogenesis of cerebral vasospasm[8].

Activation of white blood cells can be induced by chemotactic agents, such as leukotriene B[4], or peptides released from bacteria. These substances are known to attract and to activate leukocytes together with an alteration of the structure of their cell surface membrane[4]. Activated neutrophils were found attached to the venular endothelium ("stickers"), or to roll slowly along the vessel wall ("rollers"). In the current study, effects of the leukotaxin n-formyl-methionyl-leucyl-phenylalanine (fMLP) — an analogue of a chemotactic peptide released by E. coli — were analyzed on white blood cell dynamics in the rat brain surface microcirculation. Thereby, the usefulness of the closed cranial window preparation in rats was assessed to investigate pertinent microcirculatory properties of white blood cells on an quantitative scale.

Materials and Methods

Male Sprague-Dawley rats of 230–320 g *bw* were anaesthetized with α-chloralose (100 mg kg^{-1} *bw i.v.*). The animals were tracheotomized, immobilized with pancuronium bromide (1, 2 mg^{-1} *bw* h^{-1} *i.v.*) and artificially ventilated with room air supplemented with oxygen. An improved closed cranial window technique[5] was used for the present study. Following trephination the dura mater was incised and microsurgically reflected. Utmost care was taken to avoid touching or depressing of the exposed brain surface. Finally, the skull was closed again by sealing a cover glass onto a wall of dental cement surrounding the cranial window. The brain surface was continously superfused then with buffered artificial CSF at a rate of 5 ml × h^{-1}. With this preparation it is possible to investigate the cerebral microcirculation without disturbing the blood-brain barrier function.

The white blood cells were observed by intravital fluorescence microscopy after *i.v.* administration of rhodamin G, which is staining vessels and leukocytes. Measurement of white blood cell behaviour in the cerebral microcirculation was carried out off-line from the videoscreen. During the control period the brain was superfused with artificial CSF. Thereafter, fMLP dissolved in mock CSF was superfused in rising concentrations from 10^{-8} to 10^{-5} M with intervals of 20 minutes between. Arterial and venular segments of 100 µm length were observed for one minute. Determinations were made to obtain the number of rolling and sticking leukocytes and the changes of vessel diameters. The cerebral blood vessels were observed during superfusion with mock CSF and at each concentration of fMLP. Animals with massive adhering and rolling of leukocytes under control conditions were discarded from further experimentation.

Results

Under control conditions there were no significant changes, neither in the number of stickers and rollers, nor in vessel diameters. The number of rolling leukocytes increased from five to about twelve cells per segment during superfusion with fMLP (Table 1). The influence of fMLP on leukocyte adherence, however, was less dramatic. Only at concentrations of 10^{-6} and 10^{-5} M a significant increase from four under control conditions to a maximum of six cells per vascular segment were found. Significant changes were also observed in vessel diameters. Arterioles had a more pronounced dilation than venules. Arterioles dilated by about 30% at a concentration of 10^{-5}, whereas venules only by about 15% at this fMLP level in the superfusate (Table 2).

Table 1. *Effect of fMLP on Leukocyte Rolling and Sticking in Brain Surface Venules*. Stickers and rollers are given as number of cells/venular segment of 100 µm × min^{-1}

	mock CSF	fMLP			
		10^{-8} M	10^{-7} M	10^{-6} M	10^{-5} M
Stickers	3.39	3.80	4.53	5.20	6.53
Rollers	5.26	11.57	12.43	12.57	12.00

Table 2. *Effect of fMLP on Diameter Changes (Percent Control) of Cerebral Surface Vessels*

	mock CSF	fMLP			
		10^{-8} M	10^{-7} M	10^{-6} M	10^{-5} M
Arterioles	100	117	116	126	131
Venules	100	107	113	117	116

Discussion

As seen, the extravascularly adminstered chemotactic substance fMLP is able to alter vessel diameters and leukocyte behaviour in the cerebral microcirculation. These observations are remarkable, since fMLP was administered from the extravascular side of the blood-brain barrier. The leukotaxin is a hydrophilic compound which cannot be expected to penetrate freely the intact barrier to reach intravascular blood cell elements. Stimulation of leukcyte rolling and sticking along the endothelial luminal surface may, thus, have required an intermediate mechanism initiated by the leukotaxin. Or, specific transport processes for fMLP through the intact blood-brain barrier might have provided access of the compound to the intravascular compartment.

Our results are different from observations made during administration of fMLP in other organs. For example, blood vessels of the hamster cheek-pouch constrict during superfusion with fMLP[6]. Besides, induction of leukocyte sticking to the vessel wall was far more pronounced in the cheek-pouch preparation, whereas leukocyte rolling was actually found to be inhibited[7]. The current results suggest that the mechanism mediating chemotaxis across the cerebrovascular-parenchymal interface are different in the brain from those in other organs indicative of a protection of the brain against powerful mediators. Taken together, the present study demonstrates that microcirculatory properties of white blood cells, such as leukocyte rolling and sticking at the vascular endothelium can be analyzed in the brain at a quantitative level. Rolling and sticking of these inflammatory cells might be important in the pathological phenomena associated with acute disruption of the blood-brain barrier or impairment of cerebral blood flow.

References

1. Engler RL, Schmid-Schönbein GW, Pavelec RL (1983) Leukocyte capillary plugging in myocardial ischemia and reperfusion in the dog. Am J Pathol 111: 98–111
2. Ernst E, Matrai A, Paulsen F (1987) Leukocyte rheology in recent stroke. Stroke 18: 59–62
3. Hallenbeck JM, Dutka AJ, Tanishima R, Kochanek PM, Kumaroo KK, Thompson CB, Obrenovitch TP, Contreras TJ (1986) Polymorphonuclear leukocyte accumulation in brain regions with low blood flow during the early postischemic period. Stroke 17: 246–253
4. Harlan JM (1985) Leukocyte-endothelial interactions. Blood 65 (3): 513–525
5. Kawamura S, Schürer L, Goetz A, Kempski O, Schmucker B, Baethmann A (1990) An improved closed cranial window tech-

nique for investigation of blood-brain barrier function and cerebral vasomotor control in the rat. Int J Microcirc: Clin Exp (in press)

6. Lewis RE, Miller RA, Granger HJ (1989) Acute microvascular effects of the chemotactic peptide N-formyl-methionyl-leucyl-phenylalanine: Comparisons with leukotriene B_4. Microvasc Res 37: 53–69

7. Nagai K, Katori M (1988) Possible changes in the leukocyte membrane as a mechanism of leukocyte adhesion to the venular walls induced by leukotriene B_4 and fMLP in the microvasculature of the hamster cheek pouch. Int J Microcirc: Clin Exp 7: 305–314

8. Nazar GB, Neal FK, Povlishock JT, Lee J, Hudson S (1988) Subarachnoid hemorrhage causes adherence of white blood cells to the cerebral arterial luminal surface. In: Wilkins Re (ed) Cerebral vasospasm. Raven Press, New York, pp 343–356

9. Unterberg A, Wahl M, Hammersen F, Baethmann A (1987) Permeability and vasomotor response of cerebral vessels during exposure to arachidonic acid. Acta Neuropathol (Berl) 73: 209–219

Correspondence: Dr. S. Corvin, Institute for Surgical Research, Klinikum Großhadern, Ludwig-Maximilians-Universität München, Marchioninistrasse 15, D-8000 München 70, Federal Republic of Germany.

Acta Neurochirurgica, Suppl. 51, 58–60 (1990)

Leukotriene Production by Human Glia

S. R. Shepard, R. J. Hariri, R. Giannuzzi, K. Pomerantz[1], D. Hajjar[1], and J. B. G. Ghajar

Aitken Neurosurgery Laboratory, Division of Neurosurgery; Department of Surgery; and Laboratory of Vascular Biology,
[1] Department of Pathology, Cornell University Medical Center, New York, New York, U.S.A.

Summary

Elevated intracranial pressure and acute cerebrovascular changes following head injury remain the principle challenge in the management of traumatic brain injury. Recent work has demonstrated that leukotrienes can induce increases in blood brain barrier permeability and alter cerebrovascular dynamics. We investigated whether human astroglia in culture: 1. generate specific leukotrienes; 2. how they metabolize leukotrienes, and; 3. if astroglia generate leukotrienes in response to barotraumatic injury. Human astroglial cultures established from normal human brain obtained at surgery were exposed to either ionophore, exogenous ^{3}H-LTC$_4$, or barotraumatic injury. Supernatants were assayed for specific leukotrienes by one of three methods: HPLC, radioimmunoassay, or enzymeimmunoassay. Glial cells exposed to exogenous LTC$_4$ metabolized nearly all of the LTC$_4$ to LTD$_4$ and LTE$_4$ within 20 minutes. Glial cells stimulated with ionophore produced mostly LTC$_4$ at five minutes after stimulation and LTD$_4$ and LTE$_4$ at fifteen minutes after stimulation. Glial cells subject to barotraumatic injury produced LTC$_4$ in concentrations of 40–200 pg/ml 15 minutes after injury. These results demonstrate that human astroglial cells are capable of rapidly generating and degrading LTC$_4$ and this capability of glial cells may play an important role in the pathophysiology of cerebrovascular changes following head injury.

Introduction

Elevated intracranial pressure and acute cerebrovascular changes following head injury remain the principle challenge in the management of traumatic brain injury. Increased cerebrovascular permeability associated with the development of vasogenic oedema and abnormalities of vascular reactivity are frequent concomitants of traumatic brain injury as well as other conditions such as subarachnoid hemorrhage. Recently, a dose response relationship between increased blood brain barrier permeability and leukotrienes of the B, C, and E classes after direct intraparenchymal injection of these compounds has been described[1]. In addition, a vasoconstrictive effect of leukotriene D$_4$

both *in vivo* and in vitro has been demonstrated[9]. Furthermore, the presence of lipoxygenase products in whole brain preparations after experimental concussive injury has been demonstrated[5]. The vasoactive effects and oedema generating ability of these eicosanoids suggest a potential role for lipoxygenase products in cerebrovascular dynamics following head injury.

The precise cellular origin of these leukotrienes remains unclear. Recently, we have demonstrated that human astroglial cells in culture produce 5, 12, and 15 class hydroxy-eicosatetranoic acid (HETE) eicosanoids after ionophore stimulation[4]. We sought to further investigate: 1. whether human astroglia in culture produce specific leukotriene products (LTC$_4$ and LTD$_4$) after ionophore stimulation: 2. the time course for human astroglial metabolism of exogenous LTC$_4$, and; 3. to evaluate whether these astroglial cells generate specific leukotriene products after experimental mechanical injury.

Materials and Methods

Human glial cell cultures were established from normal whole brain specimens obtained at surgery as described previously[4]. Confluent human astroglial cell cultures in second, third, or fourth passage were employed in these experiments.

Labelling of cells. Glial cell cultures were grown to confluence before overnight incubation in medium containing 10% FBS supplemented with 2.0 µCi of [^{3}H] arachidonic acid, 94.5 µCi/mmol, in ethanol (New England Nuclear; Boston, MA) with a final concentration of ethanol of less than 0.2%. After removal of the media, the cells were washed three times in PBS containing 0.2% bovine serum albumin (BSA), followed by 3 washes in PBS without BSA to remove unincorporated label.

Metabolic Studies. Confluent 25 cm^2 flasks of human glial cells were incubated with arachidonic acid as above and exposed to either serum-free Waymouth's media or Waymouth's media with 20 µM calcium ionophore A23187 (Calbiochem). Supernatants were then removed 5, 10, 15 or 30 minutes and immediately frozen at −80 °C. These supernatants were characterized by high pressure liquid chromatography (HPLC) or radioimmunoassay analysis. Determination of glial cell capacity to metabolize ^{3}H-LTC$_4$ was performed in human astroglial cells. Cells were adminstered either serum-free Waymouth's media or Waymouth's media with 1 ng/ml ^{3}H-LTC$_4$. Supernatants were harvested at 5, 10, and 20 minutes and analyzed by HPLC. A third set of cells underwent mechanical injury in a fluid percussion device specially designed to apply barotrauma to cell monolayers. Confluent 55 mm petri dishes of human astroglial cells were subject to injury at pressures ranging from 1.0 to 2.5 atmospheres. Supernatants were removed after 15 minutes, centrifuged for 3 minutes at 0 °C to remove any cells and immediately frozen at −80 °C. These samples were then analyzed by enzyme-immunoassay described below. Cell protein content was determined by the method of Lowry[6].

High Performance Liquid Chromatography (HPLC). In these studies, eicosanoids were extracted by the method of Powell[8]. Recovery of leukotrienes, HET's, and arachidonic acid by this method is typically greater than 90%. The HPCL chromatographic separation of eicosanoids was performed using a Waters HPCL System: radioactivity of the column effluent was assessed using a Radiomatic Flow-One Model HS in-line scintillation spectrometer using FLOW-Scint II; DPM were quantitated at 30-seconds. Eicosanoids were separated using a 3.5 × 250 mm analytical column (C-18 microsorb reverse phase silica, 50-micron particle size, Rainin Instrument Co., Woburn, MA) maintained at 25 °C. Leukotrienes were eluted with 100% MeOH and 10% MeOH in H$_2$O (62 : 38 v/v) mixture at a flow rate of 1.0 ml/minute. Identity of the leukotrienes was confirmed by co-migration of radioactive peaks with peaks of UV absorbance at 280 nm, at equivalent retention times for LTC$_4$, LTD$_4$, and LTE$_4$.

Radioimmunoassay (RIA). Samples were prepared for RIA analysis as follows: After exposure to ionophore or mechanical injury supernatants were removed, immediately frozen in liquid nitrogen and stored at −80 °C until analysis. Samples were thawed, applied to octadecyl silane derived silica columns (Analtech, Delaware), washed twice with H$_2$O and eluted with MeOH. The eluted material was then evaporated under nitrogen and reconstituted in methanol: H$_2$O (1 : 1 v/v). Extraction efficiency ranged between 65 and 75%. Samples were then assayed using a commercial ^{3}H-LTC$_4$ RIA (New England Nuclear, Boston) with the following cross reactivities: LTD$_4$-55.3%; LTE$_4$-8.6%, 5-HETE <.1%, LTB$_4$ <0.1% and a reported detection limit of approximately 0.2 ng. Radioactivity was counted in a Beckman scintillation counter after the addition of 10 ml of Aquasol (New England Nuclear) per sample.

Enzyme-immunoassay (EIA). Samples were prepared in the same manner as for the RIA and analyzed with a commercial LTC$_4$ EIA (Cayman Chemical, Michigan). The detection limit of this system is approximately 15 pg/ml and cross reactivities at 22 °C are reported for LTD$_4$ (46%), LTE$_4$ (2%), and LTB$_4$ (<0.01%).

Results

Initial studies utilizing 20 μM ionophore to stimulate arachidonate primed glial cells for thirty minutes demonstrated a peak by HPLC with a retention time (r.t.) of 41 minutes, consistent with the retention time for LTE$_4$ and only trace peaks at 18 minutes and 37 minutes, the retention times for LTC$_4$ and LTD$_4$, respectively. This indicated that the cells were producing

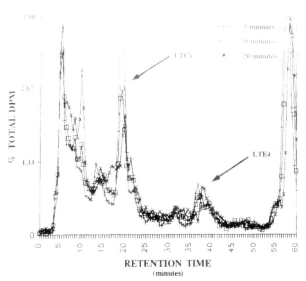

Fig. 1. LTC$_4$ Catabolism. 1 ng/ml ^{3}H-LTC$_4$ was administered to human astroglia, supernatants removed at 5 minutes, 10 minutes, and 20 minutes and analyzed by HPLC. DPM = decay per minute

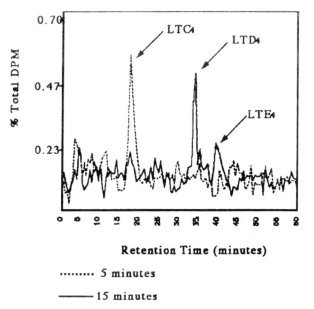

Fig. 2. LTC$_4$ Synthesis. Human astroglia were stimulated with 20 μM calcium ionophore A23187 in serum-free Waymouth's media. Supernatants were removed at 5 minutes and 15 minutes and analyzed by HPLC. DPM = decay per minute

LTC$_4$ and LTD$_4$ quickly after ionophore stimulation and were rapidly converting these products to LTE$_4$.

Exogenous LTC$_4$ was catabolized by glial cells as shown in Fig. 1. LTC$_4$ catabolism by the glial cells was evaluated by monitoring the metabolism of ^{3}H-LTC$_4$ treated cells. The LTC$_4$ peak (r.t. = 18 minutes) was large in supernatants collected 5 minutes after LTC$_4$ addition, significantly smaller in the 10 minute samples, and at 20 minutes, less than half of that measured at 5 minutes (Fig. 1). There was no appreciable LTD_4 peak, but there were two significant peaks for LTE_4 (r.t. = 41 minutes) in samples collected at 5 minutes and 20 minutes. An unidentified peak (r.t. = 11 minutes) was present in the 20 minute sample and may represent an oxidized metabolite of LTC_4.

Rapid production of leukotrienes by glial cells was then studied. Supernatants were obtained 5 and 15 minutes after glial cells were stimulated with ionophore. One significant peak (r.t. = 18 minutes), consistent with the presence of LTC_4 was seen 5 minutes after exposure to ionophore. Two peaks (r.t. = 37 minutes, 41 minutes) indicating the presence of LTD_4 and LTE_4, respectively (Fig. 2) were present at 15 minutes. There was a small peak (r.t. = 18) in the 15 minute sample suggesting the presence of only trace amounts of LTC_4.

Human glial cells stimulated with 10 µM ionophore were analyzed by RIA to quantitate the amount of LTC_4 production. After ten minutes, 25 cm^2 flasks of glial cells produced concentrations of 40 to 80 pg/ml of i-LTC_4, corresponding to a total production of 200 to 400 pg of i-LTC_4 per 5 ml of supernatant. Human glial cell undergoing mechanical injury at pressures from 1.0 to 2.5 atmospheres produced i-LTC_4 ranging from 40 to 200 pg/ml, yielding a total of 120 to 600 pg i-LTC_4. LTC_4 production after mechanical injury was maximal at 1.5 atmospheres (200 pg/ml). At a pressure of 2.0 atmospheres, the glial cells produced 180 pg/ml of LTC_4 and at a pressure of 2.5 atmospheres, the glial cells produced 130 pg/ml of LTC_4.

Discussion

The ability of arachidonic acid and leukotrienes B, C, and E to promote the development of cerebral oedema after intraventricular injection has been well established[1, 3]. Furthermore, elevated levels of leukotrienes have been demonstrated in the CSF of patients with subarachnoid haemorrhage and increasing levels of leukotrienes have been associated with increased levels of oedema in association with certain types of brain tumours[2, 7]. These findings, when considered with the knowledge that human atroglial cells produce lipoxygenase derivatives, suggest a potential role for glial cell generation of leukotrienes in the pathophysiology of cerebral oedema.

Our findings demonstrate that human astroglial cells rapidly produce as well as degrade leukotrienes in response to non-specific stimuli such as ionophore. When stimulated by ionophore, glial cells produce mostly LTC_4 within five minutes, followed by LTD_4 and LTE_4 within fifteen minutes. Furthermore, these cells rapidly degrade leukotrienes, metabolizing exogenous LTC_4 to LTD_4 and LTE_4 almost entirely within twenty minutes.

Although the demonstration that glial cells have the capacity of synthesize and metabolize LTC_4 is significant, the ability to do this does not itself define a physiological role for this phenomenon. Since cerebral oedema is a frequent complication of traumatic brain injury and leukotrienes can induce such oedema, we demonstrated that glial cells generate LTC_4 in concentrations ranging from 40 to 200 pg/ml in response to cellular barotrauma. We hypothesize that mechanical forces may result in allosteric changes to the glial cell calcium/magnesium membrane bound ATPase, resulting in transient increases in intracellular calcium levels. This, in turn, activates cellular phospholipases causing release of free arachidonic acid from cell membranes. It is likely that additional arachidonate is also released from the effects of mechanical damage to the membranes themselves. This free arachidonate is then converted by lipoxygenase to leukotrienes which in turn initiate a series of cerebrovascular changes resulting in increased blood brain barrier permeability and cerebral oedema.

References

1. Black KL, Hoff JT (1985) Leukotrienes increase blood-brain barrier permeability following intraparenchymal injections in rats. Ann Neurol 18: 349–351
2. Black KL, Hoff JT, McGillicuddy JE, Gebarski SS (1986) Increased leukotriene C_4 and vasogenic oedema surrounding brain tumours in humans. Ann Neurol 19: 592–595
3. Chan Ph, Fishmann R, Caronna J, Schmidley J, Prioleau G, Lee J (1983) Induction of brain oedema following intracerebral injection of arachidonic acid. Ann Neurol 13: 625–632
4. Hariri RJ, Ghajar JB, Pomerantz KB, Hajjar DP, Patterson RH (1989) Human glial cell production of eicosanoids: A potential role in the pathophysiology of vascular changes following traumatic brain injury. J Trauma 29: 1203–1210
5. Kiwak KJ, Moskowitz MA, Levine L (1985) Leukotriene production in gerbil brain after ischemic insult, subarachnoid haemorrhage, and concussive injury. J Neurosurg 62: 865–869
6. Lowry O, Rosebrough N, Farr A et al (1951) Protein measurement with the folin phenol reagent. J Biol Chem 193: 265–275
7. Paoletti P, Gaetani P, Grignani G, Pacchiarini L, Silvani V, Baena RRV (1988) CSF Leukotriene C_4 following subarachnoid haemorrhage. J Neurosurg 69: 488–493
8. Powell WS (1980) Rapid extraction of oxygenated metabolites of arachidonic acid from biological sampels using octadecylsilyl silica: Prostaglandins 20: 947–955
9. Tagari P, DuBoulay GH, Artken V et al (1983) Leukotriene D_4 and the cerebral vasculature *in vivo* and *in vitro*. Prostaglandins and Leukotrienes in Medicine. 11: 281–288

Correspondence: R. J. Hariri, M.D., Ph.D., Aitken Neurosurgery Laboratory, The New York Hospital-Cornell Medical Center, 1300 York Avenue, New York, NY 10021, U.S.A.

Acta Neurochirurgica, Suppl. 51, 61–64 (1990)
© by Springer-Verlag 1990

The Role of Arachidonic Acid and Oxygen Radicals on Cerebromicrovascular Endothelial Permeability

R. M. McCarron, S. Uematsu[1], N. Merkel, D. Long[1], J. Bembry, and M. Spatz[1]

Laboratory of Neuropathology and Neuroanatomical Sciences, NINDS, National Institutes of Health, Bethesda, MD
and [1] Department of Neurosurgery, The Johns Hopkins Hospital, Baltimore, MD, U.S.A.

Summary

Arachidonic acid release from tissue membranes and/or formation of free radical species have been considered to affect blood-brain barrier permeability and formation of brain oedema. To determine whether exogenous arachidonic acid or H_2O_2 may alter blood-brain barrier permeability, we examined their effect on cultured endothelium derived from cerebral microvessels of human and animals. Release of ^{51}Cr from labeled endothelium exposed to these substance was used as a main marker for the assessment of endothelial injury. The results of these studies indicate that endothelial cells (EC) are susceptible to exogenous arachidonic acid or H_2O_2 insult irrespective of their origin. However human endothelial cells are less affected than animal EC by H_2O_2-generated systems. The findings suggest that a disturbance of the existing balance between the endogenous antioxidant properties of EC and exogenous oxidant leads to EC injury.

Introduction

The release of arachidonic acid (AA) from cellular membranes and production of oxygen-derived species (O_2^-, H_2O_2, HO^-) have been considered among other tissue or blood-derived substances to play a role in altering the blood brain barrier (BBB) permeability, formation of cerebral edema and tissue injury[1, 2, 6, 7, 9, 10, 15]. In support of this contention are numerous data accumulated during experimental studies of brain ischaemia, trauma and hypertension. Even though the mechanism responsible for the development of brain oedema is multifactorial, it is generally accepted that both primary and/or secondary changes in the capillary endothelial function greatly contribute to the accumulation of water in the tissues. Studies *in vitro* demonstrated that AA alone in contrast to AA and H_2O_2 can induce a reversible change in permeability of cerebromicrovascular endothelium derived from the brains of rats[17].

Recently, we established an endothelial cell (EC) culture derived from cerebromicrovessels of human brain. The purpose of this study was to assess the susceptibility of human cerebromicrovascular EC to exogenous AA, H_2O_2 or H_2O_2-generated system and compare it with EC obtained from the brains of animals.

Methods

Cell Cultures. Human cerebral endothelial cultures were established from isolated microvessels by a modified technique of Gerhard *et al.*[5], while those from SJL/J mice were obtained by a method similar to that used for rats[14]. The purity of the cultured EC was ascertained by immunostaining for Von Willebrand (Factor VIII-related) antigen and contamination by astrocytes was assessed by staining for glial fibrillary acidic protein (GFAP).

Endothelial permeability assay. Confluent cultures of endothelial cells grown in 96-well flat-bottom microtiter plates were washed and labeled by 16 hr incubation in the presence of media containing 1% serum and ^{51}Cr-sodium chromate (1 μCi/well). After incubation, cells were washed four times with HBSS containing 25 mM Hepes and resuspended in the same buffer. In indicated experiments cells were then exposed to glucose (27 mM) in the presence or absence of various concentrations of glucose oxidase (1×10^{-5}-10 U/ml) and glucose oxidase in absence of glucose for one to four hours at 37 °C. Cell cultures incubated in the presence of boiled glucose oxidase served as controls.

Endothelial cell cultures were also incubated in the presence of various concentrations of H_2O_2 (1.8×10^8-1.8×10^{-1} nM) or arachidonic acid (5–100 μM) for 90 min at 37 °C. In indicated experiments, cells were pre-incubated in the presence of 50 mM aminotriazole (16 hr) or 2 mM sodium azide (15 min) prior to the addition of glucose oxidase or H_2O_2. Also, in indicated experiments cell cultures contained catalase (20–40 U/ml) or superoxide dismutase (60–80 U/ml). Cell cultures incubated for the designated periods of time in the presence of buffer only or buffer containing 1% H_2O_2 were utilized to obtain spontaneous release and maximum release values, respectively. All experiments were performed in quadruplicate at a final volume of 0.2 ml/well.

Permeability determination. After endothelial cell cultures were incubated for the indicated time periods, supernatants were harvested and the cells remaining in the wells were lysed by overnight incubation in the presence of 2% triton-x. All samples were counted on LKB 1282 compugamma cs gamma counter and the percent ^{51}Cr-release values were calculated by dividing the supernatant cpm values by the total counts (supernatant + cell) obtained in each well. In all experiments, the spontaneous release ranged between 5%–17% and the total release was 92%–100%. The data are presented as % specific release which was calculated as follows: [%^{51}Cr-release (supernatant) − %^{51}Cr-release (buffer only)]/(0.01)[%^{51}Cr-release(1%H_2O_2) − %^{51}Cr-release (buffer only)]. Arachidonic acid solutions and controls were prepared as described previously[17].

Enzyme assays. In designated experiments, EC cultured as monolayers in Petri dishes were treated with glucose/glucose oxidase in the presence and absence of sodium azide the same way as those prepared for ^{51}Cr-release (but without the addition of the isotope). They served for the determination of catalase, superoxide dismutase (SOD), glutathione peroxidase (GSHPx) and lactic dehydrogenase (LDH) activities[2, 12, 13]. The protein was determined according to the technique of Lowry *et al.*[11].

Reagents. Glucose oxidase, type V-S (spec. act. 1,130 U/5.3 mg/ml), catalase (spec. act. 14,300 U/mg), superoxide dismutase (spec. act. 3,360 U/mg), arachidonic acid, aminotriazole (3-amino- 1,2,4-triazole) and H_2O_2 were all obtained from Sigma Chem Co. (St. Louis, MO). Sodium azide (NaN$_3$) was obtained from Fisher Scientific Co. (Pittsburg, PA).

Results

The purity of several generations (4–22) of the cultured EC was greater than 95% as ascertained by positively staining for Factor VIII-related antigen. Only occasional cells stained with GFAP antibody.

Effect of arachidonic acid. The ^{51}Cr-release from human EC exposed to various concentrations of exogenous free AA was similar to the ^{51}Cr-release from AA-treated EC derived from rat and mice (data not shown). EC exposed to AA (50–100 μM) showed the greatest release of the marker. A lesser degree of ^{51}Cr-release from EC was observed after incubation of EC with AA doses lower than 50 μM; EC exposed to 5 μM AA did not release ^{51}Cr from the cells.

Effect of hydrogen peroxidase. A marked ^{51}Cr-release from EC was induced with a single H_2O_2 dose $(1.8 \times 10^7$ nM) in all EC irrespective of their origin (Table 1). A lesser degree of release was observed after EC exposure to 6.8×10^6 nM H_2O_2. This release was blocked by catalase. Lower concentrations of H_2O_2 did not affect the ^{51}Cr-release from human and murine EC in contrast to EC of rats. The addition of sodium azide in the presence of 6.8×10^6 nM or 1.8×10^6 nM induced an increase of ^{14}C release from these cells.

Effect of enzymatically generated hydrogen peroxidase. In general, the ^{51}Cr-release from EC exposed to glucose/glucose oxidase (representing a continuous

Table 1. *Effect of Catalase or Sodium Azide on Glucose Oxidase- and H_2O_2-induced Permeability of Cerebral Vascular Endothelial Cells*[1]

Treatment	Human	Rat	Mouse
Glucose oxidase (U/ml)			
0.1	37.6	50.8	33.4
0.1 + CAT	1.9	19.6	1.1
0.01	40.0	–	27.9
0.01 + AZ	–	–	49.1
0.005	30.2	–	–
0.005 + AZ	35.5	–	–
0.001	0	–	0
0.001 + AZ	3.6	–	0.9
0.0001	–	4.1	–
0.0001 + AZ	–	17.6	–
[H_2O_2], nM			
1.8×10^7	43.5	45.8	26.3
1.8×10^7 + AZ	–	–	–
1.8×10^6	8.3	16.9	2.1
1.8×10^6 + AZ	24.2	18.0	8.8
1.8×10^5	3.6	2.1	0
1.8×10^5 + AZ	0	16.0	0

[1] EC from the indicated sources were incubated in the presence of glucose oxidase (4 hrs) or H_2O_2 (90 min) ± catalase (CAT) or sodium azide (AZ; 2 mM); Data are from a representative experiment and are expressed as percent specific ^{51}Cr release (S.E.M. values were within 5% of the mean and are not shown)

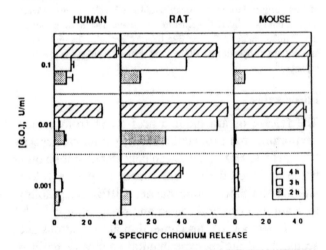

Fig. 1. Time course and dose-dependency of glucose oxidase-induced permeability changes in human, rat and murine EC. Cell cultures were incubated with glucose oxidase (0.1–0.001 U/ml) for the indicated time periods. Data presented were obtained from representative experiments and are expressed as means percent specific lysis ± S.E.M

H_2O_2-generated system) was time and dose dependent (Fig. 1). EC incubated for 1 hr with these substances had no effect on ^{51}Cr-release (data not shown). The earliest significant release was observed after EC treatment with glucose/glucose oxidase for 2 hrs. The human EC in contrast to murine and rat EC showed only a slightly greater release after 3 hrs treatment with glucose/glucose oxidase. ^{51}Cr-release from EC treated with glucose/glucose oxidase for 4 hrs was most affected. Both human and murine EC exposed to 0.001 U/ml glucose oxidase in presence of glucose showed no loss of the isotope marker, whereas the same concentration of these substances caused ^{51}Cr-release from rat EC (Table 1). Catalase blocked the release of ^{51}Cr-labeled EC exposed to all the concentrations of glucose oxidase irrespective of EC origin (Table 1). However, the catalase inhibition was less effective in EC of rats. SOD had no effect on the ^{51}Cr-release from the EC. Addition of NaN_3 or aminotriazole increased (2–7 fold) the ^{51}Cr-release from all EC.

Enzymes activities: Three anti-oxidant enzymes GSHPx, catalase and SOD were detected in all cultured cerebral EC irrespective of their origin. The level of GSHPx was the only enzyme showing a 2-fold greater activity in human EC than in animal EC. Preliminary investigation of human EC exposed to glucose/glucose oxidase (0.1 U/ml) showed a 30% reduction in the activity of catalase without changes in GSHPx and SOD. In addition, a 2-fold increase of LDH was detected in the supernatant as compared to that of controls. On the other hand murine EC treated with the same dose of glucose/glucose oxidase caused a 77% reduction in GSHPx, a 29% decrease in catalase activity and a 18-fold increase of LDH content in supernatant.

Discussion

The results of this study indicate that the cultured EC derived from human cerebral microvessels are susceptible to exogenous free AA or H_2O_2 injury. These findings are in agreement with the previously reported experimental *in vivo* and *in vitro* studies demonstrating that AA and/or H_2O_2 can induce changes in BBB permeability and formation of brain oedema[1, 2, 6, 10, 15, 17].

Based on the ^{51}Cr-release from the AA treated EC, it appears that the AA inducible damage of human EC is similar to that observed in EC derived from brain microvessels of animals. The threshold for inducing EC leakage of chromium is greater than 5 µM of free AA. On the other hand, the detected requirements of human EC for either higher concentrations of a bolus H_2O_2 or longer exposure of EC to H_2O_2-generated system to release labeled chromium strongly suggest that human cerebral EC are more resistant than animal EC to peroxidative injury. The absent or reduced peroxidative

effect on the activities of GSHPx, catalase and LDH on human EC as compared to the murine EC exposed to the same treatment supports this contention. However it remains to be clarified whether this difference is species specific or depends on other factors like number of EC passages.

The catalase protection and the NaN_3-augmented ^{51}Cr-release from all EC suggest that two enzymatic mechanisms, catalase and glutathione peroxidase may be involved in the EC detoxification of exogenous H_2O_2. A variety of experimental *in vitro* studies demonstrated catalase protection against H_2O_2 mediated cellular injury[4, 8, 16, 18]. However, the peroxidative role of catalase *in vivo* has been debatable since suitable electron donors required for its peroxidative function have not been identified. Studies in humans expressing a genetic deficiency of peroxidative enzymes suggested that glutathione peroxidase plays a more important biological function against various oxidant than catalase. The detection of these enzymes and SOD in the cultured cerebromicrovascular endothelium renders them suitable as a model for the investigation of oxidant-anti-oxidant mechanisms involved in altering BBB functions.

In conclusion, this report represents the first demonstation of human brain EC response to exogenous free AA or H_2O_2 insult. The results support the existence of a balance between the endogenous anti-oxidant properties of cerebral EC and exogenous oxidants. The disturbance of these parameters results in EC injury. Thus the altered EC function may contribute to the development and/or progression of brain oedema.

References

1. Chan PH, Fishman RA (1984) Phospholipid degradation and early release of polyunsaturated fatty acids in the evaluation of oedema. In: Go GK, Baethman A (ed) Recent progress in the study and therapy of brainoedema. Plenum Press, New York, pp 193–212
2. Chan PH, Longar S, Fishman RA (1985) Oxygen Radicals: Potentialoedema mediators in brainoedema. In: Inaba Y, Klatzo I, Spatz M (eds) Brainoedema. Springer, Berlin Heidelberg New York Tokyo, pp 317–323
3. Del Maestro RF, McDonald W (1985) Oxidative enzymes in tissue homogenates. In: Greenwald RA (ed) CRC handbook of methods for oxygen radical research. CRC Press, Boca Raton, FL, pp 291–296
4. Fantone JC, Ward PA (1982) Role of oxygen-derived radicals and metabolities in leukocytes-dependent infammatory reaction. Am J Pathol 107: 395–418
5. Gerhardt DZ, Broderius MA, Drewes LR (1988) Cultured human and canine endothelial cells from brain microvessels. Brain Res Bull 21: 785–993

6. Hall DR, Braughler JM (1987) The role of oxygen radical-induced lipid peroxidation in acute central nervous system trauma. In: Halliwell B (ed) Oxygen radicals and tissue injury. The Upjohn Co by the Fed Amer Soc for Exp Biol, Bethesda, MD, pp 92–98

7. Halliwell B (1987) Oxidants and human disease: Some new concepts. FASEB J 1: 358–364

8. Harlan JM, Levine JD, Callahan KS, Schartz BA, Harker LA (1984) Glutathione redox cycle protects cultured endothelial cells against lysis by extracellularly generated hydrogen peroxide. J Clin Invest 73: 706–713

9. Kogure K, Arai H, Abe K, Nakano M (1985) Free radical damage of the brain following ischaemia. In: Kogure R, Hossmann KA, Siesjo BK, Welsh A (eds) Progress in brain research. Elsevier-Science Publishers BV, pp 237–259

10. Kontos HA (1985) Oxygen radical in cerebral vascular injury. Circ Res 57: 508–516

11. Lowry OW, Rosebrough NJ, Farr AL, Randall RJ (1951) Protein measurement with folin phenol reagent. J Biol Chem 193: 266–275

12. Marklund S, Marklund G (1974) Involvement of superoxide anion radical in the autoxidation of pyrogallol and a convenient assay for superoxide dismutase. Eur Bioch J 47: 469–474

13. Spatz M, Bembry J, Dodson RF, Hervonen H, Murray MR (1980) Endothelial cell cultures derived from isolated cerebral microvessels. Brain Res 191: 577–582

14. Unterberg A, Wahl M, Hammersen F, Baethmann (1987) Permeability and vasomotor response of cerebral vessels during exposure to arachidonic acid. Acta Neuropathol (Berl) 73: 209–219

15. Vercellotti GM, Dolson M, Schorer AE, Moldow CR (1988) Endothelial cell heterogeneity: antioxidant profiles determine vulnerability to oxidant injury. Proc Soc Exp Biol and Med 187: 181–189

16. Villacara A, Spatz M, Dodson RF, Corn C, Bembry J (1989) Effect of arachidonic acid on cultured cerebromicrovascular endothelium: permeability, lipid peroxidation and membrane 'fluidity', Acta Neuropathol (Berl) 78: 310–316

17. Wei EP, Ellison MD, Kontos HA, Povlishock JT (1986) O_2 radicals in arachidonate-induced increased blood-brain barrier to proteins. Am J Physiol 251: H693–699

Correspondence: Dr. M. Spatz, National Institutes of Health, 9000 Rockville Pike, Building 36, Room 4D-04, Bethesda, MD 20892, U.S.A.

Acta Neurochirurgica, Suppl. 51, 65–67 (1990)
© by Springer-Verlag 1990

Possible Mechanism of Vasogenic Brain Oedema Induced by Arachidonic Acid

T. Ohnishi[1], **H. Iwasaki**[1], **T. Hayakawa**[1], and **W. R. Shapiro**[2]

[1] Department of Neurosurgery, Osaka University Medical School, Osaka, Japan, [2] Department of Neurology,
Memorial Sloan-Kettering Cancer Center, New York, U.S.A.

Summary

Free archidonic acid was infused into normal rat brains and the effect of arachidonic acid on capillary permeability was investigated by measuring the regional uptake of ^{14}C-aminoisobutyric acid with a quantitative autoradiographic method. Arachidonic acid increased capillary permeability in a dose-dependent manner up to 2 mM. A high dose of arachidonic acid (> 5 mM) produced a profound tissue destruction around the needle track and less increased capillary permeability than 2 mM arachidonic acid. Time-course study disclosed that arachidonic acid markedly increased capillary permeability within 2 hours after infusion, and continued to increase with time to 24 hours. The effect of 48 hours infusion was about a half of that at 24 hours, indicating that the effect of arachidonic acid was partially reversible. Pretreatment with dexamethasone significantly inhibited the arachidonic acid-induced increase in capillary permeability and the administration of actinomycin D 1 hour before the pretreatment with dexamethasone suppressed the inhibitory effect of dexamethasone. These results suggest that arachidonic acid, which is deposited in the extracellular space, increases brain capillary permeability by two different ways. One is the direct detergent effect of arachidonic acid, and the other is the effect of arachidonic acid that is released from the membrane by the activation of phospholipase A_2.

Introduction

Substantial evidence has suggested that alteration of membrane phospholipids and release of arachidonic acid may play an important role in the pathogenesis of vasogenic brain oedema resulting from seizure, hypoxia, and ischaemia. Chan, *et al.* demonstrated that intracerebral injection of arachidonic acid in rats induced vasogenic brain oedema as well as cellular oedema[3]. In addition, arachidonic acid and free radicals were found to produce the opening of the blood-brain barrier and affect the vasomotor response of cerebral vessels when they were extravascularly superfused[6, 7]. In these studies, however, it is unknown whether arachidonic acid itself or its metabolites are causally in-

volved in the formation of vasogenic oedema. Moreover, pathological significance of arachidonic acid in the extracellular space has never been documented.

In the present study, we report effects of arachidonic acid, which was infused into normal rat brain, on capillary permeability by using a quantitative autoradiographic (QAR) technique. We also discuss a possible mechanism underlying the formation of vasogenic brain oedema by intracerebral infusion of arachidonic acid.

Materials and Methods

Intracerebral Infusion of Arachidonic Acid

Male wistar rats were anaesthetized with N_2O-O_2-ethrane, and fixed in a stereotaxic device. Two burr holes were made symmetrically on the frontal bone. Thirty-gauge needles were inserted through these holes to a depth of 6 mm into the brain. These points corresponded to the right and left caudate nucleus-putamen. Sodium arachidonate (0.5–5 mM) in Krebs Ringer buffer solution was infused through the right needle, and the control solution (Krebs Ringer buffer) through the left needle at a rate of 1 µl/min for 50 min with a Harvard pump. After infusion, the animals were allowed to awake from anesthesia.

Assay of Capillary Permeability

To determine the effect of arachidonic acid, which was infused into the brain, on capillary permeability, regional uptake of ^{14}C-aminoisobutyric acid (AIB) was measured by QAR method. The rats were reanaesthetized and the right femoral artery and vein were cannulated. The animals were permitted to recover from anaesthesia before administration of ^{14}C-AIB. At the scheduled time interval after the intracerebral infusion began, ^{14}C-AIB (100 µCi/ml, 1 ml) was injected into the femoral vein. Timed blood samples were rapidly collected from the femoral artery. Fifteen minutes later, the rats were decapitated, and the brains were processed for QAR and histological examination. Capillary permeability was expressed as an unidirectional blood-brain transfer constant, K (µl/g/min), and calculated from the experimental data as previously described[5].

Treatment Studies with Dexamethasone and Actinomycin D

To examine the effect of glucocorticoids on the capillary permeability change produced by arachidonic acid, the rats were treated with dexamethasone (10 mg/kg, ip) 1 hour before intracerebral infusion of arachidonic acid. In addition, actinomycin D (0.5 mg/kg, iv), which inhibits de novo protein synthesis, was given to the animals 1 hour before the pretreatment with dexamethasone.

Results

The alteration of capillary permeability 6 hours after intracerebral infusion of 2 mM arachidonic acid is depicted as a computer-generated image of transfer constant (K) for ^{14}C-AIB (Fig. 1, inset). Compared to the marked increase of capillary permeability around the right needle track, into which arachidonic acid was infused, only a slight increase of capillary permeability was observed around the left needle with control solution.

To quantitatively evaluate the effect of arachidonic acid infusion on capillary permeability, mean K values were determined in sequential "rings" drawn around the needle track and these K values were plotted as a function of distance from the needle track. Figure 1 shows the distribution curves of ^{14}C-AIB for the treated and control groups. From the curves, two parameters, K_{max} and $D_{1/2}$, were defined as follows: K_{max} was the highest K value and $D_{1/2}$ was the distance (mm) from the needle track at which the mean K value was 50% of K_{max}. K_{max} indicates how much the infusate increased capillary permeability, and $D_{1/2}$ indicates the

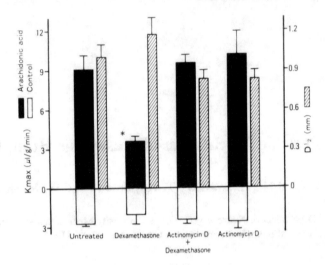

Fig. 2. Effects of various drugs on the arachidonic acid-induced increase in capillary permeability. Capillary permeability and half-distance of effect are expressed as K_{max} and $D_{1/2}$, resepectively (see text). The mean ($\pm$ standard error) for three experiments is indicated by the bar. Significantly different from the untreated group at $p < 0.01$ by t-test

distance the infusate markedly affected capillary permeability, *i.e.*, the spatial effect.

The K_{max} and $D_{1/2}$ of arachidonic acid were increased in a dose-dependent manner up to 2 mM. At the concentration of 5 mM, however, the K_{max} decreased to about 60% of K_{max} of 2 mM arachidonic acid. On the other hand, $D_{1/2}$ of 5 mM arachidonic acid was not different from that of 2 mM.

The time-course of the effect of arachidonic acid on capillary permeability was investigated by changing the interval time between the initiation of the infusion and the administration of ^{14}C-AIB. The K_{max} of arachidonic acid 2 hours after infusion was significantly higher than that of the control solution [12.43 ± 1.90 and 3.18 ± 0.32, respectively, $p < 0.001$ (t-test)]. Both K_{max} and $D_{1/2}$ of arachidonic acid increased with time up to 24 hours. At 48 hours after the infusion, the K_{max} decreased to 50% of that of 24 hours, but $D_{1/2}$ of 48 hours did not change.

The effects of various drugs on capillary permeability changes induced by arachidonic acid infusion were shown in Fig. 2. The K_{max} of arachidonic acid was significantly decreased by the pretreatment of animals with dexamethasone ($p < 0.01$). This inhibitory effect of dexamethasone, however, was almost completely suppressed by the pretreatment with actinomycin D. Actinomycin D itself did not affect the arachidonic acid-increased capillary permeability. For the group treated with dexamethasone, $D_{1/2}$ was the same or greater than $D_{1/2}$ of the untreated group.

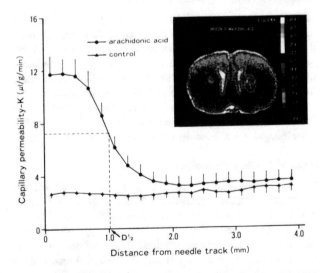

Fig. 1. Profiles of transfer constants (K) for ^{14}C-AIB. Mean K values in sequential "rings" drawn around the needle track were plotted as a function of distance from the needle track. Inset: a computer-generated image of ^{14}C-AIB distribution after arachidonic acid infusion on the right side of the brain. The control it the left side of the brain

Discussion

We have demonstrated that free arachidonic acid markedly increased brain capillary permeability when infused into the normal rat brain. The effect of arachidonic acid on capillary permeability was localized to an area within about 1.5 mm of the needle track. Histologically, arachidonic acid produced an oedematous change and a poor staining of the neuropil in the area surrounding the tissue damaged by needle insertion. Particularly, the latter change was prominent in the area 0.2–0.4 mm away from the needle track. The uptake of ^{14}C-AIB in this area was relatively low, resulting in the ring-like appearance in the autoradiogram. Although Chan *et al.*[3] reported a dose-dependent increase of brain water content up to the concentration of 5 mM arachidonic acid, our studies indicated that capillary permeability increased less with 2 mM than with 5 mM arachidonic acid. In the case of high concentration of arachidonate, the diminished effect was considered to be due to the tissue damage produced by arachidonic acid.

The mechanism of the effect of arachidonic acid, which is released by pathological insults of brains, on capillary permeability is poorly understood. Since arachidonic acid is rapidly broken down into the metabolites by the cascade reaction, the question is raised whether arachidonic acid itself or its metabolites are causally involved in the increased capillary permeability in a given pathological condition. Unterberg *et al.*[6] showed that opening of the blood-brain barrier by superfusion of the brain with arachidonic acid was not inhibited by indomethacin or by BW 755C (lipoxygenase inhibitor). Black *et al.*[2], however, reported that injection of leukotrienes and arachidonic acid into the brain parenchyma increased blood-brain barrier permeability and that the increased permeability induced by arachidonic acid could be prevented by BW 755C, but not by indomethacin. In our studies, dexamethasone significantly inhibited the increase in capillary permeability after intracerebral infusion of arachidonic acid. Moreover, the inhibitory effect of dexamethasone was suppressed by pretreatment with actinomycin D. It is well known that glucocorticoids inhibit the activation of membrane phospholipase A_2 through the

stimulation of the synthesis of an inhibitory protein[4]. Thus, it is suggested that extracellular arachidonic acid may increase brain capillary permeability by two pathways. One is the direcet detergent effect of arachidonic acid, which alters the endothelial cell membrane and immediately increases capillary permeability. The other is the effect of arachidonic acid activation of membrane-bound enzymes such as phospholipase A_2. By enzyme action, newly produced arachidonic acid is metabolized to prostaglandins, thromboxane, and leukotrienes which are responsible for the later increase in capillary permeability.

The vasogenic brain oedema produced after brain damage involves a system of mutually interacting factors, including arachidonic acid, glutamine and kinins[1]. Further biochemical studies focused on such mediators are necessary for the understanding of the molecular mechanism underlying the formation of vasogenic brain oedema and the development of new forms of treatment.

References

1. Baethmann A, Maier-Hauff K, Schürer L *et al* (1989) Release of glutamate and of free fatty acids in vasogenic brain oedema. J Neurosurg 70: 578–591
2. Black KL, Hoff JT (1985) Leukotrienes increase blood-brain barrier permeability following intraparenchymal injections in rats. Ann Neurol 18: 349–351
3. Chan PH, Fishmann RA, Caronna J *et al* (1983) Induction of brain oedema following intracerebral injection of arachidonic acid. Ann Neurol 13: 625–632
4. Hirata F (1981) The regulation of lipomodulin, a phospholipase inhibitory protein, in rabbit neutrophils by phosphorylation. J Biol Chem 256: 7730–7733
5. Ohnishi T, Sher PB, Posner JB *et al* (1990) Capillary permeability factor secreted by malignant brain tumour. Role in peritumoral brain oedema and possible mechanism for anti-oedema effect of glucocorticoids. J Neurosurg 72: 245–251
6. Unterberg A, Wahl M, Hammersen F *et al* (1987) Permeability and vasomotor response of cerebral vessels during exposure to arachidonic acid. Acta Neuropathol 73: 209–219
7. Wei EP, Ellison MD, Kontos HA *et al* (1986) O_2 radicals in arachidonate-induced increased blood-brain barrier permeability to proteins. Am J Physiol 251: H693–H699

Correspondence: T. Ohnishi, M.D., Department of Neurosurgery, Osaka University Medical School, 1-1-50 Fukushima, Fukushima, Osaka 553, Japan.

Acta Neurochirurgica, Suppl. 51, 68–70 (1990)

Effects of the Arachidonate Lipoxygenase Inhibitor BW755C on Traumatic and Peritumoural Brain Oedema

Y. Ikeda and D. M. Long

Department of Neurosurgery, Nippon Medical School, Tokyo, Japan and Johns Hopkins University School of Medicine, Baltimore, Maryland, U.S.A.

Summary

The abstract BW755C is a novel non-steroidal anti-inflammatory agent. It inhibits synthesis of prostaglandins and leukotrienes by inhibition in the cyclooxygenase and lipoxygenase pathways of arachidonic acid metabolism. Cold injury induced vasogenic oedema was produced in 18 cats. The animals were sacrificed at six and twenty-four hours. One group was treated with BW755C.

Seventeen white rabbits bearing an experimental brain tumour VX-2 carcinoma were treated for five consecutive days from the eighth day after tumour injection to the thirteenth day. Untreated tumour bearing rabbits were used as control.

Brain water content was measured by specific gravity method. BW755C did not reduce the water content following cold injury. There was no effect upon peritumoural oedema. The use of this novel blocking agent with diffuse effects in the arachidonic cascade was not beneficial for the reduction or prevention of brain oedema.

Introduction

Leukotrienes are powerful metabolites of arachidonic acid produced via the lipoxygenase pathway. Leukotrienes are important mediators of inflammation and have been shown to increase the vascular permeability of peripheral vessels[14]. Recently leukotrienes have been implicated in the development of all types of brain injury and brain oedema.

BW755C [3-amino-1-(3-(trifluoromethylphenyl)-2-pyrazoline] is a novel nonsteroidal anti-inflammatory agent. BW755C inhibits the synthesis of both prostaglandins and leukotrienes by dual inhibition of cyclooxygenase and lipoxygenase pathways of arachidonic acid metabolism *in vitro* and *in vivo*[6, 8] and is therefore considered a superior anti-inflammatory compound over the conventional cyclooxygenase inhibitors.

We investigated the effects of BW755C on two standard models of vasogenic brain oedema in an attempt to determine if leukotrienes play a role in the development of vasogenic brain oedema.

Materials and Methods

Cold Injury Oedema

Eighteen cats of either sex, weighing from 2 to 4 kg, were anesthetized intramuscularly with ketamine hydrochloride (25 mg/kg). Vasogenic brain oedema was produced by a standardized cortical freezing lesion. The cortical cold injury was produced by applying to the exposed cortex a metal probe 5 mm in diameter that had been cooled with dry ice.

Animals were separated into two groups. In first group (N = 10), the animals were cold injured and sacrificed at either 6 or 24 hours after production of the lesion. In the second group (N = 8), the animals were pretreated with BW755C (20 mg/kg) intraperitoneally 1 hour prior to cold injury and were sacrificed at either 6 or 24 hours after the production of the lesion. All animals were sacrificed by intravenous injection of a pentobarbital overdose.

Peritumoural Oedema

Seventeen New Zealand white rabbits (2 ~ 4 kg) were anesthetized with ketamine hydrochloride (25 mg/kg) and acepromazine maleate (2.5 mg/kg). Experimental brain tumours were produced by the injection of a 25 µl suspension of 3×10^5 viable VX2 carcinoma cells into the right frontoparietal lobe of the rabbit brain. The animals were separated into the following three groups: 1. (n = 5) tumour-free rabbits; 2. (n = 7) untreated tumour-bearing rabbits; 3. (n = 5) BW755treated tumour-bearing rabbits. The latter group was treated with BW755C (19 mg/kg/day) intraperitoneally for 5 consecutive days from 8th day following tumour transplantation and was sacrificed on 13th day.

Determination of Water Content

Brain water content was measured by the specific gravity (SG) method. The twelve sampling areas in cold-injured brain are illustrated in Fig. 1. The eight sampling areas in brain subjected to transplantation of VX2 carcinoma cells are illustrated in Figure 2.

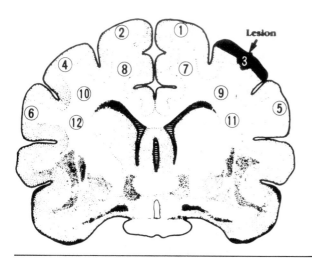

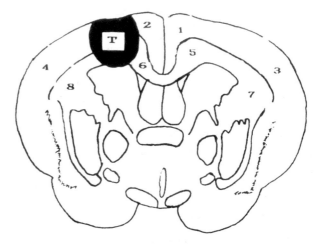

Area	SG values of 24 hours study	
	BW 755 C untreated (n = 5)	BW 755 C treated (n = 4)
1	1.0432 ± 0.0029	1.0458 ± 0.0016
3	1.0328 ± 0.0034	1.0343 ± 0.0022
5	1.0458 ± 0.0017	1.0447 ± 0.0021
7	1.0336 ± 0.0023	1.0353 ± 0.0016
9	1.0339 ± 0.0015	1.0336 ± 0.0004
11	1.0366 ± 0.0035	1.0388 ± 0.0034
2	1.0476 ± 0.0005	1.0483 ± 0.0013
4	1.0469 ± 0.0007	1.0480 ± 0.0011
6	1.0467 ± 0.0007	1.0470 ± 0.0009
8	1.0467 ± 0.0010	1.0473 ± 0.0014
10	1.0457 ± 0.0014	1.0461 ± 0.0010
12	1.0454 ± 0.0013	1.0452 ± 0.0010

Values are mean ± SD.

Fig. 1. 12 sampling areas used for measurement of water content by the specific gravity method in brain subjected to cold injury and values for specific gravity in untreated and BW755C treated groups at 24 hours after cold injury

Chemicals

BW755C was a gift of Wellcome Research Lab., Beckenham. Kent, England.

Statistical Analysis

All data quoted in this study are expressed as mean ± standard deviation. Student's test was used to assess significance and $P < 0.05$ is considered statistically significant.

Results

SG values showed that the increase in the brain water content was predominantly in the white matter in this cold injury model. BW755C did not reduce the water content of the injured side of the brain at 6 and 24 hours following cold injury (Fig. 1).

The peritumoural white matter of the affected hem-

Area	Tumour-free control rabbits (n = 5)	Untreated tumour-bearing rabbits (n = 7)	BW 755 C treated tumour-bearing rabbits (n = 4)
1	1.0460 ± 0.0010	1.0448 ± 0.0013	1.0453 ± 0.0007
2	1.0458 ± 0.0007	1.0451 ± 0.0013	1.0433 ± 0.0026
3	1.0447 ± 0.0006	1.0441 ± 0.0013	1.0445 ± 0.0009
4	1.0455 ± 0.0004	1.0450 ± 0.0022	1.0449 ± 0.0010
5	1.0440 ± 0.0010	1.0388 ± 0.0019*	1.0362 ± 0.0021*
6	1.0439 ± 0.0011	1.0363 ± 0.0019*	1.0369 ± 0.0016*
7	1.0443 ± 0.0007	1.0423 ± 0.0028	1.0423 ± 0.0025
8	1.0434 ± 0.0003	1.0426 ± 0.0035	1.0411 ± 0.0032

Data are presented as mean ± standard derivation.
Symbols indicate a significant difference from tumour-free control rabbits group at the level of: *$p < 0.05$.

Fig. 2. 8 sampling areas used for measurement of water content by the specific gravity method in brain subjected to transplantation of V × 2 carcinoma cells and values for specific gravity in tumour-free rabbits, untreated tumour-bearing rabbits and BW755C treated tumour-bearing rabbits

isphere (area 6) and the adjacent white matter of the contralateral hemisphere (area 5) in untreated tumour-bearing rabbits showed significantly lower specific gravities ($P < 0.05$).

Treatment of tumour-bearing rabbits with BW755C did not significantly alter or improve specific gravity in the peritumoural white matter of the affected hemisphere and the adjacent white matter of the contralateral hemisphere. BW755C had no effect on the other areas (Fig. 2).

Discussion

The possible role of arachidonic acid metabolism in the genesis of brain injury and of brain oedema has gained considerable attention. Arachidonic acid is the precursor of prostaglandin via the cyclooxygenase pathway and leukotriene via the lipoxygenase pathway, and oxygen free radicals are also generated by its conversion. Recent reports[1, 11, 13] have shown that the intracerebral injection of arachidonic acid causes brain oedema, which is not ameliorated by administration of indomethacin, the cyclooxygenase inhibitor. Black et al.[2, 3] reported that the intraparenchymal injection of leukotrienes in rats increased blood-brain barrier (BBB) permeability, which could be prevented by pretreatment with BW755C but not with indomethacin. These data taken together indicate that leukotrienes may play a role in the development of increased BBB permeability or the formation of vasogenic brain oedema. However, Mayhan et al.[7] demonstrated only minimal disruption of the barrier of pial vessels to Na^+-fluorescein after superfusing the cerebral cortex of hamsters with leukotriene C_4. Unterberg et al.[11] emphasized that leukotrienes do not affect BBB-permeability and do not induce vasogenic brain oedema after intraparenchymal infusion. They concluded that leukotrienes are not important mediators of brain oedema, which is supported by the additional finding that BW755C does not attenuate oedema formation after a cold injury. These reports on BBB dysfunction due to leukotrienes are still controversial.

Peritumoural brain oedema remains a significant cause of mortality and neurological deficits in patients with brain tumours. Black et al.[4] reported that there was a significant correlation between the degree of peritumoural brain oedema and leukotriene C_4 levels in brain tumour tissues. Shinonaga et al.[9] also reported that a correlation was found between the degree of macrophage infiltration and the amount of peritumoural brain oedema detected on CT scans. This study indicates that macrophage infiltration may be an important factor in the formation of peritumoural brain oedema and that arachidonate metabolites secreted by macrophages may cause peritumoural brain oedema in some cases. They speculated that the beneficial effect of glucocorticoids on peritumoural brain oedema is associated with the inhibition of the release of free arachidonic acid, probably through the inhibition of phospholipase A and reduced formation of leukotrienes[4].

This preliminary study indicates that BW755C had no beneficial effect upon the vasogenic brain oedema produced in either model, suggesting that leukotrienes do not contribute importantly to the development of vasogenic brain oedema in these two animal models. Leukotrienes involvement can not be completely ruled out, however, because BW755C is hydrophilic and may

not pass through the normal BBB nor penetrate into brain cells[12]. In addition, some reports demonstrated species difference with respect to brain response and formation of leukotrienes. Hambrecht et al.[5] showed that considerable species variability exists with respect to brain formation of 12 - HETE (12- hydroxy - 5, 8, 10, 14-eicosatetraenoic acid) from endogenous arachidonic acid. Ueno et al.[10] reported that synthetic leukotriene C_4 and D_4 increase vascular permeability to differing extents among species.

References

1. Aritake K, Wakai S, Asano T, Takakura K (1983) Peroxidation of arachidonic acid and brainoedema. Brain Nerve 35: 965–973
2. Black KL (1984) Leukotriene C_4 induces vasogenic cerebraloedema in rats. Prostaglandins Leukotrienes Med 14: 339–340
3. Black KL, Hoff JT (1985) Leukotrienes increase blood-brain barrier permeability following intraparenchymal injection in rats. Ann Neurol 18: 349–351
4. Black KL, Hoff JT, McGillicuddy JE, Gebarski SS (1986) Increased leukotriene C_4 and vasogenicoedema surrounding brain tumours in humans. Ann Neurol 19: 592–595
5. Hambrecht GS, Adesuyi SA, Holt S, Ellis EF (1987) Brain 12 - HETE formation in different species, brain regions, and in brain microvessels. Neurochem Res 12: 1029–1033
6. Higgs GA, Mugridge KG, Moncada S, Vane JR (1984) Inhibition of tissue damage by the arachidonate lipoxygenase inhibitor BW755C. Proc Natl Acad Sci USA 81: 2890–2892
7. Mayhan WG, Sahagun G, Spector R, Heistad DD (1986) Effects of leukotriene C_4 on the cerebral microvasculature. Am J Physiol 251: H471–H474
8. Mullane KM, Moncada S (1982) The salvage of ischaemic myocardium by BW755C in anaesthetized dogs. Prostaglandins 24: 255–266
9. Shinonaga M, Chang CC, Suzuki N, Sato M, Kuwabara T (1988) Immunohistological evaluation of macrophage infiltrates in brain tumours. Correlation with peritumouraloedema. J Neurosurg 68: 259–265
10. Ueno A, Tanaka K, Katori M, Hayashi MN, Arai Y (1981) Species difference in increased vascular permeability by synthetic leukotriene C_4 and D_4. Prostaglandins 21: 637–648
11. Unterberg A, Schmidt W, Polk T, Wahl M, Ellis E, Marmarou A, Baethmann A (1987) Evidence against Leukotrienes as mediators of brainoedema. J Cereb Blood Flow Metab 7 [suppl 1]: S625
12. Unterberg A, Wahl M, Hammersen F, Baethmann A (1987) Permeability and vasomotor response of cerebral vessels during exposure to arachidonic acid. Acta Neuropathol (Berl) 73: 209–219
13. Wahl M, Unterberg A, Baethmann A, Schilling L (1988) Review. Mediators of blood-brain barrier dysfunction and formation of vasogenic brainoedema. J Cereb Blood Flow Metab 8: 621–634
14. Wolfe LS (1982) Eicosanoids prostaglandins, thromboxanes, leukotrienes, and other derivatives of carbon – 20 unsaturated fatty acids. J Neurochem 38: 1–14

Correspondence: D. M. Long, M.D., Ph.D., Professor and Chairman, Department of Neurosurgery, The Johns Hopkins Hospital, 601 N. Wolfe Street, Baltimore, Maryland, U.S.A.

Acta Neurochirurgica, Suppl. 51, 71–73 (1990)

The Contribution of Secondary Mediators
to the Etiology and Pathophysiology of Brain Oedema:
Studies Using a Feline Infusion Oedema Model

I. R. Whittle, I. R. Piper, and **J. D. Miller**

University Department of Clinical Neurosciences, Western General Hospital, Edingburgh, Scotland

Summary

Secondary mediator compounds are postulated to have a role in vasogenic oedematogenesis. They may also cause focal brain dysfunction due to their neuronal, axonal and glial modulating properties. Using the feline model of infusion brain oedema the effects of right frontal intracerebral infusion (200 μl/hr for 3 hrs) of saline, bradykinin (10^{-4} to 10^{-6} M), arachidonic acid (10^{-2} to 10^{-3} M), 20% protein and four human glioma cyst fluids were evaluated. Somatosensory evoked potentials (SSEP), motor evoked potentials (MEPs), rCBF and rCBF CO_2 reactivity (Hydrogen clearance), ICP, craniospinal compliance, local brain tissue water content (microgravimetry), brain histology and BBB function (Evans Blue 2%) were measured. Brain water content increased locally from 69% to 79%, ICP increased (by mean 14 mmHg) and compliance decreased (mean 70%) and there were the histological features of brain oedema with all infusates. BBB opening occurred with Bradykinin (+), arachidonic acid (+ +), 20% protein (+ + +) and glioma cyst fluid (4 +). Polymorphic and macrophage infiltrates were seen with all infusions but rCBF and MEPs remained normal. SSEPs changed with high dose bradykinin and some glioma cyst infusates whilst CBF CO_2 reactivity was locally impaired by all infusates except saline and arachidonic acid. This study suggests that certain compounds in brain oedema fluid could mediate local brain dysfunction.

Introduction

Many disorders that cause brain oedema result in the release of kinins, eicosanoids, glutamate and free radicals from the pathological tissue[1]. These compounds, termed secondary mediators, may increase capillary hydraulic conductivity and increase oedema formation. Many of these compunds also have neuro-, glial and vasomodulating properties that could impair local homeostasis and cause local brain dysfunction[1, 10]. This study evaluates the role of certain compounds in both the etiology and pathophysiology of vasogenic brain oedema.

Methods

The feline model of infusion brain oedema was used[5]. Over a three hour period 600 μl of various compounds [sham, normal saline, bradykinin 10^{-4} to 10^{-6} M, arachidonic acid 10^{-2} to 10^{-3} M, 20% serum protein and human glioma cyst fluid from four patients (two glioblastoma, one anaplastic astrocytoma and one oligodendroglioma] were infused into the right anterior forebrain. The following parameters were monitored from the infused and control (left) hemispheres; (1) regional cerebral blood flow (rCBF) and CBF CO_2 reactivity using the hydrogen clearance technique. (2) Cortical somatosensory evoked potentials (SSEP) to contralateral forepaw stimulation. (3) Motor evoked potentials[3]. ICP, PVI, compliance and CSF outflow resistance were measured through a ventricular catheter using standard equations[4]. BBB opening was assessed using intravenous Evans Blue extravasation (1 ml/kg of 2% solution given 60 minutes prior to sacrifice). Brain oedema was assessed using microgravimetry[6] and histology (H & E, solochromecyanin).

Results

Forty cats were studied with sham[3], saline[4], bradykinin[8], arachidonic acid[8], 20% serum protein[9] and glioma cyst[8] infusion. White matter water content increased by a mean of 10 ml/100 g tissue following all the infusions, but there was no change in cortical water content (mean 80 ml/100 g tissue). ICP increased significantly, from baseline values of around 5 mmHg to 15–25 mmHg at the end of infusion with all except sham infusion. In the three hours after stopping infusion ICP started to return towards normal values. Lumped craniospinal compliance fell significantly from mean baseline values around 0.040 ml/mmHg to around 0.010 ml/mmHg at the end of infusion. PVI remained stable at around 0.4 ml in all experiments. Brain tissue hydraulic resistance fell from peak values around 10×10^3 mmHg/ml/min shortly after starting the in-

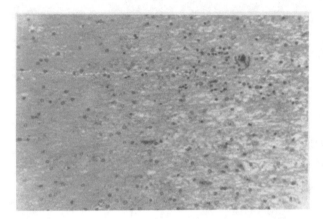

Fig. 1. Photomicrograph showing spreading oedema "front" in the gyral white matter following infusion of 20% serum protein into the forebrain white matter. Expansion and dissolution of the extracellular space (right) is apparent when compared to the normal white matter (left). There is also infiltration of the oedematous white matter with granulocytes. H & E ×100

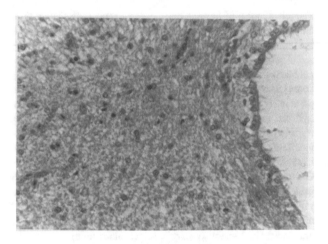

Fig. 2. Photomicrograph of oedematous periventricular white matter following bradykinin infusion into the forebrain white matter. The ependyma is intact, but there is marked expansion and vacuolation of the subependymal white matter (H &E ×250)

fusion to 4×10^3 mmHg/ml/min at 60 minutes. CSF outflow resistance was stable at around 100 mmHg/ml/min with all except the protein rich infusates where significant increases were noted.

Histological studies showed expansion of the white matter extracellular space, with dissolution and vacuolation of the eosinophilic ground substance (Fig. 1). These changes extended to the subependymal periventricular white matter (Fig. 2). There was regional perivenous and white matter infiltration with granulocytes after the bradykinin and arachidonic acid infusions. The cortex and subcortical U-fibres were normal. Granulocytic and macrophage infiltration was more extensive following 20% protein and glioma cyst infusions. BBB opening occured following all but sham and saline infusates. Blue discolouration of the white matter was more extensive and intense following both glioma cyst fluid and 20% protein infusions than bradykinin and arachidonic acid infusion.

rCBf during the infusions was stable in both hemispheres and ranged between 40–50 ml/100 g brain/min. There was a focal and unilateral decrease in rCBF during infusion with only one malignant glioma cyst fluid. rCBF CO_2 reactivity was normal (3–4% increase in CBF/mmHG increase in P_aCO_2) with saline, sham, oligodendroglioma cyst fluid and arachidonic acid infusates, but was variably impaired following all other infusates. In only three experiments was this loss of flow reserve associated with asymmetry of the cortical SSEP. A large long latency wave (P_{30}-N_{45}-P_{70}) developed with high dose bradykinin, one glioblastoma and the oligodendroglioma cyst fluid infusions. MEP remained symmetrical and normal throughout all experiments.

Discussion

This study confirms that bradykinin, arachidonic acid, components of serum proteins and glioma cyst fluid are secondary mediators of vasogenic brain oedema. BBB opening was variable but relatively mild after the bradykinin and arachidonic acid infusions. Given the pharmacological doses infused it may be that in patients neither of these compounds are clinically significant. However normal white matter capillaries have low hydraulic conductivity and high osmotic reflection coefficients[2] and kinins and eicosanoids may well be more active on damaged or neoplastic brain capillaries. In such tissue kininase levels may be reduced, and arachidonic acid preferentially converted into leukotrienes. The granulocytic infiltrates with both these infusates may underlie the BBB opening, however in leucopenic rats arachidonic acid can still open the BBB[8]. It is difficult to determine from these studies whether the BBB opening with serum protein and the glioma cyst fluids was a direct effect of some biochemical mediator[1, 7, 10] or mediated by an inflammatory response.

Electrophysiological function was preserved in the brain despite the various contents of the infused oedema fluid. This finding confirms clinical observations that brain oedema may cause no focal brain dysfunction. However some experimental findings are surprising in view of the multiplicity of effects of both bradykinin and arachidonic acid on neuronal function.

Preoperative treatment of patients with steroids may have suppressed secretion of neuromodulative compounds into the glioma cyst fluid. This may explain the lack of change in the feline SSEPs and MEPs despite profound, steroid reversed neural deficits in the patients. Stable, non-ischaemic levels of rCBF could also account for the preservation of the electrophysiological waveforms, even though infused oedematous brain has a higher CNR_{glu} than normal[9]. The mechanism underlying loss of CBF CO_2 reactivity is uncertain. It seemed to relate to protein or polypeptide infusions, however one glioma cyst infusate increased CBF CO_2 reactivity. However even when CBF CO_2 reactivity was impaired both MEPs and cortical SEPs remained normal in 20/23 cats.

These studies demonstrate the versatility of the infusion model of brain oedema and that it quantitatively and qualitatively resembles the extracellular brain oedema seen in many neuropathological situations. The brain biochemical and histological studies also confirm that the oedema fluid spreads by decreased brain tissue hydraulic resistance towards the ventricular system. Another advantage with the model is that it enables studies of brain dysfunction in the oedematous brain to be performed without the potentially confounding variable of severe raised ICP or brain tissue herniation.

Acknowledgements

This work was supported by a project grant from the Medical Research Council of the United Kingdom.

References

1. Baethmann A, Oettinger W, Rothenfusser O *et al* (1980) Brain oedema factors; current state with particular reference to glutamate and plasma constituents. Adv Neurol 28: 171–197
2. Fenstermacher JD, Patlak CS (1976) The movements of solutes and water in the brains of mammals. In: Pappius H, Feindel W (eds) Dynamics of brain oedema. Springer, Berlin Heidelberg New York, pp 87–94
3. Levy WJ, McCaffrey M, York DM *et al* (1983) Motor evoked potentials from transcranial stimulation of the motor cortex in cats. Neurosurgery 15: 214–227
4. Marmarou A, Shulman K, LaMorgese J (1975) Compartmental analysis of compliance and outflow resistance of the cerebrospinal fluid system. J Neurosurg 48: 523–534
5. Marmarou A, Takagi H, Shulman K (1980) Biomechanics of brain oedema and effects of local blood flow. Adv Neurol 28: 345–358
6. Marmarou A, Tanaka K, Shulman K (1982) An improved gravimetric measure of cerebral oedema. J Neurosurg 56: 246–253
7. Ohnishi T, Sher PB, Posner JB *et al* (1990) Capillary permeability factor secreted by malignant brain tumour. J Neurosurg 72: 245–251
8. Papadopulus S, Black KL, Hoff JT (1989) Cerebral oedema induced by arachidonic acid: Role of leukocytes and 5-lipoxygenase products. Neurosurgery 25: 369–372
9. Sutton LN, Barranco D, Greenberg J *et al* (1989) Cerebral blood flow and glucose metabolism in experimental brain oedema. J Neurosurg 71: 868–874
10. Whittle IR (1989) The contribution of secondary mediators to the etiology and pathophysiology of brain oedema. PhD thesis, University of Edinburgh, pp 186

Correspondence: Ian R. Whittle, M.D., Ph.D., FRACS, FRCSE (SN), Department of Clinical Neurosciences, Western General Hospital, Edingburgh EH4 2XU, Scotland.

Acta Neurochirurgica, Suppl. 51, 74–76 (1990)

Comparative Effects of Direct and Indirect Hydroxyl Radical Scavengers on Traumatic Brain Oedema

Y. Ikeda and D. M. Long

Department of Neurosurgery, Nippon Medical School, Tokyo, Japan and Johns Hopkins University School of Medicine, Baltimore, Maryland, U.S.A.

Summary

The effects of dimethyl sulfoxide (DMSO) and dimethylthiourea (DMTU) and deferoxamine (DFO) were investigated in cats during a standard cortical freezing method for the production of vasogenic brain oedema. Hydroxyl radical scavenging activity was studied by electron spin resonance assay using DMPO as the trap. The hydroxyl radical spin adduct was extinguished by DFO and a large dose of DMTU. Evans blue dye extravasation was reduced in DFO and DMSO treated groups as well. DFO appears to be an effective agent for the prevention of cold induced oedema. DMSO and DMTU are direct hydroxyl radical scavengers. These studies suggest that scavenging the hydroxyl radicals has a beneficial effect in the treatment of traumatic brain oedema.

Introduction

Oxygen free radicals have been implicated as very important causative factors of many pathological processes[1]. Recent studies demonstrate the importance of iron and oxygen free radicals in the genesis of brain injury and brain oedema[2]. Partial reduction of oxygen can produce three different, highly reactive species; superoxide radical anion, hydrogen peroxide, and superoxide radical anion. Moreover, hydrogen peroxide can react to form the most highly reactive species, hydroxyl radicals, via the iron-catalyzed Haber-Weiss reaction. These oxygen free radicals can damage the endothelial cells, disrupt the blood-brain barrier, and directly injure brain, causing brain oedema and structural changes in neurons and glias. In this study we investigated the effects of dimethyl sulfoxide (DMSO) and dimethylthiourea (DMTU) as direct hydroxyl radical scavengers and iron chelating agent deferoxamine (DFO) as an indirect hydroxyl radical scavenger.

Materials and Methods

Thirty-six cats, weighing 2.5 to 3.5 kg each, were anaesthetized with ketamine (25 mg/kg). Vasogenic brain oedema was produced by a cortical freezing lesion. The cortical cold injury was produced by application of a previous prepared metal probe (2 mm in diameter) cooled with dry ice for 1 min to the dura of the right parietal region. The dura was left intact. Brain water content was measured by the specific gravity (SG) method. Brain samples, approximately 1 mm³, were excised and placed onto a kerosene/monobromobenzene column. The 8 sampling areas are illustrated in Fig. 1. Competence of the blood-brain barrier was assessed by spread of Evans blue dye by planimetry. Each animal received an intravenous injection of 2.5 ml/kg of 2% Evans blue dye immediately prior to the production of the cold lesion.

Animals were separated into the following five groups.

Group 1 (n = 8): normal control group without lesion.

Group 2 (n = 8): saline-treated group.

Group 3 (n = 7): DFO (50 mg/kg)treated group.

Group 4 (n = 6): DMSO (2 g/kg in 40% solution with saline) treated group.

Group 5 (n = 7) DMTU (500 mg/kg) treated group.

Each agent was administered intravenously 15 minutes before the lesion and the animals were sacrificed at 6 hours after lesion production. The acitivity of direct and indirect hydroxyl radical scavengers DMSO, DMTU and DFO was studied by electron spin resonance assay using DMPO as a spin trapper. The Fenton system containing ferrous sulfate and hydrogen peroxide was used as detection of the hydroxyl radical.

Results

Signals of hydroxyl radical spin adduction of DMPO were extinguished by the addition of DFO and large dose of DMTU. This signal was significantly reduced by the addition of DFO (Fig. 2). SG values for area 5, 6, and 7 were higher in DFO-treated group than those for the saline-treated group. SG values for area 6 and

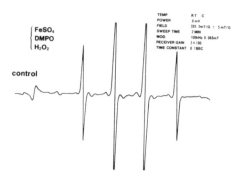

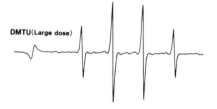

Fig. 2. The Fenton system containing ferrous sulfate and hydrogen peroxide is used as detection of the hydroxyl radical. Signal of hydroxyl radical spin adduct of DMPO was extinguished by the add of large dose of DMTU and DFO

Area	Untreated (n=8)	DFO (n=8)	DMTU (n=7)	DMSO (n=6)
1	1.0487±0.0011	1.0494±0.0006	1.0494±0.0003	1.0497±0.0005
2	1.0301±0.0020	1.0316±0.0023	1.0300±0.0025	1.0308±0.0008
3	1.0460±0.0016	1.0485±0.0007**	1.0481±0.0013*	1.0489±0.0008**
4	1.0460±0.0025	1.0492±0.0010*	1.0482±0.0010	1.0481±0.0006
5	1.0343±0.0029	1.0409±0.0050*	1.0362±0.0021	1.0399±0.0047
6	1.0374±0.0049	1.0472±0.0020***	1.0399±0.0025	1.0465±0.0015**
7	1.0420±0.0042	1.0480±0.0009**	1.0462±0.0011*	1.0477±0.0007*
8	1.0457±0.0009	1.0479±0.0007***	1.0469±0.0010	1.0477±0.0011*

* p<0.05, ** p<0.01, *** p<0.001
Untreated : untreated cats with cold injury
DFO : deferoxamine
DMTU : dimethylthiourea
DMSO : dimethyl sulfoxide

Fig. 1. Effect of hydroxy 1 radical scavengers on specific gravity values in 8 sampling areas

7 were also higher in DMSO-treated group than those for the saline-treated group.

White matter 3, 5, and 7 mm beneath the lesion cortex showed the maximum increase of water content among all sampling areas. DMTU reduced water content only in area 7. DFO and DMSO were more effective than DMTU (Fig. 1). Evans blue dye extravasation was significantly reduced in DFO and DMSO treated groups relative to saline and DMTU treated groups.

Discussion

Oxygen free radicals such as superoxide radical and iron-catalyzed hydroxyl radical generated by the superoxide system have been implicated in the pathological processes of ischaemia, trauma, and inflammation. Recent studies have also shown that oxygen free radicals are involved in the genesis of brain oedema. Among oxygen free radicals, the hydroxyl radical is one of the most highly reactive species and is produced from two active oxygen species, superoxide radical and hydrogen peroxide, via the iron-catalyzed Haber-Weiss reaction.

$$O_2 \cdot^- + Fe^{3+} \rightarrow Fe^{2+} + O_2$$
$$H_2O_2 + Fe^{2+} \rightarrow OH \cdot + Fe^{3+} + OH^-$$
$$O_2 \cdot^- + H_2O_2 \rightarrow OH \cdot + OH^- + O_2$$

The hydroxyl radical reacts at great speed with almost every molecule found in living cells including deoxyribonucleic acid, proteins, and carbohydrates. It attacks the fatty acid side chains and starts the process of lipid peroxidation. Inasmuch as the brain is especially rich in iron, it is possible that the iron-catalyzed Haber-Weiss reaction is involved in the production of hydroxyl radical. Evaluation of role of the hydroxyl radical has been hindered by the fact that there are no oxygen-free radical scavenger enzymes and some of the available chemical scavengers lack potency or specificity.

DFO has been widely used as a probe for iron-dependent radical reactions. Halliwell and Gutteridge[3] summarized the effects of DFO. DFO is a powerful chelator of ferric ion and inhibitor of iron-dependent hydroxyl radical generation. DFO is expected to reach both the extracellular and intracellular compartments and chelate intracellular and extracellular iron. Our study also demonstrates that DFO is an effective therapeutic agent in preventing cold-induced brain oedema.

Mannitol, dimethylsulfoxide (DMSO) and dimethylthiourea (DMTU) are known to be direct hydroxyl radical scavengers. Bolli et al.[5] emphasized several favorable effects of DMTU. The effects of DMTU reflect the cytotoxic action of the hydroxyl radical rather than that of superoxide radical or hydrogen peroxide. DMTU is highly permeable and may scavenge oxygen metabolites both intracellularly and extracellularly. DMTU is considerably more effective than other hydroxyl radical scavengers such as mannitol and DMSO. Pharmacokinetic data demonstrate that DMTU has a relatively longer plasma half-life, which seems to be useful in prolonged protection against oxygen-free radical-mediated damage[6]. This study showed that DMTU prevented the early development of brain oedema in the white matter adjacent to the lesion 6 hours following the lesion. In vitro evidence[7] suggests that DMTU may be most effective in scavenging hydrogen peroxide (H_2O_2), a precursor of the hydroxyl radical, while DMSO may be most effective in scavenging hydroxyl radical.

This study indicates that both direct and indirect hydroxyl radical scavengers are effective in reducing brain oedema and suggest that the iron-dependent hydroxyl radical may play a significant role in the pathogenesis of traumatic brain oedema. However, this study also demonstrated that there is difference of effectiveness among direct and indirect hydroxyl radical scavengers.

Recent critical papers[8,9] have shown that DNA damage is produced by the ferryl radical, not the free hydroxyl radical.

$$Fe^{2+} + H_2O_2 + H^+ \rightarrow Fe^{+3} - HO\cdot + H_2O$$
$$Fe^{3+} - OH\cdot \rightarrow Fe^{3+} + OH\cdot$$

$$Fe^{2+} + H_2O_2 + H^+ \rightarrow Fe^{3+} + HO\cdot + H_2O$$

These authors suggest the biologically active radical would be the ferryl radical (Fe^3-HO.), an intermediate of the reaction. This interesting possibility warrants further investigation. Other free radicals are possibly important.

References

1. Bulkley GB (1987) Free radicall-mediated reperfusion injury: a selective review: Br J Cancer [Suppl] 8: 66–73
2. Ikeda Y, Anderson JH, Long DM (1989) Oxygen free radicals in the genesis of traumatic and peritumoral brain oedema. Neurosurgery 24: 679–685
3. Halliwell B, Gutteridge JMC (1986) Oxygen free radicals and iron in relation to biology and medicine: Some problems and concepts. Arch Biochem Biophys 246: 501–514
4. Ikeda Y, Ikeda K, Long DM (1989) Protective effect of the iron chelator deferoxamine on cold-induced brain oedema. J Neurosurg 71: 233–238
5. Bolli R, Zhu W (1987) Hartley CJ, Michael LH, Repine JB, Hess ML, Kukreja RC, Roberts R. Attenuation of dysfunction in the postischemic stunned myocardium by dimethylthiourea. Circulation 76: 458–468
6. Fox RB (1984) Prevention of granulocyte-mediated oxidant injury in rats by a hydroxyl radical scavenger, dimethylthiourea. J Clin Invest 7: 1456–1464
7. Parker NB, Bergen BM, Curtis WE, Muldraw ME, Linas SL, Repine JE (1985) Hydrogen peroxidase causes dimethylthiourea consumption while hydroxyl radical casuses dimethyl sulfoxide consumption in vitro. J Free Radical Biol Med 1: 415–419
8. Imlay JA, Chin SM, Linn S (1987) Toxic DNA damage by hydrogen peroxide through the Fenton reaction in vivo and in vitro. Science 240: 640–642
9. Ikeda Y, Ikeda K, Long DM (1989) Comparative study of different iron-chelating agents in cold-induced brain oedema. Neurosurgery 24: 820–824

Correspondence: D. M. Long, M.D., Ph.D., Professor and Chairman, Department of Neurosurgery, The Johns Hopkins Hospital, 601 N. Wolfe Street, Baltimore, Maryland, U.S.A.

Acta Neurochirurgica, Suppl. 51, 77–78 (1990)
© by Springer-Verlag 1990

Pathophysiology III (Formation, Resolution)

The Effects of Cerebral Haemodynamics on the Progression of Cold-Induced Oedema

H. Kuchiwaki, S. Inao, M. Nagasaka, K. Andoh, T. Hirano, and **K. Sugita**

Department of Neurosurgery, Nagoya University School of Medicine, Nagoya, Japan

Summary

A cold-injury lesion was made on the brain of 14 of 21 adult mongrel cats. Cerebral blood flow (CBF) and cerebral blood volume (CBV) were observed in both groups during the next 6 to 9 hours using a laser Doppler flow meter. Intracranial pressure (ICP), water contents, and blood pressure were also measured. Hyperemia frequently reached a peak early in the test period in the gray matter and somewhat later in the white matter. In a few cases, hyperemia was observed to occur late in the test period in the gray matter and even later in the white matter. The least frequent pattern was a continuous decrease of CBF and CBV in gray matter over time and of CBF in white matter. Intracranial pressure rose rapidly and correlated well with the early hyperemia. The water content was significantly increased in the white matter. However, the difference in the rate of the increment of the water content was not significant from 7–9 hours after injury. Hyperemia seems to cause the severe impairment of the microcirculation during oedema formation induced by cold injury. A time difference of hyperemia was detected in the gray matter and the white matter.

Introduction

Brain oedema is frequently observed in patients with vascular injury following head trauma. In order to minimize the injury and to protect the normal brain surrounding the lesion, it is important to understand the haemodynamic factors involved in the progression of vasogenic oedema.

The aim of our study was the continuous measurement of cerebral blood flow (CBF) and cerebral blood volume (CBV) in real time in the gray matter and white matter after vascular injury and during the progression of the resulting oedema. Oedema was induced in the cerebrum of cats by cold injury.

Material and Method

Twenty-one adult mongrel cats, each weighing 3.7 ± 1.1 kg were anesthetized with intramuscular injection of Ketamine hydrochloride (30 mg/kg:im), Chloralose (20 mg/kg:iv), and Urethane (80 mg/kg:iv). The animals were then placed on a stereotaxic operation table and immobilized with Pancuronium bromide (iv) under a respirator according to the "Animal Experimentation Guide, Nagoya University School of Medicine". Expired CO_2 gas was monitored using gas monitor (1H21A, NEC-SANEI Co.). Arterial blood pressure (SAP) was measured in the abdominal aorta. Lactate Ringer's solution was continuously infused intravenously at a rate of 0.94 ml/min.

A small craniectomy was performed to make a cold lesion, to attach a probe for the laser Doppler flow meter (LDFM; ALF 2000, Advance Co.), and to measure the pressure with two catheter tip pressure transducers (Tokairika Co.). A probe of LDFM was at first attached to the brain surface. The values in cortical blood flow (CBF-g) were obtained in the control stage. Thereafter, CBF-g was continuously recorded throughout the experiments. To measure CBF in the white matter (CBF-w) a probe was inserted into the oedema front in the white matter just beneath the lesion using a micromanipulator (Sumitt Med. Co.) after making a lesion. The normal value in CBF-w was assumed to be one quarter of the CBF-g in the control stage in each animal. Tissue pressure probes were also inserted into the white matter under the lesion and into the contralateral white matter.

In 14 of the cats, a lesion was made by touching the epidural surface for 30 sec with a metal bar (diameter = 7.7 mm), which was cooled in dry ice-alcohol. After the initial freezing, the procedure was repeated.

Neuro-electrical activity (EEG) was recorded with a monopolar lead (G 1 : lt. and rt. parietal region, G 2 : nasion). Power spectrum analysis was continuously performed. The skull was sealed in most of the cases but was left open in a few cases. Water content of the oedematous area, which was visualized by Evans Blue staining (60 mg/kg, iv), was measured in the gray matter and the white matter. The observations were made for 6 to 9 hours after making the lesion.

Results

1. Changes in SAP:

Systemic arterial pressure was slightly elevated in most animals after the lesion was made. Systemic arterial pressure was considered high when diastolic pressure exceeded 110 mmHg (control mean diastolic pressure was 92 mmHg). Nine of the fourteen cats were included in this category. Diastolic pressure higher than 120 mmHg was seen in six of the nine cats.

2. Changes in Intracranial Pressure (ICP):

Changes in ICP ranged from 0 to 80 mmHg. In the control state, ICP was 3.9 ± 1.8 mmHg (mean + SD, $n = 13$). In five cats ICP increased more than 15 mmHg. In two of them, ICP was elevated more than 35 mmHg. The rate of ICP increment was greatest from 1–4 hours after cold injury. This was observed in animals that are classified as Type I (below).

3. Changes in Water Content:

Water content in the cortex ranged from $79.6 \pm 1.5(\%)$ to 81.5 ± 2.9 in the lesioned animals and was not significantly different than control values (79.4 ± 1.1, $n = 6$). On the other hand, water content in the white matter of the injured cats ranged from 74.3 ± 3.3 to 76.1 ± 3.6 and was significantly different than the control values (65.7 ± 0.8, $n = 6$) ($p < 0.001$).

4. Gray Matter CBF and CBV:

Control CBF-g was 2.2 ± 1.5 [V] ($n = 14$). Changes in CBF were classified into three patterns. For the Type I response ($n = 3$), CBF initially rose and peaked 2–3 hours after making the lesion, whereas CBV increased and reached a maximum level at 4 hours. For Type II response ($n = 1$), CBF was highest 7–9 hours after making the lesion. In this animal CBV had begun do decrease by 7–8 hours. For the Type III response ($n = 1$), CBF decreased continuously after making the lesion, but CBV did not begin to decrease until 4 hours after injury.

5. White Matter CBF and CBV:

The maximal increase in CBF-w was observed 2–3 hours after the peak in CBF-g. The data were also grouped into the three types. For the Type I response ($n = 5$), CBF-w reached a peak 3–4 hours after making the lesion, and CBV-w increased slightly and then decreased after 5 hours. For the Type II response ($n = 3$), both CBF-w and CBV-w were highest 8–9 hours after making the lesion. For the Type III response ($n = 1$), CBF-w decreased steadily but CBV rose during the first two hours and declined continuously thereafter.

6. Changes in EEG Recording:

EEG was recorded in 2 cats. Within the first three hours after lesioning the EEG's showed slow waves with lower voltage in the injured side than in the contralateral side. In this period sharp waves and spike discharges were frequently detected. After 4–5 hours the frequency increased slightly. After 6 hours the voltage decreased further and sporadic large slow waves appeared.

Discussion and Conclusion

There is frequently a transient hyperemia during the progression of cold-induced oedema, but this increase in CBF is often followed by a decrease. The interval between CBF peaks in gray and white matter was 2–3 hours. The patterns of CBV change did not match the changes in CBF in either gray or white matter. In gray matter, three different combinations of CBF and CBV changes were observed. Three different combinations of CBF and CBV changes were also found in white matter, but they were not identical to those in gray matter. This indicates that CBF-CBV responses to cold injury vary both between gray and white matter and among individual animals.

It is interesting to observe that only one peak was detected in CBF at any single location. In most experiments ICP rapidly reached a maximum in white matter. Water content increased in white matter but began to decrease after 7–9 hours.

After a period of hyperemia, CBF decreased in both gray and white matter. This may be caused by oedema formation and be partly due to focal pressure increases in the parenchyma. Although water content did not increase from 7–9 hours after cold injury, EEG activity deteriorated during this period. These observations are not fully explained by rising tissue pressure[2]. Some other factors may, therefore, be involved in these alterations of the brain microcirculation.

References

1. Overgaåd J, Tweed WA (1976) Cerebral circulation after head injury. Part 2: The effects of traumatic brain oedema. J Neurosurg 45: 292–300
2. Sutton LN, Welsh F, Bruce DA (1980) Bioenergetics of acute vasogenic oedema. J Neurosurg 53: 470–476

Correspondence: Dr. H. Kuchiwaki, Department of Neurosurgery, Nagoya University School of Medicine, 65-Tsurumai-Cho Showa-Ku, Nagoya, Showa-Ku, Nagoya, 466 Japan.

Acta Neurochirurgica, Suppl. 51, 79–81 (1990)

Cryogenic Brain Oedema: Loss of Cerebrovascular Autoregulation as a Cause of Intracranial Hypertension. Implications for Treatment

H. E. James and **S. Schneider**

Division of Neurosurgery and Department of Pediatrics, School of Medicine of the University of California, San Diego, California, U.S.A.

Summary

Experimental cryogenic brain oedema was created in albino rabbits, and intracranial pressure (ICP), cerebral blood flow (CBF), EEG, blood pressure, central venous pressure, were subsequently studied at a constant $PaCO_2$. Upon completion the brain water content of the gray and white matter was analyzed by gravimetry. The findings were compared to controls and sham-operated. Two subsets of elevated ICP following cryogenic injury were identified: one with a mean of 6.2 ± 3.3 torr (n = 7) and the other of 19.3 ± 9 (n = 5) (p < 0.005). Both these subsets had similar white and gray matter gravimetry values indicating that the magnitude of cerebral oedema was comparable and could not explain the difference in ICP. There was however a significant difference in the CBF of the left hemisphere between these subsets, with Subset A at 49.1 ± 9 ml/100 g/min and Subset B with 70 ± 9.1 (p < 0.001). We conclude that the ICP elevation in the cryogenic oedema model may be due to not only increased brain water of the injured hemisphere, but also to increased cerebral blood volume due to increased CBF.

Introduction

The oedema that develops following a cryogenic lesion (vasogenic oedema) through the intact dura in the various experimental models has been the basis for the explanation of the elevation of intracranial pressure (ICP). We here present results of our studies on the elevation of ICP following cryogenic lesions by not only documenting increased water content in the injured hemisphere but also studying the changes in cerebral blood flow (CBF).

Materials and Methods

Three groups of albino rabbits were used: controls, sham-operated, and a cryogenic lesion group (90 second cryogenic left hemisphere lesion through the intact dura)[1]. The 3 kg animals were anesthetized with halothane (2%), their scalps were infiltrated with marcaine (1%), and a sham operation or a freeze lesion was performed. The scalp was sutured. The animals recovered, were returned to their cages, and were allowed food and water ad lib. 24 hours later the experimental trials were commenced. The animals were reanesthetized with halothane, intubated and mechanically ventilated with a mixture of nitrous oxide (45%), halothane (0.5%), and oxygen, then paralyzed with pancuronium (1 mg). Arterial and venous femoral catheters were placed and continuous monitoring of blood gases, blood pressure (SAP), and central venous pressure (CVP) was performed to maintain a $PaCO_2$ 38–42 torr. A cisternal 19-gauge catheter recorded ICP continuously. Bilateral EEG subdermal electrodes permitted EEG monitoring. Bilateral platinum CBF electrodes were inserted by stereotaxic technique 2 mm into the frontal lobes (junction of gray and white matter), 1.5 mm from the sagittal suture. CBF runs were performed over 6 hours, with 5 to 6 runs per animal. At the end of the experimental period, the animals were sacrificed, and the brains were rapidly extracted and placed in kerosene for gravimetric determinations of brain water content[1]. The extent of Evans blue extravasation indicated the presence and severity of the brain lesion[1]. The site and location of the CBF electrodes were confirmed.

Results

Systemic arterial pressure (SAP) ranged from 85 to 120 torr in all groups. There was no statistical difference between the groups (Table 1). The CVP ranged from 1 to 5 torr in all groups. The EEG revealed the previously described slow wave activity over the left hemisphere in the cryogenic injury groups, with fast small wave activity on the right[1].

Intracranial Pressure

The ICP in the controls was 2.2 ± 1.2 torr. In the sham operated animals, it was 4.7 ± 2.4. This difference is not significant. In the cryogenic lesion group there were 2 subsets. One (Subset A) with next to normal or slightly elevated ICP (n = 7), with an average range of 4.1 to 9.0 torr and a mean of 6.2 ± 3.3 torr. This difference was statistically significant from controls (p < 0.005). The second one (Subset B), had a ICP average range from 11.5 to 33.3 torr and a mean of

Table 1. *Lesion Group*

	Subset A	Subset B	Sham-op	Controls
CBF (left hemisphere)	49.1±9.0	70±9.1	51.8±3.1	65.7±4.9
CBF (right hemisphere)	41.8±9.4	45.2±7.9	48.3±5.8	71.1±6.0
SAP	98.7±9.7	110.4±14.7	104.7±14.7	102.2±14.9
$PaCO_2$	39.2±2.3	40±2.4	39.7±1.6	39.7±1.5
ICP	6.2±3.3	19.3±9.0	4.7±2.4	2.2±1.2
	n=7	n=5	n=12	n=12

Table 2. *Water Content – Specific Gravity*

	Controls	Sham operated	Lesion groups	
			A	B
Left hemisphere				
Gray	1.0415±0.0009	1.0413±0.0089	1.0400±0.0008	1.0401±0.0038
White	1.0404±0.0009	1.0404±0.0008	1.0360±0.0036*	1.0365±0.0002[1]
Right hemisphere				
Gray	1.0418±0.0036	1.0417±0.0040	1.0415±0.0004	1.0412±0.0003
White	1.0406±0.001	1.0405±0.0009	1.0385±0.0023	1.0375±0.0033

* $p < 0.005$ from sham operated and controls.

19.3 ± 9 ($p < 0.005$ from Subset A). There was no difference in SAP or $PaCO_2$ between Subsets A or B (cryogenic lesion group). These findings are listed in Table 1.

Gravimetry

The specific gravity for the gray matter in the left hemisphere in the controls was 1.0415 ± 0.0009 (SD). In the sham-operated animals, it was 1.0413 ± 0.0089. Similar findings were obtained for the right hemisphere (Table 2). The specific gravity of the white matter in the controls was 1.0404 ± 0.0009, and in the sham-operated it was 1.0404 ± 0.0008. Similar findings were seen in the right hemisphere. The expected reduction of specific gravity, indicating an increase in water content, was seen in the cryogenic lesion group in the white matter of the left hemisphere, when compared to the controls and sham-operated rabbits. These findings are presented in Table 2. For the purpose of discussion the data from the Subset A and Subset B of the previously analyzed ICP changes, are presented. There is no significant difference between these subsets.

Cerebral Blood Flow

The CBF of the control group was 65.7 ± 4.9 ml/100 g/ min for the left hemisphere and 71.1 ± 6 for the right.

In the sham operated rabbits, there was a reduction of CBF in both hemispheres. This may be a reflection of the effect of surgery, because it was consistently seen in all animals. This reduction was seen in all the cryogenic injury animals as well, except in Subset B, but was only observed in the left hemisphere (Table 1). The elevation of CBF of the left hemisphere in Subset B when compared to sham-operated animals was significant ($p < 0.001$). The elevation of CBF when compared to the contralateral hemisphere of the same subset was also significant ($p < 0.001$) and when compared to the left hemisphere of Subset A ($p < 0.001$).

Discussion

The increase in ICP in the various preparations of cryogenic brain oedema has been well documented in the literature[1]. This is felt to be due to increased brain water and progression of brain oedema[2]. We had previously noted the wide variability of ICP in this model, and we were not able to explain this on the basis of brain oedema alone. In clinical studies of pediatric head injuries and Reyes syndrome, Swedlow et al.[3] have shown that in the face of insults of similar aetiologies, there were very divergent responses in CBF, which were associated with elevated ICP. In some patients the CBF values were well above the normal, indicating a blood flow-metabolism missmatch; in others elevated ICP was accompanied by very low blood flows and metabolism[3]. We believe that in the Subset B the high ICP values are due to the presence of cerebral oedema and the addition of increased cerebral blood volume, which is a consequence of increased CBF in that subset. The current findings in this experimental model confirm the clinical studies and indicate that in studying and treating ICP in the model of cryogenic brain oedema, brain oedema studies should be accompanied by CBF estimates, to permit a better understanding of the effects of therapy on elevated ICP.

Acknowledgement

This research was supported in part by The Foundation for Pediatric and Laser Neurosurgery, Inc. San Diego, California.

References

1. James HE, Laurin RA (1981) Intracranial hypertension and brain oedema in albino rabbits. Part 1. Experimental models. Acta Neurochir (Wien) 55: 213–226
2. Klatzo J, Wisniewski H, Steinwall O, Streicher E (1967) Dynamics of cold injury oedema. In: Klatzo I, Seitelberger F (eds) Brain oedema. Springer, Berlin Heidelberg New York, pp 554–563
3. Swedlow D, Frewen T, Watcha M, Bruce DA (1984) Cerebral Blood Flow, AJDO 2 & CMRO 2 in comatose children. In: Go KG, Baethmann A (eds) Recent progress in the study and therapy of brain oedema. Plenum Press, New York London, pp 365–371

Correspondence: H. E. James, M.D., 7930 Frost Street, Suite 304, San Diego, CA 92123, U.S.A.

Acta Neurochirurgica, Suppl. 51, 82–83 (1990)
© by Springer-Verlag 1990

An Electron Microscopic and Electron Probe Study of the Microcirculation in Cold-induced Oedema

H. Kuchiwaki, S. Inao, T. Woh, M. Nagasaka, K. Sugita, and **T. Hanaichi**[1]

Department of Neurosurgery and [1] Central Laboratory of Electro-Microscopy, Nagoya University School of Medicine, Nagoya, Japan

Summary

An ultrastructural study of adult mongrel dogs and cats was made to evaluate the changes in the microcirculation during cerebral oedema formation. Two to five cold injuries were made in one hemisphere in dogs and one lesion was made in cats. In several dogs arterial hypertension was induced with a balloon in the aorta. Intracranial pressure (ICP) and water content were measured. The specimens from the oedematous region were studied with transmission electron microscopy (TEM) and electron probe x-ray microanalysis (EPMA). The TEM data showed swelling of the endothelium and astrocytic foot processes, enlarged perivascular spaces and increased number of endothelial vesicles. The EPMA findings indicated increases in Fe and Ca content in the perivascular spaces. In some cases, the amount of chloride in red cells was increased. The altered distributions of these metals suggested tissue injury and impairment of red cell and vessel wall functions.

Introduction

An abnormal distribution of metals in injured tissue and a loss of normal function of the cerebral vessels may be important in the progression of brain oedema resulting from neural trauma. The aim of our study was to assess the ultrastructural distributions of iron and calcium in a model of cold-induced brain injury. Chloride distribution in red cells was also evaluated. The x-ray microanalyzer data are used to estimate the roles these metals play in the progression of cerebral oedema.

Materials and Methods

Thirteen adult mongrel dogs and twelve adult mongrel cats were anaesthetized with Ketamine hydrochloride (40 mg/kg, im) and allowed to breath spontaneously. The cats were anaesthetized first with Ketamine hydrochloride (25 mg/kg, im) and then with Chloralose (15 mg/kg), and Urethane (80 mg/kg); their breathing was artificially controlled. These procedures were performed according to the "Animal Experimentation Guide of the Nagoya University School of Medicine". Expired CO_2 was continuously monitored.

In dogs two models were used: 1) animals in which systemic arterial blood pressure (SAP) was artificially elevated by intermittent balloon inflation in the abdominal aorta (Model B(+), n = 8), and 2) animals without a balloon in the aorta (Model B(−), n = 2). A catheter was introduced into the brachial artery to measure SAP and into the femoral vein to infuse saline. Intracranial pressure (ICP) was measured using epidural pressure transducers in dogs and measured using small catheter tip pressure transducers (Tokairikakogyo Co.) in the brain tissue in the cats.

These animals were placed on an operating table and observed for 5 to 9 hours after the lesion was made. At the end of the experiment, the oedematous lesion was identified by an iv injection of Evans Blue dye (60 mg/kg). The water content (%) of the oedematous tissue was assayed with a wet-and-dry method.

The cold lesions were made epidurally using a cooled metal bar (7.7 mm in diameter) that was kept at −80°C in a dry ice-alcohol solution. In one cerebral hemisphere, two to five cold lesions were made in the dogs. A single cerebral lesion was made in cats. Sham operations were performed on two dogs and one cat. Additional control data were used from earlier experiments.

Samples of tissue from each oedema site were used in the study to map the distribution of the materials. Samples were analyzed by transmission electron microscope (TEM) and by electron probe x-ray microanalysis (H 800, Hitachi Co., Kevec, Kevec group) (EPMA). The samples were fixed in glutaraldehyde solution, cut with a microtome (LKB Co.), and examined without staining. The localization of the atoms was determined on a color image scanner (GT 4000, Epson Co. Ltd.) by superimposing an EPMA dot-mapped image on an EPMA photograph of the same area.

Results

1. Changes in ICP and SAP:

Mean SAP was 187.0 ± 36.2 mmHg in B(+) dogs and 177.5 in B(−) dogs and was 102.5 ± 9.0 mmHg in control dogs. In cats SAP was 145.9 ± 20.0 mmHg in the experimental group, 116.5 in controls. Mean control ICP was 3.4 ± 0.7 mmHg (dogs; n = 3) and 4.8 ± 2.3 (cats; n = 8). The maximum ICP in dogs averaged 54.6 ± 28.5 mmHg in the B(+) group (n = 8) and 33.0 mmHg in the B($-$) group (n = 2). In cold-injured cats ICP was 27.5 mmHg (n = 2). The water content in the gray matter was 82.3 ± 0.7% in B($+$) dogs, 79.9 ± 0.3 in B($-$) dogs, and 79.4 in control dogs. In the white matter of dogs, the water content was significantly greater (p < 0.05) in group B($+$) (76.1 ± 4.6%) than in controls (66.4 ± 0.3, n = 5). In cats the water content of the gray matter was 80.7% in experimental animals (n = 2) and 79.4 ± 1.1% in controls (n = 6); in the white matter it was 75.7% in the experimental group (n = 2) and 65.7 ± 0.8 in controls (n = 6).

2. Transmission Electron Microscopic (TEM) Findings:

The perivascular fluid space was prominent. The astrocytic end feet and endothelial cells were swollen in the oedematous tissue. Endothelial mitochondria were enlarged and often destroyed. The number of vesicles in the endothelium was increased in oedematous tissue.

3. X-ray Microanalysis (EPMA) Findings:

In normal tissue erythrocytes were rarely observed in the capillaries. Iron was generally seen in the red cells and seldom observed in the wall and the tissue surrounding cerebral vessels in normal animals. Calcium was mostly located in the red cells. In the capillaries chloride distribution in the red cells as well as in the luminal space surrounding the red cells was slight.

In injured tissue the frequency of red cells within capillaries was higher than in the normal group. The density of iron in capillary walls and pericapillary spaces was high and obscured the localization of this metal. In experimental animals, the amount of calcium was very high in the perivascular spaces and in the capillaries. Such findings were more common in dogs (multiple lesions) than in cats (one lesion). In some cases the amount of chloride in red cells was remarkably large.

Discussion and Conclusion

The most severe changes were observed in the cold-lesioned dogs group with high SAP. This suggests that intravascular hydrostatic pressure plays a role in altering the distribution of these metals in the progression of brain oedema during the acute stages.

In oedematous brain the metals accumulated in the perivascular spaces. Ferrous and ferric iron form free radicals[2]. Free radicals are thought to be involved in the progression of brain oedema. Our findings on metal distribution do not indicate the form or state of this iron and do not directly show that there was production of free radicals.

Calcium regulates various cellular functions[1]. We speculate that the accumulation of this metal is involved in the dysfunction of the walls of the capillary (vasoparalysis). However, the present study only showed that this material accumulates in the perivascular space during oedema progression.

Red cell function depends strongly on chloride[3]. Accumulation of chloride in red cells may cause some impairment in the function of red cells and may alter microcirculatory function during cerebral oedema.

References

1. Faber JL (1981) The role of calcium in cell death. Life Sci 29: 1289–1295
2. Koppenol WH, Butler J, Van Leeuwen JW (1978) The Harber-Weiss cycle. Photochem Photobiol 28: 655–660
3. O'Neill WC (1987) Volume-sensitive Cl-dependent K transport in human erythrocytes. Am J Physiol 22: C 883–888

Correspondence: Dr. Hiroji Kuchiwaki, Department of Neurosurgery, Nagoya University School of Medicine, 65-Tsurumai-Cho, Showa-Ku, Nagoya, 466 Japan.

Acta Neurochirurgica, Suppl. 51, 84–86 (1990)
© by Springer-Verlag 1990

Accumulation of Oedema Fluid in Deep White Matter After Cerebral Cold Injury

T. Kuroiwa, J. Yokofujita, H. Kaneko[1], and **R. Okeda**

Department of Pathology, Medical Research Institute, Tokyo Medical and Dental University, [1] Unique Medical Co., Ltd., Tokyo, Japan

Summary

The distribution of oedema fluid was examined in cats subjected to a cryogenic cortical injury. The lesion was made in the parietal cortex, and the animals were sacrificed 6 hr after the injury. The serum concentration of ^{125}I bovine serum albumin was kept constant over the 6 hr period by a programmed infusion. Autoradiograms were made from the coronal sections through the lesion and were used to quantify densitometrically the regional content of extravasated serum albumin. After autoradiographic exposure, the section was stained with luxol-fast blue (LFB), and the degree of LFB discoloration was quantified.

The maximal accumulation of extravasated serum albumin was observed in the deeper white matter under the subcortical white matter and not in the subcortical white matter of the lesion. The degree of oedema indicated by LFB discoloration showed a similar distribution pattern. This indicates that the compliance of the white matter in vasogenic oedema is regionally different. This difference of regional compliance seems to depend on the structural characteristics of each region such as the type of the fibers and the orientation of the fibers.

Introduction

The spread of oedema fluid in cryogenic brain injury is driven by the pressure gradient between the lesion and non-oedematous white matter[3]. The interstitial pressure of the white matter under the lesion decreases in proportion to the distance from the lesion[1]. These findings suggest that the regional content of oedema fluid also decreases in proportion to the distance from the lesion and that the expansion of white matter due to oedema parallels the increase of the interstitial pressure.

In this study, we examined quantitatively the distribution of oedema fluid following cryogenic injury and observed that the regional content of oedema fluid did not increase as a function of distance from the lesion but showed a maximal accumulation in the deeper white matter. This finding indicates that the regional compliance of the white matter is heterogeneous in cold lesion oedema.

Materials and Methods

Cats weighing 2.5 to 4.5 kg of either sex were used in these experiments. Anaesthesia was induced by intramuscular injection of ketamine and maintained by intermittent intravenous injection of alpha choralose. The animals were tracheostomized and artificially ventilated after a muscle relaxant injection. The left femoral artery was cannulated for arterial blood pressure monitoring and blood sampling and the left femoral vein was cannulated for the injection of anaesthetics, tracers and other substances. A cold lesion was made by applying a metal plate cooled to $-50\,°C$ with a mixture of dry ice and acetone to the suprasylvian gyrus for 1 min[4]. ^{125}I-labeled serum albumin (1 mCi) dissolved in 2% Evans blue solution was continuously infused for 6 h after the cold lesion according to a method by Patlak and Pettigrew[2] to keep the serum specific activity constant. The animal was sacrificed 6 hr after cold injury. The brain was then removed and cut coronally through the lesion. A coronal block was frozen and cut for macroscopical autoradiography. The quantitative image of the extravasated serum albumin was obtained by using an image processor (UHG-101, Unique Medical Co. Ltd., Tokyo, Japan) and the following formula:

$$rAC = Ci/SA \text{ mg/gP}$$

where rAC is the regional content of extravasated serum albumin (mg/g brain), Ci is the brain tissue concentration of tracer (mCi/g) and SA is the specific activity of tracer (mCi/mg). After autoradiographic exposure, the section was stained with LFB for the determination of LFB discoloration by densitometry. Another coronal block was immediately immersed in buffered formalin for histological examination with LFB staining and in some animals for microscopical autoradiography with basic fuchsin counter staining.

Results

Arterial blood pressure, tension of blood gases, and blood pH were within the normal range. The specific activity of ^{125}I-BSA was kept constant during the ex-

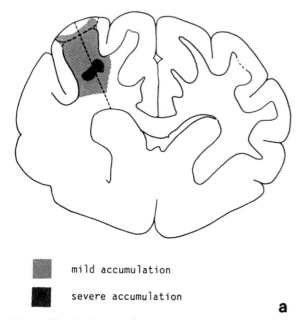

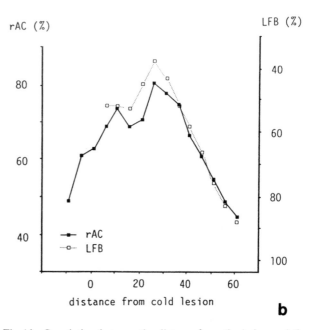

mild accumulation

severe accumulation

a

Fig. 1 a. The distribution of oedema fluid in the white matter under the cold lesion. Note the severe accumulation in the deeper white matter

b

Fig. 1 b. Correlation between the distance from the lesion and the severity of oedema assessed by the regional albumin content and LFB discoloration

Fig. 2. White matter in the control area (a) and in the oedematous white matter (b, c). The greatest accumulation of oedema fluid was in the white matter that was made up of intertwining fibers (2). Luxol fast blue staining, × 130

periment. Macroscopical autoradiography of ^{125}I-BSA showed increased optical density and radioactivity in the lesioned cortex, subcortical white matter of the lesion, and the deeper white matter closer to the lateral ventricle. The area of increased optical density corresponded to the area stained with Evans blue. The content of extravasated serum albumin gradually increased along the oedematous white matter ventral to the lesion and reached a maximum in the deeper white matter. Then it gradually decreased toward the lateral ventricle (Fig. 1 a). The microautoradiograms from the adjacent sections counter-stained with basic Fuchsin showed that most of ^{125}I-BSA in the non-oedematous area was localized to the vascular space. In the oedematous white matter the ^{125}I-BSA diffusely distributed in the extravascular space, especially in the deeper white matter under the subcortical white matter. Densitometric analysis of LFB discoloration showed that the degree of LFB discoloration parallels the regional content of extravasated serum albumin (Fig. 1 b).

Discussion

To analyze quantitatively the distribution of oedema fluid in vasogenic oedema, the regional albumin content was autoradiographically visualized in the presented study. Water content has been shown to correlate closely with the content of extravasated serum albumin in the early phase of vasogenic brain oedema[5, 6]. Constant level of the tracer specific activity, which simplifies the quantification of distribution, was achieved by employing a method described by Patlak and Pettigrew[2]. Providing that the oedematous expansion of white matter is proportional to interstitial pressure, a gradual decrease in oedema fluid content along the white matter toward the periphery of the oedematous tissue could be anticipated since a cold lesion develops a pressure gradient in the white matter be-

tween the lesion and non-oedematous white matter[1]. However, the present study showed that oedema fluid around the lesion did not decrease in proportion to the distance from the lesion but rather accumulated in the deeper white matter under the lesion. This indicates that the compliance of the white matter in vasogenic oedema varies. The cause of this regional variation in compliance is unknown. Structural differences of the white matter such as the type of fibers and the orientation of the fibers, for example, whether they are parallel or intermingle, may be important.

References

1. Marmarou A, Shulman K, Shapiro K, Poll W (1976) The time course of brain tissue pressure and local CBF in vasogenic oedema. In: Pappius H, Feindel W (eds) Dynamics of brain oedema. Springer, Berlin Heidelberg New York, pp 113–121
2. Patlak C, Pettigrew K (1976) A method to obtain infusion schedules for prescribed blood concentration time courses. J Appl Physiol 40: 458–463
3. Reulen HJ, Graham R, Fenske A, Tsuyumu M, Klatzo I (1976) The role of tissue pressure and bulk flow in the formation and resolution of cold-induced oedema. In: Pappius H, Feindel W (eds) Dynamics of brain oedema. Springer, Berlin Heidelberg New York, pp 103–112
4. Klatzo I, Chui E, Fujiwara K, Spatz M (1980) Resolution of vasogenic brain oedema. In: Cervós-Navarro J, Ferszt R (eds) Advances in neurology, Vol 28: Brain oedema: Pathology, diagnosis and therapy. Raven Press, New York, pp 359–373
5. Kuroiwa T, Cahn R, Juhler M, Goping G, Campbell G, Klatzo I (1985) Role of extracellular proteins in the dynamics of vasogenic brain oedema. Acta Neuropathol (Berl) 66: 3–11
6. Kuroiwa T, Shibutani M, Tajima T, Hirasawa H, Okeda R (1990) Hydrostatic pressure versus osmotic pressure in the development of vasogenic brain oedema induced by cold injury. In: Long D *et al* (eds) Advances in neurology, Vol 52. Brain oedema: Pathogenesis imaging and therapy. Raven Press, New York (in press)

Correspondence: T. Kuroiwa, M.D., Medical Research Institute, Tokyo Medical and Dental University, 1-5-45, Yushima, Bunkyo-Ku, Tokyo 113, Japan.

Acta Neurochirurgica, Suppl. 51, 87–89 (1990)
© by Springer-Verlag 1990

A Microstructural Study of Oedema Resolution

H. Naruse, K. Tanaka, S. Nishimura, and **K. Fugimoto**[1]

Department of Neurosurgery, Osaka City University Medical School, Osaka, [1] The 1st Department of Anatomy, Simane Medical University, Osaka, Japan

Summary

The infusion model was used to investigate the resolution mechanism of vasogenic brain oedema. HRP dissolved in autoserum was infused into the white matter of cats. The infused protein in the oedema fluid seemed not only to be absorbed by pinocytosis or phagocytosis, as has been reported previously, but also to be quickly washed out via the perivascular space of the post-capillary venules. Smaller molecules such as HRP moved faster than serum albumin in these spaces after the infusion pump was turned off. Rapid flow of extracellular fluid via the perivascular space may play an important role on the resolution process of vasogenic brain oedema by accelerating diffusion

Introduction

It has been postulated that the resolution of vasogenic brain oedema closely correlates with the disappearance of extravasated serum proteins[2, 3]. Reulen *et al.* demonstrated that two different mechanisms, bulk flow and diffusion, may be operating at different stages of oedema resolution[5]. Rennels *et al.* reported the presence of rapid fluid circulation throughout the brain via paravascular pathways[4]. In order to investigate the resolution process of brain oedema, the direct infusion model was used. In this model, the blood-brain barrier (BBB) is thought to be intact; if true, then the oedema fluid comes solely from the infused fluid, and the resolution process of oedema can be separated from its formation process[6]. Horseradish peroxidase (HRP) dissolved in autoserum was used as a tracer of fluid movement, and the fine structural features were studied sequentially.

Materials and Methods

Adult cats were anaesthetized with intraperitoneal injection of pentobarbital sodium (30 mg/kg), intubated and mechanically ventilated. Infusion oedema was produced by inserting a 25 gauge needle into the right centrum semiovale using stereotaxic coordinates (AP: + 19.0, LAT: 9.0, V: + 19.0). The needle was connected by stiff polyethylene tubing to a syringe mounted in a Harvard dual infusion pump. Three mg of HRP was dissolved in 0.25 ml of autoserum and infused at an average inflow rate of 0.1 ml/hr.

For the microstructural study, animals were sacrificed by perfusion immediately, 2 hours and 3 days after infusion. After 0.2 ml of heparin was administered intravenously, the animals were perfused with 4% paraformaldehyde in phosphate buffer for 5 min, then with a mixture of 2% paraformaldehyde and 2.5% glutaraldehyde in phosphate buffer for 10 min, and finally with 2.5% glutaraldehyde in phosphate buffer for 10 min. Two hours after perfusion, the brain was removed from the skull and sliced in the coronal plane with a vibratome. Areas of interest were sampled from the vibratomed sections.

1. Macroscopic Study

The slices were incubated for 30 min in 10 ml of 0.05 M Tris-HCl buffer containing 5.0 mg of 3,3'-diaminobenzidine (DAB) and 3 ml of 0.3% hydrogen peroxide, then air dried on gelatin coated glass slides.

2. Electron Microscopic Study

Small pieces of tissue were cut and incubated with DAB for 30 min. These tissues were postfixed in 1% osmium tetroxide in phosphate buffer for 2 hours and embedded in Epon after dehydration. Thin sections were prepared, but not stained, and examined with a transmission electron microscope (H-300, Hitachi).

Results

1. Macroscopic Observations

Dark staining indicative of the presence of HRP reaction product was observed on the slices prepared immediately and 2 hours after infusion. The staining was noted in the infused white matter, cortex and periventricular region immediately after infusion. After two hours the HRP had spread away from the infusion site, as indicated by a decrease in the intensity of staining. HRP reaction product could hardly be seen on the slices 3 days after infusion (Fig. 1).

2. Electron microscopic observations

a) Changes immediately after infusion

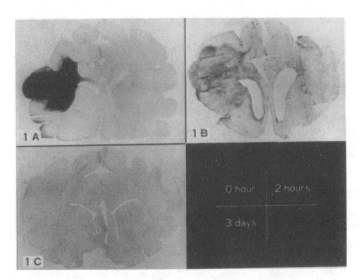

Fig. 1. Coronal sections of brain immediately, 2 hours, and 3 days after infusion. HRP reaction product is observed immediately and 2 hours after infusion (Fig. 1 A, B). Immediately after infusion, the staining is observed in the vicinity of the infused site, in the cortex, and in the periventricular region (A). Two hours after infusion, the staining has spread to more remote areas and become less intense (B). Three days after infusion, staining can hardly be seen (C)

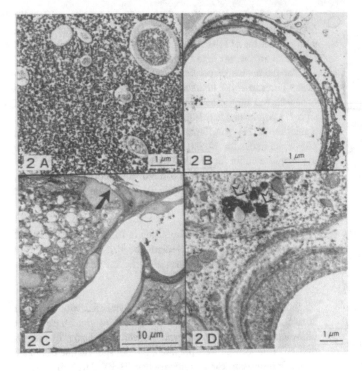

Fig. 2. Electron micrograph of infused white matter immediately after infusion (A). The distended extracellular space is filled with HRP reaction products. Axons also contain dense material. The perivascular space is greatly expanded and filled with HRP reaction products (B). The endothelial cells are stained with reaction products, too. Low magnification view of a blood vessel at the surface of the brain immediately after infusion (C). The oedema fluid has apparently reached the brain surface via the perivascular space (solid arrow) and been deposited in the subpial space. In the vicinity of the infused site, the extracellular space and the perivascular space around post-capillary venules are still expanded two hours after infusion (D). Less dense material indicating serum albumin is seen in these regions, but HRP reaction product has almost disappeared. Dense material was also visible in the astrocyte (open arrows)

The extracellular space was markedly distended in the infused white matter. In the distant subependymal and subpial regions, interstitial distension was minimal. The expanded extracellular and perivascular spaces were filled with dense HRP reaction products (Fig. 2 A, B). In the infused white matter, the perivascular space was expanded more around post-capillary venules than around arterioles and capillaries. Pinocytotic vesicles containing HRP reaction product were seen within endothelial cells of these microvessels and a thin dense layer of the materials was observed at the luminal surface of the endothelium. In the subependymal extracellular space, large deposits of dense materials were noted which led to the ventricle via the clefts between ependymal cells. Large deposits of dense materials were also observed even in the remote cortex of the infused hemisphere. The distended perivascular space extended to the subpial space where electron dense materials seemed to be deposited (Fig. 2 C). Perivascular macrophages were apparently phagocytosing the dense material. Nerve cells also contained large amounts of dense material, which resembled that in oedema fluid.

b) Changes 2 Hours After Infusion

The extracellular and the perivascular spaces around post-capillary venules were still distended and filled with somewhat less dense material. HRP reaction product was greatly reduced at the infused site. Active phagocytosis by astrocytes and perivascular macrophages was noted (Fig. 2 D).

c) Changes 3 Days After Infusion

In the infused white matter, the extracellular and the perivascular spaces of post-capillary venules were still as widely expanded as immediately after infusion, but the amount of dense material was markedly diminished. Macrophages containing dense materials were also visible. Many lysosomes were seen in the glial cells.

Discussion

Resolution of vasogenic brain oedema may involve various mechanisms. Klatzo *et al.* contended that intraglial uptake of extravasated serum proteins is primarily responsible for the resolution of vasogenic brain oedema[2]. Reulen *et al.* divided the resolution process into two phases; 1. the initial phase where the BBB is leaky and oedema fluid is being formed, and 2. the later phase where fluid leakage across the BBB has stopped. They presumed that resolution would be accomplished by bulk flow in the initial phase and by diffusion in the later phase[5]. With the present model of oedema, bulk fluid movement ceases after the infusion pump is turned off. Our macroscopic observation at 2 hours after infusion revealed that HRP reaction product had spread away from the site of infusion, which was now less intensely stained. This movement of HRP reaction product might be far more rapid than can be accomplished by diffusion. Rennels *et al.* reported the presence of rapid fluid circulation throughout the brain via paravascular pathways[4]. Two hours after infusion, the distension of extracellular space and the perivascular space around post-capillary venules was similar to that immediately after infusion. HRP reaction product was mostly seen in macrophages, and the expanded extracellular space and the perivascular space were filled with less dense material, which was probably serum albumin. These microscopic features that the smaller particle as HRP moved faster than the larger particle as serum protein support the idea that the process of diffusion is operating in the later phase of oedema resolution. Three days after infusion, the extracellular space and the perivascular

space were still similarly distended, but the density of the fluid in these spaces markedly decreased.

Bodsch *et al.* reported that extravasated protein undergoes enzymatic breakdown, yields degradation products of smaller molecular weight[1], and increases the number of particles in the extracellular space and the colloid osmotic pressure. The results obtained here show that HRP of 40,000 molecular weight has mostly been washed out in the early phase of the resolution process (2 hours post-infusion) and indicate that protein degradation products such as peptides might be rapidly washed out. We have previously reported that 3 days is necessary for the oedema, which was produced by the infusion of protein-free mock CSF, to be resolved[6]. We, therefore, propose that there is a wash-out mechanism that continuously operates throughout the period of oedema resolution.

References

1. Bodsch W, Hossmann K-A (1983) ^{125}I-antibody autoradiography and peptide fragments of albumin in cerebral edema. J Neurochem 41: 239–243
2. Klatzo I, Chui E, Fujiwara K, Spatz M (1980) Resolution of vasogenic brain edema. In: Cervós-Navarro J, Ferszt R (eds) Advances in neurology 28: Brain edema. Raven Press, New York, pp 359–373
3. Marmarou A, Tanaka K, Shulman K (1982) The brain response to infusion edema: Dynamics of fluid resolution. In: Hartmann A, Brock M (eds) Treatment of cerebral edema. Springer, Berlin Heidelberg New York, pp 11–18
4. Rennels ML, Gregory TF, Blaumanis OR, Fujimoto K, Grady PA (1985) Evidence for a "paravascular" fluid circulation in the mammalian central nervous system, provided by the rapid distribution of tracer protein throughout the brain from the subarachnoid space. Brain Res 326: 47–63
5. Reulen HJ, Tsuyumu M, Tack A, Fenske AR, Prioleau GR (1978) Clearance of edema fluid into cerebrospinal fluid. A mechanism for resolution of vasogenic brain edema. J Neurosurg 48: 754–764
6. Tanaka K, Ohata K, Katsuyama J, Nishimura S, Marmarou A (1986) Effect of steroids on the resolution process of edema. In: Miller JD, Teasdale GM, Rowan JO, Galbraith SL, Mendelow AD (eds) Intracranial pressure. VI. Springer, Berlin Heidelberg New York, pp 611–619

Correspondence: H. Naruse, M.D., Department of Neurosurgery, Osaka City University Medical School, 1-5-7 Asahi-machi, abeno-ku, Osaka 545, Japan.

Acta Neurochirurgica, Suppl. 51, 90–92 (1990)

Routed Protein Migration After Protein Extravasation and Water Leakage Caused by Cold Injury

Y. Shinohara, H. Ohsuga, S. Ohsuga, S. Takizawa, and **M. Haida**

Department of Neurology, Tokai University School of Medicine, Isehara, Kanagawa, Japan

Summary

The movement of extravasated endogenous protein induced by cold injury was investigated in the rat. Extravasated proteins were observed around the cold injury immediately after injury; within 3 hr they had moved into the ipsi- and contralateral hemispheres along the nerve fibers (routed protein migration). The water content in the areas where the extravasated protein was observed was increased, and NMR-CT scans (TR 2000 msec, TE 90 msec) showed high intensity patterns whose distribution and progression coincided with those of extravasated protein. NMR relaxation time, T_1, showed a slight increase in the area where the extravasated protein was observed, but T_2 value did not show any significant changes. In the opposite hemisphere, CBF decreased within 3 to 6 hours after injury. Local cerebral glucose utilization was reduced, but this change occurred more than 6 hours after injury. These results indicate that water leakage and protein extravasation occur simultaneously after the injury without any significant change of water state or protein conformation. Subsequently, a reduction of CBF is induced, which is followed by changes in cerebral metabolism.

Introduction

In 1981 we reported that after blood-brain barrier destruction, extravasated protein was observed not only in the marginal areas of the cold injury in the brain, but also in areas remote from the site of the injury[2]. Extravasated protein such as albumin or globulin seemed to move from the damaged region into the ipsilateral and even into the contralateral hemispheres, mainly through the corpus callosum. We designated this protein movement as "routed protein migration" because the route of protein movement along the nerve fibers was generally similar when the injury was produced in the same place[3, 5].

The purpose of this study was to examine whether the presence of extravasated proteins in areas remote from the site of direct brain damage is correlated with changes of water content and water states, to establish more precisely the electron microscopic changes in the remote areas, and to assess the coupling between cerebral blood flow and metabolism in these areas.

Materials and Methods

Seventy Wistar rats weighing 250 to 400 grams were studied. Blood-brain barrier (BBB) function was investigated by means of the anti-horseradish peroxidase (HRP) immunostaining method or the direct HRP-labelled antibody method[1, 4, 6]. Using these methods the extravasated serum protein, which is an endogenous protein, and not a foreign protein, can be observed macroscopically, microscopically, and also electron microscopically.

The water content of the brain was measured by a conventional vacuum-drying method. Water states were estimated from the values of proton relaxation times, T_1 and T_2, by means of 100 MHZ NMR equipment (Nihon Denshi, JNR-100), and the water distribution was obtained from the MRI (Nihon Denshi JHM-CTX 270). Cerebral blood flow (CBF) values was quantitated by the autoradiographic C-14-iodoantipyrine method. Local cerebral glucose utilization was determined by the 2-deoxyglucose method.

BBB destruction was produced by cold injury under anaesthesia (intraperitoneal administration of chloralose 50 mg/kg and urethan 500 mg/kg). Cold injury was inflicted by opening the skull on the left and applying dry ice gently to the left cerebral cortex for 60 seconds.

Results and Discussion

1. The Movement of Extravasated Protein

Extravasated proteins (immunoglobulin for the anti-HRP immunostaining method, and albumin and globulin for the direct HRP-labeled antibody method) were observed around the lesion immediately after the cold injury. In most cases these proteins then moved from the damaged region to the surrounding area and moved into the opposite hemisphere via the corpus callosum from 3 to 5 hours after injury. Twenty-four hours after injury, extravasated proteins were observed in the op-

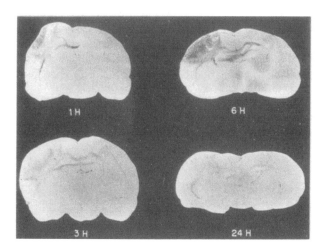

Fig. 1. Protein distribution in brain at 1, 3, 6, and 24 hr after cold injury

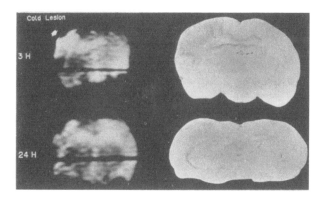

Fig. 2. Left side, magnetic resonance images obtained by the long spin echo technique at 3 and 24 hr after cortical injury. Right side, protein distribution in brain at 3 and 24 hr after cortical injury as indicated by antibody – HRP labelling

posite hemisphere in all cases (Fig. 1). In the case of middle cerebral artery ligation with focal damage during operation, the extravasated protein also migrates along the nerve fibers as in the case following cold injury. The distribution of extravasated protein was far wider than that of Evans blue seen with the conventional method.

2. Microscopic and Electron-microscopic Studies

At higher magnification, a number of damaged and swollen nerve cells were visualized. The cytoplasm of these cells in the areas adjacent to the necrotic lesions was intensely stained by both methods. This may be due to the passage of the proteins into the damaged cells. Electron microscopic findings showed that extravasated protein was observed only along the nerve

fibers and not in nerve cells or nerve fibers in the opposite hemisphere.

3. Changes of Water Content and States

In the injured hemisphere, the water content was elevated by about 2% within 1 hour after cold injury. In the opposite hemisphere a slight but definite transient increase of the water content was observed around 3 hours after the injury, suggesting that "routed protein migration" coincides with changes of water content in areas remote from the damage. The T_1 value of biopsied brain determined by the spin-echo method was elevated around the area of cold injury as well as in the contralateral hemisphere, indicating increased water contents in these tissues. However, the T_2 value did not show any significant change at 3 hours, suggesting that in these areas extravasated protein is accompanied with water but that a significant change of the bound state of water or conformational changes of protein may not occur until at least 3 hours after injury.

The left side of Fig. 2 shows magnetic resonance images obtained by the long spin echo method with the pulse sequence of TR 2000 msec and TE 90 msec.

High intensity signals were observed around the lesion immediately after injury; after about 3 hr, they appeared in the opposite hemisphere. The time course of these MRI changes was identical with that of protein distribution. These findings confirmed the postulation that the extravasated proteins move along the nerve fibers in the brain together with water.

4. CBF and Metabolism Change

Cerebral blood flow and glucose utilization were measured in 24 rats. Cerebral blood flow in the contralateral hemisphere was decreased within 3 to 6 hours after cold injury, and a reduction of glucose utilization appeared a little later.

References

1. Houthoff HJ, Go KG (1980) Endogenous versus exogenous protein tracer passage in blood-brain barrier damage. In: Cervos-Navarro J, Ferszt R (eds) Advances in neurology, Vol 28: Brain oedema. Raven Press, New York, pp 75–80
2. Shinohara Y, Izumi S, Watanabe K (1981) Evaluation of BBB function and movement of protein in cold injury by anti-peroxidase immunostaining method. J Cereb Blood Flow Metab 1 [Suppl 1]: pp 393–394
3. Shinohara Y, Takagi S, Yoshii F, Tanaka M, Izumi S, Watanabe K (1983) Movement of extravasated protein and changes of CBF and cerebral glucose utilization in BBB dysfunction in rats. J Cereb Blood Flow Metab 3 [Suppl 1]: pp 437–438

4. Shinohara Y, Ohsuga H, Takizawa S, Tanaka M, Izumi S, Watanabe K (1985) Evaluation of BBB function by anti-peroxidase immunostaining method and comparison with enzyme labeled antibody method. Acta Neurol Scand 72: 112–113
5. Shinohara Y, Ohsuga H, Takizawa S, Haida M, Saitoh S, Taniguchi R (1987) Routed protein migration followed by changes of water content and water states. In: Meyer JS *et al* (eds) Cerebral vascular disease 6. Elsevier Science Publishers B.V., Excerpta Medica, Amsterdam New York Oxford
6. Shinohara Y, Ohsuga H, Tanaka M, Takizawa S, Izumi S, Watanabe K (to be submitted) Evaluation of blood-brain barrier function by anti-horseradish peroxidase immunostaining method and comparison with horseradish peroxidase-labelled antibody method

Correspondence: Y. Shinohara, M.D., Department of Neurology, Tokai University School of Medicine, Isehara, Kanagawa, Japan.

Acta Neurochirurgica, Suppl. 51, 93–95 (1990)
© by Springer-Verlag 1990

Immunocytochemical Studies of Oedema Protein Clearance in the Rat

K. Ohata[1], **A. Marmarou**[1], and **J. T. Povlishock**[2]

Richard Roland Reynolds Neurosurgical Research Laboratories, [1] Division of Neurosurgery and
[2] Department of Anatomy, Medical College of Virginia, Richmond, Virginia, U.S.A.

Summary

Twenty µl of rat albumin solution was infused into the caudate nucleus of anaesthetized rats and the distribution of the albumin was followed using immunocytochemical methods with LM and EM at 15 min, 24 hr, and 48 hr post-infusion. Fifteen min post-infusion, the albumin was distributed in the extracellular space of the white matter and in the overlying deep cortical layers. At 24 hr post-infusion, the albumin was detected in the extracellular space around the glia limitans. At the surface of the ventricular wall of the 48 hr post-infusion animals, most of the albumin had been cleared from the extracellular space (ECS) of the subependymal white matter and the ependymal clefts, although a large amount of albumin had been observed in these areas at 15 min after infusion. At the temporobasal area of the cortex, there was continuity of the labelled perivascular space of the venous vessel from the deep oedematous area to the cortical surface not only immediately after infusion but also during the chronic phase. In conclusion, oedema fluid and protein migrate not only to the ventricle but upward toward the cortical surfaces to reach the subarachnoid spaces for eventual clearance into CSF. This seemingly occurs in the absence of significant pressure gradients.

Introduction

The three possible mechanisms that have been proposed for the clearance of the oedema fluid are: 1. passage into the CSF by bulk flow in the presence of pressure gradients[3], 2. glial and neuronal uptake of the protein compounds[1], and 3. reverse vesicular transport from the ECS to the blood via transendothelial passage[4]. In this study, an attempt was made to elucidate the mechanisms responsible for the clearance of the oedema fluid in the late stage of resolution when large pressure gradients have subsided. More specifically, we utilized the infusion oedema method and targeted this study to describe the resolution of oedema protein by its passage into the CSF.

Materials and Methods

Male adult Sprague-Dawley rats were anaesthetized by intraperitoneal injection of pentobarbital (60 mg/kg). According to the preliminary study, all burr holes were drilled 12 days before infusion to eliminate the effect of contusion in subsequent studies. Then, the animals were reanaesthetized, and a 1 cc syringe was filled with the infusate and connected via PE-10 tubing to a 30 gauge needle. The infusion needle was inserted into the brain according to the following stereotaxic coordinates relative to the bregma: + 1.0 mm forward, 3 mm lateral, 4 mm deep. Rat albumin was diluted to 65 mg/ml with mock CSF and 20 µl of albumin solution was infused in one hour. Fifteen minutes after the infusion, the infusion needle was carefully removed. After survival periods of 15 min (n = 3), 24 hr (n = 3), and 48 hr (n = 3), the animals were anaesthetized and transcardially perfused with 4% paraformaldehyde and 0.2% glutaraldehyde. The brain was cut into 50 µm thick coronal sections using a vibratome.

Rat albumin was stained by the avidin-biotin peroxidase complex method. Tissues were immersed in goat anti-rat albumin (diluted 1 : 10 000), biotinylated rabbit anti-goat immunoglobulin G antibody and avidin-biotinylated peroxidase complex, in turn. After the visualization of peroxidase, sections were osmicated and embedded in plastic. The cortical white matter, the cortical surface and the ventricular wall were extensively examined by LM and EM.

The specificity of immunocytochemical reaction in this method was confirmed. In two shun-operated animals the needle was inserted without infusion. Tissues from these sham animals were observed at 13 days after drilling and at 24 hrs after drilling.

Results

In animals sacrificed 15 minutes after infusion, immunoreactivity was intense within the interstices of the brain parenchyma around the infusion site (Fig. 1). The immunocytochemical reaction product distributed over a wide expanse of the white matter and caudate-putamen, extending to the ventricular wall of the infused hemisphere. At this time, albumin had migrated from the cortical white matter infusion site to the deep cortical gray, which was a distance of about 600 µm. As

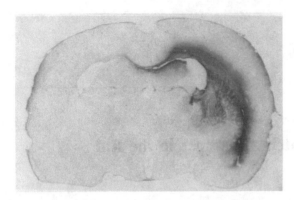

Fig. 1. Photograph of a coronal section 1.5 mm posterior to the needle site immediately after infusion. Reaction product is distributed in the caudate-putamen, the cortical white matter and the deep cortical gray, and extends to the ventricular wall in the infused hemisphere (50 µm thick sections, no counterstaining)

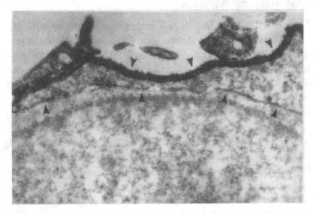

Fig. 2. Electron micrograph of the cortical surface at the infused site 24 hours post-infusion. Note that reaction product (arrowheads) is observed around the glia limitans (GL) (no staining, × 48,000)

time progressed, the albumin spread radially, and the intensity was decreased.

At 1 hr and 48 hrs after infusion, the reaction product was observed spreading to the cortical surface. Ultrastructurally, the reaction product was recognized throughout the ECS and reached the superficial cortex (Fig. 2). When the albumin reached the glia limitans, it passed through this lamina to reach the subarachnoid space. In general, within cortical gray the reaction product moved through the extracellular compartment, which was not overtly expanded. Furthermore, on the cortical surface, strong immunoreactivity was exclusively observed at the infusion site. No immunoreactivity was found on the contralateral side of the same section nor along the needle insertion site of the sham-operated animals.

Immediately after the infusion considerable reaction

product had distributed in the ECS of the subependymal white matter, and spread to the CSF space via interependymal clefts. At 48 hr after infusion reaction product was difficult to visualize within the tissue adjacent to the ventricular surface. However, deep brain areas remained heavily stained with reaction product. Some reaction product could be seen in the ependymal clefts. It, thus, appeared that albumin around the ventricular wall continued to move into the CSF space despite normalization of pressure gradients from the lesion site to the CSF.

Light microscopy indicated intense perivascular immunoreactivity around the walls of vessels both in the oedematous white matter and in the temporobasal cortex surrounding the infusion site throughout the two day period of observation. There was continuity of vascular wall labeling from the deep oedematous area to the cortical surface. This finding was particularly prominent in the temporobasal area. Electron microscopy indicated that the involved vessels were typically venules. In the ultrastructural analysis of the white matter and the cortex of the infused hemisphere, the reaction product was found to be localized to the basal laminae of microvessels and venules. In every case, endothelial cell tight junctions remained intact.

Discussion

It is now well established that bulk flow processes are involved in the acute stage of brain oedema clearance when pressure gradients are available as a driving force[3]. We suggest that the resolution process occurs in two distinct phases. During the first stage, oedema fluid continues to form and the concurrent rise in tissue pressure in and around the site of the lesion is sufficient to drive fluid clearance by bulk flow. However, when fluid formation ceases, the large pressure gradients dissipate quickly (stage two), and the clearance of oedema fluid must take place by some other mechanisms. Klatzo and co-workers postulated that the intracellular uptake of extravasated serum proteins by glial cells constituted the principal mechanism for vasogenic oedema resolution in the late stages of the oedema fluid clearance[1].

This study indicates that the CSF pathway remains the major route for resolution of infused proteinaceous fluid in the late stages of the clearance process when large tissue pressure gradients have subsided. Fluid was observed to pass both toward the ventricles and also toward the cortical surface. The conduits through which the infused albumin cleared the brain parenchyma were primarily the extracellular and cerebrospinal fluid compartments.

In the temporobasal area of the cortex, immunoreactivity was prominent in the perivascular spaces of the venules and veins both in the early and late stages of clearance, suggesting that this route is another path for albumin passage to the CSF space. Recent morphological investigations have suggested that a convective influx of the fluid from the CSF space to the brain parenchyma may occur via the perivascular space of the penetrating arteries and arterioles[2]. The present study suggests that the venules and veins are a pathway for the flow of solutes from the brain parenchyma to the CSF space.

Acknowledgements

This work was supported in part by grants NS-19235 and NS-12587 from the National Institutes of Health.

Additional facilities and support were provided by the Richard Roland Reynolds Neurosurgical Research Laboratories.

References

1. Klatzo I, Chui E, Fujiwara K, Spatz M (1980) Resolution of vasogenic brain oedema. In: Cervos-Navaro J, Ferszt R (eds) Brain oedema. Advances in neurology, Vol 28. Raven Press, New York, pp 359–373
2. Rennels M, Gregory TF, Blaumanis OR, Fujimoto K, Grady P (1985) Evidence for a "paravascular" fluid circulation in the mammalian central nervous system, provided by the rapid distribution of tracer protein throughout the brain from the subarachnoid space. Brain Res 326: 47–63
3. Reulen HJ, Tsuyumu M, Tack A, Fenske AR, Prioleau GR (1978) Clearance of oedema fluid into cerebrospinal fluid: a mechanism for resolution of vasogenic brain oedema. J Neurosurg 48: 754–764
4. Vorbrodt AW, Lossinsky AS, Wisniewski HM, Suzuki R, Yamaguchi T, Masaoka H, Klatzo I (1985) Ultrastructural observations on the transvascular route of protein removal in vasogenic brain oedema. Acta Neuropathol (Berl) 66: 265–273

Correspondence: A. Marmarou, Ph.D., Division of Neurosurgery, Medical College of Virginia, MCV Station, Box 508, Richmond, VA 23298, U.S.A.

Acta Neurochirurgica, Suppl. 51, 97–99 (1990)

Pathophysiology IV (Treatment)

Effects of Anaesthetic Agents on Cerebral Blood Flow and Brain Oedema from a Focal Lesion in Rabbit Brain

R. Murr[1], L. Schürer[2], S. Berger[2], R. Enzenbach[1], and A. Baethmann[2]

[1] Institute Anesthesiology and [2] Institute Surgical Research, LMU Munich, Klinikum Großhadern, München, Federal Republic of Germany

Summary

Anaesthetic agents reduce cerebral metabolism and may impair coupling of cerebral blood flow and metabolism. We analyzed the effects of isoflurane (I) (1 MAC), fentanyl (F), thiopental (T) (32.5 mg/kg × hr) and α-chloralose (C) on rCBF and brain oedema formation after a focal cerebral injury (cold lesion) in rabbits (n = 6 per group). In the isoflurane group, angiotensin II (0.15 μg/kg × min) was given to maintain blood pressure. rCBF of cerebral cortex was measured 3 times per hr by H_2-clearance with needle electrodes placed at different distances to the lesion during 6 hrs after induction of trauma. Thereafter, samples of white matter were obtained near the focal lesions and from corresponding areas of the contralateral hemisphere for measurement of specific gravity (SG) by a linear density column (Percoll R). Blood pressure was 78, 86, 72, and 88 mmHg for groups I, F, T, and C, respectively. After induction of the lesion, hyperemia of approximately 1 hr was observed in all groups. This was most pronounced distant to the lesion. Close to the lesion rCBF remained unchanged in groups C and T, but fell significantly below control in I and F. The blood flow response distant to the trauma was characterized by a moderate increase (C), or no alteration (T), while isoflurane animals had a pronounced secondary hyperemia for about 3 hrs. With fentanyl, however, rCBF was markedly reduced in this area. SG of white matter close to the lesion decreased significantly to values of 1.032 g/cm^3 (I, F, T), or 1.031 (C), indicative of oedema. Specific gravity was 1.034 in the contralateral hemisphere (control). The differences in SG adjacent to the lesion between groups I, F, and T were statistically not significant. Further, no significant differences were observed between the specific gravity of group C and the other groups. It is concluded, that with an open skull preparation formation of posttraumatic brain oedema from a focal cerebral lesion does not seem to be markedly affected by either hyperemia, or a blood flow reduction from various anaesthetic agents, at least during the first hours after trauma.

Introduction

Brain oedema is a major cause of morbidity and mortality in patients with severe head injury. Anaesthetic agents, which often have to be administered in these patients for diagnostic or operative procedures usually decrease cerebral metabolism[7]. On the other hand, inhalation anaesthetics may increase cerebral blood flow (CBF) by vasodilation of brain vessels[1]. Therefore, anaesthesia may either protect the injured brain by reduction of metabolism, or contribute to the spread of brain oedema by cerebral hyperemia. Until now, only little information is available about the effect of anaesthetic agents in traumatized brain. Therefore, we performed experimental investigations to analyze the influence of isoflurane, fentanyl, thiopental, and α-chloralose on regional cerebral blood flow (rCBF) and formation of brain oedema from a focal injury to the brain.

Methods

4 groups of 6 New Zealand White rabbits each were studied. In all groups, anaesthesia was induced with thiopental. After tracheostomy and start of artificial ventilation (30% O_2/70% N_2, $pCO_2 = 35$–37 mmHg), the following anaesthetic agents were administered: isofluran (I) 1 MAC (in rabbits 2.1 vol%), fentanyl (F) (per kg b. w.: 5 μg bolus, 90 min: 1 μg/min, thereafter: 0.5 μg/min), thiopental (T) (32.5 mg/kg b. w. × h), and α-chloralose (C) (50 mg/kg b. w.). In the isoflurane group, angiotensin II in a mean dosis of

0.15 µg/kg b.w. × min was given to support the blood pressure. Arterial and venous catheters were placed into femoral vessels for monitoring (mean arterial blood pressure, blood gas, hematocrit) and for administration of fluids and drugs. After fixation of the head, the left hemisphere was completely exposed by a rectangular trephination. Four platinum needle electrodes (75 µm Φ) were placed then into the cortical gray matter at the lateral border of the skull window for measurement of rCBF by H_2-clearance. The electrodes were placed at different distances from the lesion, *e.g.* 5, 9, 13, and 17 mm from the center of trauma. The brain surface was protected by a thick layer of paraffin oil. After control measurements a focal cold lesion according to Klatzo[6] was induced in the occipital region of the brain. Measurements of rCBF were taken at intervals of 15 min (first hr) and 20 min during the remaining experimental period, respectively until termination of the experiment 6 hrs after trauma. The brain was rapidly removed then and frozen. For assessment of specific gravity (SG), the brain was sectioned in 7 coronar slices. Small tissue samples taken from standardized areas were immersed into a linear density column (Percoll R) for assessment of SG.

Results

pCO_2 and Hct remained largely unchanged throughout the experimental course and did not differ between the groups. Arterial pressure was 78, 86, 72, and 88 mmHg in groups I, F, T, and C. Control values of rCBF in the 4 groups ranged from 40 to 65 ml/100 g × min. After induction of the lesion, marked hyperemia of 30 to 60 min duration was observed with fentanyl (increase to 127%) and α-chloralose (increase to 204%) in the vicinity of the focus, while blood flow was only moderately altered with isoflurane- (115%), or thiopental anaesthesia (86%). In the later course of the experiment, rCBF remained largely unchanged in this area in animals receiving α-chloralose. On the other hand, regional perfusion close to the lesion decreased markedly below control with isoflurane, fentanyl, and thiopental anaesthesia (−24%, −23%, −17%) 3 to 4 hrs after trauma. Also in areas distant to the lesion, a hyperemic response was observed in all groups subjected to different forms of anaesthesia. Later, blood flow remained largely unchanged with thiopental, or increased moderately with α-chloralose. With fentanyl anaesthesia, rCBF decreased also in areas distant to the lesion (−11%). Animals with isoflurane anaesthesia, however, had then a marked and long lasting second hyperemic response with increases of flow to 122%. Further details of the cerebral blood flow response have been published elsewhere[8].

Values for SG averaged from two sections in the region of the lesion and the corresponding contralateral site are given in Table 1. Decreases of SG indicate an increased water content. In the lesioned hemisphere, a marked and significant ($p < 0.01$) decrease of SG was found as compared to the corresponding region of the

Table 1. *Specific Gravity of White Matter in the Vicinity of a Focal Lesion and in the Corresponding Regions of the Contralateral Hemisphere (SG in g/cm³)*

Group	Lesion		Contralateral	
Isofluran	1.0327 ± 0.0004	n = 33	1.0347 ± 0.0004	n = 24
Fentanyl	1.0321 ± 0.0005	n = 28	1.0341 ± 0.0004	n = 22
Thiopental	1.0319 ± 0.0005	n = 27	1.0341 ± 0.0004	n = 18
a-Chloral.	1.0306 ± 0.0006	n = 29	1.0332 ± 0.0004	n = 18

contralateral hemisphere. The lowest values for SG were found in α-chloralose anaesthesia. This was significantly different from animals with isoflurane- or fentanyl anaesthesia. On the other hand, no significant differences in SG between groups I, F, and T were obtained.

Discussion

The immediate and pronounced hyperemia after focal cold injury as observed with isoflurane and fentanyl anaesthesia in our experiments is in contrast to other studies, where depression of blood flow was found in an inverse relationship with the distance from the lesion[3, 9]. However, in those studies, rCBF was not measured in the first hour after trauma. Further, a closed skull preparation was employed then allowing the ICP to rise quickly after trauma. This may have blunted the posttraumatic hyperemic response. Recent studies focussing on microcirculatory alterations from cerebral lesions have found brief hyperemia associated with pial vasodilation after fluid percussion injury, or cold lesion[2, 11]. Since posttraumatic hyperemia may enhance formation of brain oedema, its extent and duration is of considerable clinical importance. While in our experiments isoflurane led to a marked early and late hyperemia, with fentanyl hyperemia was limited to a period of 60–90 min after trauma. This was followed by a flow decrease in all regions during the later course. Animals with thiopental had only a moderate increase of blood flow after induction of the lesion. Blood flow remained largely unchanged during the later experimental observation period.

In spite of the rather variable flow pattern, measurements of SG revealed no specific influence of anaesthesia on the water content in groups C, F, and T. Inhalation anaesthetics, such as isoflurane, are considered less suitable for neurosurgical procedures, since with a traumatized brain the agents may increase ICP and support formation of brain oedema[4, 10]. Recently,

Kaieda[5] has analyzed this question again and found, however, no differences between inhalation anaesthetics and barbiturates in the formation of brain oedema from a focal lesion. Besides of a hyperemic blood flow response, an increase of the systemic blood pressure may also contribute to the formation of vasogenic brain oedema in the presence of a cerebral lesion. The comparatively high blood pressure observed in animals with α-chloralose may be responsible for a somewhat more pronounced extravasation of oedema in this group. However, animals receiving fentanyl had an even higher blood pressure than animals with thiopental, but brain oedema formation was not differently affected in those groups.

In conclusion, in spite of a markedly variable blood flow response to focal trauma observed in animals subjected to different forms of anaesthesia, the development of vasogenic brain oedema was obviously not affected by the anaesthetic procedures. However, this interpretation should be restricted for conditions with an open skull. Then, brain swelling or an increased cerebral blood flow is unlikely to result in an increase of intracranial pressure.

Acknowledgement

Supported by Deutsche Forschungsgemeinschaft Ba 452/6–7.

References

1. Cucchiara RF, Theye RA, Michenfelder JD (1974) The effects of isoflurane on canine cerebral metabolism and blood flow. Anaesthesiology 40: 571–574

2. DeWitt DW, Jenkins LW, Wei EP, Lutz HJ, Becker DP, Kontos HA (1986) Effects of fluid percussion brain injury on regional cerebral blood flow and pial arterial diameter. J Neurosurg 64: 787–794

3. Frei HJ, Wallenfang W, Pöll H, Reulen HJ, Schubert R, Brock M (1973) Regional cerebral blood flow and regional metabolism in cold induced oedema. Acta Neurochir (Wien) 29: 15–28

4. Grosslight K, Foster R, Colohan AR, Bedford RF (1985) Isoflurane for neuroanaesthesia: risk factors for increases in intracranial pressure. Anaesthesiology 63: 533–536

5. Kaieda R, Todd MM, Weeks JB, Warner DS (1989) A comparison of the effects of halothane, isoflurane, and pentobarbital anaesthesia on intracranial pressure and cerebral oedema formation following brain injury in rabbits. Anaesthesiology 71: 571–579

6. Klatzo I, Piraux A, Laskowski EJ (1959) The relationship between oedema, blood-brain barrier and tissue elements in a local brain injury. J Neuropath Exp Neurol 17: 548–564

7. Michenfelder JD (1980) The cerebral circulation. In: Prys-Roberts C (ed) The circulation in anaesthesia. Blackwell Scient Publ, Oxford

8. Murr R, Berger S, Schürer L, Baethmann A (1989) Regional cerebral blood flow and tissue pO₂ after focal trauma: effects of isoflurane and fentanyl. In: Hammersen F, Messmer K (eds) Cerebral microcirculation. Progr Appl Microcirc. Karger, Basel, pp 61–70

9. Pappius HM (1981) Local cerebral glucose utilization in thermally traumatized brain. Ann Neurol 9: 484–491

10. Smith AL, Marque JJ (1976) Anaesthetics and cerebral oedema. Anaesthesiology 45: 64–72

11. Yamamoto L, Soejima T, Meyer E, Feindel W (1976) Early haemodynamic changes at the microcirculatory level following cryogenic injury over the cortex. In: Pappius HM, Feindel W (eds) Dynamics of brain oedema. Springer, Berlin Heidelberg New York, pp 59–62

Correspondence: Dr. R. Murr, Institute Anaesthesiology, LMU Munich, Klinikum Großhadern, Marchioninistraße 15, D-8000 München 70, Federal Republic of Germany.

Acta Neurochirurgica, Suppl. 51, 100–103 (1990)

Dose- and Time-dependent Effects of Dexamethasone on Rat Brain Following Cold-injury Oedema

G. Meinig and **K. Deisenroth**

Department of Neurosurgery and Neurotraumatology, Berufsgenossenschaftliche Unfallklinik, Frankfurt/Main, Federal Republic of Germany

Summary

Experiments were carried out with a rat model of brain-injury oedema to establish the most efficacious dose and administration schedule for dexamethasone treatment. The results indicate that there are statistically significant dose- and time-dependent effects for dexamethasone treatment of cold-injury oedema. The equivalent of a 500 mg dose of dexamethasone had the highest anti-oedematous effect. With higher doses no further improvement could be achieved and the potential of hazardous effects increased. As expected, pretraumatic drug treatment had the greatest therapeutic effect. Dexamethasone administration up to about 30 min after cold injury led to measurable beneficial results. Drug injection from 90 min or longer after injury only slightly reduced cerebral oedema. No therapeutic effects were found when dexamethasone was administered more than 21 hours after inducing cerebral oedema. If similar results are obtained in corresponding clinical studies, the recommended dexamethasone dose and schedule for treating traumatic cerebral oedema would be: 1) high doses of drug (*e.g.* 500 mg); 2) drug administration to begin as early as possible, preferably within the first 2–3 hours after head injury; and 3) treatment should be terminated within 2–3 days to avoid major side effects.

Introduction

Brain oedema is one of the most significant complications following head injury. A very wide range of therapeutic measures are used in an attempt to treat it. The glucocorticoids, especially dexamethasone, have in the past been the subject of extensive investigations, but their effect on brain oedema following head injury has not been definitively assessed[4, 5, 10, 11, 17].

Animal studies have usually shown steroid therapy to be effective in treating cold-injury oedema. In many cases, however, the drug was administered before injury (reviews in 10 and 11). As for clinical studies, considerable sampling difficulties such as overly severe injuries, too few patients, and too heterogeneous a patient group have limited their value.

In addition, little attention has actually been paid to optimizing corticosteroid administration. Excessively high doses and inadequate durations of administration are evident in the literature[9, 10, 12, 14]. At present, neither the optimal dose nor the efficacious period after head injury has been established. Information on the appropriate duration of administration is also lacking. In view of these shortcomings, a dose- and time-response study of steroid therapy in brain-injury oedema has been begun[5, 18].

Material and Methods

To determine the dose-response relationship, 20 mg, 100 mg, 500 mg or 2500 mg of dexamethasone (Fortecortin, E. Merck, Darmstadt) was administered intraperitoneally 24 hours before, 12 hours before, during and 12 hours after lesion induction[4].

To determine the time response relationship, 500 mg dexamethasone was administered and followed by 4 doses of 200 mg. The first dose of dexamethasone was administered either 24 hours before lesion induction or 10, 20, 40 or 80 minutes after lesion induction[5]. All the dosages given are the human dosages adjusted to the body weight of the test animals.

Brain-injury oedema was induced using the cold lesion model of Klatzo[16]. Details concerning methods, calculation of water and electrolyte content, and statistical evaluation are given in previous communications[4, 5, 17, 18].

Results

Water content rose as a result of brain injury in untreated animals (controls, Fig. 1). Dexamethasone reduced this "brain swelling" in a dose-dependent manner. The optimal dosage was 500 mg, which reduced the oedematous fluid content by about 50%[5, 18]. The reduction was less when a higher dose (2500 mg) was used. As expected, the changes in sodium content paralleled those of water content, whereas the changes in

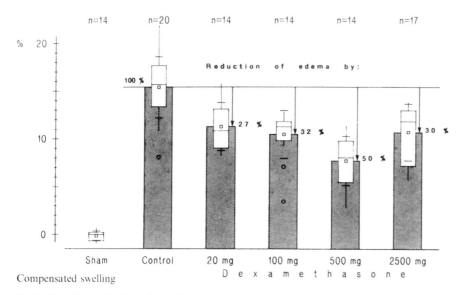

Fig. 1. The effects of various doses of dexamethasone on reducing brain swelling. The best response was achieved with 500 mg dexamethasone. At higher doses there are less favourable or even detrimental effects. The slope of the log-linear dose-response curve for 20 mg, 100 mg, and 500 mg is statistically significantly different from 0 (p = 0.0006)

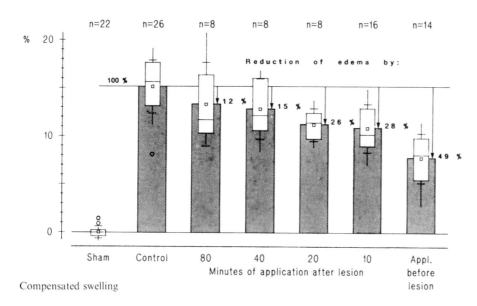

Fig. 2. The effects of various starting times for drug administration on reducing brain swelling. The strongest effects were obtained with pretraumatic administration. Significant response was also found when treatment was begun up to about 30 min after injury. The effect was minimal when the drug was given more than 30 min after injury. The slope of the log-linear time-response curve for 10, 20, 40, and 80 minutes is statistically significantly different from 0 (p = 0.03)

potassium levels were the opposite of those of sodium and water[17, 18].

In the determination of the time-response relationship it was presupposed that the pretraumatically initiated administration of 500 mg dexamethasone would bring about maximum oedema reduction[4, 17, 18]. A reduction in brain swelling of about 50% was, indeed, achieved with this dose (Fig. 2); administering dexa-

methasone 10, 20, 40, or 80 min after cold injury was less effective in lowering brain water content (Fig. 2). An appreciable reduction in brain oedema was achieved by starting dexamethasone treatment within 30 min of injury. After 90 min dexamethasone at this dose has little effect on cerebral oedema.

Again sodium and water content changes were similar. The effects on brain sodium increase were greatest

(44% reduction) when treatment was initiated within 30 min of injury. After 90 min, dexamethasone administration had little effect on brain sodium relative to untreated (control) animals[5].

There is considerable scatter in the potassium data, nonetheless strongly time-dependent changes were observed. Drug administration at times up to 30 min after cold-injury tended to normalize, that is, raise, brain potassium content. This affect was no longer found when dexamethasone was given 90 min after freeze injury[5].

Discussion

The time- and dose-dependent affects of dexamethasone on cerebral oedema induced in the rat by cold injury can be clearly seen in three different parameters, brain water, sodium, and potassium content. This gives the overall data some degree of plausibility. In untreated animals, brain water and sodium rose, whereas brain potassium fell. Dexamethasone brought about time-dependent improvements in these changes, namely lowering of oedematous brain water and sodium and raising of potassium.

One noteworthy finding of the dose-response relationship (Fig. 1) is the outcome following the ultra-high dose (2500 mg) dexamethasone. In various emergency situations including septic and haemorrhagic shock as well as head injury, ultra-high doses of corticosteroids have been used[1, 2, 7, 8, 14, 15, 20]. It may be concluded from these experiments that, in terms of human doses of dexamethasone, optimal effects following head injury are recorded at about 500 mg, while at doses of about 1000 mg and upwards the effects must be expected to diminish or even be damaging[5, 18].

In addition, our own results have been confirmed by two new experimental investigations[6, 13]. A similar dose-response relationship was found in the epidural tumor compression model of the vertebral column for doses corresponding to human doses of 7 mg, 70 mg, and 700 mg dexamethasone. The most effective dose in the treatment of this model was 700 mg of dexamethasone. Groger et al. have found that the most effective dose in the treatment of a model of extracellular oedema is 500 mg dexamethasone (Fortecortin, E. Merck, Darmstadt)[13]. Furthermore, high-dose corticosteroid treatment has been reported to be most effective in a double-blind clinical study of spinal cord injury[3].

Extending the rat time-response data (Fig. 2) to man is difficult. Since the metabolic activity of the rat is three times higher than that of man[9], high-dose cor-

ticosteroid therapy may actually be effective in humans when initiated up to 2–3 hrs after head injury[5]. Moreover, because dexamethasone therapy in the rat is successful only during the first 21 hrs after injury, the period of beneficial treatment in man may be up to 2–3 days after brain injury[5, 10, 11].

Corticosteroid therapy rational duration and dosage is to be recommended all the more strongly in view of the fact that in short-term administration the usual side effects of corticosteroids, mainly those of Cushing's syndrome, are generally able to be controlled or do not even occur[5, 10, 11]. If corticosteroids are administered for no longer than 2–3 days the suppression of the endocrine system (hypothalamus-pituitary anterior lobe – adrenal cortex axis) also plays no clinically relevant role. High dose corticosteroid therapy of this type may, therefore, be carried out without the usual consideration of the circadian rhythm and may be discontinued abruptly without tailing off.

If similar results are obtained in corresponding clinical studies, the practical recommendation – at least from the animal experimental viewpoint – would be to administer high cortisone doses (e.g. 500 mg of dexamethasone) as early as possible, preferably within the first 2–3 hours after head injury. Cortisone therapy period of more than 2 or 3 days does not appear appropriate[4, 5, 10, 11, 14, 17, 18].

References

1. Altura BM et al (1974) Peripheral vascular actions of glucocorticoids and their relationship to protection in circulatory shock. J Pharmacol Exp Ther 190: 300
2. Altura BM (1975) Glucocorticoid-induced protection in circulatory shock: Role of reticuloendothelial system function. Exper Biol Med 150: 202
3. Bracken MB et al (1990) A randomized, controlled trial of methylprednisolone or naloxone in the treatment of acute spinal-cord injury. New Engl J Med 322: 1405
4. Deisenroth KW (1986) Tierexperimentelle Erstellung einer Dosis-Wirkungsbeziehung von Dexamethason am durch Kälteläsion induzierten Hirnödem der Ratte. Inaug.-Diss., Mainz
5. Deisenroth KW, Meinig G, Schürmann K (1990) Dosis und zeitabhängige Dexamethasonwirkungen beim Kältelesion-induzierten Hirnödem der Ratte. Neurchirurgia 33: 1–6
6. Delattre JY, Arbit E, Thaler HT et al (1989) A dose-response study of dexamethasone in a model of spinal cord compression caused by epidural tumor. J Neurosurg 70: 920
7. Dietrich KA (1986) Der Einsatz von Glukokortikoiden im Endotoxinschock – eine experimentelle Studie an der Ratte. Inaug.-Diss., Marburg
8. Dietz W et al (1986) Steroide in der Therapie des septischen Schocks: Probleme „äquieffektiver" Dosen verschiedener Glukokortikoide und Applikation unter Berücksichtigung von Serumsteroidspiegeln. Anaesthesist 35: 441

9. Faden AI, Jacbos Th P, Patrick DH *et al* (1984) Megadose corticosteroid therapy following experimental traumatic spinal injury. J Neurosurg 60: 712

10. Gaab MR (1986) Kortikosteroid-Therapie beim Schädel-Hirn-Trauma? In: Walter W, Krenkel W (Hrsg) Jahrbuch der Neurochirurgie. Regensberg & Biermann, Münster, p 179

11. Gaab MR, Dietz H (1989) Ultrahohe Dexamethason-Kurzzeittherapie bei Schädel-Hirn-Trauma. Rationale und Design einer Multicenter-Studie. Neurochirurgia 32: 93

12. Giannotta SL, Weiss MH, Apuzzo MLJ *et al* (1984) High dose glucocorticoids in the management of severe head injury. Neurosurgery 15: 497

13. Gröger U, Keller S, Reulen H-J (1989) Besteht eine Dosis-Wirkungs-Beziehung von Steroiden auf die Resolution des extrazellulären Hirnödems? Arbeitstagung der Arbeitsgruppe „Intrakranieller Druck, Hirnödem und Hirndurchblutung" der Deutschen Gesellschaft für Neurochirurgie, Essen – Duisburg, 26. – 28. Oktober 1989

14. Hall ED (1985) High-dose glucocorticoid treatment improves neurological recovery in head-injured mice. J Neurosurg 62: 882

15. Horeyseck G *et al* (1985) Dosis-Wirkungsbeziehung verschiedener Glucocorticoide im Endotoxinschock der Ratte: Einfluß auf Überlebenszeiten und Histaminneubildung in verschiedenen Organen. In: Doenicke A, Lorenz W (Hrsg) Histamin und Histamin-Rezeptor-Antagonisten. Springer, Berlin Heidelberg New York, p 185

16. Klatzo I, Wisniewski H, Steinwall O *et al* (1967) Dynamics of cold injury oedema. In: Klatzo I, Seitelberger F (eds) Brain oedema. Springer, Wien, p 554

17. Meinig G, Deisenroth KW, Behl M *et al* (1986) Dosis-Wirkungs-Beziehung von Dexamethason beim Kältelesion-induzierten Hirnödem der Ratte. In: Walter W, Krenkel W (Hrsg) Jahrbuch der Neurochirurgie. Regensberg & Biermann, Münster, p 177

18. Meinig G, Deisenroth KW (1990) Dose-response relation for dexamethasone in cold lesion-induced brain oedema in rats. Adv Neurol 52: 295–300

19. Prosser CL (1988) Comparative animal physiology, 4th ed. Saunders, Philadelphia

20. Vargish T *et al* (1977) Dose-response relationship in steroid therapy for haemorrhagic shock. Am Surg 43: 30

Correspondence: G. Meinig, M.D., Department of Neurosurgery, Berufsgenossenschaftliche Unfallklinik, Friedberger Landstraße 430, D-6000 Frankfurt/Main 60, Federal Republic of Germany.

Acta Neurochirurgica, Suppl. 51, 104–106 (1990)

The Effect of Various Steroid Treatment Regimens on Cold-induced Brain Swelling

A. Unterberg, W. Schmidt, C. Dautermann, and **A. Baethmann**

Department Neurosurgery, Universitätsklinikum Rudolf Virchow, Free University of Berlin, and Institute Surgical Research, Klinikum Großhadern, University of Munich, Federal Republic of Germany

Summary

The role of steroid therapy in brain oedema following acute cerebral lesions is still unsolved. This study was conducted to compare the efficacy of dexamethasone and triamcinolone, and to analyze the influence of timing and duration of treatment on cold-induced brain swelling.

In rabbits, a cryogenic lesion of the left parietal cortex was induced. 24, or 48 hrs after trauma, hemispheric swelling, water- and electrolyte-contents were measured. A first series of animals received dexamethasone, triamcinolone or saline for 24 hrs, starting treatment 10 min after trauma. In a second series, steroid treatment lasted 48 hrs and in a third series the animals were additionally pretreated for 24 hrs.

Dexamethasone and triamcinolone slightly decreased posttraumatic hemispheric swelling, from 7.7% in controls to 7.0% in treated animals. There was no significant difference between dexamethasone and triamcinolone. Reduction of swelling was most pronounced in animals with 48 hrs treatment. Pretreatment with steroids was not superior to early posttraumatic treatment. On the other hand, dexamethasone and triamcinolone significantly decreased cerebral water content in the traumatized and contralateral hemisphere, as well as in non-traumatized animals.

The unspecific reduction of water content by steroids in rabbits might explain the moderate therapeutical effect on brain swelling. This effect might be beneficial, nevertheless, with respect to an improvement of the intracranial compliance.

Introduction

Steroids have been used since decades to treat brain oedema arising from various cerebral lesions. There is, however, an ongoing debate on the efficacy of steroid treatment under these conditions[2, 6, 8]. While many groups reported beneficial effects of steroids in experimentally induced acute cerebral lesions, beneficial clinical effects are restricted to chronic processes, like tumor-associated oedema[2, 3, 4, 5, 6, 8]. Treatment of the sequelae of cold injury, esp. oedema, with steroids has

intensively been explored for many years[3, 4, 5, 6, 7, 8]. In 1980, Sugiura[8] reviewed some 25 experimental studies on this topic performed in rats. The analysis revealed that brain oedema was effectively reduced by steroids only, if treatment commenced immediately following trauma, or if the animals were pretreated.

Since steroid treatment in acute cerebral lesions remains an unsolved clinical question, this study was conducted to further analyse the efficacy of steroid therapy on brain oedema due to a cortical freezing lesion. Three questions were addressed. First, dexamethasone and triamcinolone were compared, secondly the influence of timing, that is pre- vs. posttreatment, and thirdly the influence of duration of steroid treatment on brain oedema generation was analyzed.

Material and Methods

The experiments were performed in rabbits weighing 3.2 ± 0.3 kg, anaesthetized with ketamine and xylazine. A circular craniectomy was performed over the left hemisphere for induction of cold injury by means of a copper cylinder (diameter 6 mm) cooled with a dry-ice-acetone mixture. This cylinder was stereotactically attached onto the intact dura for 15 seconds.

The steroids dexamethasone and triamcinolone were i.v. administered. The daily dosage of dexamethasone was 2 mg/kg and 4 mg/kg of triamcinolone. Three experimental series were made. In series A steroid treatment commenced 10 min post trauma. The animals survived for 24 hrs. In series B the animals were in addition pretreated for one day. In series C the rabbits were posttreated, starting 10 min after trauma, for 48 hrs. In each series an additional control group of rabbits received physiological saline i.v. instead of a steroid. At termination, i.e. 24 or 48 hrs after trauma, the animals were again anaesthetized for exsanguination.

The brain was then removed for meticulous separation of the experimental (left) and the control (right) hemisphere in the median plane. Hemispheric swelling was determined as the difference of

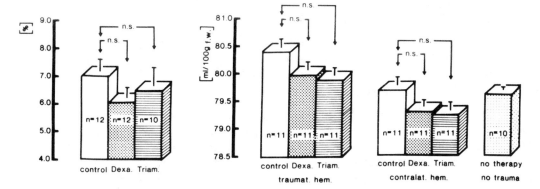

Fig. 1. Posttraumatic hemispheric swelling (top) and hemispheric water content (bottom) of dexamethasone, triamcinolone, or saline treated rabbits with cold injury. The animals were posttreated for 24 hrs (Series A). Steroids reduce swelling by app. 10% (n.s.). They induce an unspecific reduction of tissue water content in both hemispheres

weights of the traumatized and contralateral hemispheres. Cerebral water content of both hemispheres was obtained by dry/wet-weight measurement. Tissue electrolyte contents were thereafter determined in desiccated brain specimen.

Results

In series A – posttreatment for 24 hrs – steroid treatment reduced brain swelling marginally: In saline treated control animals swelling was 7.1% ± 0.5, in dexamethasone treated rabbits 6.3% ± 0.4 and in tri-

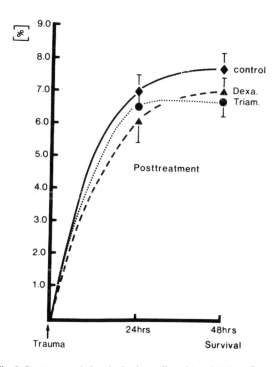

Fig. 2. Posttraumatic hemispheric swelling 24 and 48 hrs after trauma of dexamethasone, triamcinolone, or saline treated rabbits with cold injury. The additional increase in swelling between 24 and 48 hrs p.tr. is moderate. Steroids reduce swelling at both time intervals to the same amount, *i.e.* 10%

amcinolone infused rabbits 6.7% ± 0.6 (Figs. 1 and 2). Cerebral water content was more markedly affected. Both steroids reduced water content not only in the traumatized, but also in the contralateral hemisphere. This is also supported by the findings in untreated non-traumatized controls (Fig. 1).

In series B – posttreatment for 48 hrs – it should be clarified whether a prolonged treatment is more effective. 48 hrs post trauma swelling had increased to 7.7% ± 0.5 in saline infused controls. Again both steroids reduced posttraumatic swelling to 7.0% ± 0.4 (Fig. 2). Still, the difference between treated and untreated animals was statistically not significant. Thus prolonged treatment was not superior.

In series C the animals were pretreated for 24 hrs and survived one day. Again, steroid treatment affected hemispheric swelling moderately. It decreased from 7.0% ± 0.4 in controls to 6.7% ± 0.4 in triamcinolone treated rabbits. The cerebral water content was reduced in traumatized as well as non-traumatized hemispheres.

Discussion

The finding that extensive pretreatment or prolonged posttreatment with steroids only moderately decreased posttraumatic brain swelling is not surprising. The compilation of experimental studies on treatment of cold-induced oedema by Sugiura[8] stresses that even 5 of 14 pretreatment studies revealed no beneficial effects of steroids.

The fact that rabbits were used in the present study cannot account for the moderate efficacy of steroids. In another study of our group using the same experimental set-up, aprotinin significantly reduced hemispheric swelling in rabbits[9]. Moreover, other investigators using rabbits reported significant effects of dexamethasone and triamcinolone in cold-induced oedema[3, 7].

The global, *i.e.* unspecific, reduction of cerebral water content by steroids, on the other hand, is a novel finding. To our knowledge, there is no convincing explanation for this phenomenon among the numerous effects of steroids in the central nervous system[1]. Such a mechanism might be operative among others in the reversible "atrophy" of the brain seen in patients with long-term steroid treatment. Nevertheless, global reduction of cerebral water content is expected to ameliorate cerebral compliance.

Acknowledgements

The technical and secreterial assistance of Mrs. U. Goerke and Mrs. I. Brookes is gratefully acknowledged. The study was supported by a grant of the Deutsche Forschungsgemeinschaft.

References

1. Baethmann A (1985) Steroids and brain function. In: James HE, Anas NG, Perkin RM (eds) Brain insult in infants and children: Pathophysiology and management. Grune & Stratton, Orlando, pp 3–17
2. Braakman R, Schouten HJA, Blaauw-van Dishoeck M, Minderhoud JM (1983) Megadose steroids in severe head injury. Results of a prospective double-blind clinical trial. J Neurosurg 58: 326–330
3. Herrmann HD, Neuenfeldt D, Dittmann J, Palleske H (1972) The influence of dexamethasone on water content, electrolytes, blood-brain barrier and glucose metabolism in cold injury oedema. In: Reulen HJ, Schürmann K (eds) Steroids and brain oedema. Springer, Berlin Heidelberg New York, pp 77–85
4. Pappius HM (1969) Effects of steroids on cold injury oedema. In: Reulen HJ, Schürmann K (eds) Steroids and brain oedema. Springer, Berlin Heidelberg New York, pp 57–63
5. Raimondi AJ, Clasen RA, Beattie EJ, Taylor CB (1959) The effect of hypothermia and steroid therapy on experimental cerebral injury. Surg Gynecol Obstet 108: 333–338
6. Reulen HJ, Schürmann K (1972) Steroids and brain oedema. Springer, Berlin Heidelberg New York
7. Schilling L (1984) Wirkung von Di-Natrium-Dexamethason-Phosphat und Di-Kalium-Triamcinolon-Acetonid-Phosphat auf das Hirnödem nach Kälteläsion. Thesis, Mainz
8. Sugiura K, Kanazawa C, Muraoka K, Yoshino Y (1980) Effect of steroid therapy on cerebral cold injury oedema in the rat. The optimal dosage. Surg Neurol 13: 301–305
9. Unterberg A, Dautermann C, Müller-Esterl W, Baethmann A (1986) The kallikrein-kinin system as mediator in vasogenic brain oedema. III: Inhibition of the kallikrein-kinin system in traumatic brain swelling. J Neurosurg 64: 269–279

Correspondence: A. W. Unterberg, M.D., Department Neurosurgery, Rudolf Virchow Medical Center, Free University of Berlin, Augustenburger Platz 1, D-1000 Berlin 65.

Acta Neurochirurgica, Suppl. 51, 107–109 (1990)

Steroid and Hyperosmotic Diuretic Effects on the Early Phase of Experimental Vasogenic Oedema

Y. Handa[1], **H. Kobayashi**[1], **H. Kawano**, **H. Ishii**, **Y. Naguchi**, and **M. Hayashi**

Department of Neurosurgery, Fukui Medical School, Fukui, Japan

Summary

The effects of either dexamethasone or mannitol on the early phase of vasogenic oedema were studied by evaluating proton relaxation behaviour with *in vitro* NMR spectroscopy. Vasogenic cerebral oedema was induced by cold injury in cats. Significant ($p < 0.05$) decreases in both longitudinal relaxation rate ($R_1 = 1/T_1$) and transverse relaxation rate ($R_2 = 1/T_2$) and increases in water content were observed in the white matter beneath the cold lesions four hours after injury. In the dexamethasone treated group, the decrease in R_2 in the oedematous white matter was significantly ($p < 0.05$) less than that in the non-treated control group and the mannitol group. The R_2 in the oedematous white matter exhibited a two-component decay. The slow component of R_2 in the dexamethasone-treated group was significantly ($p < 0.05$) smaller than that in the control group. Our results indicated that dexamethasone has a prophylactic effect on the early phase of vasogenic oedema by preventing the accumulation of fluid in the extracellular space and that mannitol had no prophylactic effect on oedema formation.

Introduction

Vasogenic brain oedema strongly influences the morbidity of patients with intracranial disorders such as head injuries and brain tumors. In these cases, development of brain oedema worsens the neurological condition by causing intracranial hypertension. Although magnetic resonance imaging (MRI) can clearly demonstrate the degree and spread of brain oedema at both early as well as later stages[2, 5], it is still difficult to evaluate the progress or resolution and the effects of anti-oedema agents.

In clinical situations, mainly dexamethasone and hyperosmotic diuretics have been used widely for the treatment of intracranial hypertension caused by brain oedema. In the present study, we have investigated the ability of NMR to assess the effects of dexamethasone and mannitol on the resolution of early-stage brain oedema.

Materials and Methods

Eighteen mongrel cats were used in this study. The animals were divided into a control group, a dexamethasone-treated (steroid) group, and a mannitol-treated group. Vasogenic oedema was induced by cold injury in all animals. Under pentobarbital anaesthesia, a small craniectomy was performed over the right parietal bone and a cold injury was produced by applying a rod of dry ice (5 cm diameter) for 30 seconds on the surface of the dura mater. The animals were reanaesthetized four hours after cold injury and sacrificed by exsanguination. The tissue samples were taken from the injured gray matter (GM), the white matter beneath (WM 1) and remote from (WM 2) the lesion, and the corresponding regions in the contralateral hemisphere. After the measurements of proton relaxation rates, the tissue samples were kept in the oven at 80 °C for 2 days. The water content of brain tissue was calculated by the wet/dry weight method.

For the measurements of proton relaxation rates, a 30 cm bore magnet operating at 2.0 T was employed. A small tube containing the tissue sample was placed in a double-turn coil of 1.6 cm diameter. The inversion recovery pulse sequence was used for the measurements of longitudinal relaxation rate ($R_1 = 1/T_1$). For the measurement of transverse relaxation rate ($R_2 = 1/T_2$), the Curr-Purcell-Meiboom-Gill sequence was employed, and 16 spin echoes with echo times of 25 ms were calculated by curve-fitting procedure.

In the steroid group, dexamethasone (2.5 mg/kg) was administered intraperitoneally 12 hours prior to the induction of injury and an equivalent dose was administered a few minutes after injury. In the mannitol group, a solution of 20% mannitol (1.5 mg/kg) was infused intravenously for 60 minutes; the infusion began 90 minutes before sacrifice.

The data were analyzed by Student's t-test. P values of 0.05 or less were considered statistically significant.

Results

In all three groups, there were significant ($p < 0.05$) increases in water content and decreases in R_1 in each region of the injured hemisphere compared to contralateral hemisphere. However, no statistically significant

Table 1. *Water Content, Relaxation Rates and Components of Two-Exponential Transverse Relaxation Decay in Oedematous White Matter Beneath the Cold Lesion*

	Control group	Steroid group	Mannitol group
Water content (%)	84.4 ± 1.3	82.8 ± 2.6	86.0 ± 2.6
R_1 (1/S)	0.97 ± 0.04	1.00 ± 0.03	0.91 ± 0.06
R_2 (weighted average) (1/S)	4.54 ± 0.63	$5.95 \pm 0.58^*$	4.53 ± 0.65
R_{2f} (fast component) (1/S)	8.98 ± 0.71	10.20 ± 0.51	9.67 ± 0.65
R_{2s} (slow component) (1/S)	3.25 ± 0.36	$4.10 \pm 0.40^*$	3.76 ± 0.41
%S (amount of signal contributing to R_{2s}) (%)	22 ± 5	$11 \pm 4^*$	19 ± 6

Values are mean $\pm$ SD.
  * $p < 0.05$ compared to the control group.

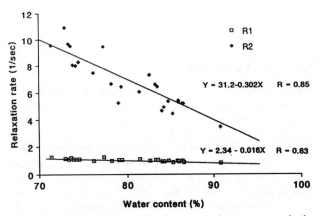

Fig. 1. Correlation between relaxation rates and water content in the oedematous white matter. *R 1*: longitudinal relaxation rate, *R 2*: transverse relaxation rate

differences in water content and R_1 value were measured between the three groups.

The weighted average of R_2 values in the injured gray matter and white matter was significantly reduced with respect to the contralateral side ($p < 0.01$). The mean decrease in R_2 values measured in the white matter just beneath the cold lesion (WM 1) of the steroid group was significantly ($p < 0.05$) less than that in the control and mannitol groups. When bi-exponential analysis was applied, T_2 decay, but not R_1 decay, in the WM 1 of the injured side showed two-compartments in all three groups. The mean value of the slow component of R_2 (R_{2s}) was significantly ($p < 0.05$) greater in the steroid group than that in the control group, and the amount of signal contributing to R_{2s} was significantly smaller ($p < 0.05$) in the steroid group ($11 \pm 4\%$) than in the control ($22 \pm 5\%$) and the mannitol groups ($19 \pm 6\%$) (Table 1).

A highly significant correlation between both R_1 and R_2, and water content, were observed in the oedematous white matter. A consistent feature of these relations was that the slope of R_2 versus water content was approximately 20 times greater than for R_1 versus water content (Fig. 1).

Discussion

The present results indicate that proton relaxation is correlated with the early changes of oedema formation and that transverse relaxation is correlated with the prophylactic effect of dexamethasone on the early phase of vasogenic oedema. Some studies have found the prolongation of relaxation times (T_1, T_2) and the separation of T_2 into two compartments occurs during the early stages of oedema formation and becomes more marked as the oedema progresses[1, 4, 7]. Analysis of relaxation behaviour, water content and pathology of the oedematous brain tissue have suggested that the change of R_1 refers to the amount of increased water. On the other hand, R_2 reflects the distribution and the content of the oedema fluid[1, 4, 7], is strongly affected by state of water molecules, and becomes much shorter when water molecules bind to macromolecules. At the beginning of vasogenic oedema, there is fairly little extravasation of macromolecules[8], and R_2 in thus, relatively unaffected. Concerning the double exponential decay of R_2, the slow component of transverse relaxation decay is probably related to the extracellular fraction of water[3].

Corticosteroids seemingly reduce abnormal vascular permeability, extravasation of macromolecules, and water content in the extracellular space[6, 8]. Long *et al.*[6] found that the volume of extracellular space in the oedematous white matter was approximately 40% when oedema developed and decreased to 10% after dexamethasone treatment. The present results observed by NMR demonstrated that mannitol did not prevent oedema formation but that dexamethasone affected the initial phase of the vasogenic oedema process by reducing the accumulation of fluid in the extracellular space.

References

1. Bakay L, Kurland RJ, Parrish RG, Lee JC, Peng RJ, Bartkowski HM (1975) Nuclear magnetic resonance studies in normal and oedematous brain tissue. Exp Brain Res 23: 241–248
2. Bartkowski HM, Pitts LH, Nishimura M, Brant-Zawadzki M, Moseley M, Young G (1985) NMR imaging and spectroscopy of experimental brain oedema. J Trauma 25: 192–196
3. Belton PS, Jackson RR, Packer KJ (1972) Pulsed NMR studies of water in striated muscle. I. Transverse nuclear spin relaxation times and freezing effects. Biochim Biophys Acta 286: 16–25
4. Go KG, Edzes HT (1975) Water in brain oedema. Observation by the pulsed nuclear magnetic resonance technique. Arch Neurol 32: 462–465
5. Go KG, van Dijk P, Luiten AL, Teelken AW (1984) Proton spin tomography in brain oedema. In: Go KG, Baethmann A (eds) Recent progress in the study and therapy of brain oedema. Plenum, New York London, pp 283–291
6. Long DM, Maxwell RE, French LA (1972) The effect of glucocorticoids upon experimental brain oedema. In: Reulen HJ, Schürmann K (eds) Steroids and brain oedema. Springer, Berlin Heidelberg New York, pp 65–76
7. Naruse S, Horikawa Y, Tanaka C, Hirakawa K, Nishikawa H, Yoshizaki K (1982) Proton nuclear magnetic resonance studies on brain oedema. J Neurosurg 56: 747–752
8. Pappius HM, McCann WP (1969) Effects of steroids on cerebral oedema in cats. Arch Neurol 20: 207–216

Correspondence: Dr. Y. Handa, Department of Neurosurgery, Fukui Medical School, Matsuoka-chou, Yoshida-gun, Fukui 910-11, Japan.

Acta Neurochirurgica, Suppl. 51, 110–112 (1990)

The Ambivalent Effects of Early and Late Administration of Mannitol in Cold-induced Brain Oedema

E. Reichenthal[1], T. Kaspi[2], M. L. Cohen[2], I. Shevach[2], E. Shalmon[2], Y. Bar-Ziv[3], Z. Feldman[1], and G. Zucker[1]

Departments of Neurosurgery, [1] Soroka Medical Center & the Ben Gurion University, Beer Sheva; [2] Beilinson Medical Center Petach-Tikva, and [3] "Hadassah" Medical Center Jerusalem, Israel

Summary

This study was undertaken in order to determine whether early administration of mannitol is different from late administration in its effect on brain oedema. Cold-induced brain oedema, which was confirmed by high resolution CT scan, was produced in 2 groups of cats. In group one mannitol was given early (90 minutes after injury); in group two 3–4 hours after the injury (late). Repeated CT scans following mannitol administration showed that the early group exhibited significantly greater dehydration (p < 0.0001) while the late group showed significant hydration, in the lesioned hemisphere. The contralateral control hemisphere responded to mannitol with similar dehydration effect in both groups.

Introduction

The dehydrating effect of osmotherapy on neural tissue has been employed for many years to decrease elevated intracranial pressure, with mannitol most commonly utilized. It is widely accepted that osmotic agents serve to dehydrate only the healthy neural tissue, while the water content of the injured brain remains unchanged or may even increase[2, 7]. Several reports however suggest that osmotherapy may dehydrate both healthy and injured brain tissue[1, 6]. The following study addresses this apparent controversy.

Material and Methods

Unilateral brain oedema was produced in craniectomized adult cats utilizing a custom-made cold-lesion probe. Intraperitoneal sodium thiopenthone was used to induce anaesthesia, which was maintained by intravenous administration of phenobarbital. Blood pressure, heart rate, arterial blood gases, and body temperature were monitored throughout most of the experiment. The animals were held in a stereotactic head-holder while a 2 × 2 cm parietal craniectomy was performed bilaterally. The cold probe, cooled to −60 °C with dry ice, was applied to the intact dura for 10 minutes on the left side only; the right side served as a control. Ten minutes prior to mannitol administration, all the animals had an initial CT scan taken on a high resolution scanner, covering the entire supratentorial area with 2 mm slices. Density measurements were collected from both the experimental and the control sites for comparison. The animals were randomly divided into two groups: in the "early" group mannitol was administered intravenously. 65–100 minutes after induction of the cold-lesion and a second CT scan taken 30 minutes thereafter; in the "late" group the mannitol was administered 150–270 minutes after the induction of the cold-lesion. The animals were then sacrificed, their brains removed and sliced after fixation, for histopathological studies.

Analysis of variance were performed on the percentage density changes from pre- to post mannitol administration comparing the "early" with the "late" group and in each group comparing the side of the lesion with the contralateral side. A similar analysis was performed comparing the percentage density changes at the site of the cold lesion (the area directly underlying the cold probe; also called focal area), with the changes occurring in the penumbral region (the remaining surrounding tissue).

Results

Percentage density changes in the lesioned area and in the contralateral control area are depicted in Fig. 1. Positive density changes reflect areas of relative dehydration, while negative density changes indicate relative water gain. Clearly, early mannitol administration in the lesioned area results in dehydration while late administration to a similar area produces marked hydration (F 91,14) + 28.789, p < 0.0001). The effects of late mannitol administration in the healthy contralateral tissue are indistinguishable from those of the early group (F(1,14) = 0.739, ns).

At the induction site, the area underlying the cold probe, and in the contralateral hemisphere, no differ-

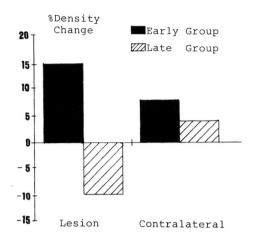

Fig. 1. Depicts the mean percentage density changes on CT for the early and late groups in the lesioned region (left) compared to the contralateral control region (right)

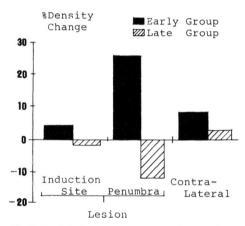

Fig. 2. Depicts the mean percentage density changes on CT for the early and late groups. The lesioned area is divided into an induction area and a surrounding penumbral region

ences were found between "early" and "late" groups. In the penumbral area, however, the 2 groups differed significantly in their responses ($F(1,11) = 16.57$, $p < 0.002$) (Fig. 2). The "early" group demonstrates a significant increase in tissue density (23.73%) reflecting a reduction in brain water, while the "late" group demonstrated a significant decrease in tissue density (− 12.03) indicating brain water gain.

Discussion

The results of many experimental studies on the effects of osmotherapy on cerebral oedema are conflicting. Some studies demonstrated either no change or even water gain in the hemisphere ipsilateral to the experimental lesion[2, 7], while the contralateral hemisphere

shows significant dehydration. Other studies showed that dehydration also occurred at the site of the injured tissue[1, 5, 6]. These apparently paradox responses of neural tissue to hyperosmolar agents can be explained by the differences between the time of injury and the time of the drug administration. This is also supported by the present study. Experiments in which the therapy was started early (in our study within 90 minutes of injury) showed evidence of oedema resolution regardless of the manner in which oedema was induced. Experimental animals given hyperosmolar agents late (more than 90 minutes after injury in our study) generally demonstrated dehydration of the intact hemisphere only.

The blood vessels in a cold-induced focal brain oedema show markedly increased permeability. The oedema spreads beyond the focal area affecting a larger perifocal territory (penumbra)[4]. The blood brain barrier in this area is still unimpaired and retains the potential to develop an osmolar gradient[8]. It is in this area that hyperosmolar agents can be highly effective provided that they are administered prior to the development of secondary damage to the microvasculature. After damage to the microvasculature has occurred with the passage of time, the blood brain barrier is disrupted[3] and at this juncture the hyperosmolar agent diffuses out of the intravascular compartment into the penumbral area, thus increasing water content of that tissue[3].

The time elapsing from the injury is the most meaningful determinant for the multiple pathophysiological processes involved in the development of cerebral oedema. This critical period in cats should not exceed two hours after injury, and the administration of hyperosmolar agents within this time limit may reverse the development of oedema.

References

1. Albright AL, Lachtaw RE, Robinson AG (1984) Intracranial and systemic effects of osmotic and oncotic therapy in experimental cerebral oedema. J Neurosurg 60: 481–489
2. Clasen RA, Cooke PM, Pandolfi S (1965) Hypertonic urea in experimental cerebral oedema. Arch Neurol 12: 424–434
3. Cooper PR (1985) Delayed brain injury: secondary insults. In: Becker DP, Povlishock JT (eds) Central nervous system trauma report. NINCDS p. 219 NIH Edition, Bethesda, Maryland
4. Fenstermacher J (1984) Volume regulation of the central nervous system. In: Staub NC, Taylor AE (eds) Oedema. Raven Press, New York, pp 383–404
5. Joyner J, Freeman LW (1963) Urea and spinal cord trauma. Neurology 13: 69–72

6. Kaupp Jr HA, Lazarus RE, Wetzel N (1960) The role of cerebral oedema in ischaemic cerebral neuropathy after cardiac arrest in dogs and monkeys and its treatment with hypertonic urea. Surgery 18: 404–460

7. Pappius HM, Dayes LA (1960) Hypertonic urea. Its effect on the distribution of water and electrolytes in normal and oedematous brain tissues. Arch Neurol 13: 395–402

8. Reulen HJ, Tsuyumu M (1981) Pathophysiology of formation and natural resolution of vasogenic brain oedema. In: De Vlieger M, deLange S, Beks JWF (eds) Brain oedema. Wiley, New York, p 32

Correspondence: E. Reichenthal, M.D., Department of Neurosurgery, P.O. Box 151, Beer Sheva 84101, Israel.

Acta Neurochirurgica, Suppl. 51, 113–115 (1990)
© by Springer-Verlag 1990

Magnetic Resonance Study of Brain Oedema Induced by Cold Injury – Changes in Relaxation Times Before and After the Administration of Glycerol

K. Yatsushiro, M. Niiro, T. Asakura, M. Sasahira, K. Terada, K. Uchimura, and **T. Fujimoto**

Department of Neurosurgery, Faculty of Medicine, University of Kagoshima, Kagoshima City, Japan

Summary

Experiments were carried out to determine the ability of magnetic resonance systems to assess change in relaxation times following the induction of experimental brain oedema and subsequent administration of hypertonic glycerol. Nine small mongrel dogs were used for these experiments. Twenty-four hours after producing a cold lesion, magnetic resonance (MR) studies were performed and physiological data were measured. Thirty min after beginning the administration of glycerol, serum osmotic pressure was raised 88 mOsm/l. After administering glycerol a small reduction in long spin echo images was observed in the periventricular high intensity area. The T_1 and T_2 values appeared to be decreased from 30 min after glycerol administration. This decrease continued until the period of observation ended. Statistically significant changes in the T_2 values, especially in the white matter of the opposite side, were seen. The changes in T_1 were generally not statistically significant. We conclude that the changes in water content of the oedematous brain following the administration of glycerol can be detected by magnetic resonance systems, that these changes become appreciable 30 min after administration of glycerol and continue at least two more hours, and that the changes in T_2 were larger than those in T_1.

Introduction

We know that hyperosmotic agents reduce intracranial pressure (ICP) but do not know how they act. Real time observations of brain water content *in vivo* following the administration of hyperosmotic agents would provide useful information about this subject, but such measurements have been difficult to make. Recently it has become possible to observe the brains of experimental animals by magnetic resonance systems that may be able to quantitate the changes in water content of the brain[2, 3]. Experiments were, therefore, carried out to ascertain whether or not we could determine changes in relaxation time and MR images following the production of experimental cerebral oedema and the administration of glycerol.

Materials and Methods

Nine adult mongrel dogs each weighing approximately 10 kg were used. Before each operation or MR study, anaesthesia was induced with ketamine hydrochloride and maintained with sodium pentobarbital. A cold lesion was produced by contact with a copper cylinder cooled in liquid nitrogen and placed on the dural surface at the right temporo-parietal portion for either five sec (the mildly injured or MI group) or 45 sec (the severely injured or SI group). Twenty-four hours after the cold lesion was produced, MR studies on the injured brain commenced. Initially, control long spin echo (SE) and inversion recovery (IR) images and calculated T_1 and T_2 images of the injured brain were determined in each dog. Following this 4 g/kg of 10% glycerol was infused for 30 min. Fifteen min after beginning the glycerol infusion a calculated T_2 image was made. Subsequently, calculated T_1 plus long SE and IR images were obtained. The same sequence of measurements was repeated over the next two hours. The MR system used in these experiments was the ASAHI Mark-J 0.1 Tesla. The T_1 value was measured by the IR method and the T_2 value by the Carr-Purcell method. We measured the relaxation times on the injured side at a set of four regions of interests (ROIs) in the periventricular white matter, which was an area of high intensity. The mean of these estimates was used as the injured side value. We also measured the relaxation time in the periventricular white matter on the opposite side. At the same time serum osmotic pressure, electrolytes, and hematocrit were measured. These values were compared by paired t-test. Statistical differences of $p < 0.05$ were considered to be significant.

Results

Physiological data are presented in Table 1. The osmotic pressure of the plasma after glycerol infusion was 87.6 mOsm/l higher than control. Serum sodium concentration and hematocrit increased after glycerol administration. These differences were statistically significant.

After glycerol infusion, there was only a slight reduction in the long SE images from the periventricular

Table 1. *Change in Physiological Data with Time*

	Serum osmotic pressure (mOsm/l)	Serum Na (mEq/l)	hematocrit (%)
Control	298.1 ± 3.7	146.7 ± 2.8	37.2 ± 8.8
30 minutes	385.7 ± 23.9**	146.9 ± 2.6	39.5 ± 6.9
1 hour	370.1 ± 15.5**	153.9 ± 2.6**	43.3 ± 7.6*
2 hours	355.8 ± 14.3**	157.3 ± 2.9**	48.8 ± 4.1*
3 hours	348.5 ± 18.1**	157.0 ± 2.9**	48.1 ± 5.7*

Values are means ± standard deviation.
Significantly different from control ** $p < 0.01$, * $p < 0.05$.

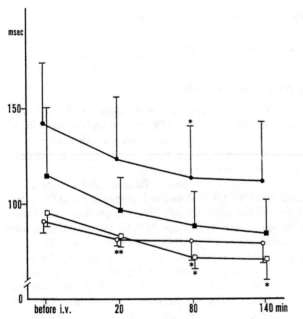

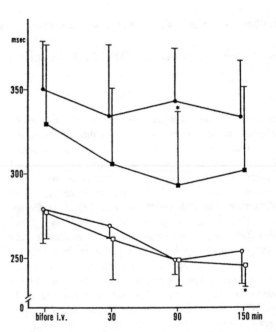

Fig. 1. The time course of mean T_2 values before and after the i.v. administration of glycerol. * Statistical significance $p < 0.05$. Severe injured group: ● mean value of point 5 to 8, ○ opposite white matter. Mild injured group: ■ mean value of point 5 to 8, □ opposite white matter

Fig. 2. The time course of mean T_1 values before and after the i.v. administration of glycerol. * Statistical significance $p < 0.05$. Legend see Fig. 1

high intensity area in some dogs, and no changes in the IR images. Reduction in the IR images was first noted 40 min after the administration of glycerol and continued thereafter.

The mean T_2 values of the brain on the injured side appeared to be reduced 20 min after beginning glycerol administration in both the SI and MI groups and to fall until the experiments were ended (Fig. 1). These reductions were only statistically significant in the SI groups at 80 min, however. Significant reductions in the T_2 values were found in the white matter of the opposite side at 20 and 80 min in the SI group. In the MI group, on the other hand, the change in T_2 was significant from 20 to 140 min after glycerol administration (Fig. 1).

The pattern of T_1 values (Fig. 2) resembled that of T_2 values. However, statistically significant differences were found only 90 min after administration on the injured side and 150 min after administration on the opposite side in the MI group (Fig. 2). On the injured side both the mean T_1 and T_2 values in the SI group were higher than those in the MI group (Figs. 1 and 2).

Discussion

The main purpose of these experiments was to see if changes in the relaxation time and in MR images could be detected in the oedematous brain following administration of glycerol. For this reason, we infused an unusually large amount of glycerol for a 10 kg animal.

Brain dehydration seemingly requires an increase in serum osmotic pressure of 30 to 45 mOsm/l[4,5]. Because serum osmotic pressure was increased, on the average, by 88 mOsm/l after the administration of glycerol, we assume that the water content of the brain was reduced.

Naruse *et al.* performed an MR study of the rat brain *in vitro*[3]. According to their results, T_2 values were reduced significantly by the administration of 1 g/kg of 10% glycerol and this reduction was correlated with the water content of the brain. On the other hand, Albright and Latchaw reported that osmotic therapy using mannitol reduced CT density *in vivo*, and its reduction continued during osmotic therapy[1]. Moreover, they indicated that mannitol lowers ICP by dehydrating normal brain. Takagi *et al.* found that brain dehydration does not correlate with the reduction in ICP[5]. We found T_2 values in vivo were reduced somewhat on the lesion side and significantly in the white matter of the opposite side from 20 min to 2 hr after glycerol administration. These results suggest that the brain was dehydrated by glycerol infusion for at least two hours.

From these findings, we conclude that MR systems are useful for studying brain oedema and the glycerol induced changes in brain water content.

References

1. Albright AL, Latchaw RE (1985) Effects of osmotic and oncodiuretic therapy on CT-brain density and intracranial pressure. Acta Neurochir (Wien) 78: 119–122
2. Bederson JB, Bartkowski HM, Moon K, Halks-Miller M, Nishimura MC, Brant-Zawadski M, Pitts LH (1986) Nuclear magnetic resonance imaging and spectroscopy in experimental brain oedema in a rat model. J Neurosurg 64: 795–802
3. Naruse S, Horikawa Y, Tanaka C, Hirakawa K, Nishikawa H (1982) Nuclear magnetic resonance studies of effects of glycerol on brain oedema. Brain Nerve 34: 805–809
4. Stern WE, Coxon RV (1964) Osmolality of brain tissue and its relation to brain bulk. Am J Physiol 204: 1–7
5. Takagi H, Saitoh T, Kitahara T, Ohwada T, Yada K (1984) The mechanism of ICP reducing effect of mannitol. Brain Nerve 36: 1095–1102

Correspondence: Dr. K. Yatsushiro, Department of Neurosurgery, Faculty of Medicine, University of Kagoshima, 1208-1 Usuki-cho, Kagoshima City, 890 Japan.

Acta Neurochirurgica, Suppl. 51, 116–117 (1990)

Transient Effect of Mannitol on Cerebral Blood Flow Following Brain Injury

E. Shalmon, E. Reichenthal, and **T. Kaspi**

Department of Neurosurgery, Experimental Research Laboratory, Beilinson Medical Center, Petah Tiqva, Israel

Summary

CBF and ICP were measured in cats following cerebral cold injury and mannitol infusion. Mannitol was found to reduce the intracranial hypertension caused by the injury. The restoration of CBF and ICP was of short duration and was followed by a reduction of CBF and elevation of ICP. A repeated restoration of CBF by a second dose of mannitol was followed by a more severe impairment of CBF. The prolonged beneficial effect of mannitol on CBF after brain injury has to be reassessed.

Introduction

Severe head injury is usually associated with brain oedema. This phenomenon is a major clinical problem, since it causes increased intracranial pressure (ICP) and decreased cerebral perfusion pressure (CPP), which result in oedema and ischaemic brain damage[1, 4, 5].

Mannitol is an osmotic agent widely used to reduce intracranial hypertension and resolve brain oedema. It has been found to affect CBF, CPP, and CBV in normal and in injured brains of humans and experimental animals. However, different patterns of response were observed in various experimental conditions[2, 3, 6].

The present study was undertaken to evaluate the effect of mannitol on CBF and ICP following a cerebral cold injury in the cat.

Material and Methods

Six cats weighing 3.0–3.5 kg were anaesthetized by an intraperitoneal injection of sodium pentobarbital (30 mg/kg), tracheotomized and mechanically ventilated with room air. The femoral vein and artery were cannulated for fluid and drug administration and for continuous monitoring of arterial blood pressure (BP), respectively.

ICP was continuously monitored through a cannula from the cisterna magna. The left ectosylvian gyrus was exposed by frontoparietal craniectomy and four platinum electrodes were inserted 2.5 mm deep into the cortex.

Local CBF was measured by the hydrogen clearance technique, utilizing a computer-combined system for on-line calculations of CBF. After several control CBF determinations, the electrodes were removed and a cortical cold injury was induced by dry ice. The electrodes were then reinserted into their original locations, and the bone flap was replaced and cemented to the skull.

Measurements of CBF were performed during 2 hours prior to, and during 3 hours after the cold injury. Mannitol 20% (1 g/kg) was infused over 10 minutes, and CBF was measured 30, 60, and 120 minutes after the infusion was completed. A second dose of mannitol was administered and was followed by an additional series of CBF measurements.

Results

Mean arterial blood pressure was 110 ± 8.3 mmHg. Changes in mean ICP are shown in the top of Fig. 1. Mean pre-injury ICP was 5 mmHg. After injury there was a gradual increase in ICP, up to 3-fold. Shortly after the beginning of mannitol infusion, ICP decreased

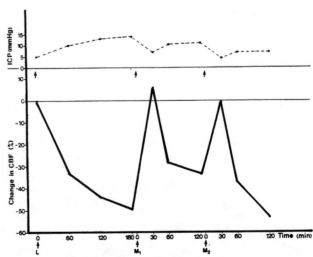

Fig. 1. Mean changes of cerebral blood flow (percentages of pre-injury values) and intracranial pressure following brain cold injury in the cat. — Ipsilateral hemisphere, ----- intracranial pressure, *L* lesion, M_1, M_2 Mannitol

and returned to baseline levels within 30 minutes, but later increased again up to twice the pre-injury value. Following the second mannitol dose ICP remained near normal levels.

Mean baseline (pre-injury) CBF was 33.8 ± 9.3 ml/min/100 g tissue. Mean changes in CBF (percentage of baseline values) are presented in Fig. 1. Following cold injury, CBF decreased gradually by 50%, 3 hours after injury.

Each of the two doses of mannitol administered brought about a transient restoration of CBF, which was followed by a repeated decrease of CBF 60 minutes after infusion. Following the first restoration, CBF decreased by 35%, while after the second time the restored CBF dropped by 55% of pre-injury level.

Discussion

Our results show that cerebral cold injury in cats causes an increase in ICP and a reduction in CBF in the lesioned cortex. This is in agreement with several experimental and clinical studies. DeWitt *et al.*[1] showed a short-lasting elevation of CBF one minute after a fluid percussion brain injury in cats; thirty minutes later baseline CBF values were observed. Pfenninger *et al.*[5] studied the changes in ICP, CPP, and CBF in piglets following fluid percussion injury. They found that ICP tripled after the trauma, while CBF and CPP decreased by 50%. In 55% of severe comatose closed-head injured patients, Obrist *et al.*[4] found an elevated CBF, which was usually associated with increased ICP; the remaining patients demonstrated reduced CBF, usually associated with low ICP.

In our study, restoration of the impaired CBF, 30 minutes after administration of mannitol, was demonstrated. Thirty minutes later CBF dropped again, although to a lesser extent. This is in agreement with the findings of Kassel[2] in normal dogs, where a short-lasting increase in CBF occurred following mannitol administration. The second dose of mannitol ameliorated CBF again for a short period, but the following reduction of CBF was more severe. An aggravation in brain oedema following mannitol administration 3 hours after brain injury was also shown by Reichenthal *et al.*[7], in tissue density measurements by computerized tomography.

Muizelaar and co-workers[3] described the changes in ICP and CBF after mannitol administration in severely head-injured patients. In patients with intact pressure auto-regulation mannitol infusion reduced ICP but did not change CBF, while in patients with defective autoregulation CBF increased but ICP remained normal. They postulated that the increased CBF was due to the lack of autoregulatory vasoconstriction in the low-flow areas. Ravussin[6] found that administration of mannitol to normal dogs and healthy human subjects caused an immediate significant increase in ICP and in CBV for 5 minutes, followed by a slight decrease of both parameters to below baseline values.

In view of the diverse effects of mannitol on CBF and of brain oedema demonstrated in various studies, it can be postulated that several mechanisms participate in determining the specific effect of the drug in different pathophysiological conditions.

References

1. DeWitt DS, Jenkins LW, Wei EP *et al* (1986) Effects of fluid-percussion brain injury on regional cerebral blood flow and pial arteriolar diameter. J Neurosurg 64: 787–794
2. Kassell NF, Baumann KW, Kitchon PW *et al* (1982) The effects of high dose mannitol on CBF in dogs with normal ICP. Stroke 13: 59–61
3. Muizelaar JP, Lutz HA, Becker DP (1984) Effect of mannitol on ICP and CBF and correlation with pressure autoregulation in severely head-injured patients. J Neurosurg 61: 700–706
4. Obrist WD, Langfitt TW, Jaggi JL *et al* (1984) Cerebral blood flow and metabolism in comatose patients with acute head injury. J Neurosurg 61: 241–253
5. Pfenninger EG, Reith A, Breitig O, Grunert AD, Ahnefeld FW (1989) Early changes of intracranial pressure, perfusion pressure and blood flow after acute head injury. J Neurosurg 70: 774–779
6. Ravussin P, Archer DP, Tyler JL *et al* (1986) Effects of rapid mannitol infusion on cerebral blood volume. A positron emission tomographic study in dogs and man. J Neurosurg 64: 104–113
7. Reichenthal E, Cohen ML, Shevach I, Shalmon E, Bar-Ziv Y, Kaspi T (1990) The ambivalent effects of early and late administration of mannitol in cold induced brain oedema. Brain Oedema (in this volume)

Correspondence: Dr. T. Kaspi, Experimental Research Laboratory, Department of Neurosurgery, Beilinson Medical Center, Petah Tiqva, 49 100, Israel.

Acta Neurochirurgica, Suppl. 51, 118–121 (1990)

Effects of Atrial Natriuretic Peptide on Brain Oedema:
The Change of Water, Sodium, and Potassium Contents in the Brain

S. Naruse[1], R. Takei[1], Y. Horikawa[1], C. Tanaka[1], T. Higuchi[1], T. Ebisu[1], S. Ueda[1], S. Sugahara[2], S. Kondo[2], T. Kiyota[2], and H. Hayashi[2]

[1] Department of Neurosurgery, Kyoto Prefectural University of Medicine, Kyoto, Japan and [2] Pharmaceuticals Lab., Asahi Chemical Industry, Co., Ltd., Kyoto, Japan

Summary

We examined the effect of atrial natriuretic peptide (ANP) administration on cerebral oedema in rats. Intravenous ANP infusion with total dose of 120 µg/kg and 100 µg/kg suppressed the elevation of water and Na contents in left middle cerebral artery (MCA) occluded and cold injured brain tissue, indicating that ANP has a suppressive effect on cerebral oedema. Similar ANP infusion at a low dose of 1 µg/kg/h for 6 h also resulted in observation of the anti-oedematous effect in both models, with no observable occurrence of the known systemic effects of ANP on systolic blood pressure (SBP), heart rate (HR), hematocrit, or serum electrolyte ion (Na^+, K^+, Cl^-) concentrations. The results thus suggest that the anti-oedematous effect of ANP is attributable to water and Na content control by ANP specific to the damaged tissue, possibly through inhibition of sodium transport. Taken together with a recent study in which it was shown that ANP might inhibit sodium transport in cerebral microvessel, our results suggest that ANP suppresses the development of brain oedema by inhibiting sodium transport and the coupled water influx.

Introduction

Since its initial discovery as a substance with diuretic effect[1], ANP has been shown to be a unique hormone or neuropeptide which regulates blood pressure and fluid homeostasis[1,3]. Recent studies have disclosed that ANP has the effect of regulating sodium transport in vascular endothelial cells[7] and in other cells[6] through activation of guanylate cyclase, thus suggesting that ANP may regulate water content in tissues.

We inferred that ANP may have a pharmacological effect on brain oedema. The effects of ANP on ischaemic and vasogenic brain oedema induced in rats were examined, by monitoring the change in water, Na, and K content.

Materials and Methods

Anti-oedematous Effect

For the model of ischaemic brain oedema, Wistar rats (male) aged 8–11 weeks were used. After anaesthetization with sodium pentobarbital at 50 mg/kg, the main trunk of the MCA was occluded according to the method described by Tamura et al.[9]. Immediately after the operation, infusion of ANP (Rat-ANP; 28 amino acids) at a dose of 120 µg/kg or of saline was begun and then continued for 24 h.

For the model of vasogenic brain oedema, Sprague-Dawley (SD) rats (male) aged 7–9 weeks were used. After anaesthetization with ketamine at 25 mg/kg, a burr-hole (4 mm in diameter) was made at the left parietal lobe. A lesion was made by applying a copper column (3 mm in diameter) cooled with liquid nitrogen through the burr-hole for 40 sec. ANP at a dose of 100 µg/kg was administered by intravenous infusion starting at 1 h before and continuing to 30 min after lesion production.

With both models, the rats were sacrificed at 24 h after the operation. The brain was immediately excised and divided into two hemispheres. Determination of the water content was performed by the dry-wet method. Na and K were extracted from the dried hemisphere by soaking in 60% nitric acid solution, and their concentrations were determined by absorption spectrophotometer and flame photometer, respectively.

Relation of Systemic Effects

SD rats (male) aged 8–10 weeks were used. Two of three groups were submitted to investigation for ANP anti-oedematous effect, one as an ischaemic and the other as a vasogenic brain oedema model, in the manner described above but at an ANP dosage of 6 µg/kg. In these two groups, either ANP solution or saline was intravenously infused from 2 h before until to 4 h after the operation. The third group underwent the same regime of ANP administration in the absence of the operation and was examined for the systemic effects of ANP. In this group arterial systolic blood pressure (SBP), heart rate (HR), electrolyte (Na^+, K^+, Cl^-) concentrations, and hemat-

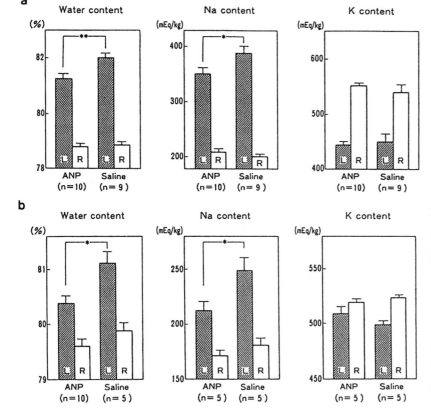

Fig. 1. Effects of intravenous infusion of ANP on brain water, Na, and K content in rats 24 hrs after MCA occlusion or cold injury. (a): ANP (120 µg/kg) or saline only was infused during 24 hrs into rats after left MCA occlusion. (b): ANP (100 µg/kg) or saline only was infused during 1.5 h into rats subjected to cold injury on left side of brain. Left, middle, and right panels represent brain water, Na, and K contents, respectively. Shaded column, left hemisphere (*L*); open column, right hemisphere (*R*). Values are represented as mean ± standard error. Number of rats in each group shown under column. * p < 0.05, ** p < 0.01

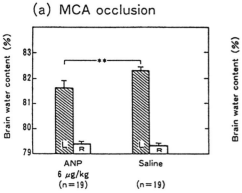
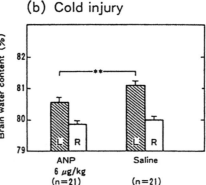

Fig. 2. Effects of intravenous infusion of 6 µg/kg (1 µg/kg/h) of ANP on brain water content in rats 24 hrs after MCA occlusion or cold injury. ANP or saline only was infused during 6 hrs into rats subjected to left MCA occlusion (a) or cold injury on left side of brain (b). Shaded column, left hemisphere (*L*); open column, right hemisphere (*R*). Values are represented as mean ± standard error. Number of rats in each group shown under column. ** p < 0.01

ocrit were measured before injection and at 2, 4, and 6 hrs after infusion. Statistical comparisons were performed by Student's t-test.

Results

Anti-oedematous Effects of ANP

The water, Na, and K contents of the hemispheres 24 hrs after left MCA occlusion in the ANP (120 µg/kg) and saline (control) group are shown in Fig. 1 a. In the left hemisphere (the occluded side), the water and Na contents were lower in the group receiving ANP than in the control group, while the K content was not significantly different. In the right hemisphere (the non-occluded side), no significant difference was observed between the two groups in water, Na, and K content.

The same results were observed for water, Na, and K contents of the hemispheres 24 hrs after cold lesioning of the left parietal lobe in the ANP (100 µg/kg) and saline (control groups) (Fig. 1 b). In the left hemisphere (the injured side), the water and Na contents were lower in the group receiving ANP than in the control group, whereas the K content was not significantly different. In the right hemisphere (the non-injured side), no significant difference was observed between the two groups in water, Na, and K content.

Relation of Systemic Effects

To investigate whether the observed anti-oedematous effect of ANP might be a result of the known systemic

		SBP[a] (mmHg)	HR[a] (beats/min)	Hct. (%)	Concentration of serum electrolyte ion[d]		
					Na$^+$ (mEq/l)	K$^+$ (mEq/l)	Cl$^-$ (mEq/l)
0hr	ANP	133±3	411±10	46.3±0.5[b]	140.1±0.6	5.6±0.3	100.9±1.1
	Saline	136±3	404±17	46.9±0.3[c]	140.3±0.7	6.0±0.5	100.5±0.7
1hr	ANP	132±2	391±7	47.1±0.5[b]	139.3±0.4	5.5±0.5	101.5±1.1
	Saline	126±3	367±15	47.6±0.4[c]	138.2±1.0	6.2±0.3	98.4±1.0
2hr	ANP	123±3	394±18	47.5±0.6[b]	138.2±1.2	5.8±0.3	102.3±1.8
	Saline	124±4	375±18	48.2±0.3[c]	139.7±1.6	6.4±0.2	100.5±0.5
6hr	ANP	112±3	390±37	45.8±0.7[b]	140.2±1.4	5.4±0.5	105.3±0.5
	Saline	119±2	374±26	46.6±0.5[c]	139.7±0.3	5.3±0.1	104.1±0.6

(a):n=4   (b):n=9   (c):n=8   (d):n=5   Mean±SE

effects of ANP such as decreased blood pressure and plasma volume we examined the effect of ANP on brain oedema at a dose presumably low enough to avoid the occurrence of ANP systemic effects.

The absence of any observable systemic effect by ANP at this dosage was confirmed (Table 1). No significant difference in systolic blood pressure, heart rate, hematocrit, or concentration of serum Na$^+$, K$^+$, and Cl$^-$ during the 6 hrs of infusion was observed between the group receiving ANP and that receiving saline.

At the same low dosage, on the other hand, the anti-oedematous effect of ANP was evident, as indicated by the water contents for the excised hemispheres 24 hrs after MCA occlusion or cold injury (Fig. 2). In both the MCA occlusion and the cold lesion groups, the water content in the left hemisphere was significantly lower with ANP than with saline, whereas that in the right hemisphere was not.

Discussion

We examined the effect of intravenous administration of ANP on ischaemic oedema with occluded left MCA and vasogenic oedema with left-hemisphere cold-injury in rats. In the occluded and lesioned hemisphere, suppression of the elevation of water and Na contents was observed in groups receiving ANP at a total dose of 120 µg/kg (5 µg/kg/h) and at 100 µg/kg (67 µg/kg/h), while no such effect was observed in the opposite hemisphere. These results suggest that ANP may have a pharmacological effect on ischaemic and vasogenic brain oedema by regulating the water and Na contents.

In basic and clinical studies, it has been reported that the characteristics of ANP administration include diuretic and hypotensive action[1, 10]. In the present study, we observed no apparent effect of ANP at the dose of 120 µg/kg on systemic blood pressure and serum

electrolyte level, except for a moderate transient drop in blood pressure at 1 h of ANP infusion. To investigate further the systemic contribution to its anti-oedematous effect, ANP at the dose of 6 µg/kg (1 µg/kg/h, 6 hrs) was studied. As shown in Table 1 and Fig. 2, ANP at this dose had no apparent systemic effect, but nevertheless significantly suppressed the elevation of water content in the oedematous tissue. These results suggest that the anti-oedematous effect of ANP results from its regulatory effect on the water and Na contents of oedematous region and not from its systemic effects.

The present results also showed that ANP has a suppressive effect on elevation of Na content, which accompanies progressive brain oedema. Like various other ANP target cells, cells in brain such as cerebral microvessel cells[8] or astrocytes[4] reportedly also contain ANP-specific receptors[2] that are involved in the regulation of guanylate cyclase, which increases the intracellular level of cGMP. Further, it has been reported that ANP has a specific inhibitory effect on amiloride-sensitive Na uptake into isolated cerebral capillaries[5]. It therefore appears that ANP may interrupt Na influx not only from blood to brain tissue across the blood-brain-barrier but also from extracellular fluid into astrocytes. This inhibiting effect of ANP on Na influx may in turn suppress water retention in brain tissue and consequently reduce the growth of brain oedema.

References

1. de Bold AJ (1985) Atrial natriuretic factor: a hormone produced by the heart. Science 230: 767–770
2. Chinkers M, Garbers DL, Chang MS, Lowe DG, Chin H, Goeddel DV, Schulz S (1989) A membrane form of guanylate cyclase is an atrial natriuretic peptide receptor. Nature 338: 78–83
3. Dóczi T, Joó F, Szerdahelyi P, Bodosi M (1987) Regulation of brain water and electrolyte contents: The possible involvement of central atrial natriuretic factor. Neurosurgery 21: 454–458
4. Friedl A, Harmening C, Schmalz F, Schuricht B, Schiller M, Hamprecht B (1989) Elevation by atrial natriuretic factors of cyclic GMP levels in astroglia-rich cultures from murine brain. J Neurochem 52: 589–597
5. Ibaragi M, Niwa M, Ozaki M (1989) Atrial natriuretic peptide modulates amiloride-sensitive Na$^+$ transport across the blood-brain barrier. J Neurochem 53: 1802–1806
6. Light DB, Schwiebert EM, Karlson KH, Stanton BA (1989) Atrial natriuretic peptide inhibits a cation channel in renal inner medullary collecting duct cells. Science 243: 383–385
7. O'Donnell ME (1989) Regulation of Na-K-Cl cotransport in endothelial cells by atrial natriuretic factor. Am J Physiol 257: C36–C44
8. Steardo L, Nathanson JA (1987) Brain barrier tissues: End organs for atriopeptins. Science 235: 470–473

9. Tamura A, Graham DI, McCulloch J, Teasdale GM (1981) Focal cerebral ischaemia in the rat: 1. Description of technique and early neuropathological consequences following middle cerebral artery occlusion. J Cereb Blood Flow Metab 1: 53–60

10. Weidmann P, Hasler L, Gnadinger MP, Lang RE, Uehlinger DE (1986) Blood levels and renal effects of atrial natriuretic peptide in normal man. J Clin Invest 77: 734–742

Correspondence: S. Naruse, M.D., Department of Neurosurgery, Kyoto Prefectural University of Medicine, Kawaramachi-Hirokoji, Kamigyo-Ku, Kyoto 602, Japan.

Acta Neurochirurgica, Suppl. 51, 123–124 (1990)

Tumour and Brain Oedema (I)

In vivo Measurement of Brain Water by MRI

A. Marmarou, P. Fatouros[1], J. Ward, A. Appley, and **H. Young**

Medical College of Virginia, Division of Neurosurgery and [1] Radiology,
Richmond, Virginia, U.S.A.

Summary

A new method for determining brain tissue water using MRI, developed in the laboratory, has been tested and applied to the clinical setting. To evaluate the accuracy of the technique, samples of human brain tissue were harvested from patients scheduled to undergo surgical removal of tumour and in whom biopsies were required for clinical management. MR determined values from imaging water maps compared favourably with gravimetric measures of samples. From these data, we conclude that accurate non-invasive measures of brain oedema in man are now possible.

Introduction

A measure of brain oedema is of critical importance in understanding the underlying mechanisms responsible for oedema formation and resolution in conditions where brain swelling occurs. Several methods are available in the experimental laboratory however there currently is no reliable method available for measuring oedema in patients. Previous work by Penn[1] utilized computerized tomography (CT) data to estimate brain water content. The difficulty in calibration among CT devices detracted from this technique although estimates of water content from Hounsfield change were reported.

The introduction of magnetic resonance imaging (MRI) and the dramatic sensitivity of MR to water change has generated new interest among researchers for developing a non-invasive means of oedema measurement[2,3]. In this report, we document our efforts in developing a technique for determining the spatial distribution of brain water non-invasively utilizing MRI. More specifically, we applied and tested a method which has been validated in our laboratories with phantom and animal studies. Our goal was to compare samples of brain tissue harvested from surgically treated tumour patients with standard laboratory measures of brain water.

Methods

Eight patients admitted to the Medical College of Virginia for surgical removal of tumour were entered into the study. The T1 measures were made in a standard commercial imaging unit (Siemens-Magnetom) employing inversion recovery imaging sequences with several inversion time intervals T1. The validity of the imaging methodology was verified with simultaneous measurements of calibration standards. From the sequential images, a neurosurgeon selected small regions of tissue for biopsy. The coordinates of the region were identified and inserted into a computer for computation of stereotaxic coordinates. At surgery, the biopsy sample was divided into two segments, one for pathologic analysis and the second for gravimetric determination of brain water. A gravimetric column was placed in a small room adjacent to the OR for rapid assessment of tissue water. The MRI water content of the tissue located sterotaxically was determined from the equation $[1/W_M = 0.949 + 0.274 (1/T1)]$ where W equals the water content of the tissue expressed in % gm H_2O/100 g t and T1 equals the relaxation time. The development of this equation is covered in greater detail in a companion article. (See Fatouros *et al.* this volume.) In addition to the patients cohort, normal volunteers were selected for MR scanning to obtain normal values.

Results

There was good agreement of brain tissue water measured from MR and gravimetric methods. The equation relating MRI determined water and gravimetry deter-

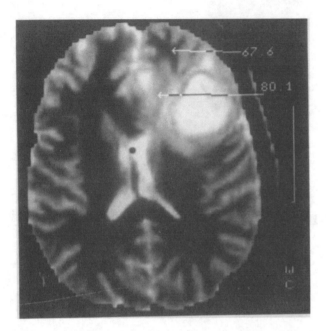

Fig. 1. Brain water map of patient with glioma. Intensity is directly proportional to water content. Enhanced area indicates increased water. Individual region can be selected by computer and brain water content determined automatically

Table 1

Region	White matter	Thalamus	Caudate nucleus
gm H$_2$O	69.0	75.5	80.5
S.D.	0.6	7.06	1.0

mined water equaled (MRI = 1.71 + 0.98 Grav) r = 0.98. The actual water maps were computed from programs converting T1 MR images to images in which the intensity was proportional to water content (Fig. 1). The example of Figure 1 shows the comparative measure of the biopsy sample and the MR determined value. A sample obtained distant from the tumour registered normal water content (67.6%) while a sample in the peritumoural area 80.1% water. These values compared to gravimetric levels of 67.3% and 80.5% respectively.

Several computerized approaches were developed for depicting the water content from the MR image. Individual areas can be selected as shown in Figure 1 or regions can be identified and the average water content and standard deviation appear directly on the screen. Values of water content determined by this method in normal subjects appear in Table 1.

An alternate method, and one which we favoured was the expression of water distribution using a histogram. In this technique, the number of pixels at a given water level were displayed. This provided a continuous profile of the water distribution which was characterized by two distinct peaks, one for white matter and one for cortex. The location of peaks corresponded to the average water content of white matter and cortex respectively. The development of oedema is characterized by more pixels at higher water content which is equivalent to a shift of the profile to the right.

Discussion

This study demonstrates that measures of water content in human brain samples using MR technique are in good agreement with values of brain tissue determined gravimetrically. The newly developed MR method has been tested and applied in the clinical setting. Our preliminary group data indicated that brain water varies 1% in white matter of normal subjects. This infers that although absolute measures are now possible by this technique, serial measures may be more meaningful for a more quantitative estimate of oedema formation over time.

Acknowledgement

This work was supported in part by grants NS 19235 and NS 12587 from the National Institutes of Health. Additional facilities and support were provided by the Richard Reynolds Neurosurgical Research Laboratories.

References

1. Penn RD, Bacus JW (1984) The brain as a sponge: A computed tomographic look at Hakim's hypothesis. Neurosurgery 14: 670–675
2. MacDonald HL, Bell BA, Smith MA, Kean DM, Tocher JL, Douglas RHB, Miller JD, Best JJK (1986) Correlation of human NMR P1 values measured *in vivo* and brain water content. Br J Radiol 59: 355
3. Bell BA, Kean DM, MacDonald HL, Barnett GH, Douglas RHB, Smith MA, McGhee CNJ, Miller JD, Tocher JL, Best JJK (1987) Brain water measured by magnetic resonance imaging. Lancet i: 66

Correspondence: A. Marmarou, Ph.D., Medical College of Virginia, Division of Neurosurgery, MCV Station, Box 508, Richmond, VA 23298, U.S.A.

Acta Neurochirurgica, Suppl. 51, 125–127 (1990)

¹H-NMR Studies on Water Structures in the Rat Brain Tissues with Cerebral Tumour or Ischaemia

T. Iwama[1], T. Andoh, N. Sakai, H, Yamada, S. Era, K. Kuwata, M. Sogami, and **H. Watari**

[1] Department of Neurosurgery, Gifu University School of Medicine, Gifu, Japan

Summary

Spin-lattice relaxation times (T_1) of water protons and cross-relaxation times (T_{IS}) between irradiated protein protons and observed water protons were measured to study water structure in various rat brain tissue. A tumour-bearing rat model and an incomplete forebrain ischaemia model of stroke prone spontaneously hypertensive rats (SHRSP) were used for the experiments. T_{IS} and T_1 were measured by the inversion recovery FT method with and without f_2-irradiation at 7.13 p.p.m. and $\gamma H_2/2\pi \sim 69$ Hz, respectively, using a 360 MHz FT-NMR spectrometer. All of experiments on T_1 and T_{IS} yielded a single kind of component. In spite of small changes in T_1 and water content, T_{IS} values in peritumoural oedematous and ischaemic brain tissue were significantly shorter than those of normal brain and far shorter than those of brain tumour, indicating a large amount of hydration of marco-molecules in the oedematous and/or ischaemic rat brain tissue. T_{IS} might be sensitive parameter for studying the water structure in various brain tissue.

Introduction

The authors reported that the change of T_{IS} between irradiated protein protons and observed water protons indicated the difference in water structure of bovine plasma albumin between in the solution state and in the gel state more sensitively than those of T_1 and T_2[6, 7]. The purpose of this study is to obtain information regarding water structures in peritumoural oedematous and ischaemic rat brain tissue by measuring T_1 and T_{IS}.

Materials and Methods

Tumour-bearing rat model was produced by semistereotaxic intracerebral injection of 1×10^5 cultured tumour cells (T9-glioma) to male F344 rats aged 6 weeks. NMR studies were conducted on the 10–12th day after the injections. Male F344 rats, aged 8 weeks, were used as controls.

Incomplete forebrain ischaemia model was made by bilateral common carotid artery occlusions (BCAO) of SHRSPs aged 10–22 weeks. Age-matched sham-operated SHRSPs were used as controls.

For ¹H-NMR experiments, each experimental rat was decapitated while awake, and the brain was removed quickly. A small volume of peritumoural oedematous brain, tumour or ischaemic brain tissue was packed into glass capillaries. ¹H-NMR experiments were conducted at 25 °C using a Bruker WM 360 wb NMR spectrometer operating at 360 MHz. T_1 was measured by the inversion recovery FT method with a pulse sequence of $(180°\text{-}\tau - 90°\text{C-acquisition-}T)_n$, using an acquisition time of 1.024 sec and T more than 5 times T_1. For T_{IS} measurements by the inversion recovery method, f_2-irradiation was applied at 7.13 p.p.m. of spectra at a power of 69 Hz in units of $\gamma H_2/2\pi$ during τ and T in the pulse sequence of T_1 measurement. After long time f_2-irradiation on the S spin system, the proton magnetization of the observed spin system (I) will reach an equilibrium value (Ioo). Longitudinal relaxation in the presence of f_2-irradiation is given by Akasaka[1, 2]

$$I = \text{Ioo} \left[1 - 2\exp\left(-\tau/T_1^*\right) \right]$$

where $1/T_1^* = 1/T_1 + 1/T_{IS}$.

After ¹H-NMR measurements, each specimen was immediately weighed and then placed in an electric oven at 60 °C for 12 hours; the dry weight was measured after further drying at 110 °C for 24 hours.

Results

Mean values of T_1 and water content for the peritumoural oedematous brain tissue in the present tumour-bearing rat model showed slight but not significant increases compared to those of the control. Nevertheless, the mean T_{IS} value of the oedematous brain tissue was significantly shorter than that of the control. On the other hand, the mean T_1 value and water content of T9-glioma were far longer and far higher than those of the control, respectively. The mean T_{IS} value of the tumour showed a slight but not significant increase (Fig. 1).

The T_1, T_{IS} values and water content of the sham-operated SHRSP brain tissue were found to be changed with age. Therefore, to evaluate the changes of those values after BCAO, the absolute values of T_1, T_{IS} and

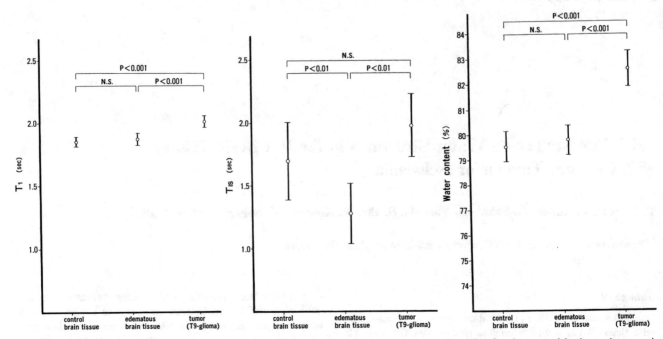

Fig. 1. Mean values and standard deviations (bars) of T_1 (left), T_{IS} (middle) and water content (right) for the control brain, peritumoural oedematous brain and T9-glioma tissues. N.S., not significant

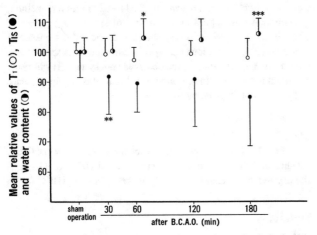

Fig. 2. Changes in mean relative values and standard deviations (bars) of T_1 (○), T_{IS} (●) and water content (◑) for the ischaemic SHRSP brain tissue after B.C.A.O. * $P < 0.05$; ** $P < 0.01$; *** $P < 0.001$

Discussion

[1] H-NMR techniques have been used to study water structures in brain oedema[3, 5, 8, 9, 11]. T_1 and T_2 values are prolonged in various kinds of brain oedema[3, 5, 8, 9, 11]; moreover, T_2 separates into two components during a course of oedema formation[3, 8, 9, 11]. Water content is elevated in oedematous brain tissues; prolongations of T_1 and T_2 are associated with increases in water content in a variety of brain tissues[3, 5, 8, 9, 11]. However, [1] H-NMR relaxation times of water proton were reported to be affected by factors other than water content, such as water-membrane and/or water-protein interaction[4, 5] and by interactions with salt ions and paramagnetic compounds[3, 10]. On the other hand, we reported that T_1, T_2 and T_{IS} from protein protons to water protons are reduced, with large effect of T_{IS} values, on the solution-gel transition of plasma albumin[6, 7]. Moreover, we found that T_{IS} measurements were useful in studying water structure in rat liver tissues[6]. The above reports might indicate that T_{IS} values are sensitive to the structural condition of water[6, 7]. Thus, we carried out T_{IS} measurements on the oedematous brain and normal brain tissue, using the inversion recovery FT method with f_2-irradiation at 7.13 p.p.m. (aromatic protons) and $\gamma H_2/2\pi$ of 69 Hz.

In the present studies, the T_1 value and water content in the peritumoural oedematous and ischaemic brain tissue showed slight but not significant increases, com-

water content in the ischaemic brain tissue were converted into relative values, as the mean values of T_1, T_{IS} and water content for the age-matched sham-operated SHRSP brains were 100. The T_1 value of the SHRSP brain tissue did not change by BCAO for 180 min, and the water content was gradually elevated after BCAO. However, the T_{IS} value was found to shorten from 30 min after BCAO (Fig. 2).

pared to those of the controls. Nevertheless, significant reduction of T_{IS} values was observed in the peritumoural oedematous and ischaemic brain tissue, indicating a large amount of hydration of macro-molecules in the peritumoural oedematous and/or ischaemic rat brain tissue. We suppose that T_{IS} measurements might be useful for water structure studying in various cerebral disorders. And it might be worth to note that T_{IS} could be applied to MRI.

References

1. Akasaka K (1981) Longitudinal relaxation of protons under cross saturation and spin diffusion. J Magn Reson 45: 337–343
2. Akasaka K (1983) Spin diffusion and the dynamic structure of a protein. Streptomyces subtilisin inhibitor. J Magn Reson 51: 14–25
3. Bakay L, Kurland RJ, Parrish RG, Lee JC, Peng RJ, Bartkowski HM (1975) Nuclear magnetic resonance studies in normal and oedematous brain tissue. Exp Brain Res 23: 241–248
4. Beall PT, Hazlewood CF, Rao PN (1976) Nuclear magnetic resonance patterns of intracellular water as a function of HeLa cell cycle. Science 192: 904–907
5. Bederson JB, Bartkowski HM, Moon K, Halks-Miller M, Nishimura MC, Brant-Zawadski M, Pitts LH (1986) Nuclear magnetic resonance imaging and spectroscopy in experimental brain oedema in a rat model. J Neurosurg 64: 795–802
6. Era S, Kato K, Sogami M, Nagaoka S, Kuwata K, Takahashi M, Suzuki E, Miura K, Watari H, Akasaka K (1986) Comparative ^{1}H NMR studies on water structure in hepatic tissues and protein gel. Biomed Res 7 [Suppl 2]: 41–46
7. Era S, Sogami M, Kuwata K, Fujii H, Suzuki E, Miura K, Kato K, Watari H (1989) [1]H-n.m.r. studies on cross-relaxation phenomena in bovine mercaptalbumin (BMA) solution and partially hydrolyzed bovine plasma albumin (BPA*) gel. Int J Peptide Protein Res 33: 214–222
8. Go KG, Edzes HT (1975) Water in brain oedema. Observations by the pulsed nuclear magnetic resonance technique. Arch Neurol 32: 462–465
9. Horikawa Y, Naruse S, Tanaka C, Hirakawa K, Nishikawa H (1986) Proton NMR relaxation times in ischaemic brain oedema. Stroke 17: 1149–1152
10. Kuntz ID, Zipp A (1977) Water in biological systems. N Eng J Med 297: 262–266
11. Naruse S, Horikawa Y, Tanaka C, Hirakawa K, Nishikawa H, Yoshizaki K (1982) Proton nuclear magnetic resonance studies on brain oedema. J Neurosurg 56: 747–752

Correspondence: T. Iwama, M.D., Department of Neurosurgery, Gifu University School of Medicine, 40 Tsukasamachi, Gifu 500, Japan.

Acta Neurochirurgica, Suppl. 51, 128–130 (1990)

Magnetic Resonance Imaging and Regional Biochemical Analysis of Experimental Brain Tumours in Cats

Y. Okada, M. Hoehn-Berlage, K. Bockhorst, T. Tolxdorff[1], and K.-A. Hossmann

Max-Planck-Institut für neurologische Forschung, Abteilung für experimentelle Neurologie, Cologne, Federal Republic of Germany, and
[1] Rheinisch-Westfälische Technische Hochschule, Medizinische Statistik und Dokumentation, Aachen, Federal Republik of Germany

Summary

Multiexponential evaluation of *in vivo* multi-echo T2 measurements at 4.7T in cat brain with implanted tumour led to monoexponential results in all tissues. T2 of normal brain tissue, tumour and oedema was 67 ms, 10–100 ms and 90–180 ms, respectively. In the ventricles biexponential solutions were observed with T2 values and relative contributions of these two components differing for ipsi- and contralateral side. Best discrimination between tumour and oedema was achieved by the magnetization value M (O), and between oedema and normal brain tissue by T2. T2 values were compared with *in vitro* biochemical analyses for ATP, lactate, glucose, tissue electrolytes and water content. NMR parameters provide reliable predictions about the water content and metabolic state of oedema, but not of the tumour.

Introduction

Conventional T1- or T2-weighted NMR images frequently fail to provide a reliable diagnosis of pathological processes in the brain. Based on results of *in vitro* studies, the application of NMR relaxation measurements for tissue characterization has, therefore, become of increasing interest[1,5]. In this study *in vivo* NMR T2 relaxation measurements were carried out at 4.7 T in order to explore the possibility of differentiating between brain tumours and peripheral brain oedema without the aid of contrast enhancing agents. For this purpose experimental brain tumours of cats were submitted to *in vivo* multicompartmental analysis of T2 relaxation times, followed by *in vitro* biochemical analysis.

Methods

Brain tumours were produced in cats by stereotactical xenotransplantation of a rat glioma clone (cell line F98)[4]. The animals were studied at various times (1–4 weeks) after the inoculation. T2 was determined with a multiecho CPMG sequence, resulting in 16 echo images of a transversal section of the brain with an interecho time TE = 28 ms. The series of 16 echo images was analyzed for multiexponential behaviour of the relaxation decay curves using the software system RAMSES[2]. This analysis provided the magnetization value M(O) at time t = O, the compartmental T2 values and their relative contributions alpha in the individual pixels in the image matrix. At the end of the NMR experiment brains were frozen in liquid nitrogen and the regional distribution of ATP, glucose, lactate and tissue pH were determined on cryostat sections (for references see[3]). The regional water and electrolyte contents were measured in tissue samples by drying and atomic absorption spectroscopy, respectively.

Results and Discussion

Figure 1 shows the initial six echos of a series of sixteen images used for T2 analysis by RAMSES software. The peritumourous oedematous tissue can be easily identified by its brightness in the later echo images but the tumour is difficult to demarcate. After processing of this stack of echo images the NMR parameters T2 and magnetization M(O) wre determined in the tumour, in the peritumourous white matter, and the non-oedematous white and grey matter of the ipsi- and contralateral hemisphere. Relaxation parameters in the ventricles were also analyzed.

At interecho time T = 28 ms, the T2 decay curves showed a monoexponential behavior for all tissue compartments. Only in the ventricular fluid were biexponential solutions found with a different distribution of T2 and alpha values on both sides. This was attributed to the outflow of oedematous fluid into the ipsilateral ventricle. Partial volume effects were unlikely because M(O) values were similar in the two ventricles.

The T2 values of grey and white matter were ap-

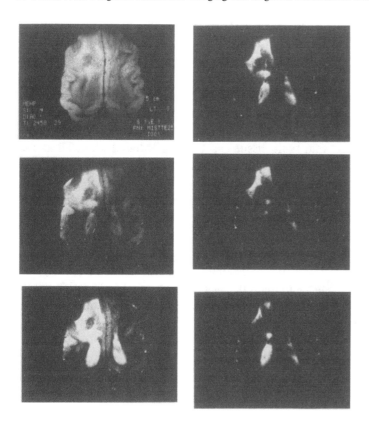

Fig. 1. Transversal nuclear magnetic resonance images of a cat brain with a three week old implantation tumour in the frontal part of the left hemisphere. Shown are the initial six echoes of a CPMG echo train of sixteen images with TE = 28 ms and TR = 2500 ms. In the later echoes peritumourous oedema and ventricles are clearly visible because of the longer T2 values

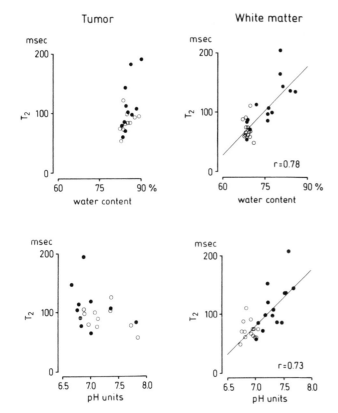

2. Correlation of T2 with water content (above) and pH (below) in brain tumour (left) and peritumorous white matter (right) of cat. Open circles indicate solid tumour regions and normal white matter, closed circles indicate necrotic tumour regions and oedematous white matter, respectively

proximately 67 ms. In the tumour T2 varied between 70 and 100 ms. A comparison with the regional distribution of ATP showed that T2 values above 100 ms consistently corresponded to necrotic tumour areas. T2 values below 100 ms were associated with either solid or necrotic tumour regions.

In the oedematous white matter the T2 values ranged between 90 and 180 ms. The magnetization (MO) was consistently higher in oedema than in tumour, irrespective of the water content. Thus, a differentiation between tumour and oedema is possible using M(O) as a discrimination variable. M(O) differed only slightly between oedema and non-oedematous normal tissue. Here T2 allowed a better differentiation (67 ms in normal tissue versus 90–180 ms in oedema).

Values of water and electrolyte content in non-oedematous grey and white matter were equivalent to those of previously reported control animals. In the tumour, water content amounted to 83% during the first week and increased to 87% during the fourth week. Electrolyte content of the tumour was similar to that of cortical tissue up to the third week. After four weeks a decrease of the intracellular electrolytes K and Mg and an increase of the extracellular electrolytes Ca and Na occurred, reflecting the onset of necrosis in the tumour tissue.

In the peritumourous oedema the water content increased from 69 to 83% during the first two weeks of tumour growth, followed by a slight regression. Water content correlated will with the increase of Na and Ca, while Mg and K showed no significant change.

The regional content of metabolites and tissue pH in tumour and peritumourous oedema correlated with water in a different way. In the tumour ATP and pH decreased with increasing water content, while glucose and lactate showed no correlation. In the peritumourous oedematous tissue a direct correlation of water content with glucose, lactate and pH, and an indirect correlation with ATP was observed. The main difference between tumour and oedema, in consequence, was pH which increased in oedema but decreased in the tumour with increasing water content.

This difference is explained by the different mechanisms of water accumulation. The tumour oedema is of the cytotoxic type, *i.e.* intracellular water content increases as a consequence of cellular injury. Since tumour necrosis is caused by inadequate blood flow, it is also associated with acidosis. In the peritumourous white matter, in contrast, oedema is of the vasogenic type, and, therefore, accumulates in the extracellular compartment. Since extracellular fluid is more alkaline than in the intracellular compartment, the alkalization of tissue with increasing water content reflects the relative increase of the extracellular volume. The normal content of intracellular electrolytes suggests that the intra-/extracellular ion homeostasis is not disturbed, and the high lactate is explained by an efflux of lactate salts from the tumour into the surrounding tissue and not as a consequence of a disturbed tissue metabolism.

Correlations between T2 and tissue constituents in peritumourous oedema revealed a linear relationship to water content, lactate, pH, Ca and Na, and an inverse relationship to ATP, K and Mg. Figure 2 (right side) shows the relationship between T2 and water content (above) and pH (below).

In the solid parts of the tumour T2 was independent of electrolyte and metabolite concentration. In the necrotic part, on the other hand, T2 correlated directly with Ca, and inversely with ATP, glucose, K and Mg. No correlation of T2 with water content and pH was found, irrespective of the viability of the tumour tissue (Fig. 2, left side).

In conclusion tissue characterization by NMR parameters allow differentiation between tumour, peritumoural oedema and normal brain tissue without the aid of contrast enhancement. Tumour and oedema can be demarcated using M(O), while oedematous and normal brain tissue can be differentiated using T2 as the discrimination variable. The comparison of NMR tissue parameters with biochemical changes demonstrates that reliable predictions about the water content and the metabolic state can be made in the peritumourous oedematous tissue but not in the tumour. Further progress, therefore, has to await the use of shorter echo times for the evaluation of fast T2 components, and measurement of self-diffusion and T1 relaxation times. Such an extension of tissue characterization is presently under investigation.

Acknowledgements

This investigation was supported by the Deutsche Forschungsgemeinschaft (SFB 200/D6).

References

1. Gersonde K, Felsberg L, Tolxdorff T, Ratzel D, Ströbel B (1984) Analysis of multiple T2 proton relaxation processes in human head and imaging on the basis of selective and assigned T2 values. Magn Reson Med 1: 463–477
2. Gersonde K, Tolxdorff T, Felsberg L (1985) Identification and characterization of tissues by T2-selective whole-body proton NMR imaging. Magn Reson Med 2: 390–401
3. Hoehn-Berlage M, Okada Y, Kloiber O, Hossmann K-A (1989) Imaging of brain tissue pH and metabolites. A new approach for the validation of volume-selective NMR spectroscopy. NMR in Biomed 2: 240–245
4. Hossmann K-A, Mies G, Paschen W, Szabo L, Dolan E, Wechsler W (1986) Regional metabolism of experimental brain tumours. Acta Neuropathol (Berl) 69: 139–147
5. Kovalikova K, Hoehn-Berlage M, Gersonde K, Porschen R, Mittermayer C, Franke R-P (1987) Age-dependent variation of T1 and T2 relaxation times of adenocarcinoma in mice. Radiology 164: 543–548

Correspondence: Dr. M. Hoehn-Berlage, Max-Planck-Institut für neurologische Forschung, Abteilung für experimentelle Neurologie, Gleueler Strasse 50, D-5000 Köln 41, Federal Republic of Germany.

Acta Neurochirurgica, Suppl. 51, 131–133 (1990)

Magnetic Resonance Studies in Human Brain Oedema Following Administration of Hyperosmotic Agents

Special References to Relaxation Times and Proton MRS

M. Niiro, T. Asakura, K. Yatsushiro, M. Sasahira[1], K. Terada[1], and **T. Fujimoto[2]**

Departments of Neurosurgery, Faculty of Medicine, Kagoshima University; [1] Neurosurgery and [2] Psychiatry, Fujimoto Hospital, Japan

Summary

Changes of proton relaxation times (T_1 and T_2) and proton Magnetic Resonance Spectroscopy (MRS) were studied in patients with brain oedema following administration of hyperosmotic agents. Relaxation times of oedema tended to decrease following infusion of hyperosmotic agents. In most patients examined, changes of relaxation times tended to achieve their lowest value at 30–60 minutes after infusion. However, the changes of relaxation times were not uniform. In some patients, relaxation times continued to decrease for more than 2 hours, while in other patients relaxation times which had earlier decreased subsequently had increased at 2 hours. The peak of water components, obtained by SIDAC (Spectroscopic Imaging by Dephasing Amplitude Changing) method was observed to change as did relaxation times. Changes of relaxation times and the peak of water component may vary depending upon factors including the kinds of lesions causing oedema, phase of oedema (acute or chronic), etc. Proton relaxation times and the peak of water component obtained by proton MRS were useful in evaluating the changes of oedematous area.

Introduction

The mechanism by which hyperosmotic agents such as Mannitol or Glycerol decrease intracranial pressure is still not clearly known[7, 10] but their dehydrating effect on the brain parenchyma is considered well-established[1, 3, 8, 11]. Proton relaxation times (T_1 and T_2) are thought to be closely related to tissue water content. We performed magnetic resonance studies in patients with brain oedema following administration of hyperosmotic agents and examined changes of relaxation times of oedematous brain and those of water components using proton magnetic resonance spectroscopy (MRS).

Materials and Methods

Fifteen patients shown in Table 1 were examined, [(6 gliomas, 3 metastatic brain tumours, 2 meningiomas, 2 hypertensive intracerebral haematomas (ICH), one malignant lymphoma, one cerebellopontine (C-P) angle tumour). Patients were selected, who showed a widely spread oedematous area and where the oedema could be relatively clearly distinguished from the tumour on the CT scan. 500 ml of 10% Glycerol or 20% Mannitol was rapidly infused intravenously. Scanning was done before, immediately after as well as 30, 60, and 120 min after the infusion. Most of the patients remained lying on the bed of the MR system motionless during the entire period of examination.

Relaxation times and the peak of water components of the oedematous area were obtained from the ROIs of the calculated images or water images corresponding to the peritumoural low density area on the CT scan. The ROIs were placed usually at the margin of the oedema with the side to the white matter. Several ROIs were examined in each patient and average values were calculated.

MR system used were a 0.1 T ASAHI Mark-J, 0.5 T YMS RESONA and 2.0 T ASAHI Super 200. Relaxation times were calculated by Inversion Recovery (IR) method for T_1 values and by Carr-Purcell method for T_2 values. For the purpose of examining the water content of oedematous area, we made use of the proton chemical shift imaging by SIDAC (Spectroscopic Imaging by Dephasing Amplitude Changing) method. The ratio of peak of oedema to the peak of the homologous contralateral white matter was used to express the changes of water components.

Results

Proton Relaxation times are known to vary according to the strength of the magnetic field of the MR system. Seven of the 15 patients were examined with the 0.1 T MR system and their relaxation times are shown in Fig. 1. In general, T_1 and T_2 relaxation times tended to decrease immediately after infusion, and most

Table 1. *Summary of the Cases*

Case	Age and sex	Diagnosis	Location	Parameter and tesla		Clinical feature
1	72 y F	glioblastoma	lt. parietal	T_1 and T_2	0.1	rapidly progressive
2	81 y M	astrocytoma (G-2)	bifrontal	T_1 and T_2	0.5	slowly progressive
3	22 y M	astrocytoma (G-2)	lt. T-P-O	T_1 and T_2	0.5	slowly progressive
4	41 y M	astrocytoma (G-2)	lt. F-T-P	SIDAC	2.0	slowly progressive
5	64 y M	glioblastoma	rt. parietal	SIDAC	2.0	intratumoural bleeding
6	64 y M	glioblastoma	rt. frontal	SIDAC	2.0	intratumoural bleeding
7	72 y F	metastatic BT	multiple	T_1 and T_2	0.1	rapidly progressive
8	51 y F	metastatic BT	multiple	T_1 and T_2	0.1	rapidly progressive
9	78 y M	metastatic BT	multiple	T_1 and T_2	0.5	slowly progressive
10	73 y F	meningioma	lt. sphenoidal	T_1 and T_2	0.1	slowly
11	34 y M	meningioma	rt. olfactory	image	2.0	slowly
12	59 y M	malignant lymphoma	rt. paraventicle	SIDAC	2.0	rapidly progressive
13	56 y F	C-P angle tumour	rt. C-P angle	T_1	0.1	slowly
14	60 y F	hypertensive ICH	lt. putaminal	T_1 and T_2	0.1	subacute stage
15	52 y F	hypertensive ICH	lt. putaminal	T_1 and T_2	0.1	acute stage

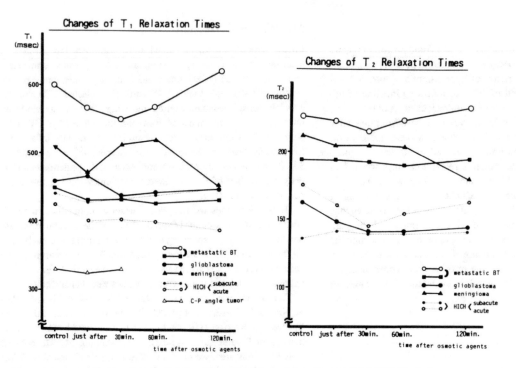

Fig. 1. Changes of relaxation times. The right is T_1 and the left is T_2. (0.1 T ASAHI Mark-J)

reached their lowest value at 30–60 minutes after infusion. However the changes of relaxation times were not uniform. In some cases, relaxation times remained reduced for more than 2 hours, while in other cases relaxation times returned toward pre-infusion values after 2 hours. In a patient with a meningioma, T_1 relaxation time initially decreased following infusion, but then increased at 30–60 minutes after infusion. In a patient with a glioblastoma multiforme and a patient with an acute ICH, the decrease in relaxation times was more obviously than in other cases. A few patients showed no changes in the relaxation times at all. The average of the maximal decreases in the relaxation times from control values was calculated and amounted to 4% for T_1 values and about 7% for T_2 values. Most of these changes were within the range of the corresponding standard deviations. In the case of the 0.5 T MR system, which was used in only 3 of the 15 patients, changes of relaxation times were not so obvious. T_1 relaxation times were slightly decreased at 60 minutes after infusion. T_2 relaxation times were also slightly decreased at 60–120 minutes after injection. Two patients with low grade astrocytoma and one with a slowly progressing metastatic brain tumour were studied with this MR-system.

Changes of water components were studied by proton chemical shift imaging obtained by SIDAC method. Subjects included in this group were two high grade and one low grade astrocytomas and one malignant lymphoma. The changes observed in these cases seemed to be similar in pattern. Water components were maximally decreased about 60 minutes after injection and tended to gradually increase about 2 hours after injection. Just after injection, water components were decreased in two patients but were nearly unchanged or increased in the others.

Discussion

Hyperosmotic agents are thought to have a dehydrating action on the brain[1, 3, 8, 9, 11]. In an experimental study, the water content (measured by wet-dry weight method) of a cold injured lesion was reduced by 2–3% in the mannitol-treated group as compared to the control group[1, 12]. After head injury in man, mannitol was reported to decrease brain water content by about 7% as measured by specific gravimetric technique[9]. Magnetic resonance study of brain oedema in brain tumours revealed that T_1 value was increased by 26% and T_2 value was increased by 78% while water content of oedema increased 10%. In the ischemic brain, T_1 value was increased by 7.1–9.2%, and T_2 value by 8.8–13.0% while water content was increased 2.4–3.4%[5]. It has been shown, therefore, that even a small change of water content is accompanied by an obvious change of relaxation times[4, 5]. In ischemic models of rats, Naruse[2, 8] reported that glycerol administration induced a significant decrease of T_2 relaxation times after 30 minutes. The ratio of decrease in T_2 value to control value was calculated to be about 7.5%. This is a similar value to our data obtained by the 0.1 T MR system. In our clinical study, changes of proton relaxation times and peak of water components were obvious in cases of highly malignant tumours such as high grade gliomas or metastatic brain tumours and an acute stage of intracerebral haematoma. They tended to decrease maximally about 30–60 minutes after injection of hyperosmotic agents. Changes of water components obtained by proton MRS were similar to those of relaxation times and the changes were relatively uniform.

When discussing the small changes of water content occurring after administration of hyperosmotic agents, a standard error in measuring relaxation times or the peak of water content should be taken into consideration. In the current MR system, because there are large standard deviations and inequalities in the magnetic field, the reasons why the changes of relaxation times are not uniform and the reasons why the time of maximum decrease is not constant, remain unclear. So, further investigations are required. By using MR system, however, it is possible to observe small changes occurring in the region of oedema following administration of hyperosmotic agents. As the MR systems are improved, more precise study of the water content of brain oedema will be possible.

References

1. Albright AL, Latchaw RE, Robinson AG (1984) Intracranial and systemic effects of osmotic and oncotic therapy in experimental cerebral oedema. J Neurosurg 60: 481–489
2. Horikawa Y, Naruse S, Tanaka C et al (1986) Proton NMR relaxation times in ischaemic brain oedema. Stroke 17: 1149–1152
3. Kamiura T (1987) Effects of mannitol on ischaemic brain oedema. Teikyo Med J 10: 11–18
4. Kamman RL, Go KG et al (1988) Nuclear magnetic resonance relaxation in experimental brain oedema. Magn Reson Med 6: 265–274
5. Kato H, Kogure H, Otomo H et al (1986) Magnetic resonance imaging of experimental cerebral ischaemia. Brain Nerve 38: 295–302
6. Miyazaki T, Yamamoto T, Iriguchi N (1989) Spectroscopic imaging by dephasing amplitude changing (SIDAC). Radiat Med 7: 1–15
7. Muizelaar JP, Wei EP, Kontos HA et al (1983) Mannitol causes compensatory cerebral vasoconstriction and vasodilation in response to blood viscosity changes. J Neurosurg 59: 822–828
8. Naruse S, Horikawa Y, Tanaka C et al (1982) Nuclear magnetic resonance studies of effects of glycerol on brain oedema. Brain Nerve 34: 805–809
9. Nath F, Galbraith S (1986) The effects of mannitol on cerebral white matter content. J Neurosurg 65: 42–43
10. Takagi H, Saito T, Kitahara T et al (1984) The mechanism of ICP reducing effect of mannitol. Brain Nerve 36: 1905–1102
11. Wilkinson HA, Rosenfeld S (1983) Furosemide and mannitol in the treatment of acute experimental intracranial hypertension. Neurosurgery 12: 405–410

Correspondence: M. Niiro, M.D., Department of Neurosurgery, Faculty of Medicine, University of Kagoshima, 1208-1, Usuki-cho, Kagoshima City, Japan, 890.

Acta Neurochirurgica, Suppl. 51, 134–136 (1990)
© by Springer-Verlag 1990

MRI Contrast Enhancement by MnTPPS of Experimental Brain Tumours in Rats

K. Bockhorst, M. Hoehn-Berlage, M. Kocher, and K.-A. Hossmann

Max-Planck-Institut für neurologische Forschung, Abteilung für experimentelle Neurologie, Cologne, Federal Republic of Germany

Summary

In the experimental F98 rat glioma model the nuclear magnetic resonance contrast agent MnTPPS was used to increase the contrast between tumour and peritumourous brain tissue. By evaluating pre- and postcontrast CPMG echo trains after different dose and circulation times of MnTPPS, the transverse relaxation time T2, the magnetization extrapolated to the time zero M (0), and the contrast ratio Rc of the magnetizations in neoplastic and normal tissues were determined. According to these data a maximum contrast is obtained a few hours after injection of 0.25 mmol MnTPPS/kg bw, using spin echo pulse sequences with short repetition times TR (1,100 ms), and short spin echo time TE (25 ms).

Introduction

Magnetic resonance images of brain tumours often fail to show a clear demarcation between neoplastic and normal brain tissue. The application of gadolinium DTPA (GD-DTPA) facilitates the detection of tumours but not the precise demarcation from the peritumourous oedema because the contrast agent does not selectively bind to neoplastic tissue. Therefore, after passing the blood-tumour barrier it spreads from the tumour into the peritumourous oedematous tissue, blurring the delineation of the tumour border.

The need for a tumour specific NMR contrast agent led to the development of manganese TPPS (MnTPPS), a synthetic metal porphyrin complex that binds to neoplastic cells and, because of its paramagnetic properties, causes a shortening of water relaxation times. Its efficiency has been documented in peripheral tumour models[5] but not yet in brain tumours. In the present communication we explore the effect of MnTPPS on contrast enhancement in experimental gliomas of rat brain.

Material and Methods

Manganese TPPS was synthesized by a condensation reaction between pyrrole and benzaldehyde (Rothemund reaction), sulfonation by sulfuric acid and metallization by manganese acetate. The product was characterized by its optical absorption spectra (450–700 nm).

The tumours were induced by stereotactical inoculation of cloned glioma cells (10,000 in 10 µl tissue culture, cell strain F98) into the right striate nucleus of Fischer rats. Two to three weeks after the inoculation the animals were placed into the NMR magnet (4.7 T Biospec System, Bruker, FRG) under halothane/N_2O/O_2 (0.4–1.0%/70%/rest) anaesthesia. Images in sagittal orientations were recorded by Gradient Echo Fast Imaging (GEFI) to define the position of the brain in the magnet. Multi Slice Multi Echo imaging was performed in coronal slices passing through the tumour center. Slice thickness was 1 mm. CPMG echo trains were evaluated with the RAMSES software package[2] that allows deconvolution of the transversal relaxation decay curve into different components and the calculation of their relative contribution alpha, the corresponding relaxation time T 2 and the magnetization M (0) extrapolated to time t = 0. The contrast enhancing effect of MnTPPS was quantified by calculating the ratio Rc of the magnetization values in neoplastic and contralateral normal tissue. Rc (0) represents the contrast ratio evaluated for the back-extrapolated time t = 0 and Rc (25) the contrast of the spin-echo image at TE = 25 ms.

After recording of precontrast images MnTPPS was injected intraperitoneally in doses varying from 0.06 to 0.38 mmol/kg bw (in 1 ml Ringers solution, pH 6–8). Postcontrast images were recorded directly and up to four days after the application. After the final imaging experiment the brains were removed and frozen in isopentane at −50 °C. They were cut in a cryostat to obtain histological sections (24 µ slices) which were then stained with cresyl violet for localization of neoplastic tissue. Further details are given elsewhere (1).

Results and Discussion

Inoculation of tumour cells into the right striate nucleus did not cause any immediate neurological deficits. However, during the third week after implantation the

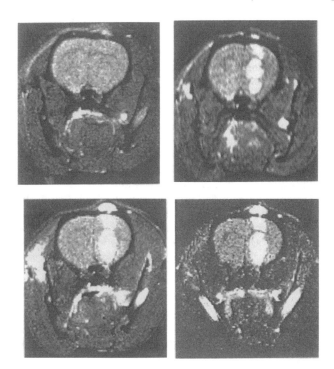

Fig. 1. Contrast enhancement of experimental glioma in rat brain with manganese TPPS (0.25 mmol/kg bw). Top left: precontrast image; top right: immediately after application; bottom left: after 1 day; bottom right: after 2 days. Note persistent contrast enhancement and absence of spread into peritumorous brain tissue (SE image, TR = 1,100 ms, TE = 25 ms)

Figure 1 represents MR images of a large tumour before and at various times after MnTPPS application. Without contrast enhancement the intracerebral mass is reflected by a shift of the midline and an asymmetrical configuration of the white matter but the tumour itself is not detectable. After MnTPPS injection the tumour becomes clearly visible and, during the four days observation time, increases in volume. At the end of the experiment the enhanced area conforms precisely to the histological localisation of tumour tissue.

Precontrast T2 was 100 ± 6 ms and Rc(0) was 0.92 ± 0.06 for the tumour. T2 of brain tissue contralateral to the tumour was 87 ± 5 ms. MnTPPS caused a dose-dependent selective decrease of T2 in tumour tissue which reached its minimum of 74 ± 8 ms at 0.25 mmol/kg body weight (Table 1). Higher doses did not cause a further shortening of T2. The greatest effect on contrast, however, was reached at a dose of 0.38 mmol/kg bw with Rc(0) = 1.36 ± 0.04. This suggests that T1, but not T2, continues to decrease at doses above 0.25 mmol/kg.

Rc(25) was dependent on the repitition time TR. At long TR (3,500 ms) MnTPPS induced only minor changes on Rc(25) (1.24 ± 0.09) but at TR = 1,100 ms Rc significantly increased (1.77 ± 0.27). Precontrast values did not depend on TR [Rc(25) = 0.96 ± 0.16 at 1,100 ms, 0.98 ± 0.08 at 3,500 ms]. The TR dependence of Rc indicates a shorter T1 in the tumour than in contralateral normal brain tissue after MnTPPS injection.

The dependence of contrast enhancement on the size of tumours was studied in two groups with different tumour age, 12–13 or 17–19 days, respectively. Although tumour size more than doubled, no difference of contrast ratio was observed, either pre- or postcontrast.

The investigation of the dependence of Rc(25) on circulation time of MnTPPS revealed an increase within two hours to a level which did not change significantly during the next four days. The persistence of the enhancement suggests tight binding of MnTPPS to the tumour. Recently, porphyrins have been described as

animals began to lose body weight, and their locomotor activity decreased. Exophthalamus was a common symptom, and some animals died at this time. The histological investigation after 2–3 weeks revealed clearly demarcated neoplasms with polymorphous tumour cells and occasional mitoses. The classification was that of a malignant polymorphous glioma[3].

Application of MnTPPS did not produce symptoms of general or central nervous toxicity. Within two hours after the application the eyes and the skin of the animals were stained green, and the urine was dark brown. The optical absorption spectrum of urine was identical with that of an aqueous solution of MnTPPS. Within two days after application the color of skin and urine returned to normal.

Table 1. *Dependence of T 1 and of Magnetization Ratio Rc(0) at TR = 1,100 ms on MnTPPS Dose (means ± SD)*

Dose (mmol MnTPPS/kg bw)	0	0.06	0.13	0.25	0.38
T 2 tumour (ms)	100 ± 6	95 ± 4	77 ± 3	74 ± 8	74 ± 6
T 2 brain (ms)	87 ± 5	90 ± 4	83 ± 3	82 ± 2	87 ± 5
Rc(0)	0.92 ± 0.06	1.11 ± 0.08	1.22 ± 0.02	1.28 ± 0.16	1.36 ± 0.04

ligands of peripheral benzodiazepine receptors at the mitochondrial membrane[7]. In the normal brain the density of these receptors is much lower than in tumours, as shown in both experimental and clinical studies by PET[4] and autoradiography[6]. We, therefore, suggest that the interaction between MnTPPS and peripheral benzodiazepine receptors is responsible for the NMR enhancement of tumours.

Acknowledgements

This investigation was supported by the Deutsche Forschungsgemeinschaft (SFB 200/D6).

References

1. Bockhorst K, Höhn-Berlage M, Kocher M, Hossmann KA (1990) Proton relaxation enhancement in experimental brain tumours. *In vivo* NMR study of manganese (III) TPPS in rat brain gliomas. Magn Reson Imaging (in press)
2. Gersonde K, Tolxdorff T, Felsberg L (1985) Identification and characterization of tissues by T2-selective whole body proton NMR imaging. Magn Reson Med 2: 390–401
3. Hossmann KA, Mies G, Paschen W, Szabo L, Dolan E, Wechsler W (1986) Regional metabolism of experimental brain tumours. Acta Neuropathol (Berlin) 69: 139–147
4. Junck L, Olson JMM, Ciliax BJ, Koeppe RA, Watkins GL, Jewett DM, McKeever PE, Wieland DM, Kilbourn MR, Starosta-Rubinstein S, Mancini WR, Kuhl DE, Greenberg HS, Young AB (1989) PET imaging of human gliomas with ligands for the peripheral benzodiazepine binding site. Ann Neurol 26: 752–758
5. Patronas NJ, Cohen JS, Knop RH, Dwyer AJ, Colcher D, Lundy J, Mornex F, Hambright P, Sohn M, Myers CE (1986) Metalloporphyrin contrast agents for magnetic resonance imaging of human tumours in mice. Cancer Treat Rev 70: 391–395
6. Starosta-Rubinstein S, Ciliax BJ, Penney JB, McKeever PE, Young AB (1987) Imaging of glioma using peripheral benzodiazepine receptor ligands. Proc Natl Acad Sci USA 84: 891–895
7. Verma A, Nye JS, Synder SH (1987) Porphyrins are endogenous ligands for the mitochondrial (peripheral-type) benzodiazepine receptor. Proc Natl Acad Sci USA 84: 2256–2260

Correspondence: Dr. M. Hoehn-Berlage, Max-Planck-Institut für neurologische Forschung, Abteilung für experimentelle Neurologie, Gleueler Strasse 50, D-5000 Köln 41, Federal Republic of Germany.

Acta Neurochirurgica, Suppl. 51, 137–139 (1990)
© by Springer-Verlag 1990

Human Malignant Gliomas Secrete a Factor that Increases Brain Capillary Permeability: Role in Peritumoural Brain Oedema

T. Ohnishi[1], **T. Hayakawa**[1], and **W. R. Shapiro**[2]

[1] Department of Neurosurgery, Osaka University Medical School, Osaka, Japan, [2] Department of Neurology, Memorial Sloan-Kettering Cancer Center, New York, U.S.A.

Summary

The effect of conditioned media obtained from two human malignant gliomas and normal human glia on rat brain capillary permeability was investigated by measuring the entry of ^{14}C-aminoisobutyric acid by a quantitative autoradiographic method. Conditioned media were concentrated 50-fold to create SUP-C. The SUP-C contained proteins with a molecular weight greater than 10 kD. The SUP-C from glioma cells markedly increased brain capillary permeability, whereas that from normal glial cells did not. The activity of capillary permeability factor in the SUP-C was significantly inhibited by pretreatment of animals with dexamethasone or BW755C (lipoxygenase inhibitor), but not with indomethacin. On the other hand, coincubation of glioma cells with dexamethasone produced SUP-C whose capillary permeability activity was about one and a half times greater than that without dexamethasone. These results indicate that human malignant glioma cells secrete a protein factor that increases brain capillary permeability. Glucocorticoids inhibit the effect of the factor by directly acting on capillary endothelial cells, possibly through the inhibition of phospholipase A_2 activity, resulting in a decrease of lipoxygenase rather than cyclo-oxygenase products.

Introduction

Peritumoural brain oedema is one of the main cause of neurological deficits and death of patients with malignant gliomas. Although peritumoural brain oedema is vasogenic, *i.e.*, characterized by increased permeability of brain capillary endothelial cells, the etiology and molecular mechanisms underlying oedema formation are poorly understood.

Our quantitative autoradiographic (QAR) studies in experimental brain tumours, demonstrating that capillary permeability in brain tissue adjacent to tumours and brain tissue surrounding tumours was increased[3], led us to a hypothesis that the tumour must produce a diffusible factor that alters capillary permeability of normal brain vessels. We have recently reported that rat C_6 glioma cells in culture secreted a protein factor that increases capillary permeability[7]. In the present report, cultured cells from human malignant gliomas are shown to secrete the same kind of factor as produced by C_6 glioma, whereas normal human glial cells do not. Also, a possible mechanism for the anti-oedema action of glucocorticoids is described.

Materials and Methods

Cells and Cell Culture

Two human malignant glioma cell lines (EI and HFA) and normal human glial cells were established and maintained in monolayer cell culture with Waymouth's MAB medium supplemented with 25% fetal bovine serum (FBS) at 37 °C in a humidified atmosphere with 5% CO_2. The feature of the normal human glial cells was previously described[1].

Preparation of Conditioned Medium

Cells in confluence were harvested by trypsinization, washed with phosphate-buffered saline (PBS), and resuspended in Eagle's minimum essential medium (MEM) without serum. Four or 20 hours after cells were incubated in serum-free MEM, the culture supernatants were collected, centrifuged, and passed through a 0.22 μm filter. All supernatants were concentrated 50-fold by a dialysis-concentrator, using a membrane with a 10,000 MW cut-off. The concentrated supernatants (SUP-C) were assayed for their activity to alter brain capillary permeability. MEM was concentrated in the same way and used as a control.

Measurement of Capillary Permeability

To determine the effect of the conditioned medium on capillary permeability, SUP-C solutions were infused into normal rat brains, and then rats were given ^{14}C-aminoisobutyric acid (AIB), which was used as a tracer for brain capillary permeability. Male Wistar rats were anesthetized with N_2O-O_2-ethrane and two burr holes were made symmetrically on the frontal bone. 30-gauge needles were inserted through these holes to a depth of 6 mm into the brain. SUP-C was infused through the right needle, and the control MEM solution through the left needle at a rate of 1 μl/min for 50 min with a Harvard pump. Six hours after the intracerebral infusion, 1 ml of ^{14}C-AIB (100 μCi/ml) was injected into the femoral vein. Timed blood samples were rapidly collected from the femoral artery. Fifteen minutes later, the rats were decapitated, and the brains were processed for QAR and histological examination.

Regional measurements of ^{14}C-AIB uptake were determined by QAR using an image digitizer and VAX 750 computer[8]. Capillary permeability was expressed as a unidirectional blood-brain transfer constant, K (μl/g/min), and was calculated from the experimental data.

Treatment Studies

To clarify the anti-oedema effect of glucocorticoids, the rats were treated with dexamethasone, indomethacin, or BW755C (lipoxygenase inhibitor) 1 hour before intracerebral infusion of SUP-C. In addition, SUP-C from EI cells was assayed for capillary permeability activity after EI cells were coincubated with dexamethasone for 4 hours.

Results

Autoradiograms from rat brains infused with SUP-C (right hemisphere) or MEM solution (left hemisphere), which were depicted as computer-generated images of transfer constant (K) for ^{14}C-AIB, are shown in Fig. 1. Supernatants from both glioma cells markedly increased capillary permeability around the inserted needles, whereas that from normal glia had little effect on

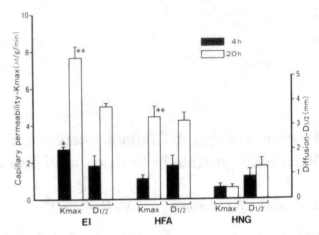

Fig. 2. Effects of capillary permeability activity in SUP-C from glioma cells (*EI, HFA*) and normal human glial cells (*HNG*). The SUP-C was obtained from two different incubation times (4 and 20 hours) in these cells. Capillary permeability is expressed as K_{max} and $D_{1/2}$. K_{max} is represented as the difference of these values between SUP-C and the corresponding control. Bars indicate the mean ± standard error of the mean for three experiments. Significantly different from the SUP-C of HNG at *p<0.01, **p<0.001

capillary permeability. Capillary permeability in the contralateral hemispheres, into which control solutions had been infused simultaneously, showed no remarkable change.

To quantify the effect of the infusion materials on capillary permeability, mean K values were determined in sequential "rings" drawn around the needle track every 0.2 mm and these K values were plotted as a function of distance from the needle track to obtain distribution curves of K for ^{14}C-AIB. From the curves, K_{max} and $D_{1/2}$ were defined as follows: K_{max} was the highest K value and $D_{1/2}$ was the distance (mm) from the needle track at which the mean K values showed a half of K_{max}. K_{max} indicates the maximum effect of the infusion materials, and $D_{1/2}$ indicates the spatial extent of the effect of the infusion materials.

The effect of the SUP-C that was expressed in terms of K_{max} and $D_{1/2}$ is shown in Fig. 2. Compared to the SUP-C from normal glial cells, those from two glioma cells showed a higher value in K_{max} (p<0.01 at EI cells). In addition, the effect of the SUP-C from 20-hour incubations of glioma cells (EI, HFA) was two to three fold greater than that of the SUP-C from 4-hours incubations. The SUP-C from the normal human glia, however, showed no difference in activity between the two incubation times.

Pretreatment of animals with dexamethasone or BW755C significantly inhibited the capillary permeability activity of SUP-C from EI cells (p<0.001 and

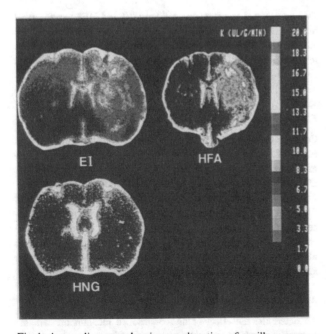

Fig. 1. Autoradiograms showing an alteration of capillary permeability after the supernatants from glioma cells (*EI, FAH*) and normal glia (*HNG*) were infused into normal rat brains (right side of each brain). (Depicted as computer-generated images of transfer constant (*K*) for ^{14}C-AIB)

$p < 0.05$, respectively). However, indomethacin had no effect when it was used at a relatively low dose (6 mg/kg). On the other hand, the SUP-C obtained from EI cells with the coincubation of dexamethasone showed capillary permeability activity about 1.5 fold greater than that without dexamethasone.

Discussion

Our recent studies have demonstrated that rat C_6 glioma cells in culture secrete diffusible factors that increase capillary permeability of normal rat brain[7, 8]. Among the factors, a protein factor, which had a molecular weight greater than 10 kD and was inactivated by heating (70 °C, 40 min), was much potent in the activity. We now report that human malignant glioma cells also secrete the same kind of a protein factor as C_6 gliomas, while normal human glial cells do not secrete the factor.

The capillary permeability activity of the protein factor in the SUP-C increased with time. No difference in cell viability and little difference in total protein contents of supernatants between the two incubation times (4 and 20 hours) indicate that the higher activity in SUP-C after longer incubation was not due to the release of intracellular components by cell lysis, instead, glioma cells actively secreted more factor with time. On the other hand, the SUP-C from normal glial cells did not show such a time-dependent increase in the activity as seen in glioma cells.

It is well known that glucocorticoids can reduce peritumoural brain oedema[5, 6], but the exact mechanism of the action is poorly understood. Our pretreatment studies suggested that glucocorticoids can reduce capillary permeability by directly acting on the capillary endothelial cells. The anti-oedema effect of glucocorticoids may function by inhibiting the activity of phospholipase A_2[4]. Moreover, lipoxygenase products may be responsible for the effect of the protein factor on capillary permeability.

Capillary permeability factor that malignant gliomas produce is likely to be the source of peritumoural brain oedema. On the other hand, the factor may also be essential for tumour growth, because generation of tumour stroma always requires host plasma proteins which usually cannot cross normal blood-brain barrier[2]. Thus, identification of capillary permeability factors in malignant gliomas may not only present a new modality in the treatment of peritumoural brain oedema, but also promote a further understanding in tumour biology.

References

1. Asch AS, Leung LLK, Shapiro JR *et al* (1986) Human brain glial cells synthesize thrombospondin. Proc Natl Acad Sci USA 83: 2904–2908
2. Dvorak HF (1986) Tumours: wounds that do not heal. Similarities between tumour stroma generation and wound healing. N Engl J Med 315: 1650–1659
3. Hiesiger EM, Voorhies RM, Basler GA *et al* (1986) Opening the blood-brain and blood-tumour barriers in experimental rat brain tumours: the effect of intracarotid hyperosmolar mannitol on capillary permeability and blood flow. Ann Neurol 19: 50–59
4. Hirata F (1918) The regulation of lipomodulin, a phospholipase inhibitory protein, in rabbit neutrophils by phosphorylation. J Biol Chem 256: 7730–7733
5. Leenders KL, Beaney RP, Brooks DJ *et al* (1985) Dexamethasone treatment of brain tumour patients: effects on regional cerebral blood flow, blood volume, and oxygen utilization. Neurology 35: 1610–1616
6. Matsuoka Y, Hossman KA (1981) Corticosteroid therapy of experimental tumour oedema. Neurosurg Rev 4: 185–190
7. Ohnishi T, Shapiro WR (1987) Vascular permeability factors produced by brain tumours: possible role in peritumoural brain oedema. J Neuro-Oncol 5: 179
8. Ohnishi T, Sher PB, Posner JB *et al* (1990) Increased capillary permeability in rat brain induced by factors secreted by cultured C_6 glioma cells: role in peritumoural brain oedema. J Neuro-Oncol (in press)

Correspondence: T. Ohnishi, M.D., Department of Neurosurgery, Osaka University Medical School, 1-1-50 Fukushima, Fukushima, Osaka 553, Japan.

Acta Neurochirurgica, Suppl. 51, 140–141 (1990)

Selective Opening of the Blood-Tumour Barrier by Intracarotid Infusion of Leukotriene C_4

K. L. Black, W. A. King, and **K. Ikezaki**

The Brain Research Institute, The Jonsson Cancer Center and The Division of Neurosurgery, UCLA Medical Center, Los Angeles, California, U.S.A.

Summary

Intracarotid infusions of leukotriene C_4 (LTC_4) were used to selectively open the blood-tumour barrier in rats with RG-2 gliomas. Blood-brain and blood-tumour barrier permeability was determined by quantitative autoradiography using ^{14}C aminoisobutyric acid. LTC_4 ($4 \mu g$ total dose) infused into the carotid artery ipsilateral to the tumour increased the unidirectional transfer constant for permeability, Ki, two-fold within the tumour while no effect on permeability was seen in the normal brain. No gamma glutamyl transpeptidase (γ-GTP) activity was seen in tumour capillaries in contrast to high γ-GTP in normal brain capillaries. These findings suggest that normal brain capillaries may resist the vasogenic effects of LTC_4 while LTC_4 will increase permeability in tumour capillaries. This could relate to the ability of γ-GTP to act as an enzymatic barrier and inactivate leukotrienes in normal brain capillaries. Intracarotid LTC_4 infusion may be a useful tool to selectively open the blood-tumour barrier for delivery of antineoplastic compounds.

Introduction

Capillaries within primary brain tumours contain a blood-tumour barrier (BTB) which impairs the delivery of many antitumour moieties to tumour cells within the brain[6, 7, 9]. One approach to open the barrier has utilized intracarotid infusions of mannitol to osmotically disrupt the tight junctions of brain endothelial cells in patients with brain tumour[9, 11]. Osmotic disruption of the blood-brain barrier (BBB) however, results in only a modest increase in drug levels within the actual tumour compared to the large increase in drug levels in normal brain[8]. In this paper we describe an experimental method that circumvents this problem. This technique selectively opens the barrier within the tumour but leaves the normal barrier intact.

In previous studies, we demonstrated a significant correlation between tissue levels of leukotriene C_4 (LTC_4) measured in brain tumours and the amount of vasogenic oedema surrounding these tumours in man[3]. This, combined with our observation that brain capillaries have high affinity binding sites for LTC_4[4], led us to speculate that LTC_4 might modulate capillary permeability in brain tumours[2]. Here we report on the ability of intracarotid infusions of LTC_4 to selectively open the tumour barrier in an experimental glial tumour, and discuss both the biological and clinical implications of this observation.

Materials and Methods

Aliquots of LTC_4 were diluted to experimental concentrations in PBS (Gibco) and the final ethanol concentration was adjusted to 2.5%. Vials containing LTC_4 were sealed under nitrogen and stored at $-80\,°C$. $[14,15-^3 H(N)]-LTC_4$ (38.4 Ci/mmol) and alpha-$[1-^{14}C]$-aminoisobutyric acid (AIB) (50.0 mCi/mmol) were obtained from New England Nuclear.

The RG-2 glioma cell line was maintained in a monolayer culture. Female Wistar rats were anesthetized with intramuscular ketamine (50 mg/kg) and xylazine (0.8 mg/kg). Glial tumours were implanted into the left hemisphere by intracerebral injections of 5×10^4 RG-2 glioma cells in $10 \mu l$ of serum-free F10 medium (Gibco). Thirteen days after tumour implantation, the rats were again anesthetized and a PE-10 polyethylene catheter (Clay Adams) was inserted retrograde through the external carotid artery to the common carotid artery bifurcation ipsilateral to the tumour. The external carotid artery was then ligated. Both femoral arteries and one femoral vein were also cannulated. LTC_4 (0.8 ml) at a concentration of $5 \mu g/ml$ or 0.8 ml of vehicle (2.5% ethanol in PBS, pH 7.1) was infused into the left carotid artery using a constant infusion pump at a rate of $53.3 \mu l/min$ for 15 min. Five minutes after the start of the intracarotid infusion, $100 \mu Ci/kg$ of ^{14}C AIB was injected as an intravenous bolus. Fifteen minutes after the start of intracarotid infusions animals were killed by decapitation and brains were rapidly removed and frozen.

Quantitative autoradiographic analysis was performed using a computer-assisted digital image analyzer. A unidirectional blood-to-brain transfer constant, Ki, was calculated and expressed in $\mu l/g/min$.

For gamma glutamyltranspeptidase histochemistry, rat brains with gliomas were fixed in 70% ethyl alcohol and embedded in paraffin. Six micron sections were cut, deparaffinized, acetone fixed, and incubated at 37 °C for 60 min in a reaction mixture of 0.5 mM γ-glutamyl-4-methoxy-2-naphthylamide (Vega Biochemicals, Tuscon, AZ), 15 mM glyclyglycine (Sigma), 0.05% fast blue BB (Sigma) in 25 mM phosphate buffer pH 7.2, containing 0.25% dimethylsulfoxide. Sections were then washed with distilled water, rinsed with 0.9% saline for 2 min, and placed in 0.1 M $CuSO_4$ for 2 min. Counter staining of sections was performed with hematoxylin for 1 min.

Ki values were calculated for experiments by measuring regions of interest in 3 continus sections. ANOVA analysis and student t-test were applied to the mean values from seperate experiments.

Results

Intracarotid infusion of LTC_4 (n = 9) resulted in a 2-fold increase in BTB permeability (Ki) within tumours which were 3 mm in diameter or smaller when compared to vehicle alone (n = 9) (45.58 ± 6.61 vs 22.33 ± 2.35, p < 0.003). There was no significant change in BBB permeability of the ipsilateral cortex (2.97 ± 1.37 vs 2.39 ± 1.09), contralateral cortex (2.87 ± 1.36 vs 2.80 ± 1.27) or contralateral corpus callosum (2.30 ± 0.92 vs 0.89 ± 0.53). A modest increase in permeability was seen in the basal ganglia immediately adjacent to the tumours (6.63 ± 2.05 vs 0.88 ± 0.42, p < 0.01) although no tumour infiltration was seen in this area. In tumours larger than 3 mm Ki values ranged between 85 and 95 µl/gm/min and were significantly not increased by LTC_4 infusions.

There was no γ-GTP activity in tumour capillaries in contrast to normal brain. Preliminary results suggested a slight decrease in γ-GTP levels in some areas of normal tissue surrounding tumours.

Leukotriene C_4 infusions did not alter physiologic parameters.

Discussion

Leukotrienes are pharmacologically active compounds that promote vascular permeability[2, 3, 10]. This study demonstrates that the blood-tumour barrier (BTB) in an experimental glial tumour can be selectively opened by intracarotid infusion of LTC_4 without increasing permeability in the normal cortex or white matter. Clinically, intracarotid LTC_4 infusions could therefore be an extremely useful tool to increase delivery of anti-tumour moieties within tumours without increasing the concentrations of potentially toxic compounds in normal brain. One possibility for the selective effect of LTC_4 could reside in the ability of normal brain capillaries to resist the vasogenic effects of LTC_4. LTC_4 will markedly increase permeability in a variety of (non-cerebral) systemic capillary beds[10]. In prior studies, we demonstrated intraparenchymal injections into the brain of high doses of leukotrienes would increase BBB permeability[2]. However, lower physiologic doses of LTC_4 do not dramatically increase permeability in normal brain tissue (unpublished observation). Brain capillaries, unlike systemic capillaries, are rich in gamma-glutamyltranspeptidase (γ-GTP)[5], an enzyme which metabolizes the peptidoleukotrienes LTC_4 and D_4 to LTE_4 and F_4 respectively[1]. High levels of γ-GTP are not present in the capillaries of experimental tumours. Theoretically, at physiologic concentrations, LTC_4 could be inactivated by γ-GTP in normal capillaries, while tumour capillaries, which lack γ-GTP, are susceptible to the vasogenic effects of LTC_4.

Acknowledgement

This work was supported by grants from NIH (1R29NS26523-01), The Robert Wood Johnson Foundation, and The Oxnard Foundation.

References

1. Aharony D, Dobson P (1984) Discriminative effect of gamma-glutamyl Transpeptidase inhibitors on metabolism of leukotriene C₄ in peritoneal cells. Life Sci 35: 2135–2142
2. Black KL, Hoff JT (1985) Leukotrienes increase blood-brain barrier permeability following intraparenchymal injections in rats. Ann Neurol 18: 349–351
3. Black KL, Hoff JT, McGillicuddy JE, Gebarski SS (1986) Increased leukotriene C₄ and vasogenic oedema surrounding brain tumours in humans. Ann Neurol 19: 592–595
4. Black KL, Betz AL, Ar DB (1987) Leukotriene C₄ receptors in isolated brain capillaries. Adv Prostaglandin Thromboxane Leukotriene Res 17: 508–511
5. DeBault LE (1981) Gamma-Glutamyltranspeptidase induction mediated by glial foot process-to endothelial contact in co-culture. Brain Res 220: 432–435
6. Fenstermacher JD, Cowles AL (1977) Theoretic limitations of intracarotid infusions in brain tumour chemotherapy. Cancer Treat Rep 61: 519–526
7. Groothuis DR, Fischer JR, Lapin G, Bigner DD, Vick NA (1982) Permeability of different experimental brain tumour models to horseradish peroxidase. J Neuropathol Exp Neurol 41: 164–185
8. Nakagawa H, Groothuis D, Blasberg RG (1984) The effect of graded hypertonic intracarotid infusions on drug delivery to experimental RG-2 gliomas. Neurology 34: 1571–1581
9. Neuwelt EA, Barnett PA, Bigner DD, Frenkel EP (1982) Effects of adrenal cortical steroids and osmotic blood-brain barrier opening on methotrexate delivery to gliomas in the rodent: The factor of the blood-brain barrier. Proc Natl Acad Sci USA 79: 4420–4423
10. Samuelsson B (1983) Leukotrienes: mediators of immediate hypersensitivity reactions and inflammations. Science 220: 568–575
11. Tomiwa T, Hazama F, Mikawa H (1983) Neurotoxicity of vincristine after osmotic opening of the blood-brain barrier. Neuropathol Appl Neurobiol 9: 345–354

Correspondence: K. L. Black, M.D., Division of Neurosurgery, 74-140 CHS, UCLA Medical Center, Los Angeles, CA 90024, U.S.A.

Acta Neurochirurgica, Suppl. 51, 142–144 (1990)
© by Springer-Verlag 1990

Oxygen Free Radicals in the Genesis of Peritumoural Brain Oedema in Experimental Malignant Brain Tumours

Y. Ikeda[1] and **D. M. Long**[2]

[1] Department of Neurosurgery, Nippon Medical School, Tokyo, Japan and [2] Johns Hopkins University School of Medicine, Baltimore, Maryland, U.S.A.

Summary

Disruption of blood brain barrier with increased vascular permeability is associated with genesis of peritumoural oedema. Oxygen free radicals play a role in increased vascular permeability. Recent studies have suggested that tumour cells can produce superoxide radicals and free radical scavengers such as superoxide dismutase (SOD) and catalase in tumour cells are impaired. In this study, we investigated the role of oxygen free radicals in the genesis of peritumoural brain oedema in experimental malignant brain tumours using $V \times 2$ carcinoma cells and 9L glioma cells. *In vitro* data indicate that the $V \times 2$ carcinoma cell and the 9L glioma cells produce superoxide radicals detected by nitroblue tetrazolium. Electron spin resonance spectroscopy using DMPO as a spin trap demonstrated that SOD activity was significantly lower in subcutaneous larger 9L glioma tumours than in normal brains and 9L glioma brain tumours.

In the subcutaneous tumours, SOD activity was lower in the central portion of the tumour than in the peripheral portion of the tumour.

In conclusion, we are not sure whether oxygen free radicals are major causative factors of peritumoural brain oedema, but the demonstration of oxygen free radicals in brain tumour cells needs further investigation.

Introduction

Peritumoural brain oedema is an extremely important complication constituting the main factor responsible for the secondary effects of brain tumours. Pathophysiology of peritumoural brain oedema is associated with increased vascular permeability. Increased vascular permeability has been reported to be caused in part by tumour-produced biochemical substances, including arachidonic acids, tissue plasminogen activator, bradykinin and vascular permeability factor[1]. Morphological abnormalities such as endothelial fenestration, increased cytoplasmic vesicles and disruption of tight junctions[2] are also associated with disruption of blood brain barrier with increased vascular permeability. Free radicals are also supposed to be associated with disruption of endothelial cell structure and function in cold-induced brain oedema and ischaemic brain oedema. Recent studies have suggested that tumour cells can produce superoxide radicals and the free radical scavenger systems such as superoxide dismutase (SOD) and catalase in tumour cells have been found to be impaired[3, 4]. In this study we evaluated the role of oxygen free radicals in the genesis of peritumoural brain oedema in rabbit brain tumour model using transplanted $V \times 2$ carcinoma and rat brain tumour model using transplanted 9L glioma.

Materials and Methods

Rabbit $V \times 2$ carcinoma brain tumour model

Experimental brain tumours were produced in 20 rabbits by the injection of a 25 µl suspension of 3×10^5 viable $V \times 2$ carcinoma cells into the right frontoparietal lobe of the rabbit brain. Animals were separated into three groups: 1) tumour-free rabbits; 2) untreated tumour-bearing rabbits, and 3) two tumour-bearing groups treated with PEG-SOD: a) rabbits treated with 10,000 U/kg of PEG-SOD on the first and fourth days after tumour transplantation and sacrified on the thirteenth day; and b) rabbits treated with 10,000 U/kg of PEG-SOD on the seventh and tenth day after tumour transplantation and sacrified on the thirteenth day. Brain water content was measured by specific gravimetry. The detection of superoxide radicals in tumour cells was performed by incubation with $V \times 2$ carcinoma cells and 0.2% nitroblue tetrazolium (NBT) dissolved in phosphate-buffered saline.

Rat 9L gliom brain tumour model

Experimental brain tumours were produced in 9 Fischer 344 rats by the injection of 10 µl suspension of 1×10^7 viable 9L glioma cells into the left frontoparietal lobe of the rat brain. Experimental subcutaneous tumours were produced in 10 Fischer 344 rats by the

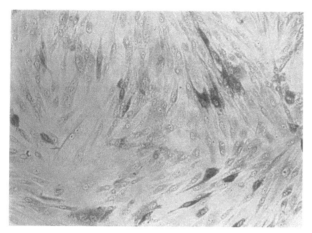

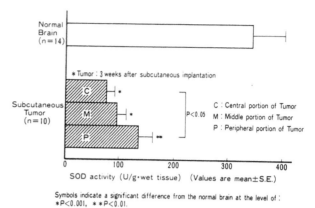

Fig. 2. Superoxide dismutase (SOD) activity in 9L gliomas. SOD activity in the subcutaneous tumour was lower in the central portion of the tumour than in the peripheral portion of the tumour

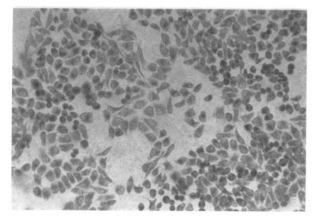

Fig. 1. Intracellular dark deposits in cytoplasm and nuclei of $V \times 2$ carcinoma cells and 9L glioma cells by nitroblue tetrazolium reduction for detection of superoxide radicals

injection of a 0.1 CC suspension of 1×10^7 viable 9L glioma cells into the back of the rat. Rats were sacrificed three weeks after tumour implantation. The detection of superoxide radicals in tumour cells was performed by incubation with 9L glioma cells and 0.2% nitroblue tetrazolium (NBT) dissolved in phosphate-buffered saline.

Measurement of SOD activity in brain tumour tissues

Determination of SOD activity was performed by electron spin resonance spectrometry using 5,5-dimethyl-1-pyrroline-1-oxide (DMPO) as a spin trap. Fifty microliters of 2 mM hypoxanthine, 35 µl of 5.5 mM DETAPAC, 50 µl of the dialyzed SOD fraction of human plasma, 15 µl of DMPO and 50 µl of xanthine oxidase were put into a test tube and mixed by an automatic mixer. Then the solution was placed in a special flat cell and DMPO-O_2-spin adduct was analyzed by electron spin resonance spectrometry. A standard curve was made using 0.8 to 100 unit/ml of SOD, and manganese oxide was used as an internal standard[5].

Results

PEG-SOD had no statistical significant effect upon peritumoural brain water content. The production of superoxide radicals was confirmed as intracellular dark

deposits by NBT reduction in the cytoplasm and nucleus of the $V \times 2$ carcinoma cells and 9L glioma cells (Fig. 1).

Electron spin resonance spectroscopy using DMPO as a spin trap demonstrated that SOD activity was significantly lower in subcutaneous larger tumours than in normal brain and brain tumours. In the subcutaneous tumour, SOD activity was lower in the central portion of the tumour than in the peripheral portion of the tumour (Fig. 2).

Discussion

Our preliminary *in vitro* data indicated that $V \times 2$ carcinoma cell and 9L glioma cell produced superoxide radicals. There seems to be a possibility that in these two models superoxide radicals may be related to the increased vascular permeability changes and production of peritumoural brain oedema. Sun *et al.*[3] have demonstrated a decrease in all the antioxidant enzymes such as MnSOD, catalase and glutathione peroxidase except $Cu - Zn$ SOD in mouse liver cells in culture. The diminished activity of these enzymes should render these tumour cells more sensitive to both superoxide radical and hydrogen peroxide compared to their normal cell counterparts. Our data indicated that SOD activity was significantly lower in subcutaneous larger 9L glioma tumours than in normal rat brains and 9L glioma brain tumours. There was no statistical significance of SOD activity in between normal brains and 9L glioma brain tumours.

In the subcutaneous large 9L glioma tumours, SOD activity was lower in the central portion of the tumour than in the peripheral portion of the tumour. These preliminary data are not in agreement with previous reports showing low levels of oxygen free radical scavengers in tumour cells. Halliwell and Gutteridge[6] postulate that areas within a large tumour mass often have a poor blood supply, and the low SOD activity could merely be a consequence of anoxia. This report was in agreement with our preliminary data about SOD activity in subcutaneous 9L glioma tumours. The precise role of oxygen free radicals and free radical scavengers in the pathogenesis of peritumoural brain oedema warrants further investigation.

References

1. Ikeda Y, Anderson JH, Long DM (1989) Oxygen free radicals in the genesis of traumatic and peritumoural brain oedema. Neurosurgery 24: 679–685
2. Long DM (1979) Capillary ultrastructure in human metastatic brain tumours. J Neurosurg 51: 53–58
3. Sun Yi, Oberley LW, Elwell JH, Sierra-Rivera E (1989) Antioxidant enzyme activities in normal and transformed mouse liver cells. Int J Cancer 44: 1028–1033
4. Oberley LW, Buettner GR (1979) Role of superoxide dismutase in cancer: A review. Cancer Res 39: 1141–1149
5. Hiramatsu M, Kohno M (1987) Determination of superoxide dismutase activity by electron spin resonance spectrometry using the spin trap method. Jeol News 23: 7–9
6. Halliwell B, Gutteridge JMC (1984) Review article. Oxygen toxicity, oxygen radicals, transition metals and disease. Biochem J 219: 1–14

Correspondence: Y. Ikeda, Department of Neurosurgery, Nippon Medical School, 1-1-5 Sendagi, Bunkyo-ku, Tokyo 113, Japan.

Acta Neurochirurgica, Suppl. 51, 145–147 (1990)
© by Springer-Verlag 1990

Effect of Recombinant Human Lipocortin I on Brain Oedema in a Rat Glioma Model

C. C. Chang[1], M. Shinonaga[1], T. Kuwabara[1], T. Mima[2], and T. Shigeno[3]

[1] Yokohama City University, [2] University Tokyo, and [3] Saitama Medical Center, Japan

Summary

Glucocorticoids have been extensively used to treat brain oedema, but little is known on the mechanisms of steroids in the prevention and resolution of tumour-induced brain oedema. Recently, the mechanism of steroid action is thought to involve synthesis of proteins with antiphospholipase activity called lipocortins. In a previous study, we have demonstrated the efficacy of dexamethasone (DEX) in resolving peritumoural oedema in a rat glioma model. Using the same model, we studied the effect of recombinant human lipocortin I on the resolution of peritumoural oedema.

Intracerebral tumours were produced in 6-week-old Wistar rats by implantation of rat glioma C6 cells. In comparison with sham-operated controls, the tumour-implanted animals showed significant increase in the cortical water content, which was reduced by DEX administration to the level in the sham-operated controls. The water content within the tumour was also significantly decreased by DEX treatment. On the other hand, there was no difference in water content between lipocortin-treated and non-treated animals. These findings suggest that tumour-induced brain oedema can be reduced by DEX treatment but not by lipocortin.

In conclusion, it is doubtful whether glucocorticoids exert their action in resolving brain oedema by inducing PLA2 inhibitory proteins named lipocortins.

Introduction

Recently, studies on the mechanisms of glucocorticoids suggest that they act by inducing the synthesis of proteins that inhibit phospholipase A2 (PLA2) and hence block mobilization of esterified arachidonic acid[7]. And these proteins with antiphospholipase activity have been named lipocortins[4]. In a previous study, we have demonstrated the efficacy of dexamethasone in resolving peritumoural oedema in a rat glioma model[1]. Using the same model, we studied the effect of recombinant human lipocortin I on the resolution of peritumoural oedema.

Materials and Methods

Intracerebral tumours were produced in 6-week-old Wistar rats by implantation of rat glioma C6 cells. For tumour cell implantation, the animals were anaesthetized and placed in a stereotactic frame. A Hamilton syringe was inserted into the right cerebral hemisphere and the suspension, containing 2,000,000 tumour cells, was injected slowly.

On day 12 after tumour implantation, the animals were sacrificed with an overdose of ether. The brain was promptly removed an dissected into the right cerebral cortex and the tumour. The water content was measured in each of these regions. In order to examine the effect of recombinant human lipocortin I in tumour-implanted animals, 0.5 mg/Kg of lipocortin was given intraperitoneally on day 11. Determination of water content was performed as follows: The tissue samples were first weighed fresh, and then weighed again after dehydration in an oven at 70 °C for 48 hours. The water percentage was calculated as:

$$\frac{\text{wet tissue weight} - \text{dry tissue weight}}{\text{wet tissue weight}} \times 100$$

Values are expressed as means ± SEM. Statistical significance was calculated by Student's t-test, and $p < 0.05$ was considered significant.

Results

The tumours grew to a diameter of about 1 cm within 10 days. Shifting of midline structures and swelling of the affected hemisphere were evident. Seven days after tumour implantation, signs of an intracerebral space-occupying lesion appeared and progressed. Some animals died before completion of the experiments.

In the tumour-implanted animals, there was a significant increase in the cortical water content, which could not be reduced by lipocortin administration (Fig. 1). The water content within the tumour also could not be decreased by lipocortin treatment (Fig.1).

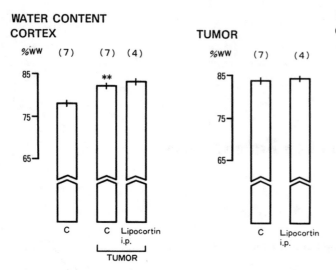

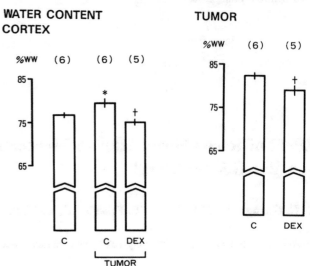

Fig. 1. Effects of lipocortin on cortical and tumoural water content in tumour-implanted animals. Vertical bars represent the means ± SEM. Numbers in parentheses denote the number of experiments. *C* controls in sham-operated or tumour-implanted animals. *Lipocortin i.p.* lipocortin-treated animals. *TUMOR* tumour-implanted animals. **p < 0.005 relative to sham-operated control

Fig. 2. Effects of DEX on cortical and tumoural water content in tumour-implanted animals. Vertical bars represent the means ± SEM. Numbers in parentheses denote the number of experiments. *C* controls in sham-operated or tumour-implanted animals. *DEX* Dex-treated animals. *TUMOR* tumour-implanted animals. *p < 0.01 relative to sham-operated control. *p < 0.005 relative to tumour-implanted control

Discussion

Glucocorticoids have been extensively used to treat cerebral oedema[6], but little is known on the mechanisms of steroids in the prevention and resolution of tumour-induced cerebral oedema.

In a previous study[1], we employed a rat glioma model and, within 7 days, the animals began to show progressive signs of intracerebral hypertension. Rats treated with DEX promptly recovered from these signs.

Vasogenic oedema, the commonest form of brain oedema, is closely related to peritumoural oedema. It is characterized by an increase in the permeability of brain capillary endothelial cells[5, 9] and is reflected by an increase in the brain water content[8]. In our previous study[1], tumour-implanted animals exhibited a significant increase in the water content of the cortex, which was reduced by DEX administration to the level in the sham-operated controls (Fig. 2). The water content within the tumour was also significantly decreased by DEX treatment (Fig. 2).

In the present study, using the same model, we examined the efficacy of lipocortin in resolving peritumoural oedema. However, we could not find any difference in water content between lipocortin-treated and nontreated animals. Published data do, however, document the ability of recombinant human lipocortin I to reduce the release of prostacyclin from suspensions of human umbilical artery rings[2] and to block the release of thromboxane from perfused guinea pig lung[3]

when triggered by leukotrien C4 but not by bradykinin. On the other hand, no data have yet emerged showing that the recombinant lipocortins can directly inhibit the cellular release of arachidonic acid or the development of an inflammatory response evoked under physiologically relevant conditions in response to specific stimuli, or that the protein designated lipocortin I can be induced by steroids in macrophage cell culture systems. Northup *et al*[10] indicate that although lipocortin I is an inhibitor of pancreatic PLA2 *in vitro*, it does not inhibit the release of arachidonate from zymosan-stimulated macrophages under conditions in which corticosteroids are effective. It also does not exhibit antiinflammatory activity in the paw oedema assay *in vivo*.

In conclusion, we could not find any efficacy of lipocortin in resolving peritumoural oedema in a rat glioma model. It is doubtful whether glucocorticoids exert their action in resolution of brain oedema by inducing PLA2 inhibitory proteins named lipocortins.

References

1. Chang CC (1989) Neurotransmitter amines in brain oedema of a rat glioma model. Neurol Med Chir (Tokyo) 29: 187–191
2. Cirino G, Flower RJ (1987) Human recombinant lipocortin I inhibits prostacyclin production by human umbilical artery *in vitro*. Prostaglandins 34: 59–62

3. Cirino G, Flower RJ, Browing JL, Sinclair LK, Pepinsky RB (1987) Recombinant human lipocortin I inhibits thromboxane release from guinea-pig isolated perfused lung. Nature (Lond) 328: 270–272

4. DiRosa M, Flower RJ, Hirata F, Parente L, Russo-Marie F (1984) Nomenclature announcement. Anti-phospholipase proteins. Prostaglandins 28: 441–442

5. Fishman RA (1975) Brain oedema. N Engl J Med 293: 706–711

6. Fishman RA (1982) Steroids in the treatment of brain oedema. N Engl J Med 306: 359–360

7. Flower RJ (1984) Macrocortin and the antiphospholipase proteins. Adv Inflamm Res 8: 1–34

8. Hossmann A, Wechsler W, Wilmes F (1979) Experimental peritumourous oedema. Acta Neuropathol (Berl) 45: 195–203

9. Klatzo I (1967) Neuropathological aspects of brain oedema. J Neuropath Exp Neurol 26: 1–14

10. Northup JK, Valentine-Braun KA, Johnson LK, Severson DL, Hollenberg MD (1988) Evaluation of the antiinflammatory and phospholipase-inhibitory activity of calpactin II/lipocortin I. J Clin Invest 82: 1347–1352

Correspondence: C. C. Chang, M.D., Department of Neurosurgery, Yokohama City University School of Medicine, 3-46 Urafune-cho, Minami-ku, Yokohama 232, Japan.

Acta Neurochirurgica, Suppl. 51, 149–151 (1990)

Tumour and Brain Oedema (II)

Formation and Resolution of White Matter Oedema in Various Types of Brain Tumours

U. Ito, H. Tomita, O. Tone, T. Shishido, and **H. Hayashi**

Department of Neurosurgery, Musashino Red-Cross Hospital, Tokyo, Japan

Summary

Following infusion of 200 ml of Iopamidol for 1 hour the propagation of extravasated contrast medium around different types of 12 brain tumours was examined and imaged via CT. Increasing volume of expanding peritumoural contrast enhanced brain tissue was measured by integrating volumes of planimetrically measured enhanced area on CT slice of 0.5 cm in thickness. So far our data failed to demonstrate differences in the peritumoural contrast expansion between the different types of tumours. Formation and resolution as well as the speed of oedema propagation were determined by calculation of the increasing volume of the enhanced peritumoural brain tissue. Average formation rate of oedema fluid from 1 cm^3 of tumour was 0.06 ml/hr, and was lower in larger tumours, while formation rate of oedema fluid from whole tumour was higher in larger tumours. Average resolution rate of oedema fluid during the passage through 1 cm^3 of the peritumoural white matter was 0.03 ml/hr, and was not affected by tumour size. Average speed of oedema propagation was 0.59 mm/hr, and was higher in larger tumours. The main therapeutic effect of steroid in peritumoural oedema was a reduction in formation rate of oedema fluid.

Introduction

It is generally accepted from experimental studies, that in brain tumours oedema fluid extravasates from leaky vessels within the tumour and secondarily spreads into the extracellular spaces of the surrounding white matter[2, 3, 9]. In a previous study[4], following 3 hours drip-infusion of contrast medium, we found a contrast-enhanced area surrounding small metastatic tumours, which expanded with time in a circular fashion into the peritumoural oedematous white matter. The expanding circular enhancement was measured planimetrically on scans, taken at various time intervals after

the infusion. This allowed a calculation of the increase in volume of contrast enhanced brain tissue assuming a spherical geometry. With these data, an estimation of the rate of oedema fluid formation[4] became possible.

In the present study, 12 patients with irregular shaped brain tumours of different histology were investigated for the same purpose.

Materials and Methods

Three metastatic brain tumours, 2 glioblastomas, 2 malignant lymphomas and 5 meningiomas were examined. All patients showed peritumoural white matter oedema of Grade II~III according to Lanksch[6, 8]. Axial CT-slices of 0.5 cm in thickness were taken. The initial scan was taken at the end of 1 hour drip infusion of 200 ml of Iopamidol. The 2nd and 3rd scans were obtained 5 and 9 hours after start of the contrast infusion. Following this study, all 12 patients received *i.v.* dexamethasone (8 mg/day) during 4 days. Thereafter, the same study was repeated.

The increasing volume of peritumoural enhanced brain tissue was determined planimetrically on each CT slice and multiplied by 0.5 cm thickness of the slice. Then, the total volume of enhanced brain tissue was determined by adding the values of the various slices.

The increasing volume from 1st to 2nd ($\triangle V_1$), and 2nd to 3rd scan ($\triangle V_2$) was calculated, repectively. From these values, the formation and resolution rate of oedema fluid were calculated with following assumptions; 1) the concentration of the contrast medium in the extravasated oedema fluid was similar to that in the plasma, 2) the contrast medium distributed only in the extracellular space, and 3) the extracellular space in the peritumoural oedematous white matter was about 30%[1, 7].

The value of newly formed oedema fluid from the tumour tissue was a summation of values of oedema fluid, which propagated into the peritumoural region ($\triangle V_1$) during interval between 1st and 2nd scan (t_1), plus volume of oedema fluid reabsorbed within $\triangle V_1$. The formation rate ($\triangle V_F$) could be calculated by following formula.

$$\Delta F_F = \frac{2\Delta V_1 - \Delta V_2}{V_1 \cdot t_1} \times 0.3 \, \text{ml/hr/cm}^3 \, \text{tumour}$$

The total tumour volume corresponded to the volume of the 1st scan (V_1) series. $\triangle V_2$ was always smaller than $\triangle V_1$. It was assumed, that this difference was due to reabsorption of oedema fluid in the white matter within the volume of $\triangle V_1$. Thus, the resolution ($\triangle V_R$) per hour during passage through 1 cm³ of white matter was calculated by following formula.

$$\Delta V_R = \frac{\Delta V_1 - \Delta V_2}{\Delta V_1 \cdot t_1} \times 0.3 \, \text{ml/hr/cm}^3 \, \text{white matter}$$

In order to estimate the speed of the propagation fo the contrast medium around the tumour, 3 slices showing the largest tumour diameter were selected in each scan series to calculate an average value. From a circle area (A) corresponding to the calculated average enhanced area, radius of the 1st scan (r_1) and 2nd scan (r_2) were calculated by following formula.

$$r = \sqrt{\frac{\pi}{A}}$$

The speed of the contrast propagation was calculated when the increase of radius from r_1 to r_2 ($\triangle r$) was divided by t_1.

Results

After 1 hour of contrast infusion, a well defined contrast enhancement of the tumour with a clear delineation of its border was observed in all 12 patients. At the second scan, 5 hours after start of contrast infusion, a halo of peritumoural contrast enhancement was observed around the tumours[5]. The contrast density in the tumour decreased at the 3rd scan, but the border of the peritumoural contrast enhancement was still well delineated. The expansion of peritumoural contrast enhancement was considerably reduced following the dexamethasone therapy[5]. Among the different types of tumours no difference of peritumoural contrast expansion was observed. Average formation rate of oe-

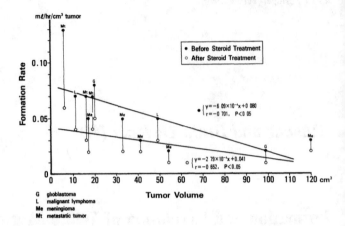

Fig. 1. Formation rate of oedema fluid from 1 cm³ of the tumour

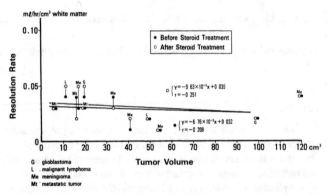

Fig. 2. Resolution rate of oedema fluid during the passage through 1 cm³ of the peritumoural white matter

dema fluid from 1 cm³ tumour was 0.06 ± 0.01 (SE) ml/hr before, and 0.03 ± 0.005 ml/hr after steroid treatment. Formation rate of oedema fluid from 1 cm³ tumour was lower in larger tumours (Fig. 1). On the other hand, formation rate of oedema fluid from the whole tumour was higher in larger tumours ($y = 1.85 \times 10^{-2}x + 0.868$, $r = 0.798$, $P < 0.01$). There was also a parallel decrease of the values after

Table 1. *Formation Rate of Oedema Fluid from Whole Tumour and Speed of Peritumoural Contrast Expansion*

	Steroid treatment	Volume of tumors (cm³)											
		6 Mt	11 L	16 Mt	17 Me	19 Mt	20 G	33 Me	40 Me	49 L	54 Me	99 G	120 Me
Formation rate	−	0.8	0.8	1.2	0.9	1.9	1.6	1.7	1.2	2.5	1.1	2.1	3.6
(ml/hr)	+	0.4	0.5	0.5	0.3	0.8	1.0	1.0	0.8	1.5	0.5	1.0	2.5
Expansion speed	−	0.51	0.45	0.77	0.34	0.45	0.45	0.58	0.25	1.10	0.49	0.47	1.21
(mm/hr)	+	0.27	0.24	0.25	0.12	0.40	0.28	0.23	0.10	0.66	0.40	0.42	0.88

Mt metastatic tumour, *L* malignant lymphoma, *Me* meningioma, *G* malignant glioma.

steroid treatment ($y = 1.30 \times 10^{-2}x + 0.363$, $r = 0.773$, $P < 0.01$) (Table). Average resolution rate of oedema fluid during the passage through 1 cm^3 of peritumoural white matter was 0.03 ± 0.003 (SE) ml/hr, and was neither affected by the tumour size nor by steroid treatment (Fig. 2). Average speed of the contrast expansion was 0.59 ± 0.09 mm/hr before, and 0.35 ± 0.06 mm/hr after steroid treatment. The speed was faster in larger tumours ($y = 4.22 \times 10^{-3} \times + 0.419$, $r = 0.517$, $P < 0.10$). There was a parallel decrease of the speed during steroid therapy ($y = 4.58 \times 10^{-3}x + 0.170$, $r = 0.735$, $P < 0.05$) (Table).

Discussion

In this study of different types of human brain tumours, Iopamidol, a contrast medium (Mol. weight 777.09), was used to label the newly formed oedema fluid in the brain tumour[4, 5]. Assumption was made that this substance moves with bulk flow of oedema fluid[7]. So far the present data have failed to demonstrate differences in the formation rate of oedema fluid between the different types of brain tumours.

Although the formation rate of oedema fluid from 1 cm^3 of tumour was lower in larger tumours, formation rate in the whole tumour was higher in larger tumours. Therefore, the speed of the contrast expansion was faster in larger tumours. In contrast, the resolution rate of oedema fluid during the passage through 1 cm^3 of peritumoural white matter was almost constant and was not related to the tumour size. The average speed of peritumoural contrast expansion was slightly slower than the speed of oedema movement in the cold injury model[7]. This could be an expression of a lower formation rate of oedema fluid in brain tumours than in the cold lesion.

Following treatment of the patients with dexamethasone, the formation rate of oedema from the tumour was reduced parallel to the tumour size. This resulted in a parallel reduction of the speed of oedema propagation. However, the resolution rate of oedema fluid during the passage through 1 cm^3 of the peritumoural white matter was not affected by the steroid treatment. The major therapeutic effect of the steroid in peritumoural brain oedema therefore is a reduction of oedema formation, probably resulting from a reduction of permeability ot the tumour vessels[4, 9, 10].

From the findings that the propagating speed of the peritumoural oedema was affected by the formation rate of oedema fluid from the tumour, it may be concluded that diffusion can not be the predominant process of the movement of the contrast medium.

Informed consent was obtained for the repeated CT scanning from all 12 patients. All had a normal renal and liver function. No adverse effects were noted in any of our patients to whom 200 ml of Iopamidol was *i.v.* infused during one hour[4].

References

1. Fenske A, Samii M, Reulen H-J *et al.* (1973) Extracellular space and electrolyte distribution in cortex and white matter of dog brain in cold induced oedema. Acta Neurochir (Wien) 28: 81–94
2. Hossmann KA, Blanik M, Wilmes F, Wechsler W (1980) Experimental peritumoural oedema of the cat brain. In: Cervos-Navaro J, Ferszt R (eds), Brain oedema. Advances in neurology, vol 28. Raven Press, New York, pp 323–340
3. Ito U, Reulen H-J, Huber P (1986) Spatial and quantitative distribution of human peritumoural brain ooedema in computerized tomography. Acta Neurochir (Wien) 81: 53–60
4. Ito U, Reulen H-J, Tomita H, Ikeda J, Saito J, Maehara T (1988) Formation and propagation of brain ooedema fluid around human brain metastases. A CT study. Acta Neurochir (Wien) 90: 35–41
5. Ito U, Tomita H, Reulen H-J, Maehara T, Kohmo Y, Ito Y (1989) A CT study on formation, propagation and resolution of brain oedema fluid in human peritumoural oedema. In: Hoff JT, Betz AL (eds), Intracranial pressure VII. Springer, Berlin Heidelberg New York, pp 959–965
6. Lanksch WR (1982) The dignosis of brain oedema by computed tomography. In: Hartmann A, Brock M (eds), Treatment of cerebral oedema. Springer, Berlin Heidelberg New York, pp 43–80
7. Reulen H-J, Graham R, Spatz M, Klatzo I (1977) Role of pressure gradients and bulk flow in dynamics of vasogenic brain oedema. J Neurosurg 46: 24–35
8. Reulen H-J, Graber S, Huber P, Ito U (1988) Factors affecting the extension of peritumoural brain ooedema. A CT-study. Acta Neurochir (Wien) 95: 19–24
9. Tamura A, Matsutani M. Nakagomi T, Nagashima T, Tsujita Y, Sano K (1988) Changes of capillary permeability in experimental brain tumour. Brain and Nerve 40: 149–156 (in Japanese with English abstract)
10. Yamada K, Bremer AM, West CR (1979) Effects of dexamethasone on tumour-induced brain oedema and its distribution in the brain of monkeys. J Neurosurg 50: 361–367

Correspondence: U. Ito, M.D., Department of Neurosurgery, Musashino Red-Cross Hospital, 1-26-24, Kyonan-cho, Musashino-city, Tokyo 180, Japan.

Acta Neurochirurgica, Suppl. 51, 152–154 (1990)

Fluid Flow Rates in Human Peritumoural Oedema

R. Aaslid[1], **U. Gröger**[1], **C. S. Patlak**[2], **J. D. Fenstermacher**[2], **P. Huber**[1], and **H.-J. Reulen**[1]

Departments of Neurosurgery and Neuroradiology, [1]University of Berne and Department of Neurological Surgery, [2]State University of New York at Stony Brook, U.S.A.

Summary

Five patients with various types of brain tumours were infused with x-ray contrast material in a schedule designed to maintain a constant plasma concentration of tracer over a period of 3 hours. CT scans from an equatorial section of the tumour were taken at frequent intervals the first hour; then at 2 and 3 hours, and when possible up to 14 hours.

Two different mathematical models – 1. simple diffusion, and 2. transport by bulk flow plus diffusion were used to analyze the changes in tracer amount along profiles placed radially from the tumour center into the oedematous white matter. We found that the simple diffusion model could not account for the spread of contrast material in 3 cases. Adding bulk flow transport gave a very good fit to the measurements, also for the late scans. This model gave bulk flow rates of 0.0005 to 0.005 ml cm^{-2} min^{-1} for the extratumoural tissue close to the tumour, and values from 0.25 to 0.55 for the extracellular space in this region.

We conclude that the peritumoural tissue is "perfused" by oedema fluid at relatively high flow rates and that this flow transports tracer and other components of plasma into the extracellular space.

Introduction

The low density area around a tumour visualized on a CT scan is an expression of the extent and degree of the oedema caused by the fluid leakage from the tumour vessels which propagates into the surrounding brain tissue. However, CT scans do not give information on the rates of flow of oedema fluid. The dynamics of oedema fluid formation within the tumour and the propagation of tracer in the peritumoural tissue can be studied by adding contrast media (tracer) to the blood and performing serial CT scanning[1].

Two processes contribute to the spread of tracer: 1. diffusion due to concentration gradients, and 2. transport by bulk flow of oedema fluid. By means of a mathematical model[2], it is possible to quantify both these mechanisms and to predict how the tracer ma-

terial would spread assuming different bulk fluid flow rates. The present study was designed to compare the predictions of the tracer transport model to data obtained from brain tumours in humans.

Patients and Procedures

Five patients with different types of brain tumours, Table 1, were studied. The patient first underwent angiography with injection of a bolus of 40 ml of X-ray contrast material (Telebrix 30-Meglumin). Blood samples were withdrawn at 1 to 5 minute intervals for at least 60 minutes after this injection. A method[3] based on impulse analysis with 3 exponentials was used to obtain infusion schedules from these data designed to achieve constant levels of tracer in plasma during the subsequent period of tracer infusion and CT scanning.

The tracer study was performed 1 to 3 days later. The calculated infusion schedule was maintained for 3 hours. Frequent blood samples were drawn to determine the actual concentration of tracer in plasma during and after infusion. Serial CT scanning was performed with a GE9800 scanner set to 3 mm slice thickness and 0.97 mm pixel distance. Individual face masks were fitted to each patient to prevent gross movements during scanning. The z-position of the image and the slice angle were adjusted to give an approximately equatorial scan plane of the tumour. The scan sequence was: one control scan before injection followed by further scans at 1, 2, 3, 4, 5, 6, 8, 10, 15, 30, 45, 60, 120, and 180 minutes during infusion. In three of the patients, we also obtained late CT scans (up to 14 hours after start of the infusion to study the long-term spread of tracer).

Results

1. Tumour tracer dynamics: The CT data were transferred to an IBM PS/2 computer for further analysis. First, the CT images were corrected for slight patient movements during the sequence by using the falx and the cranial contours as landmarks. The dynamics of tracer buildup within the tumour was estimated from the first 45 minutes of scanning. This part of the one-compartment model[2] has 3 parameters which must be determined for each case: V_p - plasma volume of the

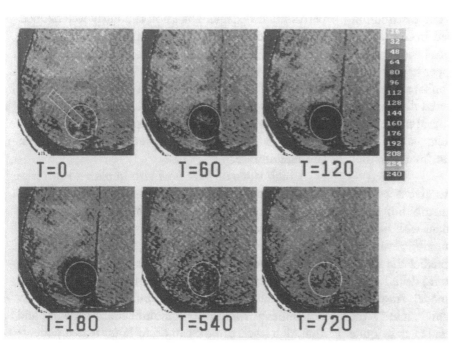

Fig. 1. Sequential CT scans of patient 4. Time is given in minutes after start of infusion of contrast material. The infusion was stopped at 180 minutes. The white circle indicates the tumour border, and the rectangular box outlines the area used for the profiles shown in Fig. 2

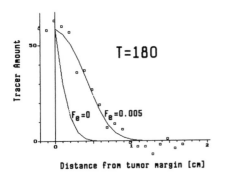

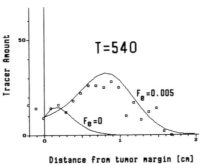

Fig. 2. Distribution space profiles of amount of tracer (Hounsfield units) in the direction shown in Fig. 1. The rectangular boxes represent the measurements taken from the CT data. The lines represent model predictions under assumption of diffusion only ($F_e = 0$); or diffusion combined with high bulk flow ($F_e = 0.005$ ml cm^{-2} min^{-1}). Clearly, diffusion only cannot explain the measured data

Table 1

No.	K_1	k_2	V_p	V_e	F_e	Tumour
1	0.0088	0.0428	0.05	0.50	0.0040	Glioblastom
2	0.0240	0.0781	0.25	0.50	0.0045	Meningiom
3	0.0065	0.0133	0.16	0.25	0.0005	Gliom
4	0.0187	0.0795	0.13	0.45	0.0050	Metastasis
5	0.0570	0.1980	0.21	0.35	0.0002	Metastasis

K_1 [ml g^{-1} min^{-1}] tumour tissue influx constant.
k_2 [min^{-1}] tumour tissue loss rate.
V_p [ml g^{-1}] plasma volume of tumour tissue.
V_e [ml g^{-1}] extracellular space of non-tumour tissue.
F_e [ml cm^{-2} min^{-1}] bulk flow per unit area of tumour border.

tumour vessels, K_1 – tumour tissue influx constant through leaky vessels, and k_2 – combined tumour loss rate through backflow of fluid into vessels and loss over the tumour border. The input to this model was the plasma concentration of tracer as measured from blood samples. The model predicts tracer amount per unit of tumour (averaged from the CT scan over a representative area of tumour tissue) as indicated by the ellipse drawn in the scan sequence shown in Fig. 1. The model parameters were determined by a least square error method to fit the model tracer amount to that measured by CT. The parameter estimates are listed in Table 1. The tumours were highly vascularized as indicated by the V_p values which were 5 to 10 times higher than those found in normal brain.

2. Peritumoural tracer dynamics: Outside the border of the one compartment comprising the tumour, we assumed that all vessels were non-leaking, that the geometry was spherically symmetrical, and that spread/removal of tracer material was by diffusion and bulk flow of oedema fluid directed radially. The model consisted of 0.5 or 1 mm thick spherical shells with a diffusion coefficient of 0.00012 cm min^{-1} [4], which was determined by multiplying a tortuosity factor of 0.40

times the agar diffusion coefficient. The distribution space profiles of tracer were calculated from the CT-scans by averaging data from 6 to 8 pixel wide rectangular regions as indicated in Fig. 1 upper left. The averaged tracer amounts (in Hounsfield units) are plotted as small rectangles to show the measured distribution space profiles in Fig. 2. The peritumoural model had two parameters: V_e – the extracellular space, and F_2 – bulk flow rate per unit area of the inner (tumour border) spherical shell.

These parameters were adjusted to give a best fit between the measured and the model distribution space profiles, as shown in Fig. 2 for a patient with a high-flow tumour. The model profiles assuming zero bulk flow (diffusion only) clearly do not predict the measured profile. The late profile (at 540 min) deviated the most markedly from diffusion only model. Assuming a bulk flow rate of $F_e = 0.005$ ml cm^{-2} min^{-1} the convection model predicted the measured data quite well as shown in Fig. 2.

The results found in the 5 patients are summarized in Table 1. In two patients (one with a glioma and one with a metastatic tumour, who was undergoing steroid treatment), the scan data fit the convective model with low bulk flow rates. The remaining 3 patients had distribution space profiles which could only be explained by high values of bulk flow according to the mathematical model.

Discussion

X-ray contrast material are known to diffuse through white and gray matter[4]. The spread of tracer by diffusion around brain tumours is almost certain; however previous reports[2] on an RG-2 glioma rat model using labelled albumin also indicate that bulk flow plays a role in tracer transport around brain tumours.

In vasogenic brain oedema induced by cold lesions, tracers with highly different diffusion constants have been shown to give almost the same distribution space profiles[5]. This is consistent with the assumption that bulk flow is prominent in such oedema.

The variance in the estimates of F_e in the present data strongly suggests a role for bulk flow in perivascular oedema. Diffusion is a fairly well defined and constant mechanism and cannot explain the large variance in the measured profiles in the 5 patients investigated. Bulk flow, in contrast, can be highly variable, as indicated by the present results. The highest values found (Patient 4) would indicate a total outflow of 0.1 ml min^{-1} or 6 ml/hour from the tumour. This value is in the same order of magnitude as the CSF production rate.

A high value of bulk flow also implies a low resistance to flow in the oedematous white matter[6]. This resistance decreases with increasing extracellular space V_e. We found a positive correlation between extracellular space and bulk flow rates ($R = 0.88$, $p < 0.05$) in our material. Therefore, we have consistency between the estimates of these two parameters.

In conclusion, we find that bulk flow seems to play a significant role in peritumoural brain oedema, and that high levels of flow can occur in untreated patients with metastases.

References

1. Ito U, Reulen HJ, Tomita H, Ikeda J, Saito J, Maehara T (1988) Formation and progapation of brain ooedema fluid around human brain metastases. A CT study. Acta Neurochir (Wien) 90: 35–41
2. Nakagawa H, Groothuis DR, Owens ES, Fenstermacher JD, Patlak CS, Blasberg RG (1987) Dexamethasone effects on [^{125}I] Albumin distribution in experimental RG-2 gliomas and adjacent brain. J Cerebr Blood Flow Metab 7: 687–701
3. Patlak CS, Pettigrew KD (1976) A method to obtain infusion schedules for prescribed blood concentration time courses. J Appl Physiol 40: 458–463
4. Fenstermacher JD, Bradbury MWB, du Boulay G, Kendall BE, Radu EW (1977) The distribution of ^{125}I-Metrizamide and ^{125}I-Diatrizoate between blood, brain and cerebrospinal fluid in the rabbit. Neuroradiology 19: 171–180
5. Reulen HJ, Graham R, Spatz M, Klatzo I (1977) Role of pressure gradients and bulk flow in dynamics of vasogenic brain oedema. J Neurosurg 46: 24–35
6. Fenstermacher JD, Patlak CS (1976) The movement of water and solutes in the brains of mammals. In: Pappius HM, Feindel W (eds) Dynamics of brain oedema. Springer, Berlin Heidelberg New York

Correspondence: Dr. A. Aaslid, Department of Neurosurgery, University of Bern, Inselspital, CH-3010 Bern, Switzerland.

Acta Neurochirurgica, Suppl. 51, 155–157 (1990)

The Finite Element Analysis of Brain Oedema Associated with Intracranial Meningiomas

T. Nagashima, Y. Tada[1], S. Hamano, M. Skakakura[1], K. Masaoka[1], N. Tamaki, and **S. Matsumoto**

Department of Neurosurgery, Kobe University School of Medicine, [1] Department of System Engineering, Faculty of Engineering, Kobe University

Summary

The mathematical model of vasogenic brain oedema, which was presented at the previous meeting in 1987[8], was applied to the analysis of peritumoural brain oedema associated with meningiomas.

Magnetic resonance images of 90 patients with intracranial meningiomas were reviewed to analyze the spatial extension of peritumoural brain oedema. It is assumed that the heterogeneous pattern of distribution of peritumoural oedema reflects the variability of the compact density of the fibers in the white matter.

A two dimensional finite element model was constructed with 786 triangular elements from a horizontal section of the human brain. The development of oedema, the change of interstitial pressure, the deformation of the brain and the absorption of oedema fluid could be simulated by the finite element method. The result of computer simulation represented interactive behaviour of the brain tissue, extracellular fluid, and cerebrospinal fluid in the clinical situation. The finite element method (FEM) may provide a new experimental tool to analyze the pathophysiology of vasogenic brain oedema.

Introduction

Many factors influence and modify the process of oedema development and resolution, their quantitative contribution, however, is not well understood. We have previously tried to establish a comprehensive mathematical model of vasogenic brain oedema which describes fluid movement across the capillary wall, fluid flow through the brain parenchyma, brain deformation and intracerebral stress distribution[7, 8, 9].

In the present study, the finite element model was extended to simulate vasogenic brain oedema associated with meningiomas. The development of this oedema was computer simulated and the result was compared with the findings of magnetic resonance imaging (MRI).

Materials and Methods

1) Mathematical model

The brain was assumed to be a continuous porous medium containing extracellular fluid. On the basis of fundamental laws of physics and constitutive equations of materials, equations were derived to describe the process of vasogenic brain oedema[9]. The model is based on the law of mass conservation and mechanical equilibrium, Hooke's law, Darcy's law, Terzaghi's principle and Starling's law.

2) MRI study

Ninety patients with intracranial meningiomas, operated on between January 1985 and July 1989, were retrospectively analyzed. All cases were examined on an Picker MR imaging system operating at 0.15 and 1.5 T. Several pulse sequences were used in each patient including Spin Echo (SE) and Inversion recovery (IR). Contrast enhancement with intravenous Gd-DTPA 0.1 mmol/kg (Shering AG) was used. The distribution of peritumoural oedema and the displacement of the brain were analyzed. For the finite element analysis, a case with a sphenoid ridge meningioma and a typical extension of peritumoural brain oedema into the frontal and temporal lobe was selected (Fig. 1).

3) Computer simulation by the FEM

The geometry of a horizontal section of human brain, 4.1 cm above the canthomeatal line, was represented by a two dimensional finite element model (Fig. 1). The computer model, which was constructed with 786 triangular elements, represents the complex anatomical geometry of the brain. Each element was assigned material properties according to estimates reported in the literature[1, 3, 4, 6, 9, 10, 11]. The right half of the brain slice was used for the simulation.

The interstitial pressure at the two points on the frontal and temporal white matter was set 10 mmHg higher than in the non-oedematous part of the brain. For the simulation, the ventricular pressure was assumed to equal 0 mmHg. The hydraulic conductivity of white matter was assumed to be 1.0×10^{-6} ml/mmHg sec, 50 times the estimated value for normal brain. The hydraulic conductivity of gray matter was assumed to be 1.0×10^{-5} ml/mmHg sec[9].

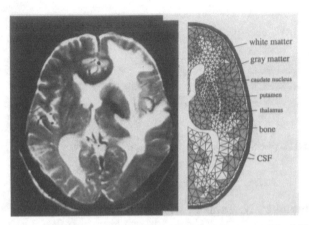

Fig. 1. MRI of a case with a sphenoid ridge meningeoma (left). MRI shows typical peritumoural brain oedema extending into the frontal and temporal white matter. The anterior limb of the internal capsule and the corpus callosum are not occupied by oedema. Finite element model constructed from a horizontal section of human brain, 4.1 cm above the cantho-meatal line. The model is constructed with 786 triangular elements (right)

Results

1) MRI study

Peritumoural brain oedema was detected by MRI in 30% of the studied patients. Edema was confined to the white matter and sharp boundaries were found at the border between white and uninvolved gray matter. The oedema occurred most frequently in the centrum semiovale (100%) and the external capsule (33.3%),

and less frequently in the corpus callosum (16.6%) and the internal capsule (6%). The anterior limb of the internal capsule was more resistant to oedema than the posterior limb. The tongue like extension of the oedema front into the internal capsule and the external capsule was a characteristic finding (Fig. 1). The oedema rarely crossed the midline along the corpus callosum.

In 24 cases (80% of patients with oedema), the oedema reached the ventricular wall. Contact with the ventricular system occurred in cases of more severe grades of oedema.

2) Computer simulation by the FEM

In the computer model, the interstitial pressure around the lesion rose gradually with time, and a pressure gradient from the lesion to the ventricle was generated. Until 6 hours, the oedema remained confined to the white matter surrounding the lesion. Between 6 and 24 hours, oedema propagated into the white matter of adjacent gyri and reached the ventricular wall. After 24 hours, oedema extended further into the surrounding white matter, the internal capsule and the external capsule (Fig. 2 left). After 40 hours, the midline shift reached 1.4 mm. The respective shift observed in the MR image was greater than the value obtained by the simulation, probably because of the mass effect by the tumour itself.

The major difference between the simulation and the MR image was the propagation of oedema into the

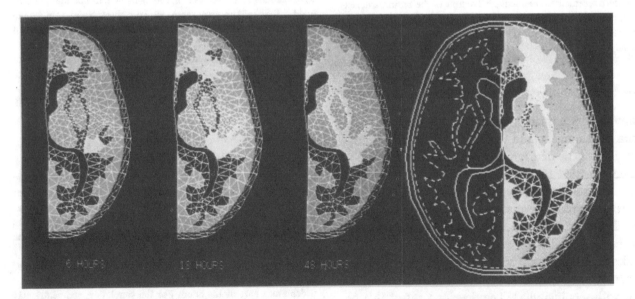

Fig. 2. Simulated brain oedema after 6, 18 and 48 hrs (left part). The interstitial pressure of two points in the temporal and frontal white matter was set 10 mmHg higher than in the other parts of the brain. After 40 hours, the oedema extends into the corpus callosum and internal capsule. Right part: The simulated brain oedema at 10 hour by reduction of the hydraulic conductivity of the corpus callosum and the internal capsule. The simulated oedema precisely represents the oedema distribution observed by MRI

internal capsule and the corpus callosum in the computer model (Fig. 1, 2-middle). If the simulation was conducted with a smaller hydraulic conductivity of the corpus callosum and the internal capsule then the distribution of the oedema precisely represented the distribution pattern observed by MRI (Fig. 2 right).

The clearance of oedema fluid into the CSF system was estimated to be not more than 40% of the fluid production. Clearance was limited by the parenchymal hydraulic conductivity and the surface area contacting the ependyma. The limited role of CSF sink action in the resolution of oedema was supported by the results of the simulation.

Discussion

The heterogeneous pattern of oedema spreading in the white matter reflects the variability in density of the fiber tracts[13]. The fiber density of areas, which offer more resistance to oedema propagation, such as the corpus callosum and internal capsule, is higher than in the other part of supratentorial white matter[2]. Also, the lower myeline density in the posterior limb as compared to the anterior limb may explain the higher incidence of oedema in the posterior limb on MRI[2]. The result of the computer simulation also supported the observations by MRI. Measurements of local tissue pressure using wick catheters revealed hydrostatic pressure gradients of 10–15 mmHg between the site of oedema and the ventricle[2]. The change in the interstitial fluid pressure was well represented by the simulation.

Elasticity and hydraulic conductivity of the brain non-linearly depend on the water content of the brain. The nonlinear relationship between physical properties and water content of the brain tissue must be determined by experiments. As the model lacks the equation to describe local osmotic pressure and the analysis was two dimensional, quantitative contribution of the "sink action" of CSF could not be determined. The simulation figured out the relative contributions of the CSF clearance and the capillary reabsorption in the resolution of brain oedema.

In considering the future of FEM analysis in brain oedema research, this method provides a new approach to a quantitative biomechanical analysis of vasogenic

brain oedema. For the next step, the greatest effort should be made to generate reliable data to put into the model. More precise formulation of material properties, boundary and loading characteristics will become a greater priority for the three dimensional, nonlinear analysis. Future progress will require a sound understanding of applied mechanics and modern nonequilibrium thermodynamics[5].

References

1. Aoyago N, Masuzawa H, Sano K, Kihira M, Kobayashi S (1980) Compliance of the brain. Brain and Nerve 32: 47–56
2. Curnes JT, Burger PC, Djang WT, Boyko OB (1988) MR imaging of compact white matter pathways. AJNR 9: 1061–1068
3. Fenstermacher JD, Patlak CS (1976) The movement of water and solutes in the brain of mammals. In: Pappius HM, Feindel W (eds) Dynamics of brain oedema. Springer, Berlin Heidelberg New York, pp 287–294
4. Gazendam J, Go KG, Zanten AK (1979) Composition of isolated oedema fluid in cold-induced brain oedema. J Neurosurg 51: 70–77
5. Glandorff P, Prigogine I (1971) Thermodynamic theory, structure, stability and fluctuation. Wiley-Interscience, London
6. Klazo I, Chui E, Fujiwara K, Spatz M (1980) Resolution of vasogenic brain oedema. Adv Neurol 28: 359–373
7. Nagashima T, Tamaki N, Matsumoto S, Seguchi Y, Tamura T (1984) Biomechanics of vasogenic brain oedema. Application of Biot's consolidation theory and finite element method. In: Inaba Y, Klazo I, Spatz M (eds) Brain oedema. Springer, Berlin Heidelberg New York, pp 92–98
8. Nagashima T, Shirakuni T, Horwitz B, Rapoport SI (1990) A mathematical model for vasogenic brain oedema. In: Long DL (ed) Advances in neurology, vol 52, Brain oedema. Raven Press, New York, pp 317–326
9. Nagashima T, Shirakuni T, Rapoport SI (1990) A two dimensional finite element analysis of vasogenic brain oedema. Neurogia Medicochirurigica 30: 1–10
10. Rapoport SI (1978) A mathematical model for vasogenic brain oedema. J Theor Biol 74: 439–467
11. Reulen HJ, Tsuyumu M, Prioleau G (1980) Further results concerning the resolution of vasogenic brain oedema. Adv Neurol 28: 375–381
12. Shulman K, Marmarou A, Weitz SN (1975) Gradients of brain interstitial fluid pressure in experimental brain infusion and compression. In: Lundberg N, Ponten U, Brock M (eds) Intracranial pressure II. Springer, Berlin Heidelberg New York, pp 221–223
13. Stevens JM, Ruitz JS, Kendall BE (1983) Observation of peritumoural ooedema in meningioma. Neuroradiology 25: 71–80

Correspondence: T. Nagashima, M.D., 7-5-1, Kusunoki-cho, Chuo-ku, Kobe, Japan 650.

Acta Neurochirurgica, Suppl. 51, 158–159 (1990)

The Effect of the Aminosteroid U-78517G on Peritumoural Brain Oedema

E. Arbit, A. Rubinstein, G. DiResta, J. Lee, F. Ali, and **J. H. Galicich**

Memorial Sloan-Kettering Cancer Center, New York, New York, U.S.A.

Summary

U-78517G belongs to a novel group of compounds-aminosteroids (AS) wich are potent inhibitors of central nervous system tissue lipid peroxidation and are devoid of classical glucocorticoid and mineralocorticoid activities. The AS have been found to possess cerebral protective properties against ischaemia, ischaemic oedema, trauma to the cerebrum and spinal cord and in sub-arachnoid hemorrhage. The possible efficacy of U-78517G in attenuating peritumoural oedema was examined in rats harbouring cerebral 9L gliosarcomas. 19 Sprague-Dawley rats were implanted intracranially with 9L gliosarcoma cells. On day 14 post implantation rats were randomized to control (no treatment n = 10) and AS treated animals (n = 9). On day 20 animals were sacrificed and six brain samples per rat were examined for the percentage of water content. Oedema was assessed with the wet/dry weight technique. The AS treated animals were found to have a significant increase in peritumoural oedema (+ 1.38%, P < 0.001) possibly related to an increase in cerebral blood flow associated with AS treatment. Further studies are, however, underway to precisely clarify this unexpected observation.

Introduction

Peritumoural oedema remains a significant source of morbidity in patients with brain tumours. Peritumoural oedema has been classified as vasogenic in origin resulting from disruption of the blood brain barrier with increased vascular permeability[10]. Recent studies have implicated oxygen free radicals in the genesis of increased vascular permeability[4]. Furthermore, tumour cells have been shown to produce superoxide radicals, and the free radical scavenger mechanism in tumours have been found to be impaired[1, 3]. A novel group of steroid compounds, 21-aminosteroids (AS) are devoid of mineralocorticoid activities and their classical side effects[2]. The aminosteroids have been shown to effectively protect cellular membranes including the blood brain barrier from a variety of insults including cerebral and spinal trauma, ischaemia and sub-arachnoid hemorrhage and reduce oedema associated with these conditions[5, 6, 7, 8, 9, 12, 13]. These facts have stimulated this study on the potential role of the AS on peritumoural brain oedema. If proven to be effective in reducing peritumoural brain oedema, the aminosteroids could become an attractive alternative of considerable clinical interest to the glucocorticosteroids.

Materials and Methods

Experiments were carried out in 19 Sprague-Dawley rats. Under general anaesthesia (Pentobarbital 40 mg/kg im), a cell suspension containing approximately 10^6 9L gliosarcoma cells was injected stereotactically into the right parietal lobe. Fourteen days after inoculation, animals were randomized either to a control (no treatment, n = 10) or aminosteroid treatment group (n = 9). The aminosteroid used was designated as U-78517G. 5 mg of the AS were dissolved in 0.5 ml water and administered orally twice a day for five days. Animals were sacrificed on day 20 and the brains removed. Six samples from each brain were sampled; one from the peritumoural region, one from an adjacent area and one from a remote area and three samples from corresponding sites on the contralateral hemisphere to serve as internal controls. Brain oedema was measured by the wet/dry weight method. Statistical analysis was done by paired and unpaired Student's t-test.

Results

The 9L gliosarcoma cell produced a spherical tumour 5 mm in diameter which was surrounded by a consistent, albeit, modest amount of oedema (0.2% increase P < 0.01) in the control group. In the AS treated group we have observed a consistent marked average increase in water content of 1.36% (from 79.71 ± 0.3% to 81.06 ± 0.9%, p < 0.001) in peritumoural regions (Fig. 1).

Discussion

The AS are compounds where an amino group is substituted for 21-carbon of the steroid moiety, a modi-

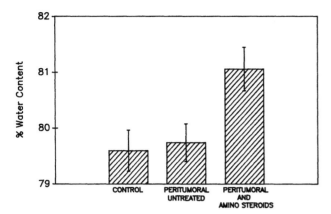

Fig. 1. The effect of the aminosteroid on peritumoural brain oedema; increase in oedema as compared to control and untreated peritumoural regions

fication that inactivates the corticosteroid receptor activity. The drugs are thus devoid of adrenocorticoid or mineralocorticoid activities and their classical side effects. *In vitro* evidence suggests that the AS are potent inhibitors of iron dependent lipid peroxidation. This action is mediated through scavenging of oxygen and lipid radicals in a mechanism which consumes the drug. Observations also suggest direct interaction with the hydrophobic core of cellular membranes, resulting in membrane stabilization. The rationale for assessing the role of AS in peritumoural oedema emanated from the fact that lipid peroxidation products are produced in excess by tumours and that tumours lack a sufficient antioxidant defense mechanism in peritumoural vessels. These free radicals in turn destabilize the blood brain barrier cellular membrane to result in a leaky barrier and oedema formation as is the case in subarachnoid haemorrhage for example[13].

Our observations did not show a reduction of peritumoural oedema in AS treated animals but in contrast showed a diffuse peritumoural increase in water content. Although no fluid balance was calculated, nor serum or urine sodium levels determined and no other report of diffuse oedema associated with the AS is known of, the mechanism remains widely speculative.

Possible explanations include: (a) water retention in AS treated animals (AS treated animals received 1 ml H_2O as drug solvent) or (b) increase in oedema as a result of enhancement of cerebral blood flow, as seen in cerebral reperfusion models[11]. That AS enhance cerebral blood flow under ischaemic conditions has been shown previously[5, 7]. While the results of this study have shown that AS did not decrease, but rather increased brain oedema in this tumour model, the hypothesis regarding the role of free radicals in peritumoural oedema may still be correct. Studies are underway with U-74006F whereby oedema, cerebral blood flow, fluid balance and serum and brain parenchyma electrolytes are measured.

References

1. Bannister WH, Frederici G, Heath JK, Bannister JV (1986) Antioxidant systems in tumour cells: The level of antioxydant enzyme, ferritine and total iron in human hepatoma cell line. Free Radic Res Commun 6: 361–367
2. Braughler J, Pregenzer JF, Chase RI *et al* (1987) Novel 21-aminosteroid as potent inhibitors of iron-dependent lipid peroxidation. J Biol Chem 262: 10438–10440
3. Cerutti PA (1985) Peroxidant states and tumour promotion. Science 227: 375–381
4. Chan PH, Schmidley JW, Fishman RA, Longar SM (1984) Brain injury, oedema and vascular permeability changes induced by oxygen derived free radicals. Neurology 34: 315–320
5. Hall ED, Yonkers PA (1988) Attenuation of postischaemic cerebral hypoperfusion by the 21-aminosteroid U74006F. Stroke 19: 340–344
6. Hall ED, Yonkers PA, McCall JM *et al.* (1988) Effects of the aminosteroid U-74006F on experimental head injury in mice. J Neurosurg 68: 456–461
7. Hall ED (1988) Effects of the 21-aminosteroid U-74006F in posttraumatic spinal cord ischaemia in cats. J Neurosurg 68: 462–465
8. Hall ED, Travis MA (1988) Inhibition of arachidonic acid induced vasogenic brain oedema by non-glucocorticoid 21-aminosteroid U74006F. Brain Res 451: 350–352
9. Hall ED, Pazara BA, Braughler MJ (1988) 21-aminosteroid lipid peroxidation inhibitor U74006F protects against cerebral ischaemia in gerbils. Stroke 19: 997–1002
10. Klazo I (1972) Pathophysiological aspects of brain oedema. In: Reulen HJ, Schurmann K (eds) Steroid and brain oedema. Springer, New York, pp 1–8
11. Kukreja RC, Kontos HA, Hess ML, Ellis EF (1986) PGH synthase and lipoxygenase generate superoxide in the presence of NADH or NADPH. Circ Res 59: 612–619
12. Young W, Wojak JC, DeCrescito V (1988) 21-aminosteroid reduces ion shifts and oedema in the rat middle cerebral artery occlusion model of regional ischaemia. Stroke 19: 91–106
13. Zuccarello M, Anderson DK (1989) Protective effect of a 21-aminosteroid on the blood brain barrier following subarachnoid haemorrhage in rats. Stroke 20: 367–371

Correspondence: E. Arbit, M.D., Memorial Sloan-Kettering Cancer Center, 1275 York Avenue, New York, NY 10021, U.S.A.

Acta Neurochirurgica, Suppl. 51, 160–162 (1990)
© by Springer-Verlag 1990

Novel 21-aminosteroids Prevent Tumour Associated Neurological Dysfunction

W. A. King, K. L. Black, K. Ikezaki, S. Conklin, and **D. P. Becker**

Brain Research Institute, Jonsson Cancer Center and Division of Neurosurgery, UCLA School of Medicine, Los Angeles, California, U.S.A.

Summary

The efficacy of the potent lipid peroxidation inhibitors U-74006F and U-78517F in the treatment of blood-tumour barrier permeability and tumour associated neurological dysfunction was evaluated. Rats with stereotactically implanted Walker 256 tumours were treated with methylprednisolone (MP), U-74006F, U-78517F, or vehicle. (0.05 N HCl) on days 6 through 10 following implantation. Neurologic function and vascular permeability was assessed on day 10. U-74006F and MP were equally effective at preventing neurologic dysfunction compared to control ($p < 0.01$). U-78517F was slightly less effective than U-74006F and MP but was significantly better than vehicle in preventing neurological dysfunction. MP resulted in a significant decrease in tumour vascular permeability ($p < 0.006$) while the lipid peroxidation inhibitors had no effect on permeability.

Introduction

Recently, two novel series of potent inhibitors of iron-catalyzed lipid peroxidation, collectively known as "lazaroids", were developed. The first series are the 21-aminosteroids which lack glucocorticoid, mineralocorticoid or other hormonal activities. One of these compounds, U-74006F, has been found to be effective in treating experimental spinal cord injury, spinal trauma, concussive head injury and cerebral ischaemia[1, 4, 7–9]. It has also been found efficacious in primate and rabbit models of vasospasm following subarachnoid haemorrhage[10–12]. The second series are the 2-methyl-amino-chromans (*e.g.* U-78517F) which are even more potent antioxidants than the 21-aminosteroids and appear to possess equivalent cerebroprotective activity[6]. In the present study a rat tumour model was used to study the effect of iron-dependent lipid peroxidation inhibition with U-74006F and U-78517F on vascular permeability and tumour associated neurologic deterioration.

Materials and Methods

1×10^3 viable Walker 256 carcinosarcoma cells were stereotactically implanted to a depth of 4 mm into the left hemisphere of 25 female Wistar rats. On the sixth day following implantation rats were randomly assigned to four treatment groups. The rats received either vehicle, methylprednisolone (MP), or U-74006F and U-78517F. Compounds were dissolved in 0.05 N HCl in phosphate buffered saline (pH 3.0). Each rat received glucocorticoid, aminosteroid, or methylamino-chroman in a dose of 10 mg/kg given as an intraperitoneal injection on a twice daily dosage. The vehicle control group received 0.2 ml of 0.05 N HCl. Neurological function and vascular permeability was assessed on day 10.

Asymptomatic rats were given a rating of 1. Those minimally symptomatic (grade 2) displayed no hemiparesis, but had diminished spontaneous activity or decreased grooming. Grade 3 rats had at least a mild hemiparesis, but were able to right themselves rapidly when placed on their sides. Rats rated as grade 4 had moderate or marked hemiparesis and righted slowly. A rating of 5 was given to those rats that appeared frankly moribund.

Permeability studies were performed on rats with neurologic rating of 4 or better. On day 10 following tumour implantation animals were anaesthetized, femoral vessels cannulated, and physiologic parameters monitored. ^{14}C-labelled α-aminoisobutyric acid (^{14}C-AIB) (100 μCi/kg) was injected into a femoral vein as a bolus. Arterial blood was continuously withdrawn for 10 min for plasma radioactivity determinations. Rats were sacrificed by decapitation 10 min after the ^{14}C-AIB bolus, their brains rapidly removed and frozen ($-40\,°C$). Twenty micron thick sections of the brains were exposed to X-ray film for 14 days. Tissue radioactivity in regions of interest were determined using computerized densitometry on the autoradiographs and a vascular permeability constant, Ki, was calculated by the method of Blasberg[2] and expressed in ml/gm/min. Maximal tumour diameter was measured by staining slides with thionin following autoradiographic exposure.

Results

Table 1 shows our results. Of the seven rats in the control group, treated with vehicle, only two rats remained asymptomatic while more than half of the rats were moderately symptomatic or worse. Six rats were tested in each of the other treatment groups. MP, U-74006F, and U-78517F significantly prevented neurologic dysfunction compared to controls ($p < 0.01$ for MP and U-74006F and $P < 0.03$ for U-78517F). All

Table 1. *Effect of Lazaroids on Rats with Experimental Brain Tumours* *

Group	Grade	Ki	Max. diameter (mm)
Vehicle	3.00 ± 0.57	0.351 ± 0.024	6.25 ± 0.214
MP	1.00 ± 0.00^a	0.241 ± 0.026^c	2.67 ± 0.167^d
U-74006 F	1.00 ± 0.00^a	0.300 ± 0.033	3.75 ± 0.403^d
U-78517 F	1.33 ± 0.21^b	0.362 ± 0.022	4.17 ± 0.380^d

* Values are means ± standard error of the means.
[a] $p < 0.01$.
[b] $p < 0.03$.
[c] $p < 0.006$.
[d] $p < 0.001$.

rats in the MP group and those treated with U-74006F were without neurologic deficits. Four rats in the U-78517F were asymptomatic while 2 had minimal symptoms.

Six rats in each group were tested for tumour vascular permeability. The MP group had a mean Ki value of 0.241 ± 0.026 ml/gm/min, which was significantly less than that of the vehicle treated rats $(0.351 \pm 0.024$ mg/gm/min) $(p < 0.006)$. Rats treated with U-74006F and U-78517F had Ki values of 0.300 ± 0.033 and 0.362 ± 0.022 ml/gm/min, respectively. These values were not significantly different from the control animals.

The average maximal tumour diameter in the vehicle treated group was 6.25 ± 0.214 mm. This was significantly larger $(p < 0.001)$ than the MP, U-74006F, and U-78517F groups which measured 2.67 ± 0.167, 3.75 ± 0.403 and 4.17 ± 0.380 mm, respectively.

Discussion

This study demonstrates the ability of the 21-aminosteroid, U-74006F and the 2-methylamino-chroman U-78517F, to prevent neurologic dysfunction in rats with experimental brain tumours. U-74006F was equally effective as MP while U-78517F was slightly less effective. In our experiments MP significantly decreased vascular permeability within the tumours which is in agreement with the findings of others. The lazaroids, on the other hand, showed no effect on the vascular permeability constants. Hence the mechanism of action of these compounds must be at least in part different than reducing vascular permeability within tumours. Perhaps they effect intracellular water content.

Tumour size was also decreased following glucocorticoid and lazaroid administration. This too could

explain the improved neurological improvement observed. Steroids have previously been implicated as playing a role in inhibiting cerebral neoplasm growth[5]. This has most extensively been investigated in meningiomas though glioma growth, at least in culture, may also be inhibited by steroid administration. Whether the decrease in the tumour size we measured is secondary to decreased water within tumours or altered tumour growth needs to be further examined.

The beneficial effects of lazaroids could also be by other mechanisms in addition to their iron-dependent lipid peroxidation inhibition capabilities. Like conventional steroids, these compounds may effect multiple biochemical pathways. U-74006F, for example, will inhibit iron or iodoacetate induced membrane release of arachidonic acid in cultured cells[3]. Other theoretical mechanisms could include a reduction in interstitial oedema by reducing cellular secretory activity or increasing interstitial resistance to bulk flow oedema. Although the mechanism of reduced neurological dysfunction with lazaroids remain unclear, current findings suggest that further studies of these compounds in the management of CNS tumours are warranted.

References

1. Anderson DK, Braughler JM, Hall ED et al (1988) Effects of treatment with U-74006F on neurological outcome following experimental spinal cord injury. J Neurosurg 69: 562–567
2. Blasberg RG, Patlak CS, Jehle JW et al (1978) An autoradiographic technique to measure the permeability of normal and abnormal brain capillaries. Neurology 28: 363
3. Braughler JM, Chase RL, Neff GL et al (1987): A new 21-aminosteroid antioxidant lacking glucocorticoid activity stimulates ACTH secretion and blocks arachidonic acid release from mouse pituitary tumour (AtT-20) cells. J Pharmacol Exp Ther 244: 423–427
4. Braughler JM, Hall ED, Jacobsen EJ et al (1989) The 21-aminosteroid, U-74006F, for CNS trauma and ischaemia. Drugs of the Future 14: 143–152
5. Giorgio B, Nadia G, Carletto Z et al (1989) Steroid hormones influence on glioma cells proliferation in vitro. Presented at the Annual Meeting of the American Association of Neurological Surgery
6. Hall ED, Braughler PA, Yonkers KE et al, (1990) U-78517F, a second generation lazaroid with potent antioxidant and cerebroprotective activity in models of CNS injury and ischaemia. J Neurotraume (in press)
7. Hall ED, Pazara KE, Braughler JM (1988) 21-aminosteroid lipid peroxidation inhibitor U-74006F protects against cerebral ischaemia in gerbils. Stroke 19: 997–1002
8. Hall ED, Yonkers PA, McCall JM et al (1988) Effects of the 21-aminosteroid U-74006F in experimental head injury in mice. J Neurosurg 68: 456–461
9. Hall ED, Travis MA (1988) Effect of the nonglucocorticoid 21-aminosteroid U-74006F on acute cerebral hypoperfusion following experimental subarachnoid haemorrhage. Exp Neurol 102: 244–248

10. Steinke DE, Weir BKA, Findlay JM *et al* (1989) A trial of the 21-aminosteroid U-74006F in a primate model of chronic cerebral vasospasm. Neurosurgery 24: 179–186

11. Vollmer DG, Kassell NF, Hongo K *et al* (1989) Effect of the nonglucocoritcoid 21-aminosteroid U-74006F on experimental cerebral vasospasm. Surg Neurol 31: 190–194

12. Zuccarello M, Marsch JT, Schmitt G *et al* (1989) Effect of the 21-aminosteroid U-74006F on cerebral vasospasm following subarachnoid haemorrhage. J Neurosurg 71: 98–104

Correspondence: A. King, M.D., Division of Neurosurgery, 74-140 CHS, UCLA School of Medicine, 10833 Le Conte Ave, Los Angeles, CA 90024, U.S.A.

Acta Neurochirurgica, Suppl. 51, 163–164 (1990)

Therapeutic Effects of Topical Dexamethasone on Experimental Brain Tumours and Peritumoural Brain Oedema

Y. Ikeda[1], B. S. Carson[2], and D. M. Long[2]

[1] Department of Neurosurgery, Nippon Medical School, Tokyo, Japan, and [2] Johns Hopkins University School of Medicine, Baltimore, Maryland, U.S.A.

Summary

Topical application of dexamethasone to brain tumour bed retains its biological effectiveness in controlling peritumoural brain oedema, possibly through the oncolytic mechanism (cytostatic effect). The question rises, whether the sytemic administration of dexamethasone can be duplicated with dexamethasone topically to the brain tumour bed.

Introduction

Peritumoural brain oedema remains a significant cause of mortality and neurological deficits in patients with brain tumours.

For the past thirty years, steroids when administered systemically have proven to be one of the few modalities sometimes effective in treating this problem[1, 2]. However, steroids are limited by many systemic complications and the mechanisms of beneficial action of steroids on peritumoural brain oedema remain controversial. The specific aim of this study is to investigate the therapeutic effect of topical application of dexamethasone to brain tumour beds in an attempt to achieve the beneficial effect and eliminate the systemic complication of systemically administered dexamethasone.

Materials and Methods

Experimental brain tumours were produced in 102 New Zealand white rabbits by the injection of a 25 μl suspension of 3×10^5 viable VX2 carcinoma cells into the right frontoparietal lobe of the rabbit brain through a 1 mm diameter cranial burr hole[3].

In peritumoural brain oedema study, 58 animals were separated into three groups; 1) untreated tumour-bearing rabbits (n = 15), 2) systemic dexamethasone treated (4 mg/kg/day) tumour-bearing rabbits (n = 18), 3) topical dexamethasone treated (2.5 μl/hr, osmotic pump) tumour-bearing rabbits (n = 25). This group was implanted with osmotic pump at the time of initial intracerebral tumour cell suspension injection. A 2 cm para-saggital incision was made and a

23 gauge burrhole was made 4 mm lateral and 3 mm posterior to the bregma. A Teflon catheter was inserted into the tumour-bed to a depth of 7 mm and glued in place.

The osmotic pump was then implanted subcutaneously and con-

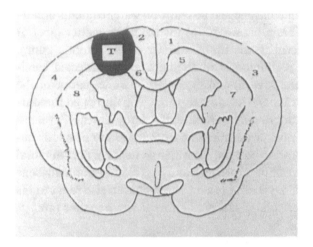

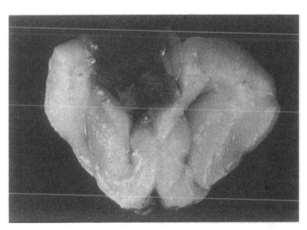

Fig. 1. Eight sampling areas used for measurement of water content by gravimetry in brain subjected to transplantation of VX 2 carcinoma cells and coronal section of the brain

nected to the Teflon catheter with P. E. 60 tubing. The incision was stapled closed and the rabbit was given antibiotics and allowed to recover. Brain water content was measured by the specific gravity (SG) method. Brain samples were excised and placed onto a kerosene/monobromobenzene column. The eight sampling areas are illustrated in Fig. 1. Systemic or topical dexamethasone was administered from day 3 or day 7 after tumour implantation or sham operation with sacrifice on day 13. In the survival study, 44 animals were separated into three groups (untreated, systemic dexa., topical dexa.) Treatments were started from day 7 after tumour implantation. The brains were removed to make sure whether experimental brain tumours were successfully produced. In addition, those rabbits with osmotic pumps had the pumps analyzed in terms of residual dexamethasone content and the integrity of the system was examined to be sure that it had remained intact. Tumour volume was estimated by measuring tumour dimensions along the long axis of the tumour and across the tumour at the widest point. Tumour volume was calculated by the equation[4]. Volume = 0.5 (length × width^2). Statistical comparisons were performed using student's t-test for independent variables, and a 95% confidence level was considered statistically significant.

Results

Systemic dexamethasone had no effect on water contents of normal brain tissues except for area 3. Topical dexamethasone had no effect on water contents of normal brain tissues except for the white matter (area 6) adjacent to the tip of catheter of the osmotic pump. The peritumoural white matter of the affected hemisphere (area 6) and the adjacent white matter of the contralateral hemisphere (area 5) in untreated tumour-bearing rabbits showed significantly lower SG values. Systemic dexamethasone from day 3 after tumour implantation reduced the water contents of peritumoural white matters (area 5, 6 and 8). Topical dexamethasone from day 3 after tumour implantation also reduced the water contents of peritumoural white matter (area 5, 7 and 8).

Systemic dexamethasone from day 7 after tumour implantation reduced the water content of peritumoural white matter, especially in the more distant oedematous region (area 5 and 8).

Topical dexamethasone from day 7 after tumour implantation had no effect on water contents of all areas (Fig. 2). Systemic and topical dexamethasone from day 3 after tumour implantation with sacrifice on day 13 showed a significant inhibition of tumour volume relative to the untreated group (31.72 ± 9.66, 26.46 ± 6.51, 135.78 ± 28.79 mm^3, mean ± S. E., P < 0.05, respectively), while systemic and topical dexamethasone from day 7 after tumour implantation with sacrifice on day 13 showed no statistical significant difference relative to the untreated group (76.18 ± 27.22, 84.25 ± 20.14, 135.78 ± 28.79 mm^3,

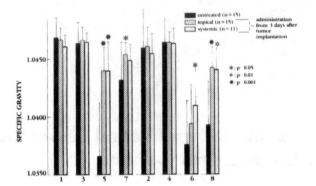

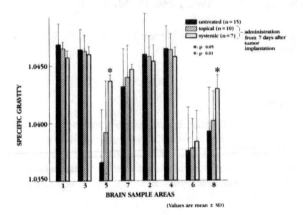

Fig. 2. Specific gravity values of eight sampling areas in tumour-bearing rabbits treated with dexamethasone topically or systemically

mean ± S.E., respectively). Systemic or topical dexamethasone resulted in a significant increase in survival relative to the untreated group (17.57 ± 2.19, 17.92 ± 2.98, 13.25 ± 4.00 days, mean ± SD, P < 0.01, respectively).

References

1. Galicich JH, French LA, Melby JC (1961) Use of dexamethasone in treatment of cerebral oedema associated with brain tumours. Lancet 81: 46–53
2. Shapiro WR, Posner JB (1974) Corticosteroid hormones. Effects in an experimental brain tumour. Arch Neurol 30: 217–221
3. Carson BS, Anderson JH, Grossman SA, Hilton J, White CL, Colvin OM, Clark AW, Grochow LB, Kahn A, Murray KJ (1982) Improved rabbit brain tumour model amenable to diagnostic radiographic procedures. Neurosurgery 11: 603–608
4. Bullard DE, Schold SC, Bigner SH, Bigner DD (1981) Growth and chemotherapeutic response in athymic mice of tumours arising from human glioma-derived cell lines. J Neuropathol Exp Neurol 40: 410–427

Correspondence: Y. Ikeda, Department of Neurosurgery, Nippon Medical School, 1-1-5 Sendagi, Bunkyo-ku, Tokyo 113, Japan.

Acta Neurochirurgica, Suppl. 51, 165–167 (1990)

Evaluation of Glucose Transport in Malignant Glioma by PET

M. Ishikawa, H. Kikuchi, S. Nishizawa[1], and Y. Yonekura[1]

Departments of Neurosurgery and [1] Nuclear Medicine Faculty of Medicine, Kyoto University, Japan

Summary

Using dynamic PET mode and [18]FDG, glucose transport in patients with gliomas were investigated. The values for transfer rate constants k1*, k2*, k3*, and glucose consumption were found to be low in the low-grade glioma as compared to those of the high-grade glioma and the contralateral cerebral cortex. The differences were statistically significant with the exception of k2*. There were no statistically significant differences between the high-grade glioma and the contralateral cerebral cortex. In contrast, the distribution volumes k1*/k2* and k1*/(k2* + k3*) were low in high-grade glioma and the difference between the high-grade glioma and the contralateral cerebral cortex was statistically significant. A difference in k1*/(k2* + k3*) was noted between the low-grade and the high-grade gliomas. Thus, the distribution volumes are most sensitive for differentiation between high-grade glioma and cerebral cortex.

Introduction

It is well known that in histologically high-grade gliomas the blood-brain-barrier (BBB) is disrupted, leading to peritumoural vasogenic oedema. In human glioma, a study using positron emission tomography (PET) and ^{18}F-labelled fluorodeoxyglucose ([18]FDG) has shown that glucose consumption increases with higher degrees of malignancy[2]. However, the mechanism of glucose uptake in human high-grade gliomas remains to be clarified. Based on a three-compartment model, influx and efflux of [18]FDG between plasma and brain through the BBB are represented by the transfer rate constants for [18]FDG of k1* and k2*. In this study, we evaluated glucose transport in patients with gliomas using dynamic PET mode and [18]FDG.

Materials and Methods

Fourteen patients with histologically confirmed gliomas were evaluated. There were five cases with a low-grade glioma and nine cases with a high-grade glioma, including four cases with recurrent glioma. All high-grade gliomas revealed peritumoural oedema on CT. A positron camera with dynamic scanning at 4-minute intervals for 60 minutes after an intravenous bolus injection of 3 to 5 mCi of [18]FDG were used to measure sequentially the radioactivity of the brain tissue. Functional images of glucose consumption were reconstructed using the autoradiographic method described by Phelps et al.[5], with a lumped constant of 0.42 and blood volume correction determined from a single inhalation of $C^{15}O$ gas. Specific regions of interest were the tumour itself and the contralateral cortex corresponding approximately with the location of the tumour. Regional transfer rate constant values for [18]FDG influx (k1*), [18]FDG efflux (k2*) and distribution volume (k1*/k2* and k1*/(k2* + k3*)), were computed. The phosphorylation rate (k3*) and glucose consumption were also computed.

Statistical comparisons were performed using one-way analysis of variance (ANOVA). If indicated, further analysis using a modified t-test with Bonferroni's correction was applied to multiple comparisons between groups.

Results

The k1* values (Fig. 1) for the area of the low-grade glioma were significantly lower than those of the high-grade glioma and the contralateral cortex. However, there was no significant difference between these latter two groups. The k2* values for the area of the low-grade glioma were significantly lower than those of the high-grade glioma but no difference was noted with the contralateral cortex. The k2* values (Fig. 1) in the high-grade gliomas were higher than those of the contralateral cortex but this difference was not statistically significant (Fig. 2).

The difference between the high-grade glioma and the contralateral cortex in terms of distribution volumes were statistically significant. Values for k3* and glucose consumption were low in the area of the low-grade glioma and high in the high-grade glioma, but the latter was not significantly different from that of the contralateral cortex.

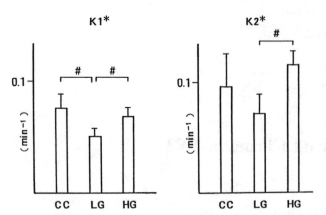

Fig. 1. Influx and efflux of ¹⁸FDG in human glioma. *k1**: ¹⁸FDG influx into brain tissue, *k2**: ¹⁸FDG efflux from brain tissue, *CC*: contralateral cortex, *LG*: low-grade glioma, *HG*: high-grade glioma, #: p<0.05

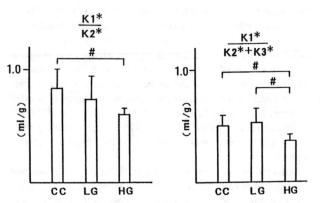

Fig. 2. Distribution volume of glucose and ¹⁸FDG in human glioma. *k1**/*k2**: glucose distribution volume (ml/mg), *k1**/(*k2** + *k3**): effective ¹⁸FDG distribution volume (ml/mg). Other abbreviations are the same as in Fig. 1

Discussion

In this study, the values for k1*, k2*, k3*, and glucose consumption were found to be low in the low-grade glioma as compared to those of the high-grade glioma and the contralateral cerebral cortex. The differences were statistically significant with the exception of k2*. There were no statistically significant differences between the high-grade glioma and the contralateral cerebral cortex. In contrast, the distribution volumes were low in high-grade glioma and the difference between the high-grade glioma and the contralateral cerebral cortex was statistically significant. A difference in k1*/(k2* + k3*) was noted between the low-grade and high-grade gliomas. Thus, k1*/k2* and k1*/(k2* + k3*), which represent the distribution volume of glucose and

¹⁸FDG, respectively, are most sensitive for differentiation between high-grade glioma and cerebral cortex. Using the same methods, Koeppe *et al.*[4] reported that the values for k1* and k2* were not significantly different between malignant glioma and cortex, while k3* typically increased more than 100% and k1*/k2* decreased more than 30%. They concluded that the parameters k3* and k1*/k2* were more sensitive than glucose consumption for distinguishing tumour from normal tissue. Although k3* was not very sensitive in this study, the low values for k1*/k2* in high-grade glioma are consistent with Koeppe's findings. Hawkins *et al.*[4] also reported a low value for k1*/k2* in brain tumours. The low values for k1*/k2* in high-grade glioma may be due to an increase in efflux, rather than a decrease in influx, as shown in Fig. 1. An increase in phosphorylation seen in high-grade gliomas may also contribute to a decrease in k1*/(k2* + k3*).

Using ¹¹[C]3-O-methyl-D-glucose (¹¹C-MeG), a D-glucose analogue not metabolized by hexokinase, as ¹⁸FDG, Brooks *et al.*[1] observed that there were no differences in the values for the rate constants k1* and k2* between tumour and the contralateral cortex. They failed to satisfactorily fit cerebral ¹¹C-MeG uptake kinetics to a two-compartment model. In a concomitant study of cerebral blood flow, they also found increased extraction of ¹¹C-MeG in gliomas. They suggested that the increased extraction of ¹¹C-MeG was due to both increased tracer free diffusion and facilitated transport.

Thus, PET is useful for *in vivo* evaluation of glucose transport in human gliomas. Further studies are necessary to elucidate the mechanism of glucose transport in malignant gliomas.

References

1. Brooks DJ, Beaney RP, Lammertsma AA, Herold S, Turton DR, Luhra SK, Franckowiak RSJ, Thomas DGT, Marshall J, Jones T (1986) Glucose transport across the blood-brain barrier in normal human subjects and patients with cerebral tumours studied using [¹¹C3-O- Methyl-D-Glucose and positron emission tomography. J Cereb Blood Flow Metab 6: 230–239
2. DiChiro G, Brooks RA, Bairamian D, Patronas NJ, Kornblith PL, Smith BH, Mansi L (1985) Diagnostic and prognostic value of positron emission tomography using (¹⁸F)-fluorodeoxyglucose in brain tumours. In: Reivich M, Alavi A (eds) Positron emission tomography. Alan R. Liss, Inc., New York, pp 291–301
3. Hawkins RA, Phelps ME, Huang SC (1986) Effects of temporal sampling, glucose metabolic rates, and disruptions of the blood-brain barrier on the FDG model with and without a vascular compartment: Studies in human brain tumours with PET. J Cereb Blood Flow Metab 6: 170–183

4. Koeppe RA, Junck L, Chen Y, Belley AT, Hutchins AT, Rodley JM, Hichwa RS (1987) Differentiation between glucose transport and phosphorylation processes in brain tumours by dynamic PET and FDG. J Cereb Blood Flow Metab 7 [Suppl 1]: S 482
5. Phelps ME, Huang SC, Hoffman EJ, Selin C, Sokoloff K, Kuhl DE (1979) Tomographic measurement of local cerebral glucose metabolic rate in humans with (^{18}-F)-L-fluoro-2-deoxy-D-glucose; validation of method. Ann Neurol 6: 371–388

Correspondence: M. Ishikawa, Department of Neurosurgery, Faculty of Medicine, Kyoto University, 54 Shogoin-Kawaramachi, Sakyo-ku, Kyoto 606, Japan.

Acta Neurochirurgica, Suppl. 51, 168–169 (1990)
© by Springer-Verlag 1990

Cerebral Blood Flow and Rheologic Alterations by Hyperosmolar Therapy in Patients with Brain Oedema

A. Hartmann, Ch. Dettmers, H. Schott, and **St. Beyenburg**

Neurologische Universitäts-Klinik, Bonn, Federal Republic of Germany

Summary

Cerebral blood flow was assessed as initial slope index by 133-Xenon inhalation in 36 patients with brain tumours subjected to osmotic dehydration. The following solutions were employed: I. 20% mannitol, II. 40% sorbitol, III. 10% glycerol. Parameters affecting blood rheologic properties as Hct, plasma viscosity, red blood cell aggregation and fluidity were simultaneously studied. CBF which was reduced in the oedematous hemisphere with brain tumour increased during infusion and thereafter by mannitol or sorbitol, respectively. The blood flow response to glycerol was more delayed, less intense, but maintained longer. Hct and plasma viscosity were significantly reduced by all osmotic agents, while red blood cell fluidity fell and aggregation rose under mannitol. It is concluded that sorbitol (40%) is superior for emergency treatment with high ICP, whereas glycerol seems to be preferable to improve cerebral blood flow in oedematous brain.

Introduction

Hyperosmolar solutions decrease tissue volume by building up an osmolar gradient between blood and tissue leading to water movement. This principle requires that the blood brain barrier is intact to prevent penetration of the osmotic substances into the tissue. Reduction of intracranial pressure (ICP) is due to the decrease of the tissue water content with resultant improvement of perfusion pressure. This may lead to an increase of cerebral blood flow (CBF), if the autoregulation is impaired. However, CBF may increase as well by the hemodiluting effect of hyperosmolar solutions, which might affect rheologic characteristics.

We studied the effect of 3 hyperosmolar agents on CBF and rheologic changes to get information on their duration.

Patients and Technique

36 measurements of CBF and rheologic parameters have been undertaken in patients with brain oedema proven by CT due to intracerebral tumours.

All patients were fully conscious and only some had mild signs of ICP increase, such as headache and nausea.

CBF was measured with the Xenon 133-inhalation technique with 32 stationary detectors arranged in a helmet like fashion with a collimation of 20 mm. CBF was calculated by the initial slope index which primarily expresses flow of the fast perfused tissue. The following rheologic parameters have been measured: Hematocrit (HCT), plasma viscosity (PV), erythrocyte aggregation (SEA), erythrocyte deformability/-fluidity (SEF), platelet aggregation (PA), and plasma osmolality (PO).

Protocol

The following substances were tested:

Group I: 125 ml of 20% Mannitol, infused over 30 min.

Group II: 250 ml of 40% Sorbitol, infused over 30 min.

Group III: 250 ml of 10% Glycerol, infused over 2 hours.

CBF and all rheologic parameters were evaluated before start of infusion, and at 30 min, 3.5, 7.5 and 11.5 hours after end of infusion.

Results

A) Rheologic parameters:

All substances reduced HCT significantly, but glycerol, which was infused more slowly than the other substances, less but longer. PV was reduced by all substances.

SEA and SEF, which are both important for the microcirculation due to their influence on erythrocyte behaviour in small caliber vessels, were not influenced by sorbitol and glycerol. Mannitol decreased SEF and increased red cell aggregation (SEA) suggesting that microcirculatory flow might be impaired. Although all

Table 1. *Effect of Hyperosmolar Solutions on Cerebral Blood Flow*

	SS	0.5	3.5	7.5	11.5
Mannitol					
il	30.6	37.2	36.9	36.3	33.9
cl	44.5	48.1	48.0	47.2	45.0
Sorbitol					
il	32.3	42.1	38.8	32.1	29.9
cl	42.0	45.8	44.1	43.8	44.1
Glycerol					
il	38.1	38.2	40.8	42.7	43.0
cl	45.2	45.5	45.9	48.6	48.9

CBF as initial slope index (sec/1). Mean of hemispheres. *SS* steady state. 0.5, 3.5, 7.5, 11.5 = time in hours after end of infusion. *il* ipsilateral (brain oedema side). *cl* contralateral.

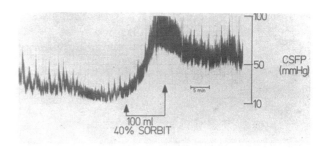

3 solutions were infused at a rather low infusion rate, the increase of PO was significant but not extensive.

B) Cerebral blood flow

Table 1 gives the actual mean ISI value for the oedematous (ipsilateral) and contralateral hemispheres.

In group I, ipsilateral CBF increased significantly shortly after start of infusion, but fell slowly back to a value slightly above steady state.

Similar observations were made with Sorbitol (group II). However, the initial CBF increase was more intense and the return to the steady state occurred faster.

With Glycerol (group III) an increase of CBF was found only after the 2nd measurement, *i.e.* at 30 min after the end of the 2-hours-infusion. The increase was not reversed within the observation time.

CBF in the contralateral hemisphere was not altered.

Discussion

A fast reduction of ICP with mannitol and sorbitol when given at the rate used in this study is well known. Fig. 1 shows an example of a steep decrease of ICP

following rapid infusion of sorbitol. Glycerol was given in this study at a much slower infusion rate due to the fact that we observed in many cases hemolysis of different degree.

In contrast to the well known fast decrease of ICP the increase of CBF was rather moderate with mannitol and sorbitol and slow with glycerol. The reason of the slow improvement of CBF, which ran parallel to the reduction of HCT, probably is that CBF is not improved by a fast decrease of ICP but predominantly by rheologic effects. This indicates preservation of autoregulation. Further, it might be suggested that an early clinical improvement after administration of hyperosmolar solutions is due to ICP reduction rather than to an increase of CBF. The only substance in this protocol which was observed to deteriorate the rheologic parameters was mannitol which impairs red cell flow behaviour. However, this did not lead to an impairment of tissue perfusion as assessed with the Xenon-133 technique.

As a clinical recommendation we suggest that for routine cases with slightly increased ICP, Glycerol should be used since it had the longest action on CBF. For emergency cases Sorbitol probably should be used, since it has the fastest action on ICP and, since under these circumstances its effect on CBF is not of primary interest.

Correspondence: Prof. Dr. med. A. Hartmann, Universitäts-Nervenklinik, D-5300 Bonn-Venusberg, Federal Republic of Germany.

Acta Neurochirurgica, Suppl. 51, 171–173 (1990)

Ischaemia and Brain Oedema I (Global Ischaemia)

Restoration of Energy Metabolism and Resolution of Oedema Following Profound Ischaemia

K. L. Allen[1], **A. L. Busza, E. Proctor, S. R. Williams, N. Van Bruggen, D. G. Gadian,** and **H. A. Crockard**[1]

[1] Institute of Neurology, London, U.K., and Royal College of Surgeons, London, U.K.

Summary

Cerebral ischaemia was produced in 2 groups of gerbils by occlusion of the common carotid arteries for 30 minutes, resulting in cerebral oedema. In group 1 cerebral oedema was measured by specific gravity microgravimetry, and in group 2 brain metabolism and blood flow were measured by ^{31}P and ^{1}H NMR spectroscopy and hydrogen clearance respectively. In group 1 the brain water content did not return to control levels by 180 minutes of reperfusion. Energy metabolism, determined by ^{31}P NMR spectroscopy returned to control by 12 minutes, intracellular pH (pH_i) by 20 minutes, and lactate, determined by ^{1}H NMR spectroscopy, by 50 minutes. There was a lag of about 10 minutes before lactate began to be cleared from the brain. We suggest that while pH_i is low, Na^+/H^+ exchange will negate the Na^+ extrusion driven by the Na^+/K^+ ATPase. When pH_i approaches normal there will be a net extrusion of Na^+, taking osmotic water with it, and presumably with passive washout of lactate. This may be the cause of the initial delay in lactate clearance.

Introduction

Oedema is a common clinical and experimental reaction to central nervous system damage, and occurs with both lethal and relatively minor stimuli. The development and resolution of ischaemic brain oedema is complex, and, to investigate some aspects of this, the present study was undertaken to examine the effects of 30 minutes bilateral carotid occlusion on brain water content in a gerbil model of cerebral ischaemia. In a separate group of animals this was correlated with cerebral blood flow (CBF), as measured by hydrogen clearance, and brain metabolism using nuclear magnetic resonance (NMR) spectroscopy.

Materials and Methods

A) Oedema Study

Twenty-eight adult male gerbils (60–70 g) were anaesthetized with halothane/oxygen and both common carotid arteries were occluded for 30 minutes using aneurysm clips. Body temperature was maintained at 36.5–37 °C using a heating lamp. Ten minutes before removal of the clips 0.1 ml 2% Evans blue was injected intraperitoneally. Animals were killed at 0, 5, 30, 60, 120, and 180 minutes of reperfusion (n = 4 for each time point). The brains were removed and dissected under kerosene into frontal, parietal and occipital cortex, and thalamus using an operating microscope. Note was made of any Evans' Blue staining. Specific gravity (SG) was determined using a linear column of organic solvents[1].

B) CBF Study

In a separate group of 6 animals CBF was measured by hydrogen clearance before, during and for a period of 1 hour after 30 minutes of ischaemia. Temperature, respiratory rate and ECG were monitored throughout the experiment. Four platinum electrodes were placed in the brain, 2 in the frontal cortex and 2 in the parietal cortex, either side of the sagittal suture. In this series of experiments, in which NMR spectroscopy was also performed, occlusion of the carotid arteries was carried out using remotely controlled nylon snares operated from outside the NMR magnet.

C) NMR Study

In the same 6 animals NMR spectroscopy was performed on a Bruker AM-360 spectrometer, using an 8.5 T vertical magnet, to measure metabolic changes in the brain during and following ischaemia. The gerbil was placed in a purpose-built probe of outer diameter 7.3 cm. Two concentric radiofrequency coils tuned to the ^{31}P and ^{1}H frequencies were placed over the skull, within which were sited the CBF electrodes. CBF and NMR measurements were taken both before and during ischaemia and at 5, 30 and 60 minutes of reperfusion. ^{31}P NMR was used to measure intracellular concentrations of phosphocreatine (PCr), adenosine triphosphate (ATP) and inorganic phosphate (P_i), and pH_i was measured from the chemical shift of the P_i resonance. ^{1}H NMR was used to measure lactate. Full experimental details for combined NMR and CBF measurments have been given elsewhere[2].

Results

A) Oedema Study

During ischaemia all four brain regions showed a fall in SG (increase in oedema) from a mean control value of 1.0500 (± 0.0002, SEM) to a mean value of 1.0480 (± 0.0005) in experimental animals. Following restoration of flow SG fell further, reaching a minimum (maximum oedema) at 30 minutes of reperfusion (1.0470, ± 0.0002). By 1 hour of reperfusion the oedema appeared to be resolving and by 3 hours of reperfusion this trend slowly continued (to a SG of 1.0482, ± 0.0003), although all areas were still significantly different from control.

No evidence of Evans' blue staining was seen until 3 hours of reperfusion, when small regions of staining were seen in the thalamus of only one of the animals.

B) CBF Study

CBF results (in ml $100 g^{-1} min^{-1}$) were as follows (± SEM); control (prior to ischaemia) 64 (± 13), during ischaemia, 12 (± 3); after 5 mins of recovery 68 (± 10), after 30 mins of recovery 39 (± 4), and after 1 hour of recovery 35 (± 5). On reperfusion flow was slightly higher than control (not statistically significant), and this was followed by post-ischaemic hypoperfusion at 30 minutes and one hour.

C) NMR Study

During ischaemia there was an increase in lactate, a decline in pH_i, ATP and PCr and a marked increase in P_i. On reperfusion there was a plateau of about 10 minutes during which the lactate concentration remained at ischaemic levels (9 mmol/kg approx.). Following this, the lactate gradually cleared over time but was not back to control levels (1.9 mmol/kg) until 50 minutes. The pH_i, on the other hand, had recovered to pre-ischaemia levels (pH_i 7.12) by 20 minutes of reperfusion, and the phosphorus metabolites had essentially recovered to control levels within 12 minutes of reperfusion.

The time course of recovery of the various parameters studied after 30 minutes ischaemia is summarized in Fig. 1.

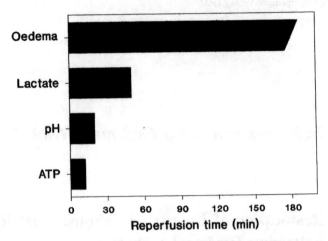

Fig. 1. Time course of recovery after 30 min ischaemia. Energy metabolism (ATP) returns earliest, within 12 minutes of reperfusion. The pH_i is back to control by 20 minutes, and lactate, after an initial plateau period of 10 minutes, returns to control by 50 minutes. Oedema is still resolving at 180 minutes of reperfusion (indicated by slashed line)

Discussion

The phosphorus spectra demonstrate that brain energy stores, as evidenced by high-energy phosphates, are depleted during ischaemia. However, on restoration of CBF following 30 minutes of ischaemia there is full recovery of phosphorus metabolites by 12 minutes of reperfusion and pH_i by 20 minutes while lactate persists for up to 50 minutes. Oedema is still resolving at 180 minutes of reperfusion.

This poses the question: why are the time scales for recovery different? As soon as blood supply is restored to the brain, water will enter the cells passively from the vascular system due to the collapse of the Na^+/K^+ gradient, and in part to an increase in perfusion pressure within the brain. At the same time ATP is recovering and so the cellular Na^+/K^+ ATPase will start to extrude Na^+, and yet a significant time is required to restore the normal intracranial volume (> 3 hours). Our results indicate that pH_i is normalized independently of lactate clearance, and therefore suggest that protons are extruded in exchange for Na^+ influx, and therefore indirectly by the Na^+/K^+ pump.

The pH_i does not return to control values until 20 minutes of reperfusion. Our hypothesis is that while there is still a significant concentration of protons in the cell, the Na^+/H^+ exchange will negate any Na^+ extrusion driven by the Na^+/K^+ ATPase. However, when the pH approaches normal there will be a net extrusion of Na^+, taking osmotic water with it — presumably with passive washout of lactate. This

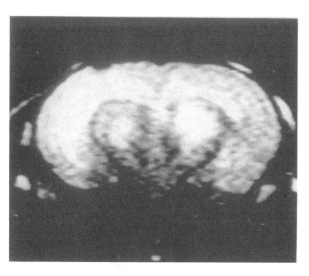

Fig. 2. Magnetic resonance image of the gerbil brain (2 mm slice thickness) acquired 5 hrs after a 60-minute period of ischaemia, using a TE 140/TR 3000 T_2-weighted spin echo sequence. An increased intensity in the thalamus (attributed to oedema) is apparent

may partially explain the lactate plateau at the beginning of reperfusion.

Evans' blue staining was only seen at 3 hrs in 1 out of 4 animals. This suggests that during the early phase of cerebral oedema most of the water is intracellular, with the possible breakdown of the blood-brain barrier occurring some time later.

To investigate further the time course of ischaemic brain oedema we are currently developing a model using magnetic resonance imaging, so that CBF and oedema can be measured in the same animal. Preliminary results using various ischaemic periods (Fig. 2) show water accumulation in the thalamus earlier than is noticeable with Evan's blue staining.

Acknowledgements

We thank Action Research for the Crippled Child, the Rank Foundation, Picker International and the Medical Research Council for their support.

References

1. Marmarou A, Poll W, Shulman K, Bhagavan H (1978) A simple gravimetric technique for measurement of cerebral oedema. J Neurosurg 49: 530–537
2. Gadian DG, Frackowiak RSJ, Crockard HA, Proctor E, Allen K, Williams SR, Ross Russell RW (1987) Acute cerebral ischaemia: concurrent changes in cerebral blood flow, energy metabolites, pH, and lactate measured with hydrogen clearance and ^{31}P and ^{1}H nuclear magnetic resonance spectroscopy. I. Methodology. J Cereb Blood Flow Metab 7: 199–206

Correspondence: K. Allen, Cerebral Oedema Research Group, Department of Clinical Neurology, Institute of Neurology, Queen Square, London WC1N 3BG, England.

Acta Neurochirurgica, Suppl. 51, 174–176 (1990)

Selective, Delayed Increase in Transfer Constants for Cerebrovascular Permeation of Blood-Borne ³H-Sucrose Following Forebrain Ischaemia in the Rat

E. Preston[1], **J. Saunders**, **N. Haas**, **M. Rydzy**, and **P. Kozlowski**

[1] Division of Biological Sciences, National Research Council of Canada, Ottawa, Canada

Summary

Experiments were conducted to explore the time course of changes in blood-brain barrier (BBB) permeability that may occur in the 2-vessel occlusion model of stroke in the rat. Anaesthetized Sprague-Dawley rats underwent 10 min of cerebral ischaemia produced by bilateral carotid occlusion plus haemorrhagic hypotension. After 6 min, or 3, 6, 18, 24, 48 h recovery and re-anaesthetization, an i.v. injection of ³H-sucrose was permitted to circulate for 30 min. Regional transfer constants (Ki) for BBB permeation of sucrose were calculated from the ratio of sucrose concentration in parenchyma relative to the time-integrated plasma concentration. In the 6-min group, all cerebral regions showed evidence of early BBB leakiness (increase in Ki above non-stroke baseline) which was maximal in forebrain cortex. This effect was diminished at subsequent time points, except in striatum and hippocampus which exhibited delayed intensification of opening, maximal in the 6 h group. Ki values had largely normalized by 24 h. Ki values were also determined 6 min, 6 h and 24 h after a 20-min stroke procedure. Early and regionally selective, delayed BBB openings were also seen, but recovery was not evident in cerebral regions at 24 h. Cortex exhibited a large increase in Ki indicating that a delayed, marked deterioration of BBB integrity had developed between the 6 h and 24 h time points. It is concluded that the combination of transfer constant measurements and the 2-vessel occlusion model could provide a sensitive means for investigating the cerebrovascular consequences and therapy of stroke.

Introduction

Recent investigations from this laboratory[5, 7] utilizing T2-weighted MRI showed image enhancement, indicative of oedematous changes, in striatum and hippocampus of rats 24 and 48 h after forebrain ischaemia induced by carotid occlusion plus hypotension. In the present work, advantage was taken of a sensitive radiotracer methodology[2] to quantitate the changes in cerebrovascular permeability that may be involved in selective regional damage and oedema occurring in this stroke model.

Materials and Methods

Male Sprague-Dawley rats, 330–400 g, were anaesthetized with pentobarbital (60 mg/kg i.v.), intubated and mechanically ventilated with 30% O_2, 70% air. The tail artery was cannulated, blood gases and pH were tested, ventilation was adjusted if required, and the carotid arteries were exposed. Stroke was induced by controlled haemorrhage to ~ 45 mm Hg for 10 min during which both carotids were clamped[6] and head and colonic temperatures maintained at 37–38 °C. Following re-perfusion and wound closures, rats were maintained normothermic and permitted to recover for 3, 6, 18, 24 or 48 h (n = 4 rats per time point). Then each rat was anaesthetized again, the right femoral vessels cannulated, and heparin was injected (280 U i.v.). Sixty microcuries of ³H-sucrose (Dupont NEN, NET 341) was injected i.v., and syringe-pump sampling of arterial blood was begun at constant rate (0.039 ml · min⁻¹) and continued for 30 min. The sampling was then stopped, and through a carotid cannula, the brain vasculature was perfused clear of blood[3] and the rat was decapitated. Routine procedures for tissue preparation and liquid scintillation counting[4] were used to obtain the following parameters: C_{paren} (tracer level in brain parenchyma, dpm · g⁻¹); $\int_0^t C_{plasma}\,dt$ (time integral of plasma tracer level, dpm · s · ml⁻¹, obtained from plasma concentration in the syringe-pump sample X 1800 s). The transfer constant for blood-to-brain diffusion of ³H-sucrose was obtained from the relationship[2]: $Ki = C_{paren}/\int_0^t C_{plasma}\,dt$.

Along with the tracer experiments 3–48 h post-stroke, there were 3 additional groups: 6 min group (tracer injected 6 min post-stroke); haemorrhage group (all stroke procedures except carotid occlusion, tracer experiment 4 h later); controls (tracer experiment alone, for baseline Ki values).

Results and Discussion

After 10 min of forebrain ischaemia, 2 temporal patterns of BBB opening were evident in certain tissue regions. In frontal cortex, apparent leakiness (increase in Ki) was most pronounced in the 6 min group (Fig. 1

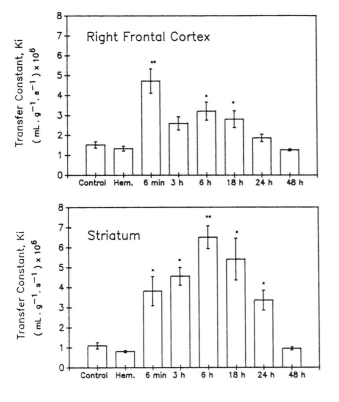

upper). Recovery (lowering of Ki) was evident at the subsequent time points so that at 24 or 48 h post-stroke there was no difference from control value (see also Table 1). The early opening of the BBB also occurred in striatum (Fig. 1 lower). However, this intensified so that at 6 h, mean Ki was significantly higher than that measured acutely post-ischaemia (6-min group). The effect was greatly diminished by 24 h post-stroke, and more so by 48 h. Hippocampus also exhibited the temporal changes in Ki similar to striatum (Table 1). Varying degrees of barrier compromise were evident in

Fig. 1. Transfer constants for ^{3}H-sucrose permeation across the rat blood-brain barrier, measured at various times after 10 min cerebral ischaemia produced by 10 min combined hypotension and carotid occlusion. n = 4 rats per group. Control group: radiotracer experiment only. Haemorrhage group (Hem.): all procedures except occlusion, radiotracer experiment 4 h later. Vertical bars indicate ± SEM. * Values different from control (p < 0.05, ANOVA and Duncan's Multiple Range Test). ** Higher than all other values (exception: striatum 18 h value)

Table 1. *Regional Transfer Constants (Ki) for ^{3}H-sucrose Permeation of the Rat Blood-Brain Barrier at Various Times After 10-min Forebrain Ischaemia*

Tissue region	Ki (ml · g^{-1} · s^{-1} × 10^6)							
	Control	6 min	3 h	6 h	18 h	24 h	48 h	Haemorrhage control
Left frontal cortex	1.64 ± 0.14	5.11 ± 0.73^{ab}	3.00 ± 0.52^{ad}	3.51 ± 0.38^{ad}	3.01 ± 0.74^d	2.17 ± 0.27^d	1.44 ± 0.11^d	1.37 ± 0.11
Right frontal cortex	1.52 ± 0.16	4.72 ± 0.61^{ab}	2.60 ± 0.33^d	3.21 ± 0.45^{ad}	2.81 ± 0.42^{ad}	1.86 ± 0.19^d	1.26 ± 0.06^d	1.33 ± 0.12
Striatum	1.10 ± 0.16	3.82 ± 0.73^a	4.56 ± 0.45^a	6.51 ± 0.57^{ac}	5.42 ± 1.05^a	3.36 ± 0.50^a	0.95 ± 0.08	0.81 ± 0.06
Hippocampus	1.15 ± 0.12	3.46 ± 0.49^a	4.51 ± 0.27^a	6.66 ± 1.16^{ac}	5.15 ± 1.21^a	2.76 ± 0.61	1.97 ± 0.27	1.18 ± 0.11
Occipital cortex	1.39 ± 0.11	3.99 ± 0.60^a	3.00 ± 0.45^{ad}	2.89 ± 0.78^{ad}	2.09 ± 0.16^d	2.18 ± 0.34^d	1.39 ± 0.11^d	1.30 ± 0.11
Diencephalon mesencephalon	1.54 ± 0.16	2.83 ± 0.52^a	2.94 ± 0.32^{ad}	4.03 ± 0.60^{ad}	3.38 ± 0.68^a	2.09 ± 0.26^d	1.29 ± 0.04^d	1.14 ± 0.07
Cerebellum	2.36 ± 0.43	2.98 ± 0.52	2.80 ± 0.37^d	2.90 ± 0.17^d	2.82 ± 0.49^d	2.26 ± 0.15^d	1.94 ± 0.14	1.66 ± 0.13
Pons-medulla	2.22 ± 0.18	2.28 ± 0.13	2.93 ± 0.33^{ad}	3.06 ± 0.08^{ad}	2.19 ± 0.24^d	2.83 ± 0.18^a	2.12 ± 0.14	2.06 ± 0.20

Values are $\bar{X}$ ± SEM, 4 rats per treatment group.
For same tissue region, between treatments (p < 0.05): ^{a} different from control group, ^{b} higher than all other values, ^{c} higher than all other values except at 18 h (ANOVA plus Duncan's Multiple Range Test).
Between tissue regions, same treatment group (p < 0.05): ^{d} lower than either striatum or hippocampus, one or both.

Table 2. *Regional Transfer Constants (Ki) for ^{3}H-sucrose Permeation of the Rat Blood-Brain Barrier at Various Times After 20-min Forebrain Ischaemia*

Tissue region	Ki (ml·g^{-1}·s^{-1} × 10^6)		
	6 min	6 h	24 h
Frontal cortex	9.38 ± 1.02	4.84 ± 0.17	28.94 ± 8.66[a]
Striatum	7.51 ± 0.68	11.44 ± 1.61	13.52 ± 1.60[b]
Hippocampus	6.91 ± 0.65	11.32 ± 0.52[b]	10.79 ± 0.52[b]
Occipital cortex	7.59 ± 0.65	5.49 ± 0.46	22.68 ± 5.79[a]
Diencephalon mesencephalon	5.13 ± 0.48	6.84 ± 0.18[a]	4.32 ± 0.48
Cerebellum	4.58 ± 0.86	4.24 ± 0.37	3.81 ± 0.44
Ponsmedulla	5.16 ± 0.99	4.42 ± 0.47	3.73 ± 0.52

Values are $\bar{X}$ ± SEM, 4 rats per treatment group. Right and left frontal cortex were combined. For same tissue region, between treatment groups ($p < 0.05$): [a] different from other 2 values, [b] different from 6-min group only (ANOVA plus Duncan's test).

other tissue regions analyzed (Table 1), but without the marked temporal changes in Ki noted above.

In additional experiments, Ki was measured 6 min, 6 h or 24 h (n = 4 rats per time point) after a more severe stroke (20 min of combined carotid occlusion and hypotension). Mean Ki values (Table 2) were higher compared to corresponding values based on 10 min stroke (Table 1). Noteworthy was the absence of recovery at 24 h in cortex, hippocampus and striatum that had been seen after 10 min strokes. Cortex showed a marked deterioration in barrier integrity between the 6 h and 24 h time points.

These data show that in the 10 min carotid occlusion model of cerebral ischaemia in the rat, two openings of the blood-brain barrier occur as seen in other stroke models (*e.g.*[1, 8]). The early BBB opening (6-min group) is attributable to reactive hyperemia and high intraluminal pressures during reperfusion of the capillaries as reported earlier[1, 8]. The delayed increase in transfer constants, evident at 6 or 18 h post-stroke, was prominent in the regions (striatum and hippocampus) that are selectively vulnerable to ischaemia. In the gerbil model[8] the second opening was associated with neuronal necrosis and oedema 3 days post-stroke. The comparatively earlier development of a second opening in the present model suggests that it may be linked to the neuronal hyperactivity and associated events

known to occur in vulnerable territories several hours post-stroke. It remains to be investigated whether these changes in the rat BBB play a significant role in the development of selective neuronal damage or oedema in this model. T 2-weighted MRI has not revealed oedematous changes 6 h after a 10-min stroke (Saunders, unpublished observations). In contrast, image enhancement in hippocampus and striatum occurred[5, 7] at a time in the present study (24 or 48 h post-stroke), when BBB opening had largely disappeared (Table 1).

In the 20-min stroke experiment cortical regions exhibited marked deterioration in BBB integrity which appeared to develop between 6 and 24 h after stroke. This phenomenon, along with the acute and regionally selective, delayed opening in the 10 min model, could prove useful for further study of the mechanisms and therapy of cerebrovascular damage associated with stroke.

Acknowledgements

The authors thank Dr. Garnette Sutherland who introduced them to the 2-vessel occlusion model, and Ariane Finsten who assisted with surgeries and data processing.

References

1. Kuroiwa T, Ting P, Martinez H, and Klatzo I (1985) The biphasic opening of the blood-brain barrier to proteins following temporary middle cerebral artery occlusion. Acta Neuropathol (Berl) 68: 122–129
2. Ohno K, Pettigrew KD, Rapoport SI (1978) Lower limits of cerebrovascular permeability to nonelectrolytes in the conscious rat. Am J Physiol 235: H 299–H 307
3. Preston E, Allen M, Haas N (1983) A modified method for measurement of radiotracer permeation across the rat blood-brain barrier: The problem of correcting brain uptake for intravascular tracer. J Neurosci Methods 9: 45–55
4. Preston E, and Haas N (1986) Defining the lower limits of blood-brain barrier permeability: factors affecting the magnitude and interpretation of permeability area products. J Neurosci Res 16: 709–719
5. Saunders JK, Smith ICP, MacTavish JC, Rydzy M, Peeling J, Sutherland E, Lesiuk H, and Sutherland GR (1989) Forebrain ischaemia studied using magnetic resonance imaging and spectroscopy. NMR in Biomedicine 2: 312–3166
6. Smith ML, Auer RN, Siesjo BK (1984) The density and distribution of ischaemic brain injury in the rat following 2–10 min of forebrain ischaemia. Acta Neuropathol (Berl) 64: 319–332
7. Sutherland G, Peeling J, Lesiuk H, Saunders J (1990) Experimental cerebral ischaemia studied using nuclear magnetic resonance imaging and spectroscopy. J Can Assoc Radiol 41: 24–31
8. Suzuki R, Yamaguchi T, Kirino T, Orzi F, Klatzo I (1983) The effects of 5 minute ischaemia in mongolian gerbils: I. Blood-brain barrier, cerebral blood flow and local cerebral glucose utilization changes. Acta Neuropathol (Berl) 60: 207–216

Correspondence: Dr. E. Preston, Division of Biological Sciences, National Research Council of Canada, Ottawa, Canada K1A OR6.

Acta Neurochirurgica, Suppl. 51, 177–179 (1990)

Electrical Impedance, rCBF, Survival and Histology in Mongolian Gerbils with Forebrain Ischaemia

W. Stummer, K. Weber, L. Schürer, A. Baethmann, and **O. Kempski***

Institute for Surgical Research, Klinikum Großhadern, Munich, Federal Republic of Germany

Summary

Survival, quantitative morphology of the hippocampus, cerebral tissue impedance and regional cerebral blood flow (rCBF) were studied in the Mongolian gerbil after 15 minutes of bilateral common carotid occlusion. A subgroup of animals was placed in cages with free access to running-wheels for two weeks preceeding ischaemia to measure voluntary locomotor activity. Survival was enhanced in the running-wheel subgroup, with 90% of the animals still alive after 14 days as compared to 48% of the non-running group. Neuronal loss was found in all animals in the hippocampus (CA 1, CA 2, CA 3 and CA 4), and was most pronounced in the CA 1 sector. In the running-wheel group, however, neuronal loss was significantly lower in sectors CA 2, CA 3 and CA 4. The increases of cerebral impedance, which indicate ischaemic cell swelling, reached 190% in both groups during ischaemia. During postischaemic recirculation, however, impedance normalized more rapidly in the running-wheel group, indicating earlier resolution of ischaemic cell swelling. In wheel-running gerbils, postischaemic hyperperfusion evolved earlier and was more pronounced as compared to non-runners. No differences in systemic blood pressure were observed during cerebral ischaemia or thereafter.

Introduction

A frequently used model for the investigation of cerebral ischaemia employs the Mongolian gerbil as experimental animal due to the uniqueness of its cerebral vascular anatomy. Ninety percent of gerbils do not have an adequate anastomosis between the carotid and basilar arteries[1]. Although the model is often employed, the differences in mortality rates with reversible ischaemia of 15 minutes duration observed by different laboratories are striking. This may be due to the fact that the determinants of outcome in forebrain ischaemia in Mongolian gerbils are not well established. An earlier investigation[5] had revealed a marked enhance-ment of survival in gerbils with access to running-wheels for 2 weeks prior to a 15 min episode of forebrain ischaemia. The current study was designed to evaluate the effect of wheel-running on physiological parameters – cortical electrical impedance and rCBF – during ischaemia and reperfusion to better understand the marked differences in outcome.

Materials and Methods

Male Mongolian gerbils were subjected to 15 min of bilateral carotid occlusion in halothane anaesthesia. A feedback-controlled heating pad ensured normothermic conditions during ischaemia and the wake-up phase. Survival and histology were studied after ischaemia with an observation period of up to 14 days. Moribund animals were perfusion-fixed earlier. A subgroup of animals was placed in cages with free access to running-wheels to measure voluntary locomotor activity prior to and after ischaemia. For quantitative histology of the hippocampus a standard slice taken 1.6 mm posterior to the bregma was chosen. In all four regions of the hippocampus (CA 1, CA 2, CA 3 and CA 4), the number of surviving neurons was assessed by projecting the microscopic image on a video screen and superimposing a standardized computer-generated counting frame.

Cerebral impedance was measured during ischaemia and 3 hrs reperfusion between two electrodes inserted into the brain, utilizing an alternating current of 10 mV at a frequency of 1 kHz. Increases in cerebral impedance indicate a reduction of the extracellular space due to fluid and electrolyte shifts into the intracellular compartment which is characteristic for ischaemia[2, 4]. The rCBF was assessed by the hydrogen clearance method. For that purpose, four 75 μm platinum wire electrodes were implanted bilaterally in the frontal and parietal cortex. The electrodes penetrated the cortical layer only without damaging deeper tissue layers. The H_2-washout curves were registered in intervals of 20 minutes over a period of 10 min for each measurement.

Results

The current study confirm earlier observations of a highly protective effect of preischaemic access to running-wheels in gerbils. Non-running gerbils had a 14-

* Present address: Institute for Neurosurgical Pathophysiology, Johannes Gutenberg-University Mainz, Langenbeckstrasse 1, D-6500 Mainz, Federal Republic of Germany.

day survival of 48% after 15 min forebrain ischaemia, whereas 90% of the animals with wheel-running prior to ischaemia were still alive at the end of two weeks ($p < 0.01$). Protection was also reflected by enhanced nerve cell counts in the hippocampus. With 15 min of ischaemia, a marked loss of neurons was observed in all four hippocampal sectors. This was particularly pronounced in the CA 1 sector in both groups. There was, however, an impressive difference between running-wheel gerbils and the animals kept in conventional cages. This was particularly noticable in the CA 3 sector of the hippocampus (runners: 683 ± 175 vs. non-runners 123 ± 36 neurons/mm^2; sham-operated: 1322 ± 43), but also in the CA 2 (286 ± 36, 141 ± 24, 1927 ± 88) and CA 4 (286 ± 99, 57 ± 13, 1629 ± 67) sectors. In the CA 1 sector no significant protection was found (120 ± 15, 90 ± 20, 3115 ± 9 neurons/mm^2).

The protective effect of preischaemic running was also reflected by the acute differences of the cerebral tissue impedance and regional CBF during and after cerebral ischaemia between runners and non-runners. Immediately after induction of ischaemia, impedance increased sharply in both groups reaching a maximum of 190% of the control level. Subsequently, the brain tissue impedance normalized very slowly in non-runners during postischaemic recirculation. After one hour impedance was still markedly elevated to 170–180%. At this time, impedance in the running-wheel group, on the other hand, had already declined to 120%.

The rCBF studies demonstrated in all animals a typical postischaemic flow pattern as already described by other authors[3], *i.e.* an impaired cerebral blood flow immediately after ischaemia followed by a period of postischaemic hyperperfusion and subsequent hypoperfusion. There were, however, pronounced differences regarding the development and degree of postischaemic hyperperfusion (Table 1). Under control conditions, both groups had a cerebral blood flow of 45 ml/100 g × min. Immediately after ischaemia, cerebral blood flow was initially depressed in both groups falling to 31 ml/100 g × min in the wheel-running animals while to 15 ml/100 g × min in animals without wheel-running (n.s.). The maximum of postischaemic blood flow in non-runners was reached at 60 minutes after ischaemia. On the other hand, animals with preischaemic wheel-running reached the maximal level of reperfusion already 27 minutes after ischaemia ($p < 0.05$). Further, the level of postischaemic hyperperfusion was far more pronounced in this group (71 ml/100 g × min in wheel-running animals vs. 47 ml/

Table 1. *Postischaemic Blood Flow in Cerebral Cortex of Gerbils Maintained in Conventional Cages (Non-Runners), or with Free Access to Running-Wheels (Runners) After 15 min of Bilateral Occlusion of the Carotid Arteries (Forebrain Ischaemia). rCBF was measured by H$_2$-clearance*

	Non-Runners	Runners
Early reflow (ml/100 g × min)	15.9 ± 1.9	31.2 ± 6.7
Hyperfusion (ml/100 g × min)	47.0 ± 5.0	70.8 ± 7.4
Time (min) after ischaemia	62.2 ± 12.3	27.4 ± 7.0
Hypoperfusion (ml/100 g × min)	30.1 ± 2.4	29.0 ± 2.4
Time (min) after ischaemia	97.3 ± 9.8	95.2 ± 11.0

100 g × min in non-running animals; $p < 0.05$). These marked differences not withstanding, the occurrence and the degree of postischaemic cerebral hypoperfusion were equal in both groups (Table 1). A different course of the arterial blood pressure can be ruled out to explain the differences in cerebral blood flow between both groups. Continuous blood pressure measurements immediately prior to, during and after ischaemia revealed an almost identical blood pressure course in both groups of animals with and without wheel-running prior to ischaemia.

Discussion

The current findings confirm that physical exercise prior to cerebral ischaemia is highly efficient to protect postischaemic survival of the animals. In addition, wheel-running was associated with a considerable reduction of delayed neuronal death in the hippocampus studied at four days postischaemia. The measurements of the electrical brain tissue impedance during forebrain ischaemia revealed a typical pattern in both experimental groups indicative of ischaemic cell swelling resulting from a shift of fluid and electrolytes from the extra- into the intracellular compartment[1, 2]. However, in animals with preischaemic wheel-running the tissue impedance normalized rapidly during postischaemic recirculation as compared to the non-running animals maintained in conventional cages. The findings suggest an earlier recovery of the normal extra- to intracellular electrolyte gradients and fluid distribution on the basis of an earlier normalization of the energy metabolism of the brain. The early and pronounced postischaemic hyperperfusion of the brain in animals with preischaemic wheel-running might be considered as a particularly critical factor providing for a better recovery and outcome in this group. On the other hand, onset and degree of postischaemic hypoperfusion were not found to be influenced by preischaemic wheel-running and, therefore, do not seem to be of significance under the current conditions. Further, the differences in postischaemic cerebral blood flow, recovery of brain tissue impedance, and survival of hippocampal nerve cells cannot be explained by an effect of physical exercise on the arterial blood pressure. As mentioned, the arterial blood pressure was largely identical in both groups of animals.

The markedly higher level of postischaemic hyperperfusion of the brain could well explain the good outcome of animals with preischaemic running, for example by a more effective wash-out of platelets and blood cells which were aggregating during the circulatory arrest of the brain. In conclusion, the current findings demonstrate that the delayed neuronal death after an episode of cerebral ischaemia is not inevitable. The present model is useful to study in the future not only mechanisms of nerve cell damage in the brain from ischaemia but also measures of protection.

References

1. Levy DE, Brierley JB (1974) Communication between Vertebro-Basilar and Carotid Arterial Circulation in the Gerbil. Expl Neurol 45: 503–508
2. Matsuoka Y, Hossmann K-A (1982) Cortical Impedance and Extracellular Volume Changes Following Middle Cerebral Artery Occlusion in Cats. J Cereb Blood Flow Metab 2: 466–474
3. Todd NV, Picozzi P, Crockard HA (1986) Quantitative Measurements of Cerebral Blood Flow and Blood Volume after Cerebral Ischaemia. J Cereb Blood Flow Metab 6: 338–341
4. Van Harreveld A, Ochs S (1956) Cerebral Impedance Changes after Circulatory Arrest. Am J Physiol 197: 180–192
5. Weber K, Tranmer B, Schürer L, Baethmann A, Kempski O (1989) Physical activity determines survival of mongolian gerbils from forebrain ischaemia. In: Hartmann A, Kuschinsky W (eds) Cerebral ischaemia and calcium. Springer, Berlin Heidelberg New York, pp 79–82

Correspondence: Dr. W. Stummer, Institute for Surgical Research, Klinikum Großhadern, Marchioninistraße 15, D-8000 Munich 70, Federal Republic of Germany.

Acta Neurochirurgica, Suppl. 51, 180–182 (1990)
© by Springer-Verlag 1990

Excitatory Amino Acid Receptors, Oxido-reductive Processes and Brain Oedema Following Transient Ischaemia in Gerbils

B. B. Mršulja[1], D. Stanimirović[2], D. V. Mićić[1], and M. Spatz[3]

[1] Institute of Biochemistry, Faculty of Medicine, Belgrade, Yugoslavia, [2] Institute for Medical Research, Military Medical Academy, Belgrade, Yugoslavia, [3] LNNS, NINCDS, National Institutes of Health, Bethesda, Maryland, U.S.A.

Summary

A key mechanism of brain injury after cerebral ischaemia is supposed to be the iron-dependent formation of highly reactive oxygen free radicals initiated by the intracellular accumulation of calcium and promoted by the excess release of glutamate. Oxido-reductive processes (formation of superoxide radicals and lipid peroxidation) are mediated through NMDA-receptors, while non-NMDA receptors, associated with (or being a part of) Na,K-ATPase, are responsible for postischaemic brain swelling. The hypothesis was put forward for consideration that release of glutamate (and other related endogenous excitatory amino acids) due to depolarization in the early minutes of ischaemia and (non)-NMDA antagonists may have roles in the development and prevention of metabolic brain impairment and cytotoxic oedema, respectively, in the ischaemic state.

Introduction

Experimental evidence indicates that excitatory neurotransmission plays a significant role in the development of ischaemic/hypoxic neuronal cell death. However, exact mechanism(s) of the excitatory amino acid (EAA)-induced neurotoxicity during ischaemia is yet to be defined. Massive calcium influx due to activation of the NMDA-receptors is supposed to mediate a prolonged cell degeneration by a number of reactions including free radical initiation and membrane lipid peroxidation[3].

It was intriguing to investigate whether or not excitatory amino acid receptors are involved in the development of ischaemic brain swelling and free radical formation. The obtained data suggest that activation of NMDA receptors leads to an oxidative stress, while blockade of non-NMDA receptors (and/or preservation of Na,K-ATPase activity) is essential for the brain to recover from swelling induced by ischaemia.

Materials and Methods

Mongolian gerbils of both sexes were subjected to 15 min bilateral ischaemia and up to 96 hours of reflow. 2-amino-5-phosphonovaleric acid (APV, 4 mg/kg b.w., i.p.) and propentofylline (HWA 285, 25 mg/kg b.w., i.p.) were administrated to the gerbils at the end of ischaemia, alone or in combination, and animals were sacrificed by decapitation 1 hour thereafter; heads were immediately frozen in the liquid nitrogen. Separate groups of gerbils were taken to estimate brain swelling[5]. Effects of postischaemia and drug application were also tested on the production of superoxide anion ($\cdot O_2^-$), superoxide dismutase (SOD) activity and lipid peroxidation (LP) in the crude mitochondrial fraction[7] of the striatum. Detection of the superoxide radicals was based upon the reduction of nitro-blue tetrazolium (NBT)[2], superoxide dismutase activity was measured as an inhibition of spontaneous epinephrine autoxidation[12] and lipid peroxidation as the content of thiobarbituric acid (TBA)-reactive material formed during the *in vitro* stimulation by Fe^{2+}-salts and ascorbic acid[13]. Protein content was determined according to Lowry *et al.*[8]. Statistical evaluation was performed using the analysis of variance (ANOVA) followed by the least squares difference testing between the means; $p < 0.05$ was considered as significant.

Results and Discussion

During ischaemia alone, only slight increase in the superoxide anion production in the gerbil striatum was observed (Table 1). However, pronounced changes were seen in the first minutes of reperfusion: marked increase in the $\cdot O_2^-$ production followed by the increase in the amount of TBA-reactive material which is indicative for enhanced lipid peroxidation (Table 1). One hour after the circulation had been reestablished, SOD activity was lowered at 73% of control values (Table 1). During the remaining period of reperfusion (until the fourth day), $\cdot O_2^-$ production was permanently enhanced, SOD activity restored and lipid per-

Table 1. *Changes of the Superoxide Anion Production ($\cdot O_2^-$), Superoxide Dismutase (SOD) Activity and Lipid Peroxidation (LP) in the Gerbil During Ischaemia, 60 min of Postischaemia, and the Effects of Drugs*

Time	$\cdot O_2^{-\,a}$	SOD[b]	LP[c]
A. Control	88.7 ± 4.9 (5)	538.1 ± 40.6 (6)	92.2 ± 7.0 (6)
A'. Isch. (15')	105.7 ± 7.5* (6)	489.1 ± 43.3 (6)	96.2 ± 7.7 (6)
B. Reperfusion			
1 min	282.2 ± 22.8* (6)	673.3 ± 52.2* (6)	211.8 ± 13.1* (6)
3 min	223.5 ± 15.5* (6)	845.2 ± 68.1* (6)	142.3 ± 4.6* (6)
15 min	199.2 ± 16.9* (5)	693.3 ± 57.7* (6)	171.1 ± 6.8* (6)
1 h	145.1 ± 5.9* (6)	392.9 ± 28.3* (6)	179.7 ± 7.9* (6)
C. Drugs after ischamia; reperfusion 1 h			
HWA 285	120.6 ± 10.7* (6)	402.8 ± 28.3* (6)	144.2 ± 6.4* (6)
APV	115.3 ± 6.3* (6)	493.6 ± 49.4 (6)	157.7 ± 7.6* (6)
HWA + APV	105.6 ± 11.2* (6)	494.3 ± 27.9 (6)	108.6 ± 8.5 (6)

Values are Means ± SEM (n); [a] nmol NBF/min/mg prot; [b] units; [c] nmol MDA/min/mg prot. *$p < 0.05$ (ANOVA) in comparison to A.

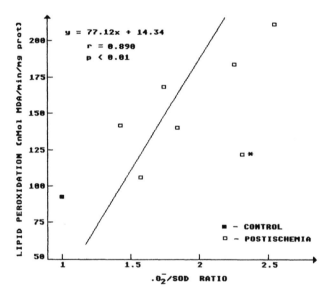

Fig. 1. Linear correlation between the $\cdot O_2^-$/SOD ratio and lipid peroxidation in the crude mitochondrial fraction of the gerbil striatum during the four days of reperfusion. * The value which is beyond the confidence level of 0.95

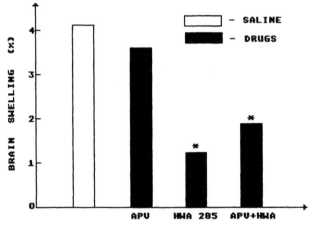

Fig. 2. Effects of APV, HWA 285 and APV + HWA 285 on the brain swelling one hour after 15-min of bilateral ischaemia in the gerbil. Brain water content in the control group was 78.29 ± 0.04; after the application of: saline – 79.14 ± 0.10; APV – 79.05 ± 0.14; HWA 285 – 78.57 ± 0.12; APV + HWA 285 – 78.67 ± 0.09. * Statistically significant difference as compared to saline-treated group (Student's t-test; $p < 0.05$)

oxidation decreased to control value (data not shown). Hence, at least in the first hour of reperfusion a disbalance exists between the enhanced superoxide anion production and the ability of SOD to quench it adequately. Since the intensity of oxidative stress to the tissue depends on its relative capacity for the protective response, it can be adequately expressed by the $\cdot O_2^-$/SOD ratio. Indeed, considering the $\cdot O_2^-$/SOD ratio, a relative incapacity for the $\cdot O_2^-$ removal was observed not only at the time of decreased SOD activity, but during the entire period of reperfusion (data not shown). $\cdot O_2^-$/SOD ratio, also, correlates well with the amount of TBA-reactive material formed during the reperfusion (r = 0.89) (Fig. 1).

Production of oxygen free radicals starts with the reperfusion in the electron transport chain. Oxygen and lipid free radicals further originate from a different metabolic reaction promoted by ischaemia-increase in cytosolic calcium concentration, activation of arachidonic acid cascade, increase in the monoamine turnover rate; all these reactions are stimulated during the *in vivo* activation of the NMDA-receptors[3]. Selective NMDA antagonist APV given immediately after 15-min bilateral ischaemia reduced $\cdot O_2^-$ production and

$\cdot O_2^-$/SOD ratio in the gerbil striatum one hour after ischaemia (Table 1). Nevertheless, it was without (significant) effect on the brain swelling (Fig. 2); obviously, receptors other than NMDA are responsible for the main sodium and water influx and cell swelling. When the HWA 285 was applied, the effects on the $\cdot O_2^-$/SOD ratio was less expressed, lipid peroxidation was diminished (Table 1) and brain swelling almost completely attenuated (Fig. 2). Finally, when the both drugs were applied together, $\cdot O_2^-$/SOD ratio and lipid peroxidation were close to control values (Table 1, Fig. 2).

HWA 285 is known to prevent postischaemic inhibition of Na,K-ATPase as well as brain swelling[9]. On the other hand non-NMDA receptors are linked to channels that allow sodium and potassium to pass through the membrane, causing acute cellular swelling; this would suggest that HWA 285 is a non-NMDA receptor antagonist or that Na,K-ATPase is linked with (or is a part of) non-NMDA receptor. Intracellular accumulation of sodium also causes depolarization and thus voltage-gated calcium channels may be activated secondarily leading to further intracellular calcium increase; new conditions develop for free radical formation.

Activation of the EAA-receptors (NMDA and non-NMDA) can be reduced either by specific alosteric inhibition using certain drugs, or indirectly, *e.g.* by the inhibition of EAA release, depolarization, Ca^{2+} entry. All these properties are intensive and may respond to endogenous adenosine or exogenous adenosine analogues[1]. HWA 285 acts as adenosine uptake inhibitor[6], decreases monoamine- and EAA-neurotransmitter release[11] and inhibits intracellular accumulation of Ca^{2+} during ischaemia[4]. Inhibition of the EAA release in addition to "indirect NMDA-blockade" also prevents the activation of non-NMDA receptors, and consequently, development of brain oedema. Since the HWA 285 is shown to prevent postischaemic reduction of Na,K-ATPase activity and increase in sodium concentration[9], direct action of HWA 285 on the non-NMDA receptors is possible. Thus, direct blockade of the NMDA-receptors by the APV and simultaneous inhibition of the non-NMDA receptors by means of HWA 285 effectively protect against lipid peroxidation and, at the same time, prevent development of postischaemic brain oedema.

References

1. Arvin B, Neville FL, Pan J, Roberts JP (1989) 2-chloroadenosine attenuates kainic acid-induced toxicity within the rat striatum: relationship to release of glutamate and Ca^{2+} influx. Br J Pharmacol 98: 225–235
2. Auclair C, Voisin E (1985) Handbook of methods for oxygen radical research. CRC Press, Florida, pp 123–132
3. Benveniste H, Jorgensen BM, Diemer HN, Hansen JA (1988) Calcium accumulation by glutamate receptor activation is involved in hippocampal cell damage after ischaemia. Acta Neurol Scand 78: 529–536
4. DeLeo J, Toth L, Schubert P, Rudolphi K, Kreutzberg WG (1987) Ischaemia-induced neuronal cell death, calcium accumulation, and glial response in the hippocampus of the mongolian gerbil and protection by propentofylline. J Cereb Blood Flow Metab 7: 745–751
5. Elliot KAC, Jasper H (1949) Measurement of experimentally induced brain swelling and shrinkage. Am J Pathol 157: 122–129
6. Fredholm BB, Linstrom K (1986) The xanthine derivative 1-(5'-oxohexyl)-3-methyl-7-propyl xanthine (HWA 285) enhances the actions of adenosine. Acta Pharmacol Toxicol 58: 187–192
7. Gurd JW, Jones LR, Mahler HR, Moore WJ (1974) Isolation and partial characterization of rat brain synaptic membrane. J Neurochem 22: 281–290
8. Lowry OH, Rosebrough NJ, Farr AL, Randall RJ (1951) Protein measurements with the folin phenol reagent. J Biol Chem 193: 952–958
9. Mršulja BB, Mićić VD, Stefanovich V (1985) Propentofylline and postischaemic brain oedema: relation to Na^+-K^+-ATPase activity. Drug Develop Res 6: 339–344
10. Murphy T, Parikh A, Schnaar R, Coyle J (1989) Arachidonic acid metabolism in glutamate neurotoxicity. In: Boland B, Culliman J, Stiefel R (eds) Ann New York Acad Sci 559: 474–477
11. Shimoyama M, Kito S, Inokawa M (1988) Effects of 3-methyl-1-(5-oxohexyl)-7-propylxanthine on the *in vivo* release of monoamines studied by intracerebral dialysis. Drug Res 38: 243–247
12. Sun M, Zigman S (1978) An improved spectrophotometric assay for superoxide dismutase based on epinephrine autoxidation. Analytical Biochem 90: 81–89
13. Villacara A, Kumami K, Yamamoto T, Mršulja BB, Spatz M (1989) Ischaemic modification of cerebrocortical membranes: 5-hydroxy-tryptamine receptors, fluidity and inducible *in vitro* lipid peroxidation. J Neurochem 53: 595–601

Correspondence: Prof. B. B. Mršulja, M.D., Institute of Biochemistry, Faculty of Medicine, Pasterova 2, YU-11000 Belgrade, Yugoslavia.

Acta Neurochirurgica, Suppl. 51, 183–185 (1990)
© by Springer-Verlag 1990

Cellular Swelling During Cerebral Ischaemia Demonstrated by Microdialysis *in vivo*: Preliminary Data Indicating the Role of Excitatory Amino Acids

Y. Katayama[1, 2], **D. P. Becker**[1], **T. Tamura**[1], and **T. Tsubokawa**[2]

[1] Division of Neurosurgery, University of California at Los Angeles, California, U.S.A., and [2] Department of Neurological Surgery, Nihon University School of Medicine, Tokyo, Japan

Summary

When rapid cellular swelling occurs, water moves from the extracellular space (ECS) into the cells and the concentration of ECS markers which do not move into the cells increases. Cellular swelling during cerebral ischaemia has therefore been demonstrated *in vivo* as an increase in ECS markers which is measurable with intracerebral electrodes. We attempted to detect the cellular swelling by brain microdialysis employing a similar principle. Dialysis probes were placed in the hippocampus, and perfused for 20 min with ^{14}C-sucrose as an ECS marker. The probes were subsequently perfused without ^{14}C-sucrose and the dialysate concentration of ^{14}C-sucrose was determined at 1-min intervals. The dialysate concentrations of ^{14}C-sucrose suddenly became elevated 1–3 min after the onset of cerebral ischaemia, indicating the occurrence of cellular swelling. The present technique is useful because it enables the mechanism of cellular swelling to be analyzed by observing the effects of pharmacological agents administered through a dialysis probe. Preliminary data indicating the role of excitatory amino acids in producing cellular swelling during cerebral ischaemia are presented as an example.

Introduction

Cellular swelling and the resultant shrinkage of the extracellular space (ECS) during cerebral ischaemia have been demonstrated *in vitro* as a change in space available for ECS marker, a substance which does not enter the cells or capillaries, such as sucrose or inulin. The time course of the cellular swelling during cerebral ischaemia has also been monitored *in vivo* from the changes in extracellular concentration of various ECS markers[2]. Thus, ECS markers are introduced into the ECS by a superfusion technique and their extracellular concentration measured with electrodes sensitive to such markers[2, 5]. When rapid cellular swelling occurs, water moves from the ECS into the cells and causes a decrease in water volume in the ECS. In contrast, the ECS marker does not move into the cells and is left behind. The cellular swelling and concomitant ECS shrinkage can thus be detected as an increase in extracellular concentration of ECS markers[2, 5].

We attempted to investigate the cellular swelling during cerebral ischaemia by brain microdialysis employing a similar principle. The method used in the present study involved the perfusion of focal brain regions with ^{14}C-sucrose, as an ECS marker, and subsequent measurement of the extracellular concentration of ^{14}C-sucrose, both performed by microdialysis. This technique is useful because it enables the mechanism of cellular swelling to be analyzed by observing the effects of pharmacological agents administered through the dialysis probe[3, 4]. Preliminary data indicating the role of excitatory amino acids (EAAs) in producing cellular swelling during cerebral ischaemia are presented as such an example.

Materials and Methods

Young female Sprague-Dawley rats weighing 180–240 g were used. They were anaesthetized with a mixture of O_2 (33%), N_2O (66%) and enflurane (1.5–2.0 ml/min), and placed in a stereotaxic frame. The rectal temperature was maintained at 37.0–38.0 °C with a heating pad. A pair of dialysis probes (O.D., 300 μ; effective length, 2 or 3 mm; cut off, 20 000 MW) were positioned vertically in the hippocampus (CA 1/dentate area) bilaterally. The probes were initially perfused with temperature-maintained, Ringer solution containing ^{14}C-sucrose (10 mM) at a rate of 5.0 μl/min. Sodium kynurenate (KYN, 0.1–10 mM) was administered through one of the two probes, chosen at random, throughout the experiment (test probe). The dialysis with the other probe served as the control (control probe).

At 20 min after the start of ^{14}C-sucrose perfusion, the perfusate for the dialysis was switched to temperature-maintained, K^+-free Ringer solution and measurements of the dialysate concentrations of K^+ and ^{14}C-sucrose ([K^+]$_d$ and [^{14}C-sucrose]$_d$) were initiated. The osmolarity of each perfusate was kept equivalent to that of normal Ringer solution by changing the concentration of sodium chloride. The dialysate fractions were collected at 1-min intervals for measurement of their K^+ concentration and radioactivity. The length of the outlet tube was shortened in order to make a dead space of 5.0 µl, so that the second dialysate fraction collected after the onset of ischaemia corresponded to the initial post-ischaemia fraction.

Global brain ischaemia was induced by decapitation. The radioactivity of each fraction was expressed in dpm and also as a percentage of the radioactivity of the original perfusate or the dialysate at the time of ischaemia induction. After completion of the experiment, autoradiograms were obtained on Kodak X-ray films.

Results

Following termination of the ^{14}C-sucrose perfusion with the dialysis probe, the value of [^{14}C-sucrose]$_d$ decreased rapidly during an initial period of a few minutes and decreased slowly thereafter. This decay curve was probably composed of two different diffusion equations[6]: diffusion within the ECS, and diffusion into the dialysis. Based on this curve, 13–18 min after the termination of the ^{14}C-sucrose perfusion was selected as the time to test the effect of ischaemia, since the slope of the curve became less steep. When typical decay curves were obtained, autoradiograms demonstrated

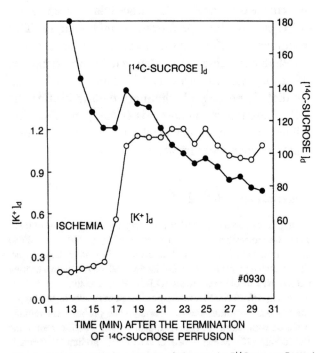

Fig. 1. A representative example of changes in [^{14}C-sucrose]$_d$ and [K^+]$_d$ during cerebral ischaemia. Dead space for dialysis system, 5 µl (1 min)

Table 1. *Effects of Kynurenic Acid* on the Increase in* [K^+]^d *and* [^{14}C-sucrose]$_d$ *During Cerebral Ischaemia*

	Delay in onset latency (min) [control probe − test probe]
[K^+]$_d$	1.45 ± 0.45**
[^{14}C-sucrose]^d	1.50 ± 0.50**

* Perfusate concentration, 10 mM; ** m $\pm$ SEM (n = 5), P < 0.02.

that a restricted area of the brain parenchyma approximately 1.5 mm distant from the probe was perfused with ^{14}C-sucrose, and ^{14}C-sucrose did not directly drain into the CSF system.

As reported previously[3, 4], [K^+]$_d$ underwent a sudden increase at 1–3 min after the onset of ischaemia (Fig. 1). An increase in [^{14}C-sucrose]$_d$ invariably occurred concomitantly with the onset of the rapid increase in [K^+]$_d$ in all animals (Fig. 1). In order to detect this change, however, it was absolutely necessary to position the probe atraumatically. The maximum level of [^{14}C-sucrose]$_d$ was 144.3 ± 10.1% (m $\pm$ SEM, n = 14, p < 0.01) of the fraction immediately before the sudden increase in [K^+]$_d$.

Administration of KYN, a broad spectrum antagonist of EAAs[1], through the dialysis probe delayed the onset of both the rapid increase in [K^+]$_d$[2, 3] and the increase in [^{14}C-sucrose]$_d$, as compared with the increases observed with the other probe perfused without KYN (Table 1).

Discussion

During cerebral ischaemia, abrupt ionic fluxes across the neuronal plasma membrane are observed with a latent period of a minute or more[2]. These ionic fluxes result in a dramatic increase in K^+, and a precipitous fall of Na^+, Ca^{2+} and Cl^- in the ECS. As a consequence of these ionic events, water moves from the ECS into the cells and cellular swelling occurs since the osmotic pressure of intracellular impermeable anions may no longer be counterbalanced by the extracellular Na^+ and Cl^-[2, 6, 7]. When water moves from the ECS into the cells, the concentration of ECS markers increases since these substances do not enter the cells[2, 5]. Sucrose is one of the ECS markers frequently used in *in vitro* experiments. The present study demonstrated that [^{14}C-sucrose]$_d$ increased concomitantly with the increase in [K^+]$_d$ employed as an indicator of the ionic flux, so that ^{14}C-sucrose behaved as an

ECS marker in these *in vivo* experiments and $[^{14}C$-sucrose$]_d$ sensitively reflected the water movement which was a direct consequence of the ionic events.

It has been reported that the concentration of ECS markers as measured with intracerebral electrodes increases to approximately 2-fold during ischaemia, and the volume of ECS is therefore estimated to decrease by approximately 50%[2]. The less marked increase observed in the present study, *i.e.*, 1.4-fold, represents a value discounted by a natural decrease in $[^{14}C$-sucrose$]_d$ due to the continuous diffusion of ^{14}C-sucrose. In addition, there may be a decrease in the effective surface area for dialysis due to the ECS shrinkage occurring during cerebral ischaemia. Thus, the absolute change in ECS volume is difficult to estimate by microdialysis. Nevertheless, this technique does provide a sensitive measure of the onset of cellular swelling.

One advantage of using microdialysis is that this technique offers a means to administer pharmacological agents in order to investigate the mechanisms underlying cellular swelling. We have demonstrated previously that the initial component of the abrupt ionic flux during cerebral ischaemia can be inhibited by the administration of EAA antagonists through the dialysis probe[3,4]. EAAs have been demonstrated to cause large ionic fluxes across the neuronal cell membrane and therefore generate cellular swelling *in vitro*. The observations made in the present study suggest that EAAs are also involved in the mechanism of early cellular swelling during cerebral ischaemia *in vivo*.

References

1. Ganong AH, Cotman CW (1986) Kynurenic acid and quinolinic acid act at N-methyl-D-aspartate receptors in the rat hippocampus. J Pharmacol Exp Therap 236: 293–299
2. Hansen AJ, Olsen CE (1980) Brain extracellular space during spreading depression and ischaemia. Acta Physiol Scand 108: 355–365
3. Katayama Y, Becker DP, Tamura T, Martin N, Tsubokawa T (1989) Inhibition of massive ionic fluxes during cerebral ischaemia (Terminal depolarization) with excitatory amino acid antagonist as demonstrated by microdialysis. J Cereb Blood Flow Metab 9 [Suppl]: 57
4. Katayama Y, Tamura T, Kawamata T, Becker DP (1989) Concomitance and dependence of massive potassium flux to early glutamate release during cerebral ischaemia *in vivo*. Neurosci Abstr 15: 358
5. Phillips JM, Nicholson C (1979) Anion permeability in spreading depression investigated with ion-sensitive microelectrodes. Brain Res 173: 567–571
6. Nicholson C, Phillips JM, Gardner-Medwin AR (1979) Diffusion from a iontophoretic point source in the rat brain: Role of tortuosity and volume fraction. Brain Res 169: 580–584
7. Van Harreveld A (1979) A mechanism for fluid shifts specific for the central nervous system. In: Wycis HT (ed) Current research in neuroscience, topical probl psychiat neurol. Karger, Basel, 10: 62–70

Correspondence: Y. Katayama, M.D., Department of Neurological Surgery, Nihon University School of Medicine, Tokyo 173, Japan, or Division of Neurosurgery, University of California at Los Angeles, LA 90024, U.S.A.

Acta Neurochirurgica, Suppl. 51, 186–188 (1990)
© by Springer-Verlag 1990

Role of Neuroexcitation in Development of Blood-Brain Barrier and Oedematous Changes Following Cerebral Ischaemia and Traumatic Brain Injury

N. Saito, C. Chang, K. Kawai, F. Joó, T. S. Nowak Jr., G. Mies, J. Ikeda, G. Nagashima, C. Ruetzler, J. Lohr, M. Spatz, and **I. Klatzo**

Laboratory of Neuropathology and Neuroanatomical Sciences, National Institute of Neurological Disorders and Stroke, National Institutes of Health, Bethesda, Maryland, U.S.A.

Summary

Potential involvement of neuroexcitatory mechanisms was studied in: 1) repetitive forebrain ischaemia in gerbils, 2) global cerebral ischaemia in rats and 3) cryogenic injury to the cerebral cortex in rats and gerbils. Uptake of ^{45}Ca was used as a marker of injury, whereas ultrastructural localization of calcium was assessed with an oxalate-pyroantimonate method. The blood-brain barrier was evaluated with immunostaining for serum albumin. Changes in extracellular glutamate were estimated by microdialysis and an enzymatic cycling assay. Changes in water content were assessed by specific gravity measurements.

Repetitive ischaemia of 3×5 min carotid occlusions produced a cumulative effect with regard to development of oedema and neuronal injury. This was associated with several-fold increments in glutamate release after repeated insults, whereas there was no apparent correlation with energy metabolism disturbances. Other studies revealed in all models a development of secondary foci distant to the primary impact of ischaemia or cold lesions, which were characterized by calcium accumulation in swollen dendrites, chronic neuronal changes and intraneuronal uptake of serum proteins, all of these changes being potentially compatible with involvement of neuroexcitatory mechanisms.

Introduction

The potential role of neuroexcitation in producing changes in the function of the blood-brain barrier (BBB) and brain oedema has become increasingly recognized. Under conditions of increased neuronal activity following seizures, Nitsch et al.[1, 2, 3] described breakdown of the BBB to protein tracers, which was characterized by uptake of extravasated horseradish peroxidase into presynaptic boutons and postsynaptic dendrites. In our studies on repetitive ischaemia[4], we observed a pronounced cumulative effect on development of oedema, where gerbils with three 5 minute carotid occlusions, spaced one hour apart, showed after 24 hours significantly lower specific gravity values than animals sacrificed after a single 15 minute insult. This cumulative effect was associated with progressive hypoxia during the intervening periods of hypoperfusion between insults[5] but not with prolonged disturbances of energy metabolism[5, 6]. A possible involvement of neuroexcitatory mechanisms has therefore been considered and has prompted the present evaluation of glutamate release following repeated insults. Moreover, we recently observed in models of global ischaemia and cryogenic injury to the cerebral cortex[8] a number of changes supporting involvement of neuroexcitatory mechanisms, and these observations are included in this study.

Materials and Methods

The three experimental models were: 1) repetitive forebrain ischaemia in gerbils by three 2-minute or 5-minute occlusions of common carotid arteries, carried out 1 hour apart, 2) global ischaemia of 10 minute duration in rats by compressing the bundle of main thoracic vessels and 3) cryogenic injury to the cerebral cortex in rats and gerbils. All surgical procedures were carried out under anaesthesia with 2% halothane in 30% O_2, 70% N_2O. Accumulations of calcium were observed with ^{45}Ca autoradiography or at the ultrastructural level by calcium visualization with an oxalate-pyroantimonate method. Calcium uptake and protein synthesis were studied simultaneously in the same sections by a double tracer method using ^{45}Ca and $[^3H]$leucine markers. The release of extracellular glutamate was evaluated by cortical microdialysis followed by an enzymatic cycling assay employing glutamate dehydrogenase and glutamate-oxaloacetate transaminase[9]. The changes in permeability of the BBB were assessed by immunostaining using antibodies specific for gerbil of rat albumin. For light microscopic observations of neuronal changes the sections were stained with cresyl violet.

Results

The pattern of calcium accumulation and albumin uptake following repeated ischaemic insults and traumatic injury have recently been described[8]. In *repetitive cerebral ischaemia* (3 × 2 min), the pattern of ^{45}Ca uptake was characterized by early (24 hrs), intense calcium uptake in the ventrolateral thalamus, which was accompanied during following days by sharply defined accumulations in the cerebral cortex, lateral portions of caudate, hippocampus, medial geniculate bodies and substantia nigra. The oxalate-pyroantimonate method revealed dense calcium precipitates located preferentially in the swollen dendrites, which otherwise contained rather well preserved mitochondria. Staining with cresyl-violet showed, in areas with ^{45}Ca accumulation, accentuated visualization of dendritic processes and occasional dark or pale and vacuolated cells, although, generally, the neurons appeared to be still viable. Areas with ^{45}Ca accumulation were associated with BBB changes which were characterized by positive albumin staining of neuronal cytoplasm. In contrast there was neither ^{45}Ca uptake nor neuronal albumin staining in central regions of thalamic foci which histologically revealed severe neuronal damage.

Evaluation of glutamate microdialysis during repeated 5 min occlusions showed a progressive increase in glutamate release after the second and third ischaemic insult (Fig. 1), often reaching concentrations of glutamate in the dialysate comparable to those observed after a single 15 minute occlusion.

Following *global ischaemia* in the rat, striking ^{45}Ca accumulation confined to sharply outlined nuclei reticularis thalami, was conspicuous in animals sacrificed after 24 hrs. After several days abnormal ^{45}Ca uptake could be seen also in ventrolateral nuclei of the thalamus, as well as in the other regions which were observed after repetitive ischaemia. In the double tracer studies inhibition of protein synthesis appeared generally to precede accumulations of ^{45}Ca, with the exception of nuclei reticularis thalami, which revealed intense ^{45}Ca accumulation with apparently normal amino acid incorporation. Areas with ^{45}Ca accumulation revealed electron microscopically calcium precipitates in swollen dendrites and were associated with neurons which showed cytoplasmic immunostaining for albumin.

In *cryogenic injury* rats with lesions in the parietal cortex revealed after 3–4 days development of secondary foci in the ipsilateral thalamus, with ^{45}Ca uptake, neuronal changes and disturbances of the BBB, as described above. Gerbils in which the cryogenic probe was applied to the entorhinal region showed bilateral changes in the CA 1 sector of the hippocampus.

Discussion

Neurotoxic effects of excitatory amino acids, and particularly of glutamate have been well established[10]. Local application of this compound results in characteristic glial and dendritic swelling, as well as frequent dark staining of the neurons[11]. Glutamate neurotoxicity has been associated with intracellular influx of Ca^{++} through membrane channels, predominantly gated by receptors of the N-methyl-D-aspartate (NMDA) subtype[12]. In cerebral ischaemia, an increase in extracellular concentrations of glutamate has been demonstrated by microdialysis[13] and is believed to play a role in neuronal injury.

In our studies on repetitive ischaemia we have correlated its cumulative effect on oedema and neuronal damage with postischaemic hypoperfusion, the temporal profile of which corresponds closely with the period of sensitivity for producing a cumulative effect[4]. Progressive hypoxia has been observed during the intervals between repeated insults[5] but neither metabolite

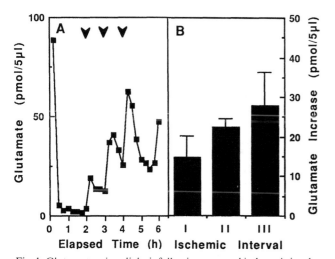

Fig. 1. Glutamate microdialysis following repeated ischaemic insults. A) Time course of dialysate glutamate levels during three occlusions of 5 min duration (arrowheads) in a representative gerbil with samples collected at 15 min intervals at a flow rate of 2 µl/min. B) Increment in dialysate glutamate levels after successive occlusions. Bars indicate the difference in glutamate concentration between samples collected before and after each occlusion (mean and standard deviation of determinations on 4 animals)

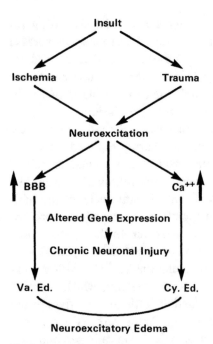

Fig. 2. Speculative involvement of neuroexcitation in development of oedema following ischaemic and traumatic insults. Va. Ed. and Cy. Ed., vasogenic and cytotoxic oedema, respectively

assays in gerbil brain extracts[5] nor NMR spectroscopic analysis in cats[6] revealed lasting energy metabolism disturbances which could account for enhancement of ischaemic damage. Our present findings indicate a significant increase of glutamate release with each consecutive insult. It remains to be determined whether this may contribute to the cumulative effect, expressed in increased severity of oedema and neuronal injury, or reflects an increasing impact of repeated insults arising by other mechanisms.

Neuroexcitatory induction of BBB opening to proteins, similar to that observed following seizures[1, 2, 3], can undoubtedly intensify the vasogenic type of oedema, whereas the effect of excitatory amino acids on glial and dendritic swelling could constitute the cytotoxic component of total oedema resulting from neuroexcitation (Fig. 2). The common features, such as uptake of calcium in swollen dendrites, characteristic changes in BBB permeability and neuronal alterations, which are observed in global ischaemia and cryogenic injury, raise the possibility that neuroexcitation may significantly contribute to the pathophysiology of cardiac arrest and head trauma.

References

1. Nitsch C, Klatzo I (1983) Regional patterns of blood-brain barrier breakdown during epileptiform seizures induced by various convulsive agents. J Neurol Sci 59: 305–322
2. Nitsch C, Goping G, Laursen H, Klatzo I (1986) The blood-brain barrier to horseradish peroxidase at the onset of biculline-induced seizures in hypothalamus, pallidum, hippocampus and other selected regions of the rabbit. Acta Neuropath (Berl) 69: 1–16
3. Nitsch C, Goping G, Klatzo I (1986) Pathophysiological aspects of blood-brain barrier permeability in epileptic seizures. In: Schwarcz R, Ben-Ari Y (eds) Excitatory amino acids and epilepsy. Plenum Publ Corp, New York, pp 175–189
4. Tomida S, Nowak TS Jr, Vass K, Lohr JM, Klatzo I (1987) Experimental model for repetitive ischaemic attacks in the gerbil: The cumulative effect of repeated ischaemic insults. J Cereb Blood Flow Metab 7: 773–782
5. Nowak TS Jr, Tomida S, Pluta R, Xu S, Kozuka M, Vass K, Wagner HG, Klatzo I (1990) Cumulative effect of repeated ischaemia on brain oedema in the gerbil. Biochemical and physiological correlates of repeated ischaemia insults. Adv Neurol 52: 1–9
6. Hossmann K-A, Nagashima G, Klatzo I (1990) Repetitive ischaemia of cat brain: Pathophysiological observations. Neurol Res (in press)
7. Nagashima G, Nowak TS Jr, Joó F, Ikeda J, Ruetzler C, Lohr J, Klatzo I (1990) The role of the blood-brain barrier in ischaemic brain lesions. In: Johanssen BB, Owman Ch, Widner H (eds) Pathophysiology of the blood-brain barrier. Elsevier Science Publishers, pp 311–321
8. Ikeda J, Nagashima G, Saito N, Nowak TS Jr, Joó F, Mies G, Lohr JM, Ruetzler CA, Klatzo I (1990) Putative neuroexcitation in cerebral ischaemia and brain injury. In: Proc 17th Princeton Conference on Cerebrovascular Disease (in press)
9. Lowry OH, Passonneau JV (1972) A flexible system of enzymatic analysis. Academic Press, San Diego
10. Olney JW (1978) Neurotoxicity of excitatory amino acids. In: McGeer EG, Olney JW, McGeer PL (eds) Kainic acid as tool in neurobiology. Raven Press, New York, pp 95–121
11. Sloviter RS, Dempster DW (1985) "Epileptic" brain damage is replicated qualitatively in the rat hippocampus by central injection of glutamate or aspartate but not by GABA or acetylcholine. Brain Res Bull 15: 39–60
12. Choi DW (1988) Calcium-mediated neurotoxicity: Relationship to specific channel types and role of ischaemic damage. Trends Neurosci II: 465–469
13. Benveniste H, Drejer J, Schousboe A, Diemer NH (1984) Elevation of extracellular concentrations of glutamate and aspartate in rat hippocampus during transient cerebral ischaemia monitored by intracerebral microdialysis. J Neurochem 43: 1369–1374

Correspondence: Dr. I. Klatzo, National Institutes of Health, NIH, Bldg 36, Room 4D04, Bethesda, MD 20892, U.S.A.

Acta Neurochirurgica, Suppl. 51, 189–191 (1990)

Inducible Ornithine Decarboxylase Expression in Brain Subject to Vasogenic Oedema After Transient Ischaemia: Relationship to C-*fos* Gene Expression

R. J. Dempsey, J. M. Carney, and **M. S. Kindy**

University of Kentucky Medical Center and Veterans Administration Hospital, Lexington, Kentucky, U.S.A.

Summary

Ornithine Decarboxylase (ODC) is the rat controlling enzyme of polyamine biosynthesis and has been shown to be produced in a delayed fashion in response to cerebral ischaemia. Its appearance has been linked to the development of vasogenic brain oedema. To understand the genetic control of this protein, Mongolian gerbils were studied for the possible expression of the ODC gene as compared to that of the inducible proto oncogenes c-fos and c-jun after transient bilateral carotid artery occlusion.

Total cellular RNA was isolated from gerbil brains by guanidine-thiocyanate extraction and characterized by northern blot analysis for c-fos, c-jun, and ODC mRNA over reperfusion times. c-fos and c-jun expression rose rapidly with peak level reached at 60 min of reperfusion ($70 \times$ control, $p \leqslant 0.01$). Peak levels of ODC mRNA induction were seen at 4 hrs reperfusion ($2.83 \times$ control, $p \leqslant 0.01$) consistent with the period of maximum of brain oedema as measured by specific gravity (1.0386 ± 0.0009, $p \leqslant 0.05$). These data indicate the differential timing of genetic expression during the reperfusion period after transient ischaemia. Such studies suggest that potential therapies may be possible by addressing the delayed ODC component of ischaemic oedema formation and allow a greater understanding of the role of gene induction in the multifaceted cerebral response to ischaemia.

Introduction

Both focal and transient cerebral ischaemia have been shown to result in the increased production of the inducible enzyme ornithine decarboxylase (ODC)[1,2]. This enzyme is the rate limiting step in the production of polyamines and is implicated in the formation of delayed vasogenic oedema[4,7,9]. ODC production is one aspect of the metabolic cascade of events that results with re-establishment of circulation after a significant period of cerebral ischaemia[3]. This cascade is energy dependent and generates several potentially harmful active metabolites. Extensive research into this process

has shown that some agents are potentially protective if given prior to the onset of cerebral ischaemia. Ornithine decarboxylase however has been particularly of interest because of its delayed induction. The enzyme possesses a short half-life and is produced in response to a stimulus such as cerebral ischaemia. This then leads to the production of its polyamine products[1,2]. This delayed nature of ODC induction, raises the possibility that pharmacologic intervention in this enzyme system during the first hours after the onset of the ischaemic event may still beneficially effect the outcome.

These early metabolic events are mediated through receptors manifested by the transcriptional products of gene expression. The techniques of molecular biology allow us to more fully understand the role that stimulation and transcription of regulator genes may have in the initiation of such metabolic events. By studying the time of onset, duration and quantity of gene expression for ornithine decarboxylase with cerebral reperfusion, a more complete understanding of the control of neuronal cell function in the face of cerebral ischaemia is obtained.

Transient cerebral ischaemia has been well studied in the Mongolian gerbil model of forebrain ischaemia. Bilateral carotid artery occlusion results in dense reversible ischaemia to the forebrain and striatal regions. With reperfusion, predictable levels of ischaemic oedema are achieved. The induction of such oedema is dependent on the duration and depth of the transient period of ischaemia[3]. We have previously shown that induction of ornithine decarboxylase is seen in a delayed fashion after cerebral ischaemia with initially very low levels of the enzyme becoming induced to progressively increasing levels between 4 and 8 hours after the onset of reperfusion[2, 7]. Due to ODC's involvement in the generation of vasogenic oedema, a further understanding of the induction of this enzyme is desired[4, 7, 9].

Methods

Animal preparation – Male Mongolian gerbils (Tumblebrook Farm, Westbrookfield, MA) were anaesthetized with Pentobarbital 40 mg/kg and bilateral carotid artery occluders were placed subcutaneously. The animals were then singly housed to prevent accidental ischaemia and allowed to recover for 2 days to completely clear anaesthetic. Transient dense cerebral ischaemia was produced by occluding both common carotid arteries. Global forebrain ischaemia was maintained for a 10 minute period of time and the occluders were removed to allow complete reperfusion. Animals were sacrificed at up to 8 hrs of reperfusion. Brains were removed and frozen in liquid nitrogen for RNA analysis. RNA was extracted from samples without thawing.

Total cellular RNA was isolated from gerbil brains by guanidine-thiocyanate extraction. The resulting mRNA was examined by Northern Blot analysis[6, 8]. Electrophoresis was done in 1% agarose, 2.2 M formaldehyde gel. The RNA was transferred to nitrocellulose and hybridized against a ^{32}P labeled DNA probe for ornithine decarboxylase. The resulting autoradiograms were then analyzed for quantity of ornithine decarboxylase message expressed.

Separate samples were hybridized against labeled probes for c-*fos* and c-*jun* proto-oncogenes[8]. The time course and quantity of this expression was then calculated and compared over the reperfusion time.

Oedema formation within the model was confirmed by specific gravity measurements for a spearate group of animals subjected to 10 minutes of forebrain ischaemia and up to 8 hours of reperfusion. Cortical samples were measured for specific gravity on a precalibrated Kerosene-bromobenzene column[5].

Results

Normal levels of c-*fos*, c-*jun* and ODC mRNA were seen in non-ischaemic controls. With reperfusion after 10 minutes of global forebrain ischaemia, c-*fos* expression rose rapidly with a peak level reached at 60 minutes of reperfusion (70 × control, p ≤ 0.01). ODC mRNA induction was seen to follow that of c-*fos*. Peak levels of ODC mRNA induction were seen at 4 hours of reperfusion (2.83 × control, p ≤ 0.01). This ODC mRNA induction returned to near control levels by 8 hours after reperfusion (see Fig. 1).

Brain Oedema Specific Gravity

Brain oedema was measured by specific gravity. Control animals frontal subcortex showed a specific gravity of 1.0393 ± 0.0005. With reperfusion, specific gravity fell at all time points measured with maximal water accumulation at 4 hours of reperfusion: (1.0368 ± 0.0009 (p ≤ 0.05) Fischer PLSD).

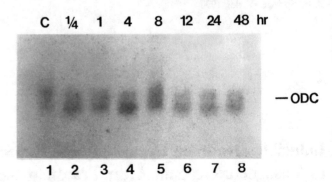

Fig. 1. Northern blot of ODC mRNA as expressed before and during cerebral ischaemia and up to 8 hours after a 10 minute ischaemic episode

Discussion

Transient cerebral ischaemia is shown to result in the generation of ischaemic oedema. In this study a model of ischaemia with known generation of ischaemic oedema was studied for the induction of 3 genes. The proto-oncogenes c-*fos* and c-*jun* were both shown to be induced to high levels within the first hour of reperfusion. The product of these proto-oncogenes, the c-*fos-jun* dimer, is involved in the activation of other gene sequences containing the AP-1 binding site[8].

An understanding of the molecular regulation of gene expression is essential to the understanding of the control of ischaemic brain oedema. Previous observations using PC 12 pheochromocytoma cell line indicated increased calcium levels altered the expression of c-*fos* mRNA[6]. Both c-*fos* and c-*jun* genes are stimulated with transient ischaemia and reperfusion. Changes in c-*fos* mRNA are seen in metrazole induced seizures. Such activation appears to occur in concert with activation of the NMDA glutamate receptor. Other studies have shown an increase in c-*fos* and c-*jun* mRNA in seizures occurring along with elevated levels of enkephalin mRNA. Enkephalin gene expression is known to be induced by the association of the *fos-jun* dimer with the AP-1 site in the promoter.

This dimer of c-*fos* and c-*jun* may be an initiator at the DNA level and part of the complex metabolic response of the brain to transient ischaemia[6]. In this study the expression of the ODC gene is delayed relative to that of c-*fos* and c-*jun*. ODC mRNA reaches maximal levels after 4 hours of reperfusion. This suggests a possible role in the later development of vasogenic oedema after an ischaemic insult. Ischaemic oedema is felt to be mixed in nature with early cytotoxic and late vasogenic components[3]. Prior work has shown the ODC enzyme to be important for the development of the vasogenic component of oedema after an ischaemic insult[4, 7]. Due to its short half-life, little ODC is present in the normal brain, but its production is induced regionally in response to ischaemia. The present study shows that the maximal induction of the ODC gene as measured by ODC mRNA is peaking at 4 hours of reperfusion. This suggests that a therapeutic window may exist in the early hours after cerebral ischaemia for treatments which would address the ODC component of the later vasogenic oedema formation. By altering polyamine production such treatments could also affect later expression of other genes. Further molecular biology studies of the expression of mRNA in response to ischaemia are planned to elucidate the role of this gene induction in the onset of the vasogenic component of multifaceted ischaemic oedema.

References

1. Dempsey RJ, Roy MW, Meyer K, Tai H, Olsen J (1985) Polyamine and prostaglandin markers in focal cerebral ischaemia. Neurosurgery 17: 635–640
2. Dempsey RJ, Combs DJ, Olson JW, Maley M (1988) Brain ornithine decarboxylase activity following transient cerebral ischaemia – Relationship to cerebral oedema development. Neurol Res 10: 175–178
3. Klatzo I, Suzuki R, Orzi F *et al* (1982) Pathomechanisms of ischaemic brain oedema. In: Go KG, Baethmann A (eds) Recent progress in the study of brain oedema. Plenum Press, New York and London
4. Koenig H, Goldstone A, Lu Cy (1983) Blood-brain barrier breakdown in brain oedema following cold injury is mediated by microvascular polyamines. Biochem Biophys Res Comm 116: 1039–1048
5. Marmarou A, Tanaka K, Shulman S (1987) An improved gravimetric measure of oedema. J Neurosurg 49: 530–537
6. Mitchell RL, Hanks SK, Verma IM (1987) Proto-oncogene *fos*: An inducible multifaceted gene. Symposium on Fundamental Cancer Research vol 39
7. Schmitz MP, Combs DJ, Dempsey RJ: Difluoromethylornithine (DFMO) inhibition of polyamine pathway synthesis decreases post-ischaemic oedema and blood-brain barrier (BBB) breakdown. Personal Communication
8. Smeal T, Angel P, Meek J, Karin M (1989) Different requirements for formation of Jun: Jun and Jun: Fos complexes. Genes & Development 3: 2091–2100
9. Trout JJ, Koening H, Goldstone A, Lu Cy (1986) Blood-brain barrier breakdown by cold injury, polyamine signals mediate acute stimulation of endocytosis, vesicular transport, microvillus formation in rat cerebral capillaries. Lab Invest 55: 622–631

Correspondence: R. J. Dempsey, M.D., Division of Neurosurgery, MS105A Chandler Medical Center, University of Kentucky, 800 Rose Street, Lexington, KY 40536-0084, U.S.A.

Acta Neurochirurgica, Suppl. 51, 192–194 (1990)

Ischaemic Brain Oedema and Xanthine-Xanthine Oxidase System

Y. Kinuta, H. Kikuchi, and **M. Ishikawa**

Department of Neurosurgery, Kyoto University, Japan

Summary

The formation of oxygen-derived free radicals in cerebral ischaemia has been implicated in altering the BBB permeability, cause oedema and tissue damage. However little attention has been paid regarding the involvement of xanthine oxidase in the cerebral ischaemic events. Recently we demonstrated that cerebral ischaemia promotes the conversion of xanthine oxidase type D (nicotinamide adenine dinucleotide – dependent dehydrogenase) to type 0 (oxygen – dependent superoxide – producing oxidase). This investigation was concerned with elucidating the relationship between the conversion of xanthide oxidase and the duration of brain ischaemia. Four vessel-occlusion served as a model for the induction of cerebral ischaemia in rats. Xanthine oxidase was assayed by high pressure liquid chromatography using ultraviolet and electrochemical detection. The enzymatic conversion of xanthine oxidase from type D to type O increased with time from 7.6–15% during 5 min ischaemia to 27% and 36% at 15 min and 30 min after ischaemia, respectively. These results support the contention that xanthine oxidase may participate in free radical-induced ischaemic brain oedema.

Introduction

It has been recognized that damage in cerebral ischaemia is at least partly due to oxidative damage caused by oxygen-derived free radical formation. Particular attention has been focused on the enzyme, xanthine oxidase, because of the ability of this enzyme to serve as a source of oxidizing agents such as superoxide radical and hydrogen peroxide. This enzyme catalyzes the conversion of hypoxanthine to xanthine and xanthine to urate. Xanthine oxidase consists of two forms in tissue: a nicotinamide adenine dinucleotide (NAD)-dependent dehydrogenase (Type D) and an oxygen-dependent superoxide-producing oxidase (Type O).

Xanthine oxidase has been shown to be located almost exclusively in the endothelial cells. The endothelial injury has been shown to cause an increase in capillary permeability. Therefore, an early involment of this pathomechanism may take part in the development of the ischaemic brain oedema.

However, no attention has been paid to the involvement of this enzyme in the free radical-induced damage of the ischaemic brain. Recently, we revealed that this enzyme exists in brain tissue and that ischaemia induces the enzymatic conversion to Type O from Type D in rat brain[3].

The present study investigated the relationship between the conversion of xanthine oxidase and the duration of ischaemia in rat brain.

Materials and Methods

Induction of Ischaemia

The experiments were performed on male Wistar rats, each weighing 250 to 300 gm. Cerebral ischaemia was produced by the four-vessel occlusion method. After the designated period of ischaemia (0, 5, 15, and 30 minutes), the rats were killed by freezing the brain *in situ* with liquid nitrogen applied for 5 minutes.

Xanthine Oxidase Assay

The xanthine oxidase assay was a modification of the method described by Della Corte and Stirpe[2]. Xanthine oxidase activities were assayed by measuring the amount of uric acid which was produced both in the presence and in the absence of exogenous NAD.

About 75 mg of frozen cortical tissue was homogenized in 1 ml of ice-cold 0.1 M Tris-HCl buffer (pH 8.1) containing 1 mM ethylenediaminetetra-acetic acid, 10 mM dithioerythritol, and 1 mM phenylmethylsulfonylfluoride. The homogenates were centrifuged at 100 000 G for 30 minutes at 4 °C. The resulting supernatant was used for the assay. The reaction mixture consisted of 0.05 mM xanthine and 0.1 M Tris-HCl buffer (pH 8.1), with or without 0.5 mM NAD. The preparation with the NAD added to the reaction mixtures allowed for the measurement of total xanthine oxidase (Type D and Type O), and the preparation without NAD allowed for the measurement of Type O alone. The reaction mixture (100 µl) was added to 100 µl of the supernatant and then incubated at 25 °C. Incubation times were set at 20, 40, or 60 minutes. The reaction was terminated by adding 20 µl of 60% perchloric acid to the mixtures. Following centrifugation at 3000 rpm for 5 minutes, 20 or 40 µl aliquots of the acid supernatants were injected into the HPLC system.

Xanthine oxidase activity was expressed as nmol urate production/gm wet weight/min at 25 °C, pH 8.1.

Table 1. *Xanthine Oxidase Activity in the Rat Brain*

Sample	Type D	Type O	Type D and O	Type O form (% of total)
Control (0 min)	0.81 ± 0.13	0.07 ± 0.01	0.87 ± 0.13	7.6 ± 1.5
Ischaemic (5 min)	0.68 ± 0.11^a	0.20 ± 0.08^b	0.88 ± 0.11^c	22.8 ± 3.9^b
Ischaemic (15 min)	0.62 ± 0.06^a	0.32 ± 0.08^b	0.93 ± 0.08^c	33.3 ± 5.4^b
Ischaemic (30 min)	0.51 ± 0.07^b	0.40 ± 0.10^b	0.91 ± 0.10^c	43.7 ± 7.6^b

Values are means ± standard deviations of five observations. Activity is expressed as nmol urate production/gm wet weight/min at 25° C, pH 8.1.

[a] Value versus control significantly different ($p < 0.05$).

[b] Value versus control significantly different ($p < 0.01$).

[c] Value versus control not significantly different.

HPLC System

The procedure for measuring uric acid was performed by the method using HPLC with ultraviolet (UV) and electrochemical (ECD) detections[5]. The two detectors were connected sequentially. Detection of peaks was at 290 nm (absorbance) for UV detection and at +0.6 V (applied potential) for ECD. A reverse-phase column (25 cm × 4 mm in inner diameter, 5 μm particle size) was used. The mobile phase composed of 0.1 M ammonium phosphate aqueous buffer brought up to pH 4.8 with ammonium hydroxide. The flow rate was 1.0 ml/min. Quantification was performed by the external standards method.

Statistics

Results were expressed as means ± standard deviations of five observations. The data were analysed using a one way analysis of variance.

Results

Table shows the results of the assay of xanthine oxidase in rat brain. There was no significant difference between the control and each of the ischaemic groups in total xanthine oxidase, of which the enzyme activity was about 0.90 nmol/gm wet weight/min at 25 °C. With regard to enzymatic forms, 92.4% was associated with the NAD-dependent dehydrogenase form (Type D) and only 7.6% with the oxygen-dependent superoxide-producing oxidase form (Type O) in the control group. However, the ratio of Type O increased with time and it reached to 22.8% at 5 minutes after ischaemia, 33.3% at 15 minutes after ischaemia, and 43.7% at 30 minutes after ischaemia, respectively. Therefore, the enzymatic conversion from Type D to Type O increased with time and was about 15% of total at 5 minutes after ischaemia, about 27% of total at 15 minutes after ischaemia, and about 36% of total at 30 minutes after ischaemia.

Discussion

The findings indicate that ischaemia induces the conversion of xanthine oxidase of Type D to Type O in the rat brain. We and other authors have observed an increase in urate production in the rat brain during ischaemia previously[4]. The increase of both xanthine oxidase Type O and of urate production in the brain tissue during ischaemia strongly suggests the possibility of increased production of superoxide anion radicals which cause the breakdown of capillary permeability. Therefore under special condition of reperfusion when the capability and reoxygenation to brain tissues is satisfied, free-radical reactions can take place and it can lead to many pathophysiological conditions including ischaemic brain oedema. Indeed, the injection of exogenous xanthine oxidase with hypoxanthine and adenosine diphosphate-Fe3+ into brain was reported to cause injury to brain capillaries and the breakdown of the blood-brain barrier[1].

There are two possibilities which cause the enzymatic conversion of xanthine oxidase *in vitro*: sulfhydryl oxidation and incubation with a protease such as trypsin. The former reaction can be readily reversed with the addition of a strong reducing agent such as dithioerythritol. However, the latter reaction, namely the proteolytic conversion, is irreversible. Since dithioerythritol was used in our homogenizing buffer, the enzymatic conversion observed in this experiment was probably the results of activated-proteases induced by ischaemia.

At 5 minutes after ischaemia, the enzymatic conversion has already begun and its rate was about 15% of total. At 30 minutes of ischaemia, the conversion rate increased to 36% of total. Therefore, it is clear that the basis for the development of brain oedema is present at an early period of ischaemia.

References

1. Chan PH, Schmidley JW, Fishman RA *et al* (1984) Brain injury, oedema, and vascular permeability changes induced by oxygen-derived free radicals. Neurology 34: 315–320
2. Della Corte E, Stirpe F (1972) The regulation of rat liver xanthine oxidase. Biochem J 126: 739–745
3. Kinuta Y, Kimura M, Itokawa Y *et al* (1989) Changes in xanthine oxidase in ischaemic rat brain. J Neurosurg 71: 417–420
4. Kinuta Y, Kikuchi H, Ishikawa M *et al* (1989) Lipid peroxidation in focal cerebral ischaemia. J Neurosurg 71: 421–429
5. Podzuweit T, Braun W, Muller A *et al* (1987) Arrhythmias and infarction in the ischaemic pig heart are not mediated by xanthine oxidase-derived free oxygen radicals. Basic Res Cardiol 82: 493–505

Correspondence: Dr. Y. Kinuta, Department of Neurosurgery, Kyoto University, Kyoto, Japan.

Acta Neurochirurgica, Suppl. 51, 195–197 (1990)

The Role of Endothelial Second Messenger's-Generating System in the Pathogenesis of Brain Oedema

F. Joó

Laboratory of Molecular Neurobiology, Institute of Biophysics, Biological Research Center, Szeged, Hungary

Summary

The role of second messengers in the regulation of protein phosphorylation was studied in microvessels isolated from rat cerebral cortex. Calcium-calmodulin (CAM)-, Ca^{2+}/phospholipid (PK C)-, cyclic GMP (cGMP)-, and cyclic AMP (cAMP)-dependent protein kinases were detected. Autophosphorylation of both the α- and β-subunits of CAM-dependent protein kinase and the proteolytic fragment of the PK C enzyme was also detected.

In other experiments, the effect of the protein kinase C enzyme inhibitor H-7 was examined on the brain oedema formation evoked by bilateral occlusion of the common carotid arteries in Sprague-Dawley rats of CFY strain.

Introduction

The first description of the presence of *adenylate cyclase* in the brain microvessels appeared in 1980[5], and the occurrence of the *guanylate cyclase* was evidenced also in 1980[6]. Histamine receptors were found to be linked to the microvascular adenylate cyclase taking part in the mediation of permeability regulation evoked by histamine[5]. Antihistamine preventing the increase of cyclic nucleotides in the cerebral endothelium were shown to prevent brain oedema formation[2].

The aims of the present study was twofold:

1. To identify in phosphorylation studies the proteins of cerebral endothelial cells, which are influenced by the elevated accumulation of different second messenger molecules.

2. To check the effect of H-7 [1,5(isoquinolinylsulfonyl)-2-methylpiperazine], a potent inhibitor of protein kinase C, on the formation of brain oedema evoked by bilateral common carotid occlusion in Sprague-Dawley CFY rats.

Material and Methods

Microvessel isolation: Microvessels were isolated from adult Wistar rats (10 in one experiments, 90 altogether) of both sexes weighing 200–300 g, according to the method described previously[5].

Phosphorylation assay: Samples were melted just before use and then suspended in 25 mM Tris-MES buffer (pH 7.4) on ice in the presence of 2 mM phenylmethylsulfonyl fluoride (PMSF). The homogenized microvessels were incubated at 25 °C for 10 min. For further details see Oláh *et al.*[13].

Treatment of brain oedema with H-7: In these studies, a closed colony of randomly bred female, normotensive Sprague-Dawley CFY rats weighing 200–220 g was used. Similarly to the results of our previous investigations[9], characteristic symptoms, like staggering and circling developed after common carotid artery occlusion in two thirds of the animals within 4 h. Treatments with different doses (1.56, 3.12, 6.25, 12.5 and 25.0 mg/kg, respectively) of H-7 (Sigma) were carried out in the form of single i.p. injections 30 min before operation. The rats were killed 4 h after surgery. Water and electrolyte contents were determined in brains of 10 rats from each experimental group.

Results

Characteristics of Phosphorylation of Proteins from Brain Microvessels and the Effects of Second Messenger Molecules

The microvessel fraction prepared from the cerebral cortex of adult rat brain always preserved some basal protein phosphorylation activity. In contrast, we were able to increase the phosphorylation of these proteins by separate addition of the putative second messengers, CAM, PK C, cGMP, and cAMP.

Protein substrates for PK C: For identification of the protein substrate for PK C, exogenous enzyme (the 45-kDa fragment of PK C) was added to the incubation medium. The exogenous PK C could recognize its ma-

Ca²⁺-calmodulin cGMP cAMP			cGMP		PK-C		cGMP cAMP	
P50	P55	P58(57)	P30	P33	P40	P45	P17	P66 P68
β- and α- subunits of CAM-dependent protein kinase II		α- and β-subunits of tubulin	DRAPP-32 (dopamine- and cAMP-regulated phosphoprotein)		Not known	Catalytic fragment of protein kinase C (PK C)	Myosin light chain	δ- and γ-subunits of nicotinic acetylcholine receptor

Fig. 1. Protein substrates of Ca²⁺-calmodulin system, Ca²⁺/phospholipide-, cGMP- and cAMP-dependent protein kinases

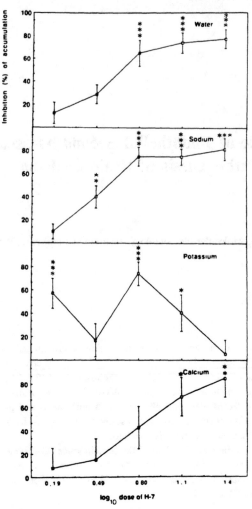

Fig. 2. Brain water and electrolyte content in the different experimental groups treated with different doses of H-7. n = 10, Mean ± S.E.M.; * $p < 0.05$; ** $p < 0.01$; *** $p < 0.001$

jor substrate in this fraction, and the reversibility of phosphorylation of the 40-kDa protein was impressive. Phosphorylation of microvessel proteins by different protein kinases in the cerebral endothelial cell is summarized in Fig. 1.

Prevention by H-7 Treatment of the Brain Oedema Formation in Sprague-Dawley CFY Rats

In comparison to the values obtained from the brains of sham-operated rats, bilateral occlusion of the common carotid arteries resulted in a significant increase in brain water, sodium and calcium contents 4 h after surgery. At the same time, the K^+ content of brains decreased considerably most likely as a result of diffusion from the extracellular space subsequent to membrane depolarization and failure of the ATP-dependent Na^+, K^+ pump. From the dose-response analysis of H-7 effect it became evident that pretreatment of rats with higher doses of H-7 prevented to a great extent the development of brain swelling and the increase in sodium and calcium (Fig. 2).

Discussion

We were able, to the best of our knowledge, to present the first evidence for the occurrence of CAM-, Ca²⁺/phospholipid-, cGMP-, and cAMP-dependent protein

phosphorylation in a well-characterized microvessel fraction isolated from rat cerebral cortex. However, in the rat brain microvessels, cGMP proved to be a better activator than cAMP, in contrast to the situation in other vessels[7].

In view of the size-dependent opening of the endothelial pores in cases of the breakdown of the blood-brain barrier[1], our present results strongly suggest that the PK C-system may be involved in the regulation of transendothelial flux of water, sodium and calcium transport through the small and intermediate pores of the tight junctions.

Acknowledgements

The author is grateful to Zoltán Oláh, Imre Lengyel, Árpád Tósaki and Mátyás Koltai for the fruitful collaboration. I thank the secretarial help of Mrs. Zsuzsanna Gonda.

References

1. Amstrong BK, Robinson PJ, Rapoport SI (1987) Size-dependent blood-brain barrier opening demonstrated with ^{14}C sucrose and a 200,000-Da ^{3}H dextran. Exp Neurol 97: 686–696
2. Dux E, Temesvári P, Szerdahelyi P, Nagy Á, Kovács J, Joó F (1987) Protective effect of antihistamines on cerebral oedema induced by experimental pneumothorax in newborn piglets. Neuroscience 22: 317–321
3. Joó F, Karnushina I (1973) A procedure for the isolation of capillaries from rat brain. Cytobios 8: 41–48
4. Joó F, Tósaki Á, Oláh Z, Koltai M (1989) Inhibition by H-7 of the protein kinase C prevents formation of brain oedema in Sprague-Dawley CFY rats. Brain Res 490: 141–143
5. Karnushina I, Palacios JM, Barbin G, Dux E, Joó F, Schwartz JC (1980) Studies on a capillary-rich fraction isolated from brain: histaminic components and characterization of the histamine receptors linked to adenylate cyclase. J Neurochem 34: 1201–1208
6. Karnushina I, Tóth I, Dux E, Joó F (1980) Presence of the guanylate cyclase in brain capillaries: histochemical and biochemical evidences. Brain Res 189: 588–592
7. Mackie K, Lai Y, Nairn AC, Greengard P, Pitt BR, Lazo JS (1986) Protein phosphorylation in cultured endothelial cells. J Cell Physiol 128: 367–374
8. Oláh Z, Novák R, Lengyel I, Dux E, Joó F (1988) Kinetics of protein phosphorylation in microvessels isolated from rat brain: modulation by second messengers. J Neurochem 51: 49–56
9. Tósaki A, Koltai M, Joó F, Ádám G, Szerdahelyi P, Leprán I, Takáts I, Szekeres L (1985) Actinomycin D suppresses the protective effect of dexamethasone in rats affected by global cerebral ischaemia. Stroke 16: 501–505

Correspondence: Dr. F. Joó, Laboratory of Molecular Neurobiology, Institute of Biophysics, Biological Research Center, P.O.B. 521, H-6701 Szeged, Hungary.

Acta Neurochirurgica, Suppl. 51, 198–200 (1990)

Effect of Glutamate and Its Antagonist on Shift of Water from Extra- to Intracellular Space After Cerebral Ischaemia

Y. Shinohara, M. Yamamoto, M. Haida, K. Yazaki, and **D. Kurita**

Department of Neurology, Tokai University, School of Medicine, Isehara, Kanagawa, Japan

Summary

The effects of glutamate and the excitatory amino acid antagonist, MK-801, were investigated on the time course of the shift of water from extracellular to intracellular space (progression or cytotoxic oedema) after total brain ischaemia in rats.

Administration of sodium glutamate intravenously before ischaemia accelerated the shift of water dose-dependently. On the contrary, preischaemic administration of MK-801, an NMDA antagonist, delayed the progression of cytotoxic oedema due to brain ischaemia.

We consider that glutamate and NMDA antagonists may have important roles in the development and prevention of cytotoxic oedema in the ischaemic state.

Introduction

It is believed that the neurotoxicity of glutamate plays not only a key role in delayed neuronal loss in hippocampus but also in acute ischaemic damage of the brain cell. Furthermore, the efficacy of the excitatory amino acid antagonist, MK-801, has recently been reported in various models of brain hypoxia or ischaemia[5, 6], though there is still some controversy especially concerning the effect of glutamate antagonists on acute ischaemic cell damage. Among the many possible methods for detecting changes of the brain tissue after exposure to a hypoxic or ischaemic insult, nuclear magnetic resonance (NMR) is particularly suitable for detecting ischaemic brain oedema or a shift of water from extracellular to intracellular space, because of its high sensitivity to the behaviour of water molecules.

The purpose of this study was to investigate by means of NMR the effects of glutamate and the excitatory amino acid antagonist, MK-801 during the time course of water shift from extracellular to intracellular space after total ischaemia of the brain.

Materials and Methods

This study comprised forty-nine 6- to 10-week-old male Wistar rats, weighing 200 to 280 g. Under light anaesthesia induced by intraperitoneal urethan (500 mg/kg body weight) and chloralose (50 mg/kg body weight), a small burr hole was drilled in the right frontoparietal area of the rat skull. A brain tissue sample, 2 mm in thickness, and consisting mainly of gray matter, was punched out with a sharp-edged Pyrex glass tube, which was sealed tightly with a water-proof material (Hematoseal), and placed in an NMR spectrometer with a special sample holder. This sample can be regarded as a highly simplified model of brain ischaemia. This method is advantageous because the total water content in the sample is constant during measurement (the brain tissue sample was sealed in the tube as mentioned above). NMR measurements were performed by using 90 MHZ FT NMR equipment (JEOL FX 90 A) at 25 °C, and were started within 1 minute after samples were obtained. The values of the longitudinal relaxation time, T_1, and the transverse relaxation time, T_2, were measured by using the inversion-recovery method and Hahn's spin-echo method, respectively. However, in this study we mainly made use of T_2, since T_2 is believed to be more sensitive to environmental changes around water molecules than T_1[1]. The magnetic field was stabilized by means of a D_2O external lock. The water signal intensity was determined in the frequency spectra obtained by Fourier transformation every 8 seconds. As we have already shown[2], the transverse magnetization decay curve of the rat brain obtained from 9 different echo times can be explained in terms of two components, and the fast component of T_2, that is T_{2f}, corresponds to the signals from the water of the intracellular space. To estimate one T_{2f} value, four echo times (40, 80, 160 and 240 msec) were used. Then, the average value of four T_{2f} was considered as mean T_{2f} during 2 minutes, and measurements were continued for approximately 60 minutes[3, 7].

Glutamate and MK-801 were injected intravenously for 3–5 minutes, and 2 minutes after administration, the brain was sampled. NMR measurements were started within 1 minute after samples were obtained.

Results and Discussion

1. Effects of Total Ischaemia on T_{2f} Change in Brain Tissue

In Fig. 1, the thick black curve is the mean of the T_{2f} changes in 21 control rat brains after total ischaemia. The value of T_{2f} increases with time, suggesting a transfer of water from extracellular space to intracellular space, because the total amount of water in the sealed tissue sample should be constant during the measurement. This increase was gradual and T_{2f} was maximally elongated by 14% within 20 minutes after total ischaemia.

The obtained value of $T_{2f}(t)$ was fitted to the following equation by the least-squares method: $T_{2f}(t) = \triangle T_{2f}(1-\exp(-t/k)) + T_{2f}(0)$, where the $\triangle T_{2f}$ is the increment of T_{2f} and $T_{2f}(0)$ is the T_{2f} value at the time of sampling. The time constant, k, may be interpreted as a parameter of the membrane permeability and/or the driving force which causes cell swelling due to change of intracellular water content[3, 7]. The time course of this change of T_{2f} was also in good agreement with that of the impedance change of the brain tissue after complete cardiac arrest, which has been explained by Hossmann[4] in terms of a shift of water molecules from extracellular to intracellular space.

2. Effects of Glutamate Administration

Sodium glutamate 6 mg/kg (n = 4) or 20 mg/kg (n = 4) was injected intravenously before ischaemia, and its effect on the various NMR parameters of the T_{2f} equation, including the time constant, k, after brain sampling was investigated. There was only a slight decrease in $T_{2f}(0)$ in the case of glutamate 20 mg/kg injection, but a marked change was observed in k value, which was significantly lower ($p < 0.001$) than that of the control (Fig. 2), suggesting that preadministration of glutamate before total ischaemia accelerates the shift of water from extra- to intracellular space in ischaemic brain cells.

3. Effects of MK-801 Administration

MK-801, 0.05, 0.1, 0.5, or 5 mg/kg, was also injected intravenously before total ischaemia in 19 rats. As shown in Fig. 2, preischaemic administration of MK-801, at more than 0.1 mg/kg, statistically significantly increased the k values without significant change of $T_{2f}(0)$ or $\triangle T_{2f}$, suggesting that MK-801 can prevent or delay the shift of water from extra- to intracellular space in ischaemic cells.

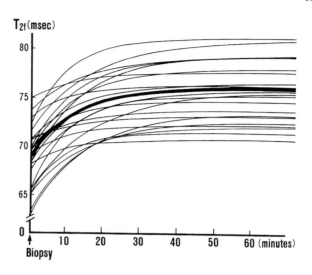

CALCULATION OF PARAMETERS

$T_{2f}(t) = \Delta T_{2f}(1-e^{(-t/k)}) + T_{2f(0)}$

$T_{2f(0)}$: Initial T_{2f} (msec)

$T_{2f\,Max}$: Maximum T_{2f} (msec)

k : Time constant (min)

$\%\Delta T_{2f}$: $\dfrac{T_{2f\,Max} - T_{2f(0)}}{T_{2f(0)}} \times 100$

Fig. 1. T_{2f} changes in control rat brain

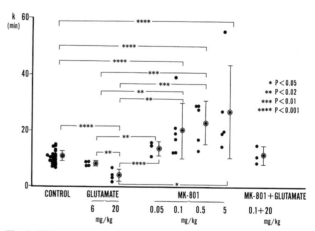

* $P < 0.05$
** $P < 0.02$
*** $P < 0.01$
**** $P < 0.001$

Fig. 2. Effects of glutamate and its antagonist on the time constant "k"

The possibility that the neuroprotection seen with MK-801 is due to a reduction in core or cerebral temperature have been denied by Swan and Meldrum[8]. Under light anaesthesia such as our experiment the effects of temperature, blood pressure and blood gas on neuroprotection was thought to be minimal.

4. Effects of MK-801 Plus Glutamate, and Other Agents

The effects of MK-801 (2 mg/kg) plus glutamate (20 mg/kg), the Ca antagonist nicardipine plus glutamate, and the competitive NMDA antagonist AP-5 or AP-7 on various NMR parameters were also examined.

MK-801 plus glutamate had essentially no effect on the cytotoxic oedema (Fig. 2), but glutamate plus nicardipine increased the k value significantly and the competitive NMDA antagonist also tended to increase in the k values although these studies were done only in small numbers of animals. It would be premature to discuss the significance of these findings until further results are available.

In conclusion, we consider that glutamate and NMDA antagonists may have an important role in the development and prevention of cytotoxic oedema in the ischaemic state.

References

1. Farrar TC, Becker ED (1971) Pulse and Fourier transform NMR. Academic Press, New York, pp 46–52
2. Haida M, Yamamoto M, Matsumura H, Shinohara Y, Fukuzaki M (1987) Intracellular and extracellular spaces of normal adult rat brain determined from the protom NMR relaxation time. J Cereb Blood Flow Metab 7: 552–556
3. Haida M, Shinohara Y, Yamamoto M, Taniguchi K, Fukuzaki M (to be submitted) A rapid shift of water from extracellular to intracellular space after total ischaemia detected by NMR-comparison with the results by impedance method
4. Hossman KA (1971) Cortical steady potential, impedance and excitability changes during and after total ischaemia of cat brain. Exp Neurol 32: 163–175
5. Kochhar A, Zivin JA, Lyden PD, Mazzarella V (1988) Glutamate antagonist therapy reduces neurologic deficits produced by focal central nervous system ischaemia. Arch Neurol 45: 148–153
6. Ozyurt E, Graham DI, Woodruff GN, McCulloch J (1988) Protective effect of the glutamate antagonist, MK-801 in focal cerebral ischaemia in the cat. J Cereb Blood Flow Metab 8: 138–143
7. Shinohara Y, Yamamoto M, Haida M, Taniguchi R (1989) Calcium antagonists and a rapid shift of water from extracellular to intracellular space after cerebral ischaemia. In: Hartmann A, Kuschinsky W (eds) Cerebral ischaemia and calcium. Springer, Berlin Heidelberg New York, pp 150–154
8. Swan JH, Meldrum BS (1990) Protection by NMDA antagonists against selective cell loss following transient ischaemia. J Cereb Blood Flow Metab 10: 343–351

Correspondence: Y. Shinohara, M.D., Department of Neurology, Tokai University, School of Medicine, Isehara, Kanagawa, Japan.

Acta Neurochirurgica, Suppl. 51, 201–203 (1990)
© by Springer-Verlag 1990

Effect of Atrial Natriuretic Peptide on Ischaemic Brain Oedema

N. Nakao, T. Itakura, H. Yokote, K. Nakai, and **N. Komai**

Department of Neurological Surgery, Wakayama Medical College, Wakayama City, Japan

Summary

We investigated the effect of intraventricularly administered atrial natriuretic peptide (ANP) on the brain water, sodium and potassium contents in ischaemic brain oedema. Global ischaemia (15 min) followed by recirculation (30 min) produced by a three vessel occlusion in the rat served as a model for induction of ischaemic cerebral oedema. Water content was measured by a drying-weighing method. Sodium and potassium contents were measured by means of flame photometry. The effect of ANP was evaluated by comparing these parameters of the ANP-treated group (5 µg/kg and 10 µg/kg atriopeptin) with those of the control group (administration of 0.9% NaCL). ANP did not significantly change the content of water, sodium and potassium in the preischaemic or ischaemic brain. Intraventricularly administered ANP (5 µg/kg and 10 µg/kg) caused significant decrease in the brain water (p < 0.02) and sodium (p < 0.01) contents after 15 min of ischaemia and 30 min of recirculation, while the brain potassium content remained unaltered. Serum osmolality, and sodium and potassium concentrations were not influenced by administration of ANP. Accordingly, ANP acts directly on the central nervous system to inhibit brain water and sodium accumulation in ischaemic brain oedema.

Introduction

Atrial natriuretic peptide (ANP) is released from the cardiac atria into the systemic circulation to regulate blood pressure and systemic fluid and electrolyte balance[6]. Recently, ANP-immunoreactive neurons and its specific binding sites have been demonstrated in the brain, suggesting that ANP may be involved in some brain functions[7, 8]. Dóczi *et al.* suggested the possible involvement of central ANP in the regulation of fluid and electrolyte balance in the brain[2]. These findings prompted us to investigate the effect of ANP on the brain water and electrolyte contents in ischaemic brain oedema where the balance of these constituents may be disturbed.

Materials and Methods

Seventy-two male Sprague-Dawley rats (250–350 gm) were used for these studies. The rats were divided into two groups: treated (n = 48) and controls (n = 24). After the rats were anaesthetized with intraperitoneal pentobarbital sodium (50 mg/kg), the following experiments were performed. Ischaemic brain oedema was produced by 15 min of global ischaemic followed by recirculation, using a three-vessel occlusion model[4]. In the treated group, ANP (5 µg/kg or 10 µg/kg atriopeptin II) in 10 µl of 0.9% NaCl was stereotaxically administered into the right lateral ventricle. In the control group, 10 µl of 0.9% NaCl was intraventricularly administered in the same manner as the treated group. In both groups, at 30 min after intraventricular administration the basilar artery was occluded and the common carotid arteries were bilaterally clipped for 15 min, and then recirculation was achieved by releasing the clips. The rats were sacrificed immediately after 15 min of ischaemia, at 15 min anf 30 min following recirculation. Some rats in each group, which underwent the same experimental procedures except for the ischaemic insult, were sacrificed at 30 min after intraventricular ANP administration, serving as ANP-treated controls.

The rats were decapitated and the brains were quickly removed. The wet weight of the brain was measured on a chemical balance. After drying the brain in an oven at 105 °C for 5 days, its dry weight was obtained. Water content was calculated as follows: water content (%) = [(wet weight − dry weight)/wet weight] × 100. The dehydrated samples were homogenized with 10 ml of 0.75 N nitric acid solution prepared with deionized water, and then sodium and potassium contents of the supernatant were measured by flame photometry and expressed as mEq/kg dry weight. Serum samples obtained at the point of sacrifice were diluted with deionized water and subjected to the determination of ion concentration (mEq/L) and osmolality (mOsm/kg, H_2O) by flame photometry and the freezing point depression method, respectively. Statistical analyses were made using Student's two-tailed unpaired t-test.

Results and Discussion

ANP had no effect on preischaemic levels of water, sodium and potassium in the brain. Ischaemia alone did not significantly affect these parameters in either

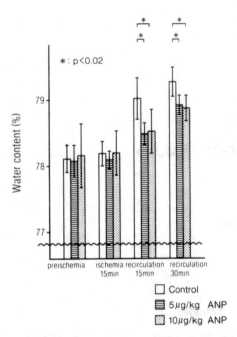

Fig. 1. Brain water content in the treated and the control groups. Each column represents the mean ± standard deviation of six animals (* $p < 0.02$, significantly different from the corresponding values in the control group)

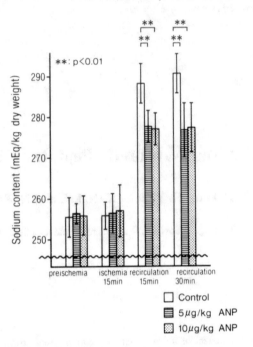

Fig. 2. Brain sodium content in the treated and the control groups. Each column represents the mean ± standard deviation of six animals (** $p < 0.01$, significantly different from the corresponding values in the control group)

untreated or ANP treated animals (Figs. 1 and 2). In the untreated group, the brain water and sodium contents increased while the potassium content decreased significantly at 15 min and 30 min after recirculation as compaired to preischaemic values ($p < 0.005$). At the same time intraventricular administrations of ANP (5 µg/kg and 10 µg/kg) significantly decreased the brain water ($p < 0.02$) and sodium ($p < 0.01$) as compared to the respective untreated postischaemic brain. However the cerebral potassium level was not affected by ANP treatment. The effects were the same irrespective of ANP concentration (Figs. 1 and 2).

Intraventricularly administered ANP did not significantly change serum osmolality or serum sodium or potassium concentrations. Therefore, the effect of ANP on the brain water and sodium is not mediated by primary changes in these parameters, but rather it is exerted by a direct action on the central nervous system.

Regarding the effect of ANP on the brain water and sodium levels in ischaemic brain oedema, the mechanism for sodium transport from blood to brain has to be considered since it may play an important role in the development of brain oedema. It has been suggested that there is an active transport system of sodium across the brain microvessels and its permeability dependent on the activity of Na^+, K^+-adenosinetriphosphatase

(ATPase) localized in the endothelial cell[3]. It was reported that reperfusion preceded by global ischaemia increased in the brain microvessel Na^+, K^+-ATPase activity[5]. In the present investigation, the change in the amount of brain water and sodium by ANP was only detected during recirculation. Therefore one of the possible mechanism responsible for the effect of ANP on the brain water and sodium in ischaemic brain oedema, may affect microvesselar Na^+, K^+-ATPase activity. Such a relationship between ANP and Na^+, K^+-ATPase localized in the brain microvascular endothelial cell has not yet been demonstrated.

Recently, specific ANP binding sites have been demonstrated in the brain microvessels that constitute the blood-brain barrier and participate in the maintenance of brain water and electrolyte homeostasis[1]. To elucidate the role of central ANP in the regulation of brain water and electrolyte balance, we need further studies, in particular do draw attention to a possible relationship between ANP and the brain microvessels.

References

1. Chabrier PE, Roubert P (1987) Specific binding of atrial natriuretic factor in brain microvessels. Proc Natl Acad Sci USA 84: 2078–2081
2. Dóczi T, Joó F, Szerdachelyi P, Bodosi M (1987) Regulation of brain water and electrolyte contents: The possible involvement of central atrial natriuretic factor. Neurosurgery 21: 454–458

3. Eisenberg HM, Suddith RL (1979) Cerebral vessels have the capacity to transport sodium and potassium. Science 28: 217–230
4. Kameyama M, Suzuki J, Shirane R, Ogawa A (1985) A new model of bilateral hemispheric ischaemia in the rat: Three-vessel occlusion model. Stroke 16: 489–493
5. Mrsulja BB, Djuricic BM, Cvejic V, Mrsulja BK, Abe K, Spatz M (1980) Biochemistry of experimental ischaemic brain oedema. Adv Neurol 28: 217–230
6. Palluk R, Gaida W, Hoefke W (1986) Minireview: Atrial natriuretic factor. Life Sci 36: 165–174
7. Quirion R, Daple M, Dam T-V (1986) Characterization and distribution of receptors for the atrial natriuretic peptides in mammalian brain. Proc Natl Acad Sci USA 83: 174–178
8. Skofitsch G, Jakobowitz DM, Eskay RL, Zamir N (1985) Distribution of atrial natriuretic factor-like immunoreactive neurons in the brain. Neuroscience 16: 917–948

Correspondence: N. Nakao, M.D., Department of Neurological Surgery, Wakayama Medical College, Wakayama City, Japan.

Acta Neurochirurgica, Suppl. 51, 204–206 (1990)
© by Springer-Verlag 1990

Suppression of Water Shift into Intracellular Space by TA-3090 (Calcium Entry Blocker) Measured with NMR

M. Yamamoto, M. Haida, R. Taniguchi, K. Yazaki, D. Kurita, M. Fukuzaki, and **Y. Shinohara**

Department of Neurology, Tokai University School of Medicine, Kanagawa, Japan

Summary

In order to clarify the effects of TA-3090 (calcium entry blocker) on the suppression of water after ischaemic events, 16 measurements of T_{2f} were continuously performed on brain biopsy (for 2–60 min) obtained from treated and control Wistar rats. The time constant (k) and NMR parameters ($T_{2f(o)}$, $\triangle T_{2f}$, $T_{2f}max(T_{2f}(o) + \triangle T_{2f})$) were obtained from 16 values of T_{2f}. The values of k in Wistar rats treated with intravenously administrated TA-3090 (0.5 mg/kg) were significantly prolonged as compared to that of control. There were no significant differences of maximum prolongation of $T_{2f}(T_{2f}max(T_{2f}(o) + \triangle T_{2f})$ among three groups. Since the prolongation of T_{2f} after biopsy reflects the water shift from extra to intracellular space, the increments of time constant indicates that TA-3090 suppresses the water shift into intracellular space. Our present results suggest that TA-3090 prevents some processes involved in irreversible cell damage and suppresses the cytotoxic brain oedema in the incomplete ischaemic area where non-competitive calcium channel is inactive.

Introduction

The influx and accumulation of calcium into the intracellular space in the ischaemic region have been reported to play a major role in producing brain oedema and several calcium entry blocker have been demonstrated to have protective effects against brain oedema[1–3]. Although the relationship between the accumulation of calcium and formation of brain oedema has been analyzed by administration of a calcium entry blocker[4], no dynamic evaluation of the protective effect of a calcium entry blocker has been made. It is well established that the interaction between water molecules and macromolecules such as proteins is a major factor determining the proton relaxation times of water measured by nuclear magnetic resonance (NMR). The dynamic changes of the amount of water molecules could be estimated by continuous measurement of relaxation times. We have previously reported that the prolongation of T_{2f} in the intracellular space of rat brain tissue after biopsy could reflect the shift of water from extracellular space to intracellular space[5–7].

The purpose of this communication is to describe the effects of the calcium entry blocker "TA-3090" (1,5-benzothiazepine derivative) on the prolongation of T_{2f}, which reflects the shift of water into intracellular space.

Materials and Methods

Wistar male rats (8–10 weeks old, weighing 200–280 g, n = 31) were anaesthetized by intraperitoneal administration of chloralose (50 mg/kg) and urethan (500 mg/kg). A small burr hole was drilled in the fronto-parietal region of the rat skull. After the intravenous administration of TA-3090 concentration of 0.1 mg/kg/0.8 ml saline (n = 5) or 0.5 mg/kg/0.8 ml saline (n = 5) in 5 minutes, a small brain tissue sample, 2 mm in thickness, was punched out with a sharp-edged glass tube (ID = 2 mm). The glass tube was sealed tightly by waterproof material, Hematoseal. Sixteen continuous measurements of T_{2f} up to 60 minutes and the calculation of the time constant (k) of prolongation of T_{2f} have been reported previously[6, 7]. We selected four different times (10, 20, 40, 60 msec) as echo times (τ) in order to calculate the value of intracellular T_{2f} by means of Hahn's spin-echo method. In order to estimate the speed of prolongation of T_{2f} after sampling, the obtained 16 values of T_{2f} were fitted to the following equation by the iteration method: $T_{2f} = \triangle T_{2f}(1-\exp(-t/k)) + T_{2f(o)}$, where $\triangle T_{2f}$ is the increment of T_{2f}, k is the time constant (min), and $T_{2f(o)}$ is the extraporated T_{2f} value at the time of sampling. The measurements of T_{2f} were done without the administration of TA-3090 in 21 Wistar rats as a control group.

Results

Figure 1 illustrates the increase of T_{2f} after sampling in control rats and rats given TA-3090 0.1 mg/kg or 0.5 mg/kg. The values of k in the 0.1 mg/kg and 0.5 mg/kg groups were 14.1 ± 7.1 min (mean $\pm$ SD) and 22.0 ± 4.5 min, respectively. The value of k in the

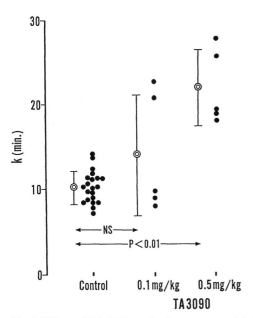

Fig. 1. Effects of TA-3090 on the time constant "k". Time constant "k" of 5 Wistar rats given TA-3090 0.5 mg/kg was significantly prolonged compared to that of control group

0.5 mg/kg group was significantly prolonged compared to that of the control group (10.5 ± 3.0 min). There are no significant differences of three parameters ($T_{2f(o)}$, $\triangle T_{2f}$, $T_{2f}max(T_{2f(o)} + \triangle T_{2f})$) among the three groups.

Discussion

1. Methodological Consideration

a) Selection of Echo-Times: It is well known that the transverse relaxation time T_2 of protons of water is more sensitive to environmental changes around the water molecules than longitudinal relaxation time T_1. We have reported that the magnetization decay curve of biopsied rat brain obtained by Hahn's spin-echo method ($\tau = 40, 80, 120, 160, 200, 240, 280, 320$ msec) could be fitted by two different T_2 components, fast T_2 and slow T_2, and we concluded that fast T_2 represented the states of water molecules in the intracellular space[5]. In our present study, we selected $\tau = 10, 20, 40, 60$ msec as echo times. T_{2f}, which was calculated by fitting the intensity obtained at these 4 different echo times, was considered to correspond mainly to fast T_2 described above. Therefore, the changes of T_{2f} could reflect mainly the changes of molecular environment around water or the amount of water in the intracellular space.

b) Prolongation of T_{2f}: In the present study, we observed the prolongation of T_{2f} of rat brain after

sampling. Since T_2 is determined from the following equation based on the fast exchange model of Zimmerman and Brittin[8]:

"$1/T_2 = x/T_{2bound} + y/T_{2free}$, T_{2bound}: T_2 of bound water, and T_{2free}: T_2 of free water, $x + y = 1$", the prolongation of T_{2f} should be accounted for by the increase of free water, the decrease of bound water, or the increase of T_{2bound} in the intracellular space. It is well known that T_2 is influenced not only by the increase of free water but also by changes in the environment of water, such as the concentration of protein. However, changes of protein structure such as proteolysis or comformational changes, the release of electrolytes, and acidosis produced in cells in an ischaemic state increase bound water and decrease T_2 though the increase of interaction between water molecules and macromolecules such as protein. Therefore, the prolongation of T_{2f} suggested the increase of free water in the intracellular space. Since in our experimental procedure, the total amount of water in the sealed biopsied tissue of rat brain should be constant, the prolongation of T_{2f} can reasonably be considered to reflect the shift of water from the extracellular space to intracellular space. In our present study, the time constant "k" is considered to be an index of the transfer speed of water into intracellular space.

2. TA-3090 and NMR Parameters

TA-3090 (a 1,5-benzothiazepine derivative) is a calcium entry blocker which acts as an inhibitor of the voltage-dependent calcium channel and has some affinity for brain tissue[9, 10]. The intravenous administration of TA-3090 caused the dose-dependent increase of the time constant k of prolongation of T_{2f}. However, there were no significant changes in the other parameters $T_{2f(o)}$, $\triangle T_{2f}$ after administration of TA-3090. No significant changes of $T_{2f(o)}$ among the three groups indicated that the administration of TA-3090 did not influence the intracellular states of water in the rat brain. Since biopsies tissue is considered to be a kind of simplified complete ischaemic model, the final intracellular state of water should be the same as that of the control group, as is suggested by the absence of significant changes of $\triangle T_{2f}$ among three groups. Other channels, such as non-competitive calcium channel, which is not inhibited by TA-3090, still operate, and many complicated events such as autolysis and membrane damage occur with the passage of time after sampling. Therefore, the final increase of T_{2f} (increase of $\triangle T_{2f}$) showed no change compared to control. Our present data suggested that administration of TA-3090 before sampling can suppress or delay the shift of water from extracellular space into intracellular space. TA-3090, as well as other calcium channel blockers such as nimodipine and nicardipine, has a protective action against brain swelling after ischaemic events[7].

The accumulation of calcium causes many events which induced an increase of permeability or/and driving force in the process leading to irreversible cell damage. Calcium entry blockers prevent some of these deteriorations and suppress cytotoxic brain oedema in the area of incomplete ischaemia where non-competitive calcium channels are inactive.

References

1. Hossmann KA, Paschen W, Cuba L (1983) Relationship between calcium accumulation and recovery of cat brain after prolonged cerebral ischaemia. J Cereb Blood Flow Metab 3: 246–253
2. Bielenberg GW, Haubruck H, Krieglstein J (1987) Effects of calcium entry blocker Emopamil on postischaemic energy metabolism of the isolated perfused rat brain. J Cereb Blood Flow Metab 7: 489–496
3. Siesjo BK (1981) Cell damage in the brain: a speculative synthesis. J Cereb Blood Flow Metab 1: 155–185
4. Abe K, Kogure K, Watanabe T (1988) Prevention of ischaemic and postischaemic brain oedema by a novel calcium antagonist (PN 200-110). J Cereb Blood Flow Metab 8: 436–439
5. Haida M, Yamamoto M, Matsumura H, Shinohara Y, Fukuzaki M (1987) Intracellular and extracellular spaces of normal adult rat brain determined from the proton nuclear magnetic resonance relaxation times. J Cereb Blood Flow Metab 7: 552–556
6. Haida M, Shinohara Y, Yamamoto M, Taniguchi R, Ohsuga H, Fukuzaki M. A rapid shift of water from extracellular to intracellular space after total ischaemia by NMR (to be submitted)
7. Shinohara Y, Yamamoto M, Haida M, Taniguchi R (1989) Calcium antagonist and a rapid shift of water from extracellular to intracellular space after cerebral ischaemia. In: Hartmann A, Kuschinsky W (eds) Cerebral ischaemia and calcium. Springer, Berlin Heidelberg New York, pp 150–154
8. Zimmerman JR, Brittin WE (1957) Nuclear magnetic resonance studies in mutiple phase systems: Lifetime of a water molecule in an absorbing phase silica gel. J Phys Chem 6: 1328–1333
9. Makamura S, Ohashi M, Suzuki T, Sugawara Y, Usuki S, Takaiti O (1989) Metabolic fate of the new Ca^{++}-channel blocking agent (+)-(2 s,3 s)-3-acetoxy-8-chloro-(2-(dimethylamino)ethyl-2,3-dihydro-2-(4-methoxyphenyl)-2,5-benzothiazepin-4-(5 H)-one maleate. Arzneim-Forsch/Drug Res 39: 1100–108
10. Murata S, Kikkawa K, Yabana H, Nagao T (1988) Cardiovascular effects of a new 1,5-benzothiazepine calcium antagonist in anaesthetized dogs. Arzneim-Forsch/Drug Res 38: 521–525

Correspondence: M. Yamamoto, M.D., Department of Neurology, Tokai University School of Medicine, Kanagawa, Japan.

Acta Neurochirurgica, Suppl. 51, 207–209 (1990)
© by Springer-Verlag 1990

Ischemia and Brain Oedema II (Focal Ischemia)

Early Detection of Ischemic Injury: Comparison of Spectroscopy, Diffusion-, T2-, and Magnetic Susceptibility-Weighted MRI in Cats

M. E. Moseley[1], J. Mintorovitch[1], Y. Cohen[2], H. S. Asgari[1], N. Derugin[1], D. Norman[1], and J. Kucharczyk[1]

Departments of [1] Radiology and [2] Pharmaceutical Chemistry, University of California, San Francisco, California, U.S.A.

Summary

Within one hour following MCA-occlusion in cats, heavily diffusion-weighted spin-echo MR images exhibited a well-defined hyperintensity in the gray matter and basal ganglia of the occluded side over normal side. This hyperintensity correlated with lactate and inorganic phosphate increases in peak areas from MR surface coil spectroscopy. T2-weighted MRI showed no significant abnormality in signal intensity from the occluded hemisphere within several hours post-occlusion. Using a paramagnetic MR contrast agent, dysprosium-DTPA-BMA together with heavily T2-weighted spin-echo or with T2*-weighted echo-planar (EPI) MR imaging, perfusion deficits resulting from MCA-occlusion were detected as a relative hyperintensity of ischaemic tissues compared to normally-perfused cerebral tissues in the contralateral hemisphere. Evidence of these deficits was observed within minutes of occlusion, and spatially correlated well with the hyperintensity seen on the diffusion-weighted images.

Diffusion- and susceptibility-weighted MRI was superior to conventional T2-weighted MRI in the detection of early ischaemic events. In contrast to surface coil spectroscopy, both techniques mapped regions of jeopardy throughout the brain, which later showed T2-weighted hyperintensity and lack of vital (TTC) staining.

Introduction

The signal intensity in proton MRI depends in T1 and T2 relaxation processes, proton density, as well as molecular motions such as diffusion. *In vivo* diffusion MRI, which can map microscopic motion of water protons (motional lengths on the order of microns), has been the subject of only a few recent studies[1–6]. New designs in gradient coils (notably the self-shielded gradients) have largely eliminated eddy current problems and can generate larger gradient strengths (up to ± 20 gauss/cm with bore sizes of 15 cm).

On potential application of diffusion-weighted MRI involves the study of cerebral ischaemia. The first minutes following a cerebral hypoxic/ischaemic insult are characterized by a progressive loss of intracellular high-energy phosphates, such as ATP and PCr, and increases in inorganic phosphate and lactate levels. This metabolic disruption results in a breakdown of the cell membrane $Na+/K+$ homeostasis, resulting in an osmotic increase in intracellular water at the expense of the water in the extracellular space[7]. Early *in vivo* sequential detection of these changes is one of the main goals of MRI and MR spectroscopy (MRS) studies of cerebral ischaemia[8]. However, problems with MRS localization and the necessity of using large volumes of tissue to obtain spectra have limited the use of MRS in producing high-resolution maps of metabolic changes in ischaemic regions. Furthermore, the use of T2-weighted MRI in the early detection of brain ischaemia has been limited to the visualization of cerebral oedema, which is not readily apparent within the first 2–3 hours following ischaemia.

These shortcomings have prompted investigators to explore the potential of diffusion-weighted MRI to detect early ischaemic injury[9–12] and to map diffusion in acute and chronic clinical infarction[13].

In addition, echo-planar MRI can acquire an image in as little as 30–70 msec[4, 6]. This speed offers tremendous advantages in MRI, namely, imaging diffusion without motion artifacts and, in particular, fast image acquisition for sequential "real time" imaging of MR contrast agent wash-in and wash-out. The acquisition of real time contrast dynamics can be used to determine blood volumes, flow transit times and relative true "perfusion" or blood flow.

Materials and Methods

Non-fasted cats weighing 2.5 to 3.5 kg were anesthetized with 1–2% isoflurane. Pavulon® (0.5 ml/kg) was given *i.v.* every 45 minutes for muscle paralysis. The temporal muscles overlying the temporo-parietal fossa were excised unilaterally to optimize extracranial external placement of the surface coil. The right middle cerebral artery (MCA) was occluded just proximal to the origin of the lateral striate arteries with bipolar electrocautery followed by surgical transection.

A General Electric CSI 2 Tesla unit equipped with GE Acustar™ self-shielded gradient coils ($\pm$ 20 gauss/cm) was used. Phosphorus-31 and H-1 MRS was performed with a 2×3 cm double-tuned, balanced-matched surface coil positioned over the MCA territory after scalp and muscle retraction. The imaging birdcage and surface coils were equipped with detuning capacitor switches to perform sequential spectroscopy and imaging without changing the position of the animal in the magnet. Successive multi-slice diffusion-weighted images, MRS spectra, echoplanar and T2-weighted images were obtained for at least six hours following occlusion. Pre- and post-contrast T2-weighted images (TR 2800, TE 80 and 160, 3 mm slices) were obtained with a field-of-view (FOV) of 80 mm in which two scans were averaged (NEX 2) for each one of the 128 phase-encoding steps (scan time = 12 minutes). Cardiac-gated and non-gated diffusion-weighted spin-echo images (TR 1000, TE 80, 3 mm slices, FOV 80 mm) were acquired with diffusion gradient pulse durations of 20 msec, a diffusion gradient separation of 40 msec and large diffusion gradients of 5.5 gauss/cm resulting in gradient b values of 1413 sec/mm² (1–2). Four scans were averaged (NEX 4) for each of 128 phase-encoding steps (scan time = 8 minutes). Echo-planar (EPI) images were acquired in a gradient-echo mode (TE = 16–64 msec) with 64×64 matrices, 60 mm FOV, 2 mm slices and image acquisition times of 65 msec. EPI diffusion was achieved by a pulsed gradient Stejskal-Tanner EPI sequences (TR 15 seconds, TE 82 msec) with 150 msec image acquisition times and b values up to 1750 sec/mm². In post-contrast studies, 0.1 and 0.25 mmol/kg dosages of DyDTPA-BMA (Salutar, Inc., Sunnyvale, CA) were delivered in 1-second bolus administrations. Up to 16 EPI images were acquired in 8–16 seconds.

Results and Discussions

The effect of increasing the diffusion gradient strength was seen as an increase in image contrast (signal intensity differences) between cerebrospinal fluid (CSF), ischaemic and non-ischaemic regions. The signal intensity difference between abnormal and normal regions of brain arises from a progressive attenuation of signal intensity from normal regions of brain with increasing gradient strength. From ROI analyses, apparent diffusion coefficients (ADC) were measured to be $0.8–0.9 \times 10^{-5}$ cm²/sec for both normal basal ganglia and cortical gray matter regions. Corresponding apparent diffusion coefficients (ADC) from the con-

tralateral abnormal, ischaemic regions were typically 0.3 to 0.6×10^{-5} cm²/sec, significantly lower than the normal regions. The ADC values in heavily myelinated white matter are greatly direction-dependent, with ADC values ranging from 0.3 to 1.2×10^{-5} cm²/sec in the internal capsule depending on the diffusion gradient direction selected. No directional dependence was found in the basal ganglia, normal or abnormal gray matter. Within the first hours following occlusion, no clear involvement of white matter has been observed.

At 1 hour following occlusion, the T2-weighted tissue signal intensity in the MCA territory ipsilateral to the occlusion was not significantly different from the contralateral normal hemisphere. In all cats studied (> 50 to date), the T2-weighted tissue signal intensity ratios ("SIR" of ischaemic ROI to normal hemisphere, contralateral ROI) in the temporal cortex average to be 1.0 ± 0.1 at the early (within 1 hour post-occlusion) timepoint. In comparison, at the same early timepoint, the SIR in the corresponding diffusion-weighted images was already significantly elevated (average SIR in the temporal cortex of 1.4–1,6). This early increase in signal intensity ratios observed in diffusion-weighted images was always observed within the vascular territory of the middle cerebral artery. The region of relative hyperintensity has also correlated well with vital staining (such as TTC). In addition, relative hyperintense regions from diffusion-weighted images has also displayed elevated levels of inorganic phosphate and lactate from regional surface coil studies done on these same animals (19–20).

Echo-planar Stejskal-Tanner diffusion-weighted MR images correlate well with the above observations. Series of progressively diffusion-weighted images (ten b values 17–2512 sec/mm² acquired) of the MCA occlusion model clearly show the enhanced attenuation of signal from normal brain. The relative hyperintensity observed in the ischaemic regions increases with b value and is the result of a lower effective ADC in these regions. Such a sequence of images can be typically acquired in 1 minute (effective TR = 6 sec) and can accurately describe regional ADC values.

The exact origin of the lower ADC value and the observed hyperintensity in the heavily diffusion-weighted images remains unclear. Two possible effects of early ischaemia events may account for the observed diffusion-weighted hyperintensity. The results from these studies suggest that the hyperintensity observed in the diffusion-weighted images may reflect accumulation of intracellular, restricted (slowed diffusion) water protons caused by the disruption of the sodium potassium transmembrane pump. Brain pulsations or micro-vibrations, created by arterial blood flow in perfused regions, may on the other hand, increase the apparent proton diffusion observed in normally-perfused cerebral tissues. A loss of perfusion in an ischaemic region may, on the other hand, attenuate brain pulsations, thereby leading to a decrease in apparent proton motion and thus to an observed diffusion-weighted relative hyperintensity. The observation that early ischaemic hyperintensity is seen only in gray matter regions may be explained by the low blood blow and denser nature of white matter tracts.

The contrast-enhanced MRI spin-echo T 2-weighted and echo-planar T 2*-weighted studies advanced the time of detection. Perfusion deficits resulting from MCA-occlusion were detected as a relative hyperintensity of ischaemic tissues compared to normally-perfused cerebral tissues in the contralateral hemisphere. Contrast differences between normally-perfused and ischaemic regions were as high as 50% in the T 2-weighted images. Evidence of these deficits was observed within minutes of occlusion, and spatially correlated well with the hyperintensity seen on the diffusion-weighted images. In the echo-planar images, dramatic decreases in perfused gray matter is seen by 6 second post-contrast. The gray-white matter tissue contrast is typically 300% on contrast-enhanced echo-planar images acquired at the height of the contrast bolus. Further, the dynamics of perfusion can be easily followed. In the MCA-occluded cats, perfused tissues loses image intensity due to the presence of T 2*-shortening contrast agent. However, by 5–9 seconds post-contrast, the ischaemic region remains hyperintense, indicating successful occlusion. In some cats, a loss of intensity in the ischaemic region after 10 seconds suggests that contrast is reaching this region later in time, perhaps due to the presence of collateral flow in the ischaemic region. The major advantage of contrast-enhanced echo-planar is that repeated fast determinations of cerebral perfusion may be obtained.

References

1. Le Bihan D, Breton E, Lallemand D, Grenier P, Cabanis E, Laval-Jeantet M (1986) MR Imaging of intravoxel incoherent motions: Application to diffusion and perfusion in neurologic disorders. Radiology 161: 401–407

2. Le Bihan D, Breton E, Lallemand D, Aubin M-I, Vignaud J, Laval-Jeantet M (1988) Separation of diffusion and perfusion in intravoxel incoherent motion MR imaging. Radiology 168: 497–505

3. Merboldt K-D, Bruhn H, Frahm J, Gyngell HM, Hanicke W, Deimling M (1989) MRI of "diffusion" in the human brain: New results using a modified CE-FAST sequence. Magn Reson Med 9: 423–429

4. Turner R, Le Bihan D, Delannoy J, Pekar J: Echo-planar diffusion and perfusion imaging at 2.0 tesla. Society of Magnetic Resonance in Medicine (SMRM), 8th Annual Meeting, 1989 (Abstract)

5. Thomsen C, Henriksen O, Ring P (1987) In vivo measurement of water self-diffusion in the human brain by magnetic resonance imaging. Acta Radiol 28: 353–361. See also: Thomsen C, Ring P, Henriksen O. In vivo measurement of water self-diffusion by magnetig resonance imaging. Society of Magnetic Resonance in Medicine (SMRM), 7th Annual Meeting, 1988 (Abstract)

6. Rosen BR, Belliveau JW, Chien D (1989) Perfusion imaging by nuclear magnetic resonance. Magn Reson Quart 5: 263–281

7. Siesjo BK (1978) "Brain energy metabolism". John Wiley and Sons, New York, chapter 15

8. Brant-Zawadski M, Weinstein PR, Bartkowski H, Moseley M (1987) Am J Rontgen 148: 579

9. Berry I, Manefe C, Gigaud M, Brueuille P. IVIM*-MRI in experimental cerebral ischaemia. Society of Magnetic Resonance in Medicine, 7th Annual Meeting, 1988 (Abstract).

10. Moseley ME, Cohen Y, Mintorovitch J, Chileuitt L, Shimizu H, Kucharczyk J, Wendland MF, Weinstein PR (1989) Early detection of regional cerebral ischaemia in cats: Comparison of diffusion- and T2-weighted MRI and spectroscopy. Magn Res Med 14: 330–346

11. Moseley ME, Kucharczyk J et al. (1990) Diffusion-weighted MR imaging of acute stroke: Correlation with T2-weighted and magnetic susceptibility-enhanced MR imaging in cats. AJNR 11: 423–429

12. Mintorovitch J, Moseley ME. Chileuitt L, Shimizu H, Cohen Y, Weinstein PR (1989) Comparison of diffusion- and T2-weighted MRI in the early detection of cerebral ischaemia and reperfusion in rats. Magn Res Med (in press)

13. Chien D, Kwong KK, Buonanno FS, Buxton RB, Gress D, MacKay BC, Kistler JP, Alpert NM, Correia JA, Ackerman RH, Brady TJ, Rosen BR. MR Diffusion of cerebral infarction. Society of Magnetic Resonance in Medicine, 7th Annual Meeting, 1988 (Abstract).

Correspondence: M. E. Moseley, Department of Radiology, Box 0628, University of California, San Francisco, CA 94143, U.S.A.

Acta Neurochirurgica, Suppl. 51, 210–212 (1990)
© by Springer-Verlag 1990

Diffusion-Weighted MR Imaging and T2-Weighted MR Imaging in Acute Cerebral Ischaemia: Comparison and Correlation with Histopathology

R. J. Sevick*, J. Kucharczyk, J. Mintorovitch, M. E. Moseley, N. Derugin, and **D. Norman**

Department of Radiology, University of California, San Francisco, California, U.S.A.

Summary

Diffusion-weighted MR imaging is a new technique which measures the microscopic motion of water protons. Signal hyperintensity on diffusion-weighted images correlates closely with evidence of ischaemic damage on histopathologic sections. Following occlusion of the middle cerebral artery (MCA), diffusion-weighted images indicate the presence of early pathophysiologic changes occurring first in the basal ganglia and, subsequently, in cortical gray matter within the MCA vascular territory. Diffusion-weighted images also better define the anatomic locus of ischaemic tissue injury than T2-weighted images. Diffusion-weighted imaging thus appears to facilitate early detection and thereby possible therapeutic intervention in patients with acute stroke.

Introduction

The increased sensitivity of magnetic resonance (MR) imaging over computed tomography in detecting pathophysiologic changes associated with acute cerebral ischaemia is well established[4]. This is primarily due to the sensitivity of T2-weighted spin echo images to changes in tissue water content. Diffusion-weighted MR imaging is a new technique which measures the microscopic motion of water protons rather than water content per se[5,6]. This is accomplished by adding strong gradients to the spin echo pulse sequence. Using this technique, diffusional motion of water protons leads to signal attenuation; thus, ischaemic regions where diffusional motion is reduced appear hyperintense.

The objective of this study was to correlate hyperintensity on diffusion-weighted MR images with histopathologic evidence of ischaemic damage obtained by staining tissue with triphenyl tetrazolium chloride (TTC)[1–2, 7, 9].

Methods

Maintained under general anaesthesia, five young adult cats had right middle cerebral artery occlusion (MCA-O) performed using a transorbital approach. Subsequently, imaging was performed using a 2 Tesla MR unit with self-shielded gradient coils. Four coronal diffusion-weighted images (DWI) (TR = 1000/1500, TE = 80, "b" (factor related to gradient strength) = 1413 sec/mm^2) and four coronal T2-weighted images (T2WI) (TR = 2800, TE = 80/160) were obtained at time points ranging from 30 minutes to 12 hours post-MCA-O.

At the conclusion of the experiment, TTC was infused transcardially and the brain was incubated in TTC. Coronal sections were analyzed planimetrically, for infarct size by measuring areas totally devoid of staining (indicating lack of viability of electron transport chain enzymes).

Planimetry and ROI analyses were also performed on T2WI and DWI obtained in closest temporal proximity to the time of sacrifice and TTC infusion.

Results

TTC-stained sections showed well-defined areas of lack of staining involving the MCA territory following MCA-O. Early changes (<2 hours) predominantly affected the basal ganglia, with later (>7 hours) involvement of frontal, temporal and parietal cortex. DWI hyperintensity corresponded closely with histopathologic changes.

Early DWI demonstrated signal hyperintensity in the head of the caudate nucleus which was poorly seen on T2WI (Fig. 1). Cortical hyperintensity was variable in extent on DWI in the early post-MCA-O period and was not visible of T2WI. Hyperintensity on DWI and T2WI showed closer correspondence at later time points.

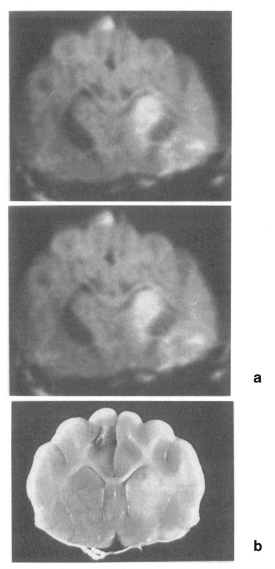

a

b

Fig. 1. 45 minute post-MCA-O coronal DWI (above), (TR = 1000, TE = 80) and 70 minute coronal T2WI (middle), (TR = 2800, TE = 160) showing clear signal hyperintensity in the head of the caudate nucleus, lentiform nuclei and a portion of cortex on DWI, poorly seen on T2WI. Corresponding TTC stained section (below) showing a lack of staining in a distribution similar to hyperintensity on the MR images

Discussion

This study demonstrates correspondence of hyperintensity on DWI with areas devoid of staining with TTC. This lack of staining implies nonviability of enzymes involved in oxidative phosphorylation and irreversible ischaemic cell damage[1]. DWI hyperintensity may have similar implications at occlusion times of 100 minutes or greater. At earlier time points it is possible that DWI hyperintensity may be reversible.

Evidence of early ischaemic damage in the basal ganglia reflected by hyperintensity on DWI may be due to the fact that the lenticulostriate arteries supplying this area are end arteries with no potential collateral flow. A previous report[3] suggested later involvement of the basal ganglia than cerebral cortex in regional cerebral ischaemia as reflected by T2WI hyperintensity. DWI clearly show early pathophysiologic changes occurring in the basal ganglia.

The better demonstration of early ischaemic injury on DWI than on T2WI is in agreement with previous reports[9]. The exact cause of hyperintensity on DWI is unknown. One possibility is that it reflects restricted diffusion in association with intracellular (cytotoxic) oedema[8]. Whatever the exact cause, DWI hyperintensity would appear to be a marker of early pathophysiologic changes associated with ischaemia.

As new pharmacologic and neurointerventional techniques are developed for early stroke treatment, an imaging modality capable of demonstrating early ischaemic changes will assume increasing importance. Preliminary examples of DWI applied to human patients with stroke have been reported[6, 11]. Unfortunately, physiologic motions and gross patient motion produce significant artifact, limiting anatomical assessment. Another factor limiting clinical implementation of DWI is the limited availability of high gradient strength coils.

DWI hyperintensity shows close correspondence with pathophysiologic changes induced by acute severe cerebral infraction. It offers advantages over T2WI in terms of early detection and accurate localization of ischaemic lesions. Clinical implementation of this technique may allow early detection of stroke and thereby earlier therapeutic intervention.

References

1. Bederson JB, Pitts LH, Germano SM, Nishimura MC, Davis RL, Bartkowski HM (1986) Evaluation of 2,3,5 triphenyltetrazolium chloride as a stain for detection and quantification of experimental cerebral infarction in rats. Stroke 17: 1304–1308
2. Bose B, Osterholm JL, Berry R (1984) A reproducible model of focal cerebral ischaemia in the cat. Brain Res 311: 385–391
3. Bose B, Jones SC, Lorig R, Friel HT, Weinstein M, Little JR (1988) Evolving focal cerebral ischaemia in cats: Spatial correlation of nuclear magnetic resonance imaging, cerebral blood flow, tetrazolium staining and histopathology. Stroke 19: 28–37
4. Brant-Zawadzki M, Pereira B, Weinstein P, Moore S, Kucharczyk W, Berry I, McNamara M, Derugin N (1986) MR imaging of acute experimental ischaemia in cats. AJNR 7: 7–11
5. LeBihan D, Breton E, Lallemand D, Grenier P, Cabanis E, Laval-Jeantet M (1986) MR imaging of intravoxel incoherant

motions: Application to diffusion and perfusion in neurologic disorders. Radiology 161: 401–407

6. LeBihan D, Breton E, Lallemand D, Aubin ML, Vignaud J, Laval-Jeantet M (1988) Separation of diffusion and perfusion in intravoxel incoherent motion MR imaging. Radiology 168: 497–505

7. Kucharczyk J, Chew W, Derugin N, Moseley M, Rollin C, Berry I, Norman D (1989) Nicardipine reduces ischaemic brain injury – Magnetic resonance imaging/spectroscopy study in cats. Stroke 20: 268–274

8. Moseley ME, Kucharczyk J, Mintorovitch J, Cohen Y, Kurhanewicz J, Derugin N, Asgari H, Norman D (1990) Diffusion-weighted MR imaging of acute stroke: Correlation with T2-weighted and magnetic susceptibility-enhanced MR imaging cats. AJNR 11: 423–429

9. Moseley M, Cohen Y, Mintorovitch J, Chileuitt L, Shimizu H, Kucharczyk J, Wendland MF, Weinstein PR (1990). Early detection of regional cerebral ischaemia in cats: Comparison of diffusion- and T2-weighted MRI and spectroscopy. Magn Reson Med 14: 330–346

10. Seidler E (1980) New nitro-monotetrazolium salts and their use in histochemistry. Histochem J 12: 619–630

11. Thomsen C, Henricksen O, Ring P (1987) *In vivo* measurement of water self diffusion in the human brain by magnetic resonance imaging. Acta Radiol 28: 353–361

Correspondence: R. J. Sevick, M.D., Department of Radiology, Neuroradiology Section, L-371, University of California, San Francisco, CA 94143-0628, U.S.A.

Acta Neurochirurgica, Suppl. 51, 213–215 (1990)

Blood-Brain Barrier Disruption Following CPR in Piglets

C. L. Schleien, R. C. Koehler, D. H. Shaffner, and **R. J. Traystman**

The Johns Hopkins Medical Institutions, Department of Anesthesiology/Critical Care Medicine, Baltimore, Maryland, U.S.A.

Summary

We studied blood-brain barrier (BBB) integrity in immature piglets during and following cardiopulmonary resuscitation (CPR). As in our previous work in the dog, there was no disruption during CPR after eight minutes of cardiac arrest, or immediately following resuscitation using a small molecule, alpha-aminoisobutyric acid. However, unlike the dog, where the BBB remained intact, we found delayed disruption of the BBB four hours after resuscitation. Young animals may be more prone to a delayed increase in BBB permeability after cardiac arrest and CPR.

Introduction

The use of adrenergic agonist drugs during CPR may cause effects on the cerebral vasculature or on cerebral metabolism if the BBB is disrupted during the following CPR. Disruption of the BBB may occur during CPR because of the large fluctuations of cerebral venous pressure which occur during chest compressions, immediately following CPR during the period of cerebral hyperemia and extreme vasodilation, or hours following resuscitation during the period of cerebral hypoperfusion. In a previous study, we reported that CPR does not result in disruption of the BBB as assessed by the transfer coefficient (Ki) of AIB (MW = 104) in an adult canine model[1]. In the present study, we determined whether AIB permeability is effected in immature swine in a different manner than in adult dogs during CPR or following defibrillation.

Methods

Two to three week old piglets (4–5.5 kg) were anesthetized with pentobarbital and catheterized. Following cardiac fibrillation, CPR was begun and epinephrine was administered which maintains perfusion pressure at levels suffient for near normal cerebral blood flow in piglets for 20 minutes of CPR. Five groups of five animals each were studied in which AIB was allowed to circulate for 10 minutes before killing the animals. Group 1, the control group, had measurements made during spontaneous circulation. In group 2, measurements were made during a 10 minute period of CPR following eight minutes of ventricular fibrillation, in order to assess any effect during CPR. In group 3, measurements were made after 40 minutes of CPR, in order to assess the effects of prolonged CPR, again following eight minutes of cardiac arrest. In group 4, following eight minutes of cardiac arrest and six minutes of CPR, the animals were defibrillated, and AIB injected three minutes later, in order to determine BBB integrity during the immediate post-ischaemic period. In group 5, following eight minutes of fibrillation, six minutes of CPR, and defibrillation, AIB was injected four hours later, in order to assess any delayed postischaemic effect.

BBB function was assessed by injecting $150\,\mu Ci$ of ^{14}C-AIB into the right atrium in order to determine its Ki into the brain. Timed arterial plasma samples were obtained in order to calculate the integrated arterial plasma concentration for AIB over the 10 minute sampling period. All of the animals received $150\,\mu Ci$ of ^3H-inulin exactly 8 minutes following the injection of AIB, in order to correct for the plasma volume space. Ten minutes following the AIB injection, the animals were killed. At the end of the experiment, following an autopsy, the brain was dissected into 14 regions. The tissue samples were dissolved, a scintillation cocktail was added, and beta counting was performed. The Ki for AIB was calculated by the following equation: $Ki = {}^{14}C\text{-tissue} - ({}^3H\text{-tissue} \times {}^{14}C\text{-plasma}_{10\,min}/{}^3H\text{-plasma}_{10\,min})/\int_0^{10} {}^{14}C\text{-plasma dt}$.

Results

Data for the 14 brain regions in the piglet is shown in the table. In the control group, Ki ranged from 0.002 to $0.005\,ml/g/min$ among regions, with a value of 0.0032 ± 0.003 for the average of the Ki from five cerebral areas. Compared to the control group, no differences were seen in Ki in groups 2, 3 or 4. In group 5, where Ki was measured 4 hours following defibrillation, the Ki was elevated in pons, midbrain, superior colliculus, diencephalon, hippocampus, caudate nucleus, anterior-middle watershed region, and the anterior, middle, and posterior cerebral arterial territo-

Table 1. K_i for AIB for 14 Brain Regions in Piglets

Region	1	2	3	4	5
Spinal cord	0.0051 ± 0.0003	0.0052 ± 0.0011	0.0037 ± 0.0005	0.0060 ± 0.0010	0.0061 ± 0.0012
Cerebellum	0.0049 ± 0.0002	0.0042 ± 0.0006	0.0035 ± 0.0004	0.0033 ± 0.0002	0.0065 ± 0.0006
Medulla	0.0035 ± 0.0003	0.0041 ± 0.0007	0.0036 ± 0.0007	0.0045 ± 0.0009	0.0066 ± 0.0008
Pons	0.0032 ± 0.0002	0.0044 ± 0.0007	0.0032 ± 0.0006	0.0037 ± 0.0003	0.0062 ± 0.0008 *
Midbrain	0.0027 ± 0.0002	0.0032 ± 0.0004	0.0024 ± 0.0003	0.0020 ± 0.0002	0.0052 ± 0.0005 *
Sup. colliculus	0.0030 ± 0.0001	0.0033 ± 0.0004	0.0026 ± 0.0005	0.0025 ± 0.0004	0.0053 ± 0.0005 *
Diencephalon	0.0022 ± 0.0001	0.0026 ± 0.0002	0.0019 ± 0.0003	0.0021 ± 0.0003	0.0046 ± 0.0006 *
Hippocampus	0.0026 ± 0.0003	0.0023 ± 0.0002	0.0021 ± 0.0002	0.0022 ± 0.0003	0.0041 ± 0.0005 *
Caudate nucleus	0.0018 ± 0.0001	0.0023 ± 0.0002	0.0016 ± 0.0003	0.0013 ± 0.0002	0.0039 ± 0.0005 *
Anterior cerebral	0.0030 ± 0.0003	0.0029 ± 0.0003	0.0031 ± 0.0007	0.0027 ± 0.0003	0.0051 ± 0.0005 *
Middle cerebral	0.0029 ± 0.0002	0.0027 ± 0.0003	0.0027 ± 0.0006	0.0025 ± 0.0002	0.0048 ± 0.0007 *
Posterior cerebral	0.0032 ± 0.0003	0.0032 ± 0.0004	0.0032 ± 0.0005	0.0026 ± 0.0001	0.0052 ± 0.0003 *
Anterior-middle watershed	0.0033 ± 0.0001	0.0032 ± 0.0004	0.0031 ± 0.0008	0.0024 ± 0.0002	0.0053 ± 0.0005 *
Posterior-middle watershed	0.0038 ± 0.0003	0.0035 ± 0.0005	0.0037 ± 0.0008	0.0025 ± 0.0003	0.0050 ± 0.0003

K_i in ml/g/min $\pm$ SEM.
* $p < 0.05$ compared to group 1 by ANOVA and Dunnett's post hoc test.

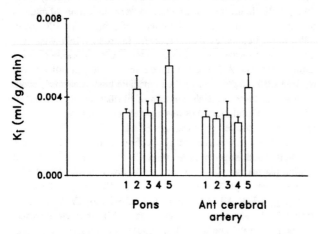

Fig. 1. K_i for AIB (ml/g/min) for 2 representative regions in piglets. Group 5, 4 hr post-resuscitation, is greater than group 1, control, by ANOVA with Dunnett's test ($p < 0.05$). 1 Control, 2–8 min ischaemia + 10 min CPR, 3–8 min ischaemia + 40 min CPR, 4–3 min post-resuscitation, 5–4 hr post-resuscitation

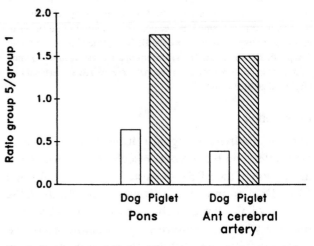

Fig. 2. Comparison of K_i for AIB ratio of group 5: group 1 dogs and piglets

ries. This is demonstrated in Fig. 1 using the pons and anterior cerebral artery region as examples.

This contrasts with the data in our canine model, where Ki was not elevated from control groups in any of the four experimental animal groups. Figure 2 demonstrates the Ki for group 5 compared to group 1, the control group, for the pons and anterior cerebral artery regions.

Discussion

A number of findings were seen in this study regarding BBB integrity during and following CPR in immature swine. As seen in the dog, the mechanical effects of vascular pressure fluctuations that occur during CPR did not disrupt the BBB. We attribute this result to the simultaneous increase in cerebrospinal fluid pressure during the compression phase of CPR, which would prevent any dynamic change in the transmural pressure. When we drained CSF in non-ischaemic dogs in our previous study, an increase in sagittal sinus pressure resulted in an increase in Ki of AIB[1]. As in dogs, there was no disruption immediately following resuscitation.

In our previous study of BBB permeability in adult dogs, we did not show disruption four hours following CPR after a cardiac arrest. In this study in piglets, using the same experimental protocol, we found evidence of BBB disruption in most of the regions studied. This may be an age related difference, which resulted in a different relationship between the ischaemic duration and severity and increase in BBB permeability.

In conclusion, BBB disruption was not evident during CPR or just following resuscitation in piglets, as in dogs. However, delayed BBB disruption was seen four hours following resuscitation in these young animals.

References

1. Schleien CL, Koehler RC, Shaffner DH, Traystman RJ (1990) Blood-brain barrier integrity during cardiopulmonary resuscitation in dogs. Stroke 21: (in press)

Correspondence: C. L. Schleien, M.D., The Johns Hopkins Medical Institutions, Department of Anesthesiology/Critical Care Medicine, 600 N. Wolfe Street, Baltimore, MD 21205, U.S.A.

Acta Neurochirurgica, Suppl. 51, 216–219 (1990)

Relationship Between Cerebral Blood Flow and Blood-Brain Barrier Permeability of Sodium and Albumin in Cerebral Infarcts of Rats

S. Ishimaru* and **K.-A. Hossmann**

Max-Planck-Institut für neurologische Forschung, Abteilung für experimentelle Neurologie, Cologne, Federal Republic of Germany

Summary

The permeability of the blood-brain barrier to sodium and albumin was investigated in rats following occlusion of the middle cerebral artery. Regional blood flow and unidirectional transfer coefficients of sodium and albumin were measured by triple tracer autoradiography, and tissue electrolyte content by atomic absorption spectroscopy. In sham-operated controls regional transfer coefficients of sodium ranged between 1.3 and $3.2 \times 10^{-3}\,\mathrm{ml \cdot g^{-1} \cdot min^{-1}}$; the transfer coefficient of albumin was below the detection limit of autoradiography. During the initial 4 h of vascular occlusion neither albumin nor sodium permeability changed although tissue sodium content increased from 56 ± 3 to $76 \pm 8\,\mathrm{\mu mol \cdot g^{-1}}$. After 24 h transfer coefficient of sodium rose to between 2.91 and $7.0 \times 10^{-3}\,\mathrm{ml \cdot g^{-1} \cdot min^{-1}}$, and tissue sodium content to $90 \pm 9\,\mathrm{\mu mol \cdot g^{-1}}$. Despite this rise the net uptake rate of sodium was 4 to 60 times lower than the unidirectional influx, indicating that permeability changes of the blood-brain barrier are without relevance for the development of stroke oedema.

Introduction

Ischaemic brain oedema is a major complicating factor of acute brain infarcts. Ng et al.[3] estimated that about 10% of stroke casualties are directly related to the mass effects of oedema. Most studies agree that the early phase of stroke oedema is of the cytotoxic type, i.e. tissue water content increases in the absence of an abnormal blood-brain barrier permeability to conventional barrier tracers such as Evan's blue or horseraddish peroxidase. However, the absence of increased barrier permeability of macromolecules does not exclude a breakdown of the blood-brain barrier for plasma electrolytes. Such a breakdown would have considerable therapeutic implications. In the presence of increased sodium permeability, an increase in blood-pressure would enhance oedema development because

of the increase in filtration pressure. In the absence of a barrier disturbance, quite contrary, an increase in blood pressure may raise blood flow above the critical threshold of ischaemic injury.

In order to assess this problem we measured blood flow and blood-brain barrier permeability of sodium and albumin in rats with middle cerebral artery occlusion using triple tracer autoradiography. As will be shown in this communication, sodium permeability did not increase before 4 h and albumin permeability before 24 h of vascular occlusion in any part of the infarct, indicating that the early phase of stroke oedema is, in fact, a purely cytotoxic type of brain swelling.

Material and Methods

Twenty-eight adult male CD Fischer rats were used. Animals were anesthetized with 1% halothane/70% nitrous oxide, and the left middle cerebral artery (MCA) was exposed by a temporal approach[2]. Twelve animals were sham-operated controls with survival times of 4 h (n = 6) and 24 h (n = 6). In the other 16 animals the MCA was electrocoagulated, followed by survival times of 4 h (n = 8) or 24 h (n = 8).

Each animal received 3 tracers: ^{125}I-bovine serum albumin (BSA 100 µCi), [22]NaCl (80 µCi), and ^{131}I-iodoantipyrine (IAP 120 µCi). [22]NaCl and ^{125}I-BSA were allowed to circulate for 5 to 60 min, followed by ramp infusion of [131]IAP for 1 min. Autoradiograms were prepared from cryostat sections, passing through the center of the MCA territory. Differentiation between the three isotopes was achieved by taking advantage of different half-lives of tracers in combination with DMP and alcohol-elution procedures.

[131]I-iodoantipyrine autoradiograms were processed for quantification of blood flow, using the algorithm of Sakurada et al.[5]. Central regions as well as the inner and outer peripheral regions of infarcts were identified on the flow images and projected onto the [22]Na and ^{125}I autoradiograms for assessment of regional sodium and albumin transfer coefficients by multiple time point analysis[4]:

$$\frac{q_{br}}{C_p(T)} = K_{in}\frac{\int_o^T C_p dt}{C_p(T)} + V_r \qquad (1)$$

---

* Present address: Department of Neurosurgery, Koshigaya City Hospital, 10-47-1 Higashikoshigaya, Koshigaya, Saitama 343, Japan.

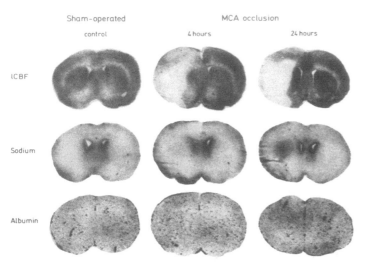

Fig. 1. Triple tracer autoradiography of blood flow (^{131}I-iodoantipyrine) and the accumulation of sodium (^{22}NaCl) and albumin (^{1235}I-bovine serum albumin) in rat brain after middle cerebral artery occlusion. Note increase of sodium (but not of albumin) in the central part of the ischemic territory after 24 h vascular occlusion

Table 1. *Local Cerebral Blood Flow and Unidirectional Transfer Coefficients for Sodium at 4 and 24 h After Middle Cerebral Artery Occlusion in Rat.* Measurements (means ± SD) were carried out in the peripheral and central parts of the infarct

	Cerebral blood flow ($ml \cdot g^{-1} \cdot min^{-1}$)		K_{in} (sodium) ($ml \cdot g^{-1} \cdot min^{-1} \cdot 10^3$)	
	4 hours	24 hours	4 hours	24 hours
Cortex				
Control (n = 6)	1.07 ± 0.15	1.07 ± 0.19	2.22 ± 0.25	2.30 ± 0.21
Outer periphery (n = 8)	0.54 ± 0.15*	0.54 ± 0.10*	2.22 ± 0.14	5.50 ± 0.33*
Inner periphery (n = 8)	0.29 ± 0.02*	0.26 ± 0.04*	2.10 ± 0.18	5.92 ± 0.54*
Central part (n = 8)	0.01 ± 0.01*	0.10 ± 0.06*	2.01 ± 0.23	3.34 ± 0.32
Basal ganglia				
Control (n = 6)	0.99 ± 0.13	1.11 ± 0.23	2.16 ± 0.25	2.21 ± 0.25
Outer periphery (n = 8)	0.58 ± 0.10*	0.56 ± 0.09*	2.10 ± 0.10	2.91 ± 0.21
Inner periphery (n = 8)	0.27 ± 0.03*	0.25 ± 0.09*	2.00 ± 0.10	6.08 ± 0.75*
Central part (n = 8)	0.01 ± 0.01*	0.08 ± 0.06*	2.20 ± 0.11	7.00 ± 0.54*

* Statistically significant difference from control (p < 0.05).

where q_{br} is the quantity of tracer in brain parenchyma (dpm/g), C_p is the tracer concentration in plasma (dpm/ml), K_{in} is the unidirectional transfer coefficient for influx into brain (ml/min per g) and V_r is the rapidly equilibrating compartment, *i.e.* mainly the plasma volume (ml/g). Measurements were carried out after increasing circulation times from 5 to 60 min, and transfer coefficients K_{in} were obtained by graphical analysis of the slope of the linear portion of the above relationship[2].

Tissue electrolyte content was measured by atomic absorption spectroscopy in samples taken from various parts of the cryostat block.

Dispersion of data is expressed as means ± SD. Statistical differences were calculated by analysis of variance (ANOVA) and a modified t test for multiple comparisons (Cochran-Cox). The significance level was set at p < 0.05.

Results and Discussion

Electrocoagulation of the middle cerebral artery resulted in focal ischaemia in the MCA territory in all animals (Fig. 1). Inspection of ^{22}Na and ^{125}I-BSA autoradiograms revealed occasionally a small area of increased radioactivity in the vicinity of the electrocoagulated MCA but little gross alterations in the rest of the brain. Only after 24 h recirculation ^{22}Na radioactivity was slightly increased in the center of the MCA territory (Fig. 1). ^{125}I-BSA radioactivity was clearly confined to the intravascular compartment at both recirculation times.

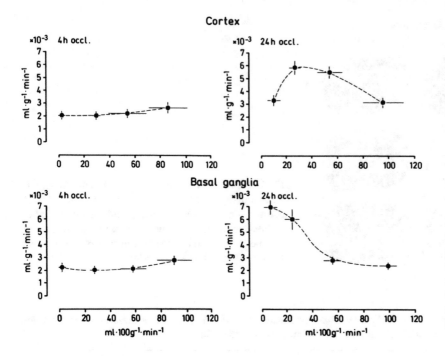

Fig. 2. Correlation between blood flow and the unidirectional transfer coefficient of sodium in cortex and basal ganglia of rat after middle cerebral artery occlusion. Measurements were carried in sham-operated controls and in the central and peripheral parts of the infarcted territory. Note the absence of permeability changes after 4 h vascular occlusion

Table 2. *Comparison of Net Increase and Unidirectional Influx of Sodium Into Cerebral Cortex After Middle Cerebral Artery Occlusion in Rats.* Rate of net increase was calculated by dividing the mean increase in tissue sodium content by the duration of ischaemia, and unidirectional influx by multiplying the mean unidirectional transfer coefficient of sodium with the plasma concentration

	Tissue content (μmol $\cdot$ g^{-1})	Net increase rate (μmol $\cdot$ g^{-1} $\cdot$ min^{-1})	Unidirectional influx (μmol $\cdot$ g^{-1} $\cdot$ min^{-1})
Control (n=6)	56±3	–	0.350
4 h ischaemia (n=6)	76±8	0.083	0.327
24 h ischaemia (n=8)	90±9	0.012	0.763

Tissue content of sodium is expressed as means ± SD.

The quantitative results are summarized in Table 1. In sham-operated animals and in non-ischaemic grey matter of experimental animals, blood flow ranged between 1.0 and 1.2 ml $\cdot$ g^{-1} $\cdot$ min^{-1}, and the unidirectional coefficient of sodium influx between 1.3 and 3.2 × 10^{-3} ml $\cdot$ g^{-1} $\cdot$ min^{-1}. Transfer coefficients of albumin were below detection limit of autoradiography. After 4 h ischaemia no significant changes of either sodium or albumin transfer coefficients could be detected in any part of the brain; after 24 h ischaemia sodium transfer coefficient - but not albumin - increased up to 7.0 × 10^{-3} ml $\cdot$ g^{-1} $\cdot$ min^{-1}. The correlation of regional sodium transfer coefficients with flow revealed that in the cortex the rise was most pronounced in the inner peripheral zone of infarct whereas in basal ganglia the central part was most severely affected (Fig. 2).

The measurement of sodium content in the infarcted region allowed the calculation of net sodium uptake (Table 2). Assuming a linear change during the two observation intervals, the net sodium uptake was 4 to 60 times lower than the unidirectional influx across the blood-brain barrier. Thus suggests that the rate of sodium accumulation and, in consequence, oedema development is not determined by the permeability state of the blood-brain barrier.

Our finding of a low sodium permeability for at least 4 h after MCA occlusion in rats corroborates similar observations in gerbils following carotid artery occlusion[1]. However, in contrast to this study, our measurements do not support the contention that the preserved blood-brain barrier function is rate limiting for oedema formation because the unilateral transfer coefficient of sodium, though low, is still far above the net sodium uptake rate. We, therefore, conclude that during the first day of infarction permeability changes of the blood-brain barrier are without relevance for the development of stroke oedema.

References

1. Betz AL, Ennis SR, Schielke GP (1989) Blood-brain barrier sodium transport limits development of brain oedema during partial ischaemia in gerbils. Stroke 20: 1253–1259
2. Blasberg RG, Patlak CS, Fenstermacher JD, Patlak CS (1983) Transport of alpha-aminoisobutyric acid across brain capillary and cellular membranes. J Cereb Blood Flow Metab 3: 8–32
3. Ng LKY, Nimmannitya J (1970) Massive cerebral infarction with severe brain swelling: A clinicopathological study. Stroke 1: 158–163
4. Patlak CS, Blasberg RG, Fenstermacher JD (1983) Graphical evaluation of blood-to-brain transfer constants from multiple-time uptake data. J Cereb Blood Flow Metab 3: 1–7
5. Sakurada O, Kennedy C, Jehle J, Brown JD, Carbin GL, Sokoloff L (1978) Measurement of local cerebral blood flow with iodo (^{14}C)antipyrine. Am J Physiol 234: H59–H66
6. Tamura A, Graham DI, McCulloch J, Teasdale GM (1981) Focal cerebral ischaemia in the rat: 1. Description of technique and early neuropathological consequences following middle cerebral artery occlusion. J Cereb Blood Flow Metab 1: 53–60

Correspondence: Prof. Dr. K.-A. Hossmann, Max-Planck-Institut für neurologische Forschung, Abteilung für experimentelle Neurologie, Gleuelerstrasse 50, D-5000 Köln 41, Federal Republic of Germany.

Acta Neurochirurgica, Suppl. 51, 220–222 (1990)

Extravasation of Albumin in Ischaemic Brain Oedema

S. A. Menzies[1], **J. T. Hoff**[1], and **A. L. Betz**[1, 2]

Departments of [1] Surgery (Neurosurgery), [2] Pediatrics, and [2] Neurology, University of Michigan, Ann Arbor, Michigan, U.S.A.

Summary

Changes in brain water, sodium, potassium, and albumin contents and blood-brain barrier (BBB) permeability were determined between 1 hr and 42 days following occlusion of the middle cerebral artery in rats. Brain oedema was maximal 24 hrs, remained high for 3 days, and was resolved by 4 weeks. These changes in water content were accompanied by a parallel increase in sodium and decrease in potassium, however, the increase in sodium always exceeded the decrease in potassium so that there was a net gain in brain cations. BBB permeability to ^{3}H-α-aminoisobutyric acid was normal for the first 4 hrs, but increased by 24 hr and returned to normal at 3 weeks. The time course for changes in brain albumin content was the same as that for BBB permeability and, at its greatest (3 days), the brain albumin was approximately 20% of the plasma albumin concentration. The relative contributions of the osmotic force produced by the increase in brain cations and the oncotic force produced by the increase in brain albumin to the observed change in water content were calculated. At all time points, the increase in brain cations accounted for nearly all of the observed brain oedema, while the increase in albumin played essentially no role in oedema development.

Introduction

During the early period following occlusion of a cerebral blood vessel, a cytotoxic type of brain oedema develops as a result of shifts in ions between blood and brain. At some later time, however, the blood-brain barrier (BBB) opens, albumin enters the brain, and the oedema, therefore, becomes vasogenic. While development of the cytotoxic oedema is entirely explained by an increase in brain ions[3, 5], the role of albumin in the vasogenic phase has not been determined. The goal of the present study was to determine the relative contributions of ions and albumin to the development of ischaemic brain oedema when the BBB is open.

Materials and Methods

Focal cerebral ischaemia was produced by occlusion of the middle cerebral artery (MCAO) in adult male Sprague-Dawley rats which were anesthetized with ketamine (50 mg/kg) and xylazine (10 mg/kg).

The method used was similar to that described by Bederson *et al.*[1] except that the initial portion of the major branches of the MCA in the vicinity of the rhinal fissure were also coagulated[7]. Animals were allowed to survive for times ranging between 1 hr and 42 days following MCAO before they were re-anesthetized and killed by decapitation. The brains were quickly removed and a 7 mm cork borer was used to obtain tissue samples from the center of the ischaemic cerebral cortex and from the corresponding area of the contralateral non-ischaemic cortex[6].

One group of animals was used for measurement of the water content as determined from the wet and dry weights and the sodium and potassium contents as determined by flame photometry[3]. The second group of animals was used for measurement of BBB permeability and brain albumin content. These animals received an intravenous injection of ^{3}H-α-aminoisobutyric acid (AIB) 10 min before termination of the experiment and a second injection of ^{14}C-inulin 7 min later[6]. Arterial blood was continuously sampled during the isotope circulation time. Brains were then removed, sampled as above, and homogenized (20% wt/vol) in water. A portion of this homogenate was used for determination of ^{3}H and ^{14}C contents while the remainder was used for measurement of the albumin content. The BBB permeability was calculated as the AIB PS product using the ^{14}C-inulin space as the brain plasma volume[6]. An enzyme-linked immunosorbent assay (ELISA)[8], was developed to measure the brain albumin content using a specific sheep anti-rat antiserum directed against rat albumin and a second sheep anti-rat antibody that was complexed to peroxidase. A Student's t-test for paired samples was used to determine whether the differences between ischaemic and non-ischaemic cortex were significant.

Results

The results shown in Table 1 are expressed as the average difference between measurements made on samples from the ischaemic and non-ischaemic cortex. Therefore, they reflect the changes due to ischaemia and the consequent brain oedema formation. The brain water content increased rapidly over the first 12 hrs and was maximal at 24–48 hrs. It subsequently decreased and returned to normal by 28 days. The brain sodium content increased and then decreased with a similar time course while the potassium content

Table 1. *Changes in Brain Water, Sodium, Potassium, Albumin, and BBB Permeability Following MCAO*

Length of ischaemia	△ water (%)	△ Na (mEq/Kg dry)	△ K (mEq/Kg dry)	△ Na + △ K (mEq/Kg dry)	△ albumin (mg/g wet)	AIB PS product (μl/g/min)
Hours						
1	1.17 ± 0.27**	91 ± 8***	-62 ± 18*	28 ± 22	0.11 ± 0.04*	
2	2.08 ± 0.34***	144 ± 22***	-90 ± 20**	54 ± 20*		0.10 ± 0.30
3	2.75 ± 0.48***	178 ± 6***	-115 ± 11***	62 ± 13**	0.11 ± 0.03*	
4	3.05 ± 0.39***	230 ± 26***	-124 ± 13***	106 ± 22**		0.06 ± 0.24
6	4.06 ± 0.36***	307 ± 33***	-155 ± 17***	152 ± 29**	0.28 ± 0.02*	
12	5.14 ± 0.55***	321 ± 37***	-109 ± 28**	204 ± 55**	0.60 ± 0.12***	
24	5.69 ± 0.37***	456 ± 53***	-193 ± 27***	264 ± 32**	1.07 ± 0.12***	1.27 ± 0.38*
Days						
2	5.67 ± 0.51***	442 ± 75***	-118 ± 27**	325 ± 79**	1.52 ± 0.13***	1.50 ± 0.30**
3	5.42 ± 0.37***	384 ± 63***	-174 ± 27***	210 ± 42**	1.90 ± 0.17***	1.78 ± 0.48**
7	3.72 ± 0.62***	232 ± 45**	-123 ± 35*	109 ± 52*	0.62 ± 0.27*	1.07 ± 0.38*
14	1.23 ± 0.89	97 ± 49	-102 ± 24	5 ± 31	1.35 ± 0.32**	2.76 ± 0.78**
21	1.47 ± 0.28	53 ± 12	-23 ± 23	30 ± 36	0.07 ± 0.01**	0.05 ± 0.06
28	0.26 ± 0.18	26 ± 22	-16 ± 24	10 ± 45	0.13 ± 0.04	0.79 ± 0.42
36	0.44 ± 0.16	22 ± 18	26 ± 10	48 ± 20		0.17 ± 0.30
42	1.26 ± 0.55	40 ± 18	-58 ± 17	17 ± 18		0.21 ± 0.06

Values are averages $\pm$ SE of 5–8 determinations.

* $p < 0.05$, ** $p < 0.01$, *** $p < 0.001$ that value is significantly different from zero.

changed reciprocally. At all times when significant oedema was present, the increase in brain sodium ($\triangle$ Na) exceeded the decrease in brain portassium ($\triangle$ K). Thus, there was a net increase in brain cations ($\triangle$ NA + $\triangle$ K).

BBB permeability to AIB was normal in the ischaemic cortex during the first 4 hr following MCAO, but by 24 hr it had doubled and by 3 days it was 2.5 times higher than the permeability of the BBB in the non-ischaemic cortex (Table 1). By 3 weeks, the BBB permeability had returned to normal. Coincident with this change in BBB permeability, the brain albumin content increased beginning about 6 hr after MCAO and peaking on the third day. It also returned to normal by 3 weeks. At its maximum, the albumin concentration of brain was approximately 20% of the plasma albumin concentration suggesting that it had equilibrated with a space similar in size to the extracellular space. Overall, the changes in BBB permeability and brain albumin content lagged behind the development and resolution of the brain oedema.

In order to determine the relative roles of ions and albumin in the development of ischaemic brain oedema, we calculated the respective osmotic and oncotic forces that would be generated by these molecules. For this comparison, the observed change in brain water (Table 1) was expressed in units of ml/g dry weight (Fig. 1). Assuming that the change in brain cations would be accompanied by an identical change in brain anions and that osmotic equilibrium with the blood could be reached, the increase in brain water which could be accounted for by ions was calculated by dividing the observed change in cations (Table 1) by the average plasma sodium concentration (145 mEq/l). The increase in brain water which could be accounted for by albumin was calculated by assuming that 5 g/100 ml of albumin exerts 20 mm Hg of oncotic pressure and that 19 mm Hg of oncotic pressure are equivalent to 1 mosmole[4]. As can be seen in Fig. 1, the change in brain water predicted on the basis of changes in ions almost totally accounted for both the formation (Fig. 1A) and resolution (Fig. 1B) of the oedema. Despite the large changes in brain albumin content, the oncotic force that it produced in brain was negligible compared to the osmotic force exerted by the ions.

Discussion

The results of this study suggest that ions are always more important determinants of ischaemic brain oedema than is albumin. Before the BBB opens, oedema

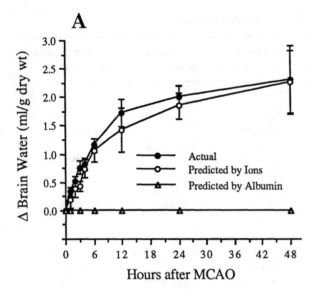

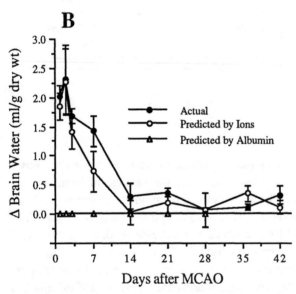

Fig. 1. Comparison of the observed and predicted changes in brain water content following MCAO. The changes in water content that could be accounted for by the observed changes in brain ions and albumin are compared during the development (A) and resolution (B) of the ischaemic oedema. Values shown are the means ± SE for 5–8 determinations at each time point

develops as sodium enters the brain in exchange for potassium, a process that is probably mediated by Na, K-ATPase in the brain capillary endothelial cell[2, 3]. During this stage, the oedema develops at a rate that appears to be limited by how fast sodium can move from blood to brain[3]. About the time that the BBB opens to low molecular weight tracers such as AIB, it also becomes permeable to albumin (Table 1). This initiates what would traditionally be called a vasogenic phase of oedema development in which the influx of

albumin should lead to a further increase in brain water. Although oedema does continue to develop, our results suggest that ions remain the principal determinants of oedema formation.

When the BBB opens to proteins, its permeability to ions also increases and oedema development may no longer be limited by how rapidly the ions can exchange across the BBB. Hydrostatic forces may further increase the influx of plasma constituents. Thus, equilibration of the sodium and potassium concentrations between interstitial fluid and plasma can occur more rapidly. Once equilibrated, brain oedema accumulation ceases despite the continued influx of proteins (Table 1). Similarly, resolution of brain oedema occurs as the ionic concentrations normalize rather than when the BBB closes.

These results suggest that the specific therapy for ischaemic brain oedema should focus on preventing or reversing ionic shifts between the intracellular and extracellular fluids. Before the BBB opens, this might be accomplished by either inhibiting BBB sodium transport or by protecting the brain cells. Once the barrier opens, however, only the latter approach is likely to be effective.

References

1. Bederson JB, Pitts LH, Tsuji M, Nishimura MC, Davis RL, Bartkowski H (1986) Rat middle cerebral artery occlusion: evaluation of the model and development of a neurologic examination. Stroke 17: 472–476
2. Betz AL (1986) Transport of ions across the blood-brain barrier. Fed Proc 45: 2050–2054
3. Betz AL, Ennis SR, Schielke GP (1989) Blood-brain barrier sodium transport limits development of brain oedema during partial ischaemia in gerbils. Stroke 20: 1253–1259
4. Fenstermacher JD (1984) Volume regulation of the central nervous system. In: Staub NC, Taylor AE (eds) Edema. Raven Press, New York, pp 383–404
5. Hatashita S, Hoff JT (1990) Brain oedema and cerebrovascular permeability during cerebral ischaemia in rats. Stroke 21: 582–588
6. Martz D, Beer M, Betz AL (1990) Dimethylthiourea reduces ischaemic brain oedema without affecting cerebral blood flow. J Cereb Blood Flow Metab 10: 352–357
7. Menzies SA, Hoff JT, Betz AL (1990) Middle cerebral artery occlusion in rats: a clinical and pathological evaluation of the model (submitted)
8. Voller A, Bidwell DE, Bartlett A (1976) Microplate enzyme immunoassay for the immunodiagnosis of virus infections. In: Rose N, Feldman H (eds) Manual of clinical immunilogy. American Society of Microbiology, Washington, D.C., pp 506

Correspondence: A. Lorris Betz, M.D., Ph.D., D3227 Medical Professional Bldg., University of Michigan, Ann Arbor, MI 48109-0718, U.S.A.

Acta Neurochirurgica, Suppl. 51, 223–225 (1990)
© by Springer-Verlag 1990

Pathophysiological Studies in the Rat Cerebral Embolization Model: Measurement of Epidural Pressure and Evaluation of Tissue pH and ATP

K. Yamane, T. Shima, Y. Okada, T. Takeda, and **T. Uozumi**[1]

Department of Neurosurgery, Chugoku Rousai Hospital, Kure, Japan, [1] Department of Neurosurgery, Hiroshima University, School of Medicine, Hiroshima, Japan

Summary

A rat embolization model was produced by occlusion of the main cerebral artery with a silicone cylinder embolus. Intracranial pressure was monitored by changes in epidural pressure (EDP), which was recorded by our designed flaccid microballoon. EDP changes could be divided into the two types: the first was moderate elevation, less than 20 mm Hg and the second was prominent, more than 20 mm Hg. Brain tissue pH on the embolized side showed heterogeneous changes composed of acidotic and alkalotic areas. Alkalotic change was frequently seen at the deep cerebrum which might be the result of plasma exudation due to disruption of blood-brain barrier (BBB). The degree of the EDP elevation was positively correlated with the extension of alkalotic area. Energy metabolism was also mainly disturbed in the deep cerebrum, but did not completely correspond to the pH change of the area. These results would indicate that embolization of the main cerebral artery could induce severe ischaemic insult on the deep cerebrum with massive brain swelling due to an early disruption of BBB.

Introduction

Brain swelling has been known as one of the most serious consequences after main cerebral artery occlusion, but its pathophysiology has been unclear. The purpose of this study was to investigate brain swelling in the embolized rat by evaluating epidural pressure (EDP), brain tissue pH and energy metabolism (ATP) and BBB function.

Materials and Methods

Fifty two male Wistar rats, aged from 20 to 25 weeks, were used. Thirty five rats served for investigating a relationship between EDP and tissue pH changes within 5 hours after embolization (Group 1) and the remaining 17 rats were used for evaluation of tissue pH and ATP changes within 24 hours occlusion (Group 2). The rats in group 1 were anaesthetized with 3% halothane, tracheotomized, mechanically ventilated and maintained with O_2 and N_2O mixture (2:3) containing 0.5% halothane. The common (CCA), external (ECA),

and internal carotid arteries (ICA) were exposed at the neck under a microscope in animals at a supine position. The ECA was cut after ligation at its distal portion. A polyethylene tube containing a silicone cylinder (400 or 500 μm in diameter and 1 mm in length) in its tip was introduced to the ICA through the ECA. The femoral artery and vein were cannulated for measuring systemic arterial blood pressure (SABP), sampling arterial blood gas and administering pancronium as necessary. Then the rat was changed to a prone position and the head was fixed in a stereotaxic frame (Narishige SR-6, Tokyo). A small hole, about 4 mm in diameter, was opened in the left parietal region for applying the microballoon. A microballoon was filled with physiological saline and connected to a pressure transducer via a polyethylene cannula. After application of the microballon on the dura, this system was opened to the air to define a zero point. Following these preparations, embolization was performed by injection of silicone cylinder embolus with 0.05 ml of physiological saline. Body temperature was maintained at about 37 °C by a heating pad.

EDP changes after embolization were measured continuously from 30 minutes to 5 hours. At the end of experiment the brain was frozen in situ with liquid nitrogen for evaluation of tissue pH. The rats in group 2 were anaesthetized for surgical preparation with mixed gas of O_2 and N_2O of the same concentration as above and maintained with 1.5–2.0% halothane using a mask, the embolization was performed by the same procedures as in the group 1. The rats were sacrificed at 1, 3 and 24 hours occlusion. The brains were frozen in situ by liquid nitrogen for evaluation of tissue pH and ATP.

For evaluation of tissue pH, umbelliferone technique reported by Csiba[2] was used. Frozen 20 μm thick coronal sections were made in a cryostat at −20 °C. This brain section was layered at −20 °C onto a strip of cellulose acetate impregnated with 0.125% umbelliferone. After thawing of the brain section at 0 °C, the strip was illuminated with ultraviolet lamp (Spectronics Corporation, model B-100, Westbury, New York) at 366 nm and recorded photographically at 450 nm using an appropriate barrier filter.

ATP was recorded by modified method reported by Kogure[3], and Paschen[6]. Frozen 20 μm thick coronal sections were freeze-dried at −20 °C for 48 hours and were overlaid with 40 μm sections of frozen solution containing all the enzymes and coenzymes required for the substrate-specific bioluminescence reaction. Light emission was recorded on photographic film. The tissue pH and ATP of the embolized hemisphere were evaluated by comparing with the homotopical region of the non-embolized hemisphere.

For evaluation of blood-brain barrier, 2 ml of 2% Evans blue (EB) was injected into the peritoneal cavity 30 minutes before freezing in 11 rats of group 2.

Results

Brain swelling was detected macroscopically in the deep cerebral structures. (*i.e.* hypothalamus, thalamus, and hippocampus) of the embolized hemisphere and a shift of midline structures to the non-embolized side was noted. Histopathological examination of the 24 hour occluded rats demonstrated infarcts in the hippocampus, thalamus and hypothalamus.

The range of changes in EDP was from 0 to 70 mmHg and mean value was 26.6 ± 22.4 mmHg (n = 27) at the end of experiment, from 30 minutes to 5 hours occlusion. The patterns of EDP changes could be divided into the following two types, one was gradual increase to less than 20 mm Hg and another was a rapid increase to over 20 mm Hg (Fig. 1). The first and second patterns were obtained in 18 and 9 rats, respectively.

The brain tissue pH of the embolized hemisphere was heterogeneously composed of acidotic and alkal-

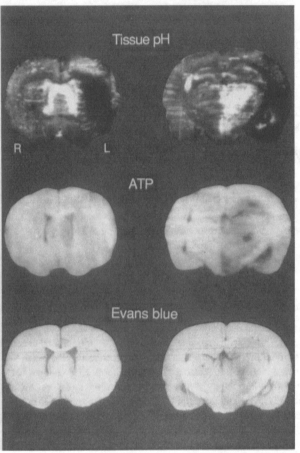

Fig. 2. Distribution of tissue pH (upper). ATP (middle) and stain of Evans blue (lower) in one hour occlusion are shown. Good correspondence between the alkalotic area and the Evans blue exudated area is noted

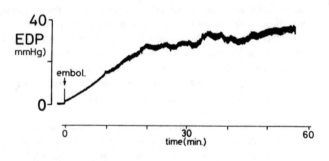

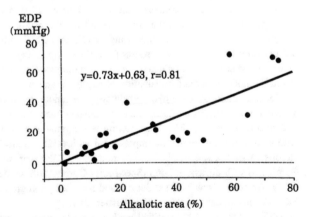

Fig. 1. Upper: One hour record of EDP after embolization demonstrates rapid elevation to near 40 mm Hg. Lower: Positive correlation between the value of EDP elevation and the extension of alkalotic area is obtained

otic areas. The cortex and caudate putamen were generally acidotic in contrast to the thalamus, hypothalamus and hippocampus which showed a mixture of alkalotic and acidotic changes (Fig. 2).

The ATP decreased area in the acidotic cortex and caudate putamen was smaller than that of the acidotic areas. The thalamus, hypothalamus and hippocampus demonstrated mild to severe decrease of ATP in the alkalotic area and severe decrease in the acidotic region. These patterns of distribution of tissue pH and ATP were almost the same after 1, 3 and 24 hours occlusion but after 24 hours occlusion the hippocampus was more alkalotic and medial thalamic region was less alkalotic than those after 1 and 3 hours occlusion.

The Evans blue staining was detected mainly in the hippocampus, thalamus and hypothalamus. The region of Evans blue-exudation corresponded well to the alkalotic area in the acute stage but after 24 hours occlusion the EB stain was not remarkable (Fig. 2).

To qualify the extension of tissue pH changed areas, the pH-changed area was calculated in the 4 coronal consequent sections at about 1.6 mm intervals, plane of the most posterior section was through near the anterior mamillary body. And the ratio of these pH-changed areas to hemispheric areas was calculated. A significant positive correlation was seen between the extension of the alkalotic area and the increase of EDP (Fig. 1).

There was no significant change in SABP during experiment. Blood gas analysis was performed in 27 rats of group 1 and pH, PaO_2 and $PaCO_2$ were 7.374 ± 0.03, 117.6 ± 18.6 mm Hg and 38.1 ± 4.0 mm Hg (mean $\pm$ SD), respectively. No significant change in arterial serum pH after injection of EB was obtained.

Discussion

In our model, the main cerebral artery could be occluded without a leakage of cerebrospinal fluid which are prerequisite for monitoring brain swelling. An embolus could not only occluded ICA but also the anterior choroidal artery and perforating arteries, which could produce serious ischaemia and massive brain swelling in the deep cerebrum. Clinicopathological studies[7] indicated that our rat embolization model was suitable for investigating massive brain swelling and its mechanism.

Elevation of intracranial pressure due to brain swelling after embolization was recorded as that of EDP which could be measured by our designed microballoon system. In the present study, the following two types of ischaemic oedema could be observed depending on the increase in EDP; the first type was a gradual increase, less than 20 mm Hg, and the second one was rapid increase, more than 20 mm Hg, within 1 hour occlusion. The degree of EDP increase may depend on the severity of the ischaemia. To clarify the mechanism of acute brain swelling due to embolism, the authors intended to investigate brain tissue pH, energy metabolism and BBB function. In the present study a close correlation was demonstrated between the extension of brain tissue alkalotic area and the degree of increase in EDP. These alkalotic areas corresponded well to BBB disrupted regions evaluated by Evans blue exudation. Further, the heterogeneous decrease of ATP was observed in the alkalotic area. And the ATP depleted area could be superimposed at the acidotic area, but the former was smaller than the later. In 24 hour occlusion the degree of EB staining was not prominent in comparison to that in earlier stage, suggesting a possible improvement of BBB function in the mild alkalotic areas.

Tissue pH is recently used as a marker for the prognosis and the effectiveness of cerebral infarction treatment[1, 4, 5]. In general, ischaemia stimulates anaerobic glycolysis and intracellular pH declines by accumulation of lactate. But some reports described an alkalotic changes in the area affected by main cerebral artery occlusion in clinical cases using magnetic resonance spectroscopy in early stage of ischaemia[4]. In this study, the observed EB exudation corresponded well with the alkalotic area indicating the passage of plasma constituents across the BBB. Normal brain tissue pH, ranged 6.93 to 7.15, is lower than pH of plasma, 7.35–7.45. Therefore, the exudated plasma fluid due to embolism would induce an alkalotic shift of brain tissue pH in the affected brain in spite of intracellular acidosis. Moreover according to the time course of brain swelling vasogenic oedema has been reported to begin at 6 hour occlusion, but in this study it occurred much earlier, within 1 hour occlusion, and disruption of BBB was observed.

In conclusion, our embolization model induced acute brain swelling mainly due to disruption of BBB in the deep cerebrum which could result in serious brain damage.

References

1. Alpert NM, Senda M, Buxton RB, Correia JA, Mackay B, Weise S, Ackerman RH, Buonanno FS (1989) Quantitative pH mapping in ischaemic disease using CO_2 and PET. J Cereb Blood Flow Metab 9: [Suppl 1]: S361
2. Csiba L, Paschen W, Hossmann KA (1983) A topographic quantitative method for measuring brain tissue pH under physiological and pathophysiological conditions. Brain Res 289: 334–337
3. Kogure K, Alonso OF (1978) A pictorial representation of endogenous brain ATP by a bioluminescent method. Brain Res 154: 273–284
4. Levine SR, Welch KMA, Gdowski JW, Chopp M, Fagan SC, Brown GG, Pajeau AK and Helpern JA (1989) The relationship of Brain pH to energy metabolism and clinical outcome in acute human cerebral ischaemia. J Cereb Blood Flow Metab 9: [Suppl 1]: S357
5. Meyer FB, Anderson RE, Sundt TM (1989) A novel dihydropyridine calcium antagonist improves CBF and brain pH in focal ischaemia. J Cereb Blood Flow Metab 9 [Suppl 1]: S219
6. Paschen W, Shima T, Hossmann KA (1984) Pial arterial pressure in cats following middle cerebral artery occlusion. II-relationship to regional disturbance of energy metabolism. Stroke 15: 686–691
7. Takeda T, Shima T, Okada Y, Yamane K, Ohta K, Uozumi T (1989) Experimental focal ischaemia produced by embolization with silicone cylinder in normotensive (NTR) and spontaneously hypertensive rats (SHR): Comparison of neurological and pathological findings. Brain Nerve (Tokyo) 41: 1119–1125

Correspondence: K. Yamane, Department of Neurosurgery, Chugoku Rousai Hospital, 1-5-1 Hirotagaya, Kure, Japan.

Acta Neurochirurgica, Suppl. 51, 226–229 (1990)
© by Springer-Verlag 1990

Determiners of Fatal Reperfusion Brain Oedema

G. M. de Courten-Myers[1], **M. Kleinholz, K. R. Wagner,** and **R. E. Myers**

[1] University of Cincinnati College of Medicine, Department of Pathology and Veterans Affairs Medical Center, Research Service, Cincinnati, Ohio, U.S.A.

Summary

Brain oedema is an important aspect of infarction from cerebrovascular occlusion. In a cat stroke model where the middle cerebral artery (MCA) was reversibly or permanently occluded, we analyzed the incidence of fatal hemispheral oedema in 35 normo- (6 mM) and 35 hyperglycaemic (20 mM for 6 hours) animals, with (N = 45) and without (N = 25) restoration of blood flow with clip release at 4 and 8 hrs of occlusion. Fatal hemispheral oedema occurred in 23% of cats (16/70) while hyperglycaemia, for one, and restoration of blood flow, for another, each quadrupled its occurrence. Further, evidence of remote oedema in the form of posterior cingulate cortical pressure atrophy from transtentorial herniation was found in animals that were allowed to survive for 2 weeks and that exhibited infarcts that affected 12 to 95% of the MCA territory. Thus, hemispheral oedema in association with MCA occlusion developed sufficiently markedly as to cause transtentorial herniation in 47% of all cats (33/70).

We carried out biochemical analyses in 14 hyper- and 10 normoglycaemic cats after 4 hrs of MCA occlusion for ATP, phosphocreatine (PCr), lactate, glucose and glycogen. The biochemical findings then were correlated with the occurrence of reperfusion oedema following clip release after 4 hrs of occlusion point-by-point in the brains. Linear regression analyses of the brain metabolic and pathologic data revealed highly significant (p < 0.001) correlations of acute oedema with brain tissue ATP and PCr reductions < 1.5 µM/g, with lactic acid accumulation > 20 µM/g and with the extents of reduction in brain tissue glucose concentrations in the ischaemic territories.

Fatal reperfusion oedema developing following clip release after 4 hrs of occlusion could be predicted with high reliability (r = 0.96, p < 0.001) from the above biochemical data obtained after 4 hrs of MCA occlusion using multivariate analyses based on partial least squares (PLS) regressions. This results suggests that MRI spectroscopy *in vivo* data may be usable to guide clinical ischaemic brain reperfusion procedures.

Introduction

Hemispheral oedema is well recognized as an aspect of cerebral infartion from cerebrovascular occlusive disease and is a leading cause of death from transtentorial herniation and brain-stem compression in man[1]. Fibrinolysis in acute focal cerebral ischaemia has not gained a similar acceptance as in the treatment of acute myocardial infarction because restoring blood flow to ischaemic brain may enhance oedema and the occurrence of hemorrhage into infarcts[2, 7]. The present experimental study of MCA occlusion in cats adds to our understanding of ischaemic brain oedema by analyzing the influence of metabolic factors (normo/hyperglycaemia) and by correlating parallel brain biochemical and pathologic studies to identify those tissue changes that correlate with subsequent oedema development and — most importantly — to make probability assessments as to this outcome prior to clip release. Our earlier results with this model have established hyperglycemia as a major influence defining infarct size both after temporary and permanent MCA occlusion[4, 5].

Materials and Methods

Details of the techniques used which are briefly summarized below have been published earlier[4, 6]. *Model*: Transorbital MCA occlusion with Yaşargil aneurysm clip placement under pentobarbital anaesthesia in adult cats. *Experimental variables*: 1) Right MCA occlusion for 4 or 8 hrs or permanently.

2) Normo/Hyperglycaemia: After 48 hrs of food deprivation, the cats were infused with saline or 10% glucose (D10W) solution for 1 hr before and 6 hr after occlusion producing mean serum glucose concentrations at occlusion of 6 mM (108 mg/dl) and 20 mM (360 mg/dl), respectively.

Physiologic Monitoring: 1) Continuously: core temperature (38.5 ± 0,5 °C); heart rate, mean arterial blood pressure and cerebral tissue pressure (CTP) using a Millar microtip pressure transducer. 2) Intermittently: arterial blood gases and pH; serum glucose and pentobarbital and hematocrit. Extensive animal physiologic monitoring for > 8 hrs after MCA occlusion and continued close clinical surveillance for 2 weeks assured group comparability.

Endpoints: a) Pathology

1) Early death (< 40 hrs) from hemispheric oedema with transtentorial herniation and brainstem compression assessed semiquan-

Animal Groups			
MCA-Occlusion	Glycemia	Endpoint	n
Permanent	6/20 mM	pathologic	13/12
8 hrs	6/20 mM	pathologic	10/10
4 hrs	6/20 mM	pathologic	12/13
4 hrs	6/20 mM	biochemical	10/14

titatively from gross and microscopic examination of the *in situ* perfusion-fixed brains.

2) Oedema/acute infarct frequencies of the same 16 brain loci as were studied biochemically in normo-and hyperglycaemic animals with 4 hrs of MCA occlusion and restoration of blood flow by clip release.

3) Morphometric (Bioquant) measurement of infarct size in animals after 2 weeks of survival from 6 standard, representative, perfusion-fixed, whole-mount, H & E stained histologic sections (expressed as % of the ischaemic MCA territory).

b) Biochemistry

The brains studied biochemically were snap-frozen *in situ* with liquid nitrogen after 4 hrs of MCA occlusion. Tissue samples were obtained topographically from 6 head slices analogous to the brain cross sections studied pathologically including 16 ipsi- and 16 contralateral sites. The tissue samples were extracted biochemically and assayed fluorometrically for ATP, phosphocreatine (PCr), lactate, glucose and glycogen. All assay results were expressed as µM/g wet tissue weight.

Data Analysis: Correlations of biochemistry data with oedema development point-by-point in the brain after 4 hrs of MCA occlusion:

1) Linear regressions of the group means for each brain site for biochemical and pathologic endpoints using different threshold values for ATP, PCr and lactate served to establish the regression closest to 1 where the tissue biochemical change parallels the tissue oedema frequency.

2) Establishment of a multivariate calibration model using partial least squares (PLS) regression enabling biochemistry based oedema predictions (Unscrambler, Giuded Wave, Inc.):

X-data (biochem) + Y-data (pathol) = calibration model

X-data (biochem) + calibration model = predictred Y-data (pathol).

We built such a calibration model with cross validation using ½ of the brain loci while we used the other ½ of the chemistry data from the reserved loci to predict oedema occurrence which was then correlated (linear regression) with the actual occurrence of oedema.

Results

Among 70 cats exposed to 4 or 6 hrs of temporary or to permanent MCA occlusion, 23% (16) died 5 to 40 hrs after clip application from acute, oedematous infarcts that involved the total MCA supply territory and that showed mass effects with midline shift and major transtentorial and vermian herniation. The time of death

from respiratory arrest occurred 1 to 5 hrs after clip release-reperfusion in 71% (10/14) while death was delayed in 4/14 with and in 2/2 without clip release by 1 to 1.5 days. Corresponding to those findings the cerebral tissue pressure (CTP) following clip release started to increase rapidly in the early fatal cases with terminal pressures > 70 mm Hg. However, in those cases where the clip remained in place, the CTP usually remained below 10 mm Hg for the initial 8 + hrs which had no predictive value as to subsequent outcome which ranged from delayed death from oedema to survival with and without infarts. Uncommonly, in animals with permanent clip application the CTP started to increase already during the initial hours to values slightly > 10 mm Hg despite the continued vascular occlusion. Such early slight tissue oedema infallibly predicted subsequent fatal oedema.

Hyperglycaemia and clip release-reperfusion each quadrupled the fatal oedema outcome (13/35 vs 3/35 hyper/normoglycaemic cats, p < 0.01; 14/45 vs 2/25 with/without clip release, p < 0.05, Fischer's exact test).

Histologic brain examination after 2 weeks of survival revealed focal pressure atrophy in the posterior cingulate gyrus on the side of MCA occlusion in those cases where the infarct affected 12 to 95% of the MCA territory. Thus, nearly half (33/70) of the MCA occluded cats developed either fatal (16) or survivable (17) oedema of the ischaemic hemisphere of sufficient severity to cause mass effect and brain tissue herniation.

The biochemical data obtained from brain tissue after 4 hrs of MCA occlusion, analyzed for ATP, PCr, lactate, glucose and glycogen, were topographically correlated point-by-point in the brain with the occurrence of reperfusion oedema from clip release also at 4 hrs of occlusion. The linear regressions revealed significant correlations between reperfusion oedema and reductions in tissue ATP, PCr and glucose and elevations in tissue lactate while no significant association was defined for glycogen. Different theresholds for ATP, PCr and lactate were tested against oedema occurrence and those with regressions close to 1 were selected and illustrated in Fig. 1. They demonstrate highly significant (p < 0.001) correlations of reperfusion oedema with threshold values for ATP and PCr of < 1.5 µM/g and of lactic acid > 20µM/g.

Predicting fatal reperfusion oedema development after clip release at 4 hrs of occlusion from the biochemical data obtained after 4 hrs of MCA occlusion using multivariate analyses based on partial least squares (PLS) regressions revealed an excellent agreement between the actual vs. predicted oedema rates

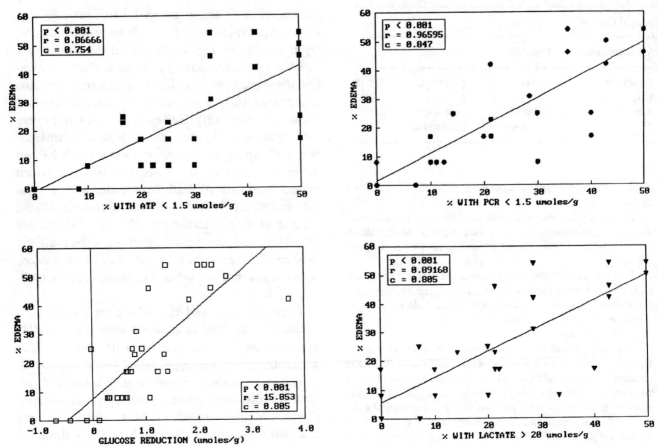

Fig. 1. Characterization of biochemical brain tissue changes indicative of later oedema development after 4 hrs of MCA occlusion

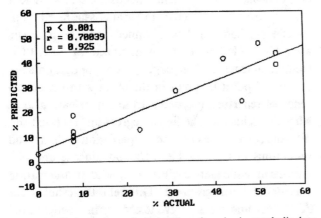

Fig. 2. Predictions of reperfusion oedema from brain metabolic data

(r = 0.96, p < 0.001). One half of the chemistry and pathology data were used to build a PLS calibration model which then was used to predict oedema in the other half of brain loci (Fig. 2).

Discussion

This experimental cerebrovascular occlusion model of temporary and permanent MCA occlusion identifies

metabolic (hyperglycaemia) risk factors that favour fatal oedema development in stroke, demonstrates the occurrence of hemispheral oedema with brain herniation in both fatal and survivable infarcts and confirms the importance of reperfustion in affecting brain pathologic outcome. Further, it defines some of the biochemical changes that underlie oedema development in the ischaemic brain after blood flow is restored and allows this malignant outcome to be predicted.

The finding that hyperglycaemia augments the frequency with which a fatal hemispheral oedema develops after temporary and permanent MCA occlusion fully agrees with a long series of experimental asphyxial and hypoxic studies from our and other laboratories[3, 9] (review). Hyperglycaemia also enlarges the mean infarct size from permanent occlusion and enhaced hemorrhagic infarct conversions[4–6]. Thus, present data suggest that hyperglycaemic strokes may not be suitable for therapeutic reperfusion procedures. Defining those biochemical changes after 4 hrs of MCA occlusion that correlate best with later fatal hemispheral oedema formation after 4 hr clip release, reveals threshold values that imply moderate reductions in high energy phosphates (ATP and PCr $<1.5\,\mu M/g$ or 2/3 and 1/3 of control, respectively) and high but not maximal accumulations of lactic acid to $>20\,\mu M/g$ or a 10-fold increase. These threshold concentrations are intermediate in their values to those that correlate with infarct development following permanent occlusion (which are less) and hemorrhagic infarct conversion (which are larger) (in preparation).

The most important contribution the present study makes is to establish a high predictability of reperfusion oedema in the ischaemic territory based on biochemical data related to energy state and carbohydrate metabolism as obtained after 4 hrs of MCA occlusion using a PLS calibration model. Indeed, these data suggests that *in vivo* MRI spectroscopy may collect the necessary information of predict ultimate stroke outcome since it can assess the same parameters in the living animal or living patient. Thus, MRI spectroscopy data may prove useful in assessing the risks that fibrinolytic or surgical reperfusion procedures in acute ischaemic strokes may enhance oedema formation. Such predictions based on *in vivo* brain chemistry may substantially expand the size of the "therapeutic window" for safe reperfusion procedures. Thus, a much larger percent of patients with ischaemic strokes may benefit from reperfusion procedures than is currently believed since symptom duration cut-off for effecting fibrinolysis is as short as 90 minutes[2, 7].

Acknowledgement

Supported by National Institutes of Health Grant R01 NS21776 (NINDS) and Department of Veterans Affairs.

References

1. Bounds JV, Wieoers DO, Whisnant JP, Okasaki H (1981) Mechanisms and timing of deaths from cerebral infarction. Stroke 12: 274–477
2. Brott T, Haley C, Levey D, Barsan W, Sheppard G, Broderick J, Reed R, Marler J (1990) Safety and potential efficacy of tissue plasminogen activator for stroke. Stroke 21: 181
3. De Courten-Myers GM, Yamaguchi S, Wagner KR, Ting P, Myers RE (1985) Brain injury from marked hypoxia in cats: Role of blood pressure and glycaemia. Stroke 16: 1016–1021
4. De Courten-Myers GM, Myers RE, Schoolfield L (1988) Hyperglycaemic enlarges in cerebrovascular occlusion. Stroke 19: 623–630
5. De Courten-Myers, Kleinholz M, Wagner KR, Myers RE (1989) Fatal strokes in hyperglycaemic cats. Stroke 20: 1707–1715
6. De Courten-Myers GM, Myers RE, Kleinholz M, Wagner KR (1990) Hemorrhagic infarct conversion in experimental stroke. Stroke 21: 357
7. Del Zoppo GJ, Zeumer H, Harker LA (1986) Thrombolytic therapy in stroke: Possibilities and hazards. Stroke 17: 595–607
8. Ito U, Go KG, Walker JT, Spatz M, Klatzo I (1976) Experimental cerebral ischaemia in Mongolian gerbils. III. Behavior of the blood brain barrier. Acta Neuropatho (Berl) 34: 1–6
9. Myers RE, Wagner KR, de Courten-Myers GM (1983) Brain metabolic and pathologic consequences of asphyxia: Role played by serum glucose concentration. In: Milunski/Friedman/Gluck (eds) Adv in Perin Med. Plenum Medical Book Co., New York London, pp 67–115
10. Wagner KR, Myers RE (1985) Topography of brain metabolite concentrations in rhesus monkeys, goats and cats. Exp Neurol 89: 146–158

Correspondence: G. M. de Courten-Myers, M.D., Assoc. Prof. Neuropathology, University of Cincinnati College of Medicine, Department of Pathology, ML 533, 231 Bethesda Avenue, Cincinnati, OH 45267, U.S.A.

Acta Neurochirurgica, Suppl. 51, 231–232 (1990)

Ischaemia and Brain Oedema III

Brain Oedema and Intracranial Pressure in Superior Sagittal Sinus Balloon Occlusion. An Experimental Study in Pigs

G. Fries, Th. Wallenfang, O. Kempski[1], J. Hennen, M. Velthaus, and A. Perneczky

Department of Neurosurgery and [1]Institute for Neurosurgical Pathophysiology, University of Mainz, Medical School, Mainz, Federal Republic of Germany

Summary

About ⅔ of all patients with thrombosis of the superior sagittal sinus (SSS) develop signs of increased ICP and/or brain oedema (BE). The time of onset and the spectrum of symptoms in SSS thrombosis vary extremely. This variability might be caused by differences in pathomechanism like BE and rise of ICP, parameters studied in the present contribution.

10 domestic pigs received a standardized occlusion of the SSS with two different balloon types (spherical and cylindrical). They were monitored for several systemic and intracranial pressures. After 4 hours of occlusion the brains were examined for BE (Evans blue, water content). They were compared with those of 4 sham-operated control animals. 4 animals underwent cerebral angiography.

Within 4 hours ICP rose to 60 mm Hg in the group with the spherical balloon. Normal ICP of 5–10 mm Hg was seen in the group with the cylindrical balloon and in the sham-operated controls. The water content of the white matter was elevated in both occlusion groups differing significantly from the control group. Haemorrhagic infarction of the frontal parts of the cerebrum occurred in animals with concomitant obliteration of bridging and cortical veins. We conclude from our experiments that SSS occlusion may initiate a multitude of possible pathomechanisms depending on the involvement of bridging veins, cortical veins and inner brain veins.

Introduction

Numerous clinical and CT-guided studies[1, 2, 5] emphasize that about ⅔ of all patients with thrombosis of the superior sagittal sinus (SSS) proven by angiography and/or autopsy develop signs of elevated intracranial pressure or brain oedema.

The time of onset and the spectrum of symptoms in SSS thrombosis vary extremely.

In the literature[1, 3, 4] differences in type and degree of brain oedema, in haemodynamic changes and in intracranial pressure are discussed as causes for the wide variety of symptoms and acuity of SSS thrombosis, though the entity of pathomechanisms is not known precisely until now.

The aim of our study was

1) using angiography to describe the intracranial circulation of the pig leading to the characterization of an animal model for SSS occlusion,

2) to determine the effect of a standardized SSS occlusion on type and degree of brain oedema and on ICP.

Materials and Methods

All animals (n = 18, b.wt. 25 kg) were generally anaesthetized, intubated and normoventilated.

They were divided into 4 groups as follows:

One group named balloon 1 (n = 5) received a spherical balloon, the other named balloon 2 (n = 5) received a long cylindrical balloon.

4 sham-operated animals without balloon occlusion served as controls.

4 pigs underwent cerebral angiography.

All animals were continuously monitored for MAP, CVP, IVP, EDP, cerebral tissue pressure, cisterna magna pressure and SSS pressure anterior (SSPa) and posterior (SSPp) to the occlusion.

After 4 hours of monitoring the brains were frozen in situ and

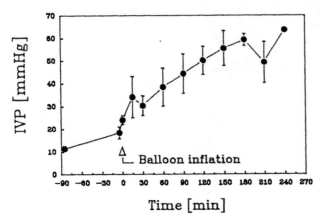

Fig. 1. IVP during balloon-occlusion of SSS (balloon 1)

Table 1. *White Matter Water Content in Pig Brain Slices Obtained 4 hrs After Sinus Sagittalis Occlusion or Sham Operation (ml/100 g fresh weight)*

	Brain slice no.		
	1	2	3
Sham-op.	71.05 ± 2.27	72.94 ± 1.46	76.26 ± 4.25
Balloon 1	$75.93 \pm 1.84*$	73.91 ± 1.55	76.91 ± 6.41
Balloon 2	$76.46 \pm 2.69*$	78.95 ± 8.80	75.94 ± 3.30

* $p < 0.01$ vs. sham-op.

cut into coronal slices. The water content of white matter was determined in slice 1 (frontal), slice 2 (parietal) and slice 3 (occipital) by drying samples of 50 mg at 100 °C.

Results

Cerebral angiography revealed the anatomy of the cerebral venous and sinus system of the pig to be very similar to the human anatomy.

Within 4 hours of SSS-occlusion with balloon 1 (spherical) a constant rise of ICP to values of about 60 mm Hg could be seen (Fig. 1).

Normal values for ICP (5–10 mm Hg) were recorded in the group with balloon 2 (cylindrical) and in the sham-operated control group throughout the experiment.

The water content of the white matter differed significantly among the groups (for values see Table 1).

One animal with a cylindrical balloon had a lesion with small dotted bleedings resembling haemorrhagic infarction of the parasagittal cortex and white matter of both hemispheres.

Discussion

In analogy to results of recent clinical studies[1, 2, 5] our experimental design with SSS balloon occlusion in pigs indicates a multitude of possible reactions.

Under clinical conditions the variation of the extension of the sinus thrombosis might be the cause for the differences in type and degree of brain oedema and ICP.

However in our experiment balloon 1 (spherical) caused a rise of ICP to about 60 mm Hg possibly due to a concomitant obstruction of the sinus confluens and the inner brain veins.

From the experimental standardized occlusion of the SSS we would suggest three possible pathomechanisms for rise of ICP and brain oedema:

a) Global reduction of cerebral venous outflow may lead to a continuous rise of ICP within 4 hours, but haemodynamic changes are not severe enough to cause infarction.

An example for this might be occlusion with balloon 1.

b) Regional obstruction of venous outflow may cause a rise of ICP, haemorrhagic infarction and brain oedema only if bridging veins are involved. Further investigations with injection of fibrin glue into the SSS anterior to the obstruction have been started to clarify this mechanism.

c) If bridging veins are not involved moderate brain oedema develops without causing a rise of ICP in the early phase. The knowledge of these mechanisms may have clinical implications[4], because very early diagnosis and treatment of SSS thrombosis might prevent the patients from fatal BE and severe elevated ICP.

References

1. D'Avella D, Greenberg RP, Mingrino S, Scanarini M, Pardatscher K (1980) Alterations in ventricular size and in intracranial pressure caused by sagittal sinus pathology in man. J Neurosurg 53: 656–661
2. Boddie HG, Banna M, Bradley WG (1974) "Benign" intracranial hypertension. Brain 97: 313–326
3. Bousser MG, Chiras J, Bories J, Castaigne P (1985) Cerebral venous thrombosis – a review of 38 cases. Stroke 16: 199–213
4. Einhäupl K, Garner C, Schmiedek P, Haberl R, Dirnagl U, Pfister HW, Franz P (1986) Reduction of intracranial pressure with anticoagulation in patients with venous sinus thrombosis. In: Miller JD *et al* (eds) Intracranial pressure VI. Springer, Berlin Heidelberg New York, pp 629–632
5. Thron A, Wessel K, Linden D, Schroth G, Dichgans J (1986) Superior sagittal sinus thrombosis: neuroradiological evaluation and clinical findings. J Neurol 233: 283–288

Correspondence: Dr. med. G. Fries, Neurochirurgische Universitätsklinik, Langebeckstrasse 1, D-6500 Mainz, Federal Republic of Germany.

Acta Neurochirurgica, Suppl. 51, 233–235 (1990)

Early Pathomorphological Changes and Intracranial Volume-Pressure: Relations Following the Experimental Saggital Sinus Occlusion

K. Tychmanowicz, Z. Czernicki, M. Czosnyka[1], G. Pawłowski, and G. Uchman

Department of Neurosurgery, Medical Research Centre, Polish Academy of Sciences, [1] Warsaw University of Technology, Warsaw, Poland

Summary

Saggital sinus occlusion was produced in cats. The intracranial volume-pressure relations were studied using the lumbar infusion tests. Right after occlusion ICP rised from 7.1 ± 1.8 to 12.4 ± 4.1 mmHg. The values of outflow resistance and volume-pressure response increased also and remained significantly unchanged with only slight decrease of the volume-pressure response. The morphometric study of the surface veins showed the dilatation of the veins draining to the occluded saggital sinus for about 10–20%. No signs of the blood-brain barrier disruption were observed.

Introduction

Superior saggital sinus (SSS) is a very important drainage for the venous blood and acts as an outflow pathway for cerebrospinal fluid (CSF) as well. The SSS occlusion produces a dangerous consequence in the clinic-severe venous congestion and intracranial hypertension. Therefore one of the major principle in surgery of the saggital region is to avoid the SSS occlusion as well as other major venous drainage routes.

The consequences of SSS occlusion in laboratory animals were studied already, but the results obtained are very different. The studies concerned dogs[2, 7], monkeys[6], rats[1] and cats[5, 8].

Material and Methods

Twenty seven cats weighing between 2800 and 4000 g were used in these experiments. The animals were anaesthetized, tracheostomized and artificially ventilated. Mean arterial blood pressure and central venous pressure were monitored. $PaCO_2$, PaO_2 and pH were checked periodically and maintained within physiological range.

The CSF pressure (ICP) was recorded from the lumbar subarachnoid space following L 4–L 5 laminectomy, dura exposure and insertion of the catheter into the subarachnoid space.

The SSS was exposed by craniectomy and ligated with Prolene 7-0 suture in two places: occipital, close to the confluens sinuum and frontal, just behind the frontal sinus. The SSS between ligatures (about 20 mm) was coagulated. Then suture sites were sealed with tissue glue. The effectiveness of occlusion was verified by digital subtraction angiography in 10 cats. The animals were divided in three groups:

Group A (8 cats): In this group the morphometric studies of surface veins were performed. The brain surface was exposed following craniotomy and dura removal over the marginal suprasylvian and ectosylvian gyri. Photos were taken before and at different times after occlusion: 1 min, 3 min, 5 min, 10 min, 15 min, 30 min, 60 min and then after every hour.

Group B (6 cats): In this group ICP changes were investigated without eventual influence of infusion tests, *i.e.* ICP monitoring was continuous and not interrupted by tests. The visual evoked potentials were studied after placing two electrodes over the visual cortex (marginal gyrus). The latency and amplitude changes were evaluated before occlusion, immediately after occlusion and every hour of experiment within 6 hours.

For the blood-brain barrier (BBB) studies Evans blue 2% was given 2 hours before the end of the experiment. The brains were removed, fixated and examined for signs of BBB disturbances. Histological examinations were performed.

Group C (13 cats): The group was used for the intracranial pressure-volume relation study by means of lumbar infusion tests[3]. The rate of infusion was 0.17 ml/min and tests were performed before occlusion – I, after craniotomy – II, immediately after occlusion – III, then in one-hour intervals until about 5 hours after occlusion. Following parameters were examined: ICP, CSF outflow resistance (R) and volume-pressure response (VPR).

Results and Discussion

The morphometric studies of the superficial veins showed a dilatation of 10–20% in diameter. The changes occurred immediately after occlusion and lasted, only slightly diminished, until the end of the experiment.

The alternations in visual evoked potentials were not very pronounced. The latency was about 5 ms shorter, but not earlier then 4–6 hours after occlusion. The amplitudes varied considerably.

Table 1. *Intracranial Pressure − Volume Relations*

	Control	After craniot.	After occls.	1 h after occls.	2 h after occls.	3 h after occls.	5 h after occls.
ICP mm Hg	7.1 ± 1.8	4.7 ± 1.5*	12.4 ± 4.1*	14.3 ± 3.8	13.7 ± 5.3	12.9 ± 5.3	15.1 ± 4.2
R mm Hg/ml/min	170 ± 42	124 ± 22*	179 ± 29*	200 ± 65	212 ± 72	189 ± 47	205 ± 80
VPR mm Hg	40 ± 16	32 ± 18*	53 ± 22*	42 ± 21*	47 ± 19	46 ± 17	45 ± 14

* Difference statistically significant.

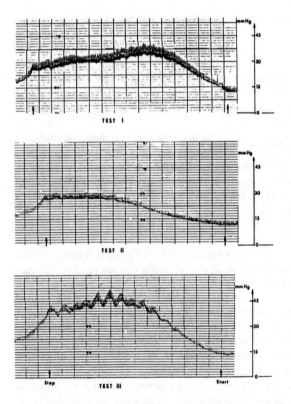

Fig. 1. Lumbar infusion tests in cat 16/88. Infusion rate 0.17 ml/min. Test I − control, Test II − after craniotomy, Test III − right after SSS occlusion

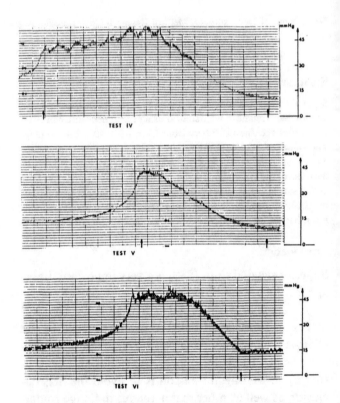

Fig. 2. Lumbar infusion tests in cat 16/88. Infusion rate 0.17 ml/min. Test IV − 1 h after occlusion, Test V − 3 h after occlusion, Test VI − 5 h after occlusion

The intracranial volume-pressure data are given in Table 1. The first significant changes were observed after craniotomy. This was a decrease in ICP, R and VPR. Immediately after occlusion a marked, significant increase of ICP (from 4.7 to 12.4 mmHg), of R (from 124 to 179 mmHg/ml/min) and of VPR (from 32 to 53 mmHg) was seen. Further changes were less pronounced and were found statistically not significant.

The results of the infusion tests in cat 16/88 are presented in Figs. 1 and 2.

The study of the BBB and histological examinations

showed no or only relative slight pathological changes following SSS occlusion in the cat. One possible reason is that SSS in some animals is developed as a big venous channel and in others is almost undeveloped. Thus, there is a big individual variability of the SSS in cats. The animals with undeveloped SSS were not included in this study.

The other explanation for the only moderate changes may be that the SSS route in the CSF outflow is less important in cats than in man. The results presented by other authors are similar[5]. Invesigations performed in the rat[1] and dog[2,7] did not show any evidence of BBB disturbances or histological signs of brain oedema. On the other hand Fujita *et al.*[2] found a significant increase of ICP following SSS occlusion in dogs and Kojima *et al.*[4] an increase of CSF outflow resistance in other species.

Our investigations confirmed the results of Kojima *et al.*[4] and Fujita *et al.*[2]. Summarizing, the early changes following SSS occlusion in the cat concern mostly the intracranial volume-pressure relations without evidence for development of brain oedema. Therefore, further investigations of the SSS role in CSF drainage are needed.

References

1. Deckert M, Freichs K, Jansen M *et al* (1987) A new experimental model of sinus thrombosis in rats. Adv Neurosurg 15: 63–66
2. Fujita K, Kojima N, Namaki N *et al* (1985) Brain oedema in intracranial venous hypertension. In: Inaba Y, Katzo J, Spatz M (eds) Brain oedema. Springer, Berlin Heidelberg New York, pp 228–234
3. Katzman R, Hussey F (1970) A simple constant infusion manometric test for measurement of CSF absorption. 1 Rationale and method. Neurology (Minneap) 20: 534–544
4. Kojima N, Fujita K, Tamaki N *et al* (1989) CSF dynamics and ventricular size in experimental saggital sinus occlusion models. In: Hoff JT, Betz AL (eds) Intracranial pressure VII. Springer, Berlin Heidelberg New York, pp 362–367
5. McComb JG, Hyman S, Weiss MH (1984) Lymphatic drainage of CSF in the cats. In: Shapiro K, Marmarou A, Portnoy H (eds) Hydrocephalus. Raven Press, New York, pp 83–98
6. Owens G, Stahlman, Capps J *et al* (1957) Experimental occlusion of dural sinuses. J Neurosurg 14: 640–647
7. Sato S, Toya S, Ohtani H *et al* (1985) The effect of saggital sinus occlusion on blood-brain barrier permeability and cerebral blood flow in the dog. In: Inaba Y, Klatzo I, Spatz M (eds) Brain oedema. Springer, Berlin Heidelberg New York, pp 235–239
8. Woolf AL (1954) Experimentally produced cerebral venous obstruction. J Pathol Bact 67: 1–16

Correspondence: K. Tychmanowicz, Department of Neurosurgery, Medical Research Centre, Polish Academy of Sciences, Barska 16, 02-325, Warsaw, Poland.

Acta Neurochirurgica, Suppl. 51, 236–238 (1990)
© by Springer-Verlag 1990

Liposome-entrapped Superoxide Dismutase Ameliorates Infarct Volume in Focal Cerebral Ischaemia

S. Imaizumi, V. Woolworth, H. Kinouchi, S. F. Chen, R. A. Fishman, and P. H. Chan

CNS Injury and Oedema Research Center, Department of Neurology, School of Medicine, San Francisco, California, U.S.A.

Summary

We studied the role of superoxide radicals in the pathogenesis of focal ischaemic brain injury using liposome-entrapped copper-zinc-superoxide dismutase which can penetrate the blood-brain barrier and cell membranes efficiently. Superoxide dismutase activities were significantly elevated in the blood and in the normal brain tissue 1, 2, 8, and 24 hours after a bolus intravenous administration of liposome-entrapped copper-zinc-superoxide dismutase. Copper-zinc-superoxide dismutase activities were also increased significantly in the ischaemic hemisphere and the contralateral cortex as well. The infarct sizes were reduced by 33%, 24% and 18%, respectively, for the anterior artery area, middle artery area, and posterior artery area by treatment at 24 hours following the injection of liposome-entrapped superoxide dismutase. These data demonstrate that superoxide radicals are important determinants of the size of an infarct following focal cerebral ischaemia, and that liposome-entrapped copper-zinc-superoxide dismutase may have pharmacological value for focal cerebral ischaemic injury.

Introduction

Oxygen radicals have been postulated to be involved in brain injury and cell death, secondary to ischaemia and traumatic injury (Kontos 1985, Chan 1988, Hall and Braughler 1989, Siesjo *et al.* 1989). Due to the transient nature of oxygen radicals and the technically inherent difficulties in accurately predicting the levels of oxygen radicals in brain, most investigators have employed indirect strategies to identify the involvement of oxygen radicals in injured brain. One experimental strategy that employs antioxidants is most often used to identify the involvement of specific oxygen radical species in ischaemic brain injury. Although free native antioxidant enzymes (superoxide dismutase, catalase) and iron chelator have been used to ameliorate brain injury in experimental models of focal ischaemia (Davis *et al.* 1987), the extremely short half-life of native su-

peroxide dismutase (SOD) and catalase in circulating blood and their inability to pass through the blood-brain barrier may limit their therapeutic usefulness in ischaemic brain injury (Turrens *et al.* 1984). These inherent technical difficulties can be circumvented by employing chemically modified antioxidative enzymes (Liu *et al.* 1989) or by liposome-entrapping techniques (Turrens *et al.* 1984, Freeman *et al.* 1985, Chan *et al.* 1987). We have previously demonstrated a strong correlation between the increased brain level SOD and the amelioration of post-traumatic brain oedema and BBB permeability to ^{125}I-bovine serum albumin (Chan *et al.* 1987). Therefore the aim of this study is to examine the effects of positively charged liposome-SOD on ischaemic brain injury, using the rat model of focal cerebral ischaemia which produces an extensive but reproducible infarct size (Chen *et al.* 1986, Liu *et al.* 1989).

Materials and Methods

One hundred and two adult male Sprague-Dawley rats weighing 300–400 g were used in the present studies. Sixty-seven animals served in the control group and thirty-five animals were used in the liposome-SOD treated group. Focal cerebral ischaemia was produced based primarily on the method of Chen *et al.* (1986). Briefly, under chloral hydrate (30 mg/kg) anaesthesia the right middle cerebral artery (MCA) was explored 1–2 mm under the junction of zygomatic bone and skull to identify the pyriform branch. The right MCA was ligated proximal to the pyriform branch with a single 10-0 suture, followed by electrical coagulation at a low power setting. The right common carotid artery was ligated at two sites and the left common carotid artery was occluded for one hour using an aneurysmal clip. Both common carotid artery occlusions were done within 5 minutes after right MCA occlusion.

In the control group, 2.5 ml of saline was injected intravenously 10 to 20 min prior to both common carotid artery occlusions. In the treatment group, 25,000 units/kg body weight of liposome-SOD diluted with saline to the total volume of 2.5 ml was administered 10 to 20 min before the common carotid artery occlusions.

The CuZn-SOD activities in the brain areas and blood were measured in both the ischaemic control group and the liposome-SOD treated group at the time intervals of preischaemia (10–20 minutes prior to the MCA occlusion), and 1 hour (immediately after the release of the clip of the left common carotid artery), 2, 8, and 24 hours after ischaemia.

At 24 hours after ischaemia, the brains were removed immediately without perfusion. Brain slices 2 mm thick at the site of division of the olfactory nerve (anterior part), at the MCA trunk (middle part) and at the posterior end of the basilar artery were obtained using a brain slicer and then immersed in a 2% 2,3,5-triphenyl-tetrazolium chloride (TTC) solution. The area of infarction was expressed as a percent area of the whole coronal section.

Results

Plasma CuZn-SOD activity in the nontreated control group did not change throughout the experiment. In the liposome-SOD treated group, the activities were significantly higher and maintained at high levels at 2 h and 8 h. There was a slight increase in the SOD level at 24 hours. In the liposome-SOD treated group, the contralateral cortex showed a significant increase of CuZn-SOD activity at 2 h and 24 h, whereas the levels of total SOD were significantly increased in the injured cortex of MCA territory at 1 h and 24 h. The SOD levels in the subcortex were also increased at 1 h and 2 h.

At 24 h following ischaemia the infarct sizes were 18.1%, 14.5%, and 9.8% for slices obtained from anterior, middle and posterior artery areas, respectively (Fig. 1). Administration of liposome-SOD reduced the infarct size in the anterior area (12.1%) and middle area (10.9%). There was no effect of liposome-SOD on the infarct size of the posterior artery area. Lipo-

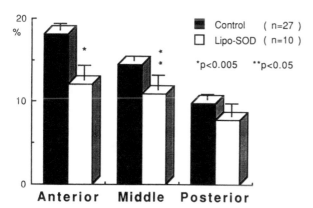

Fig. 1. Comparison of infarct size between no treatment control group and liposome-SOD treated group determined by TTC staining. Values are mean ± SEM. * $p < 0.005$, ** $p < 0.05$ versus control group. n = 27 (control group) and 10 (liposome-SOD group)

somes alone or liposomes with denatured enzymes were not effective (data not shown).

Discussion

Oxygen radicals and lipid peroxidation have been proposed to be factors involved in cerebral ischaemia and reperfusion injury. We have examined the endogenous level of antioxidant enzymes as indicators of oxidative stress in ischaemic brain injury. Our results indicate that CuZn-SOD activity in brain decreased significantly immediately following both MCA and common carotid artery occlusion (Imaizumi *et al.* 1989) and in animal model of post decapitation ischaemia (Chan *et al.* 1988). A single bolus injection of liposome-entrapped SOD 10–20 minutes prior to common carotid artery occlusion significantly elevated the level of SOD in the MCA territory. These data lead us to study the effects of liposome-entrapped SOD on infarct size following focal cerebral ischaemia. Our data indicate that liposome-SOD reduced the infarct size (as determined by TTC staining) by 33% ($p < 0.005$), 24% ($p < 0.05$), and by 18% ($p > 0.05$), respectively for anterior cerebral artery area, middle cerebral artery area and posterior cerebral artery area (Fig. 1). These results demonstrate clearly the role of superoxide radicals in the pathogenesis of an ischaemic infarct.

An important question that remains to be addressed is whether post-ischaemic treatment with liposome-SOD has beneficial effects on ischaemic infarction. In our previous studies, post-treatment with liposome-SOD reduced the severity of vasogenic brain oedema and protected blood-brain barrier permeability in cold-induced brain injury (Chan *et al.* 1987). We speculate that liposome-SOD might have therapeutic potential in ameliorating the oxidative stress associated with ischaemic brain injury. Furthermore, the usually delayed development of infarction following focal ischaemia and global cerebral ischaemia suggests that it may be possible to take advantage of a time window that would allow for therapeutic intervention using antioxidant agents.

Acknowledgements

This work was supported by NIH grants NS-25372 and NS-14543. We thank Drs. Kee-Lung Hong and Demetrios Papahadjopoulos for their assistance for liposome preparation. We also thank Marilyn Stubblebine for her editorial assistance.

References

1. Chan PH, Longar S, Fishman RA (1987) Protective effects of liposome-entrapped superoxide dismutase on posttraumatic brain oedema. Ann Neurol 21: 540–547

2. Chan PH, Chu L, Fishman RA (1988) Reduction of activities of superoxide dismutase but not of glutathione peroxidase in rat brain regions following decapitation ischaemia. Brain Res 439: 388–390

3. Chan PH (1988) The role of oxygen radicals in brain injury and oedema. In: Chow CK (ed) Cellular antioxidant defense mechanisms. CRC Press Inc., Boca Raton, FL, pp 89–109

4. Chen ST, Hsu CY, Hogan EL, Maricq H, Balentine JD (1986) A model of focal ischaemia stroke in the rat: reproducible extensive cortical infarction. Stroke 17: 738–743

5. Davis RJ, Bulkley GB, Traystman RJ (1987) Role of oxygen-free radicals in focal brain ischaemia. J Cereb Blood Flow Metab 7: S 10

6. Freeman BA, Young SL, Crapo JD (1985) Liposome mediated augmentation of superoxide dismutase in endothelial cells prevents oxygen injury. J Biol Chem 258: 12534–12542

7. Hall ED, Braughler JM (1989) Central nervous system trauma and stroke. II. Physiological and pharmacological evidence for involvement of oxygen radicals and lipid peroxidation. Free Rad Biol Med 6: 303–313

8. Imaizumi S, Woolworth V, Fishman RA, Chan PH (1989) Superoxide dismutase activities and their role in focal cerebral ischaemia. J Cereb Blood Flow Metab 9 [Suppl 1]: S 217

9. Kontos HA (1985) Oxygen radicals in cerebral vascular injury. Circ Res 57: 508–516

10. Liu TH, Beckman JS, Freeman BA, Hogan EL, Hsu CY (1989) Polyethylene glycol-conjugated superoxide dismutase and catalase reduce ischaemic brain injury. Am J Physiol 256: H 589–H 593

11. Siesjo BK, Agardh C-D, Bengtsson F (1989) Free radicals and brain damage. Cereb Brain Metab Rev 1: 165–211

12. Turrens JF, Crapo JD, Freeman BA (1984) Protection against oxygen toxicity by intravenous injection of liposome-entrapped catalase and superoxide dismutase. J Clin Invest 73: 87–95

Correspondence: Dr. P. H. Chan, Department of Neurology, Box 0114, University of California, San Francisco, CA 94143, U.S.A.

Acta Neurochirurgica, Suppl. 51, 239–241 (1990)

Effects of Ebselen (PZ 51) on Ischaemic Brain Oedema After Focal Ischaemia in Cats

H. Johshita, T. Sasaki[1], T. Matsui, T. Hanamura, H. Masayasu[2], T. Asano, and **K. Takakura**

Department of Neurosurgery, Saitama Medical Center, [1] Department of Neurosurgery, University of Tokyo, [2] Department of Medical Research Development, Daiichi Pharmaceutical Co. Ltd., Japan

Summary

Using a transorbital middle cerebral artery (MCA) occlusion model in cats, we evaluated the anti-oedema effects of a new anti-inflammatory agent, ebselen (PZ 51), on ischaemic cortical oedema caused by prolonged ischaemia and recirculation. Local cerebral blood flow was measured by the hydrogen clearance method in the MCA territory and the corresponding cortical specific gravity was assessed by a microgravimetric technique. Ebselen had no significant effect on normal and ischaemic *I*CBF, while it significantly ameliorated post-ischaemic hypoperfusion following recirculation. In the severely ischaemic regions, microgravimetry showed the beneficial effects on the ischaemic oedema caused by prolonged ischaemia and recirculation as well. Although the exact site of action is undetermined in this study, the observed effects of ebselen may be ascribed to this agent's broad-spectrum of anti-inflammatory activities.

Introduction

Rational treatment for ischaemic brain oedema which occurs during and after focal ischaemia is of the utmost importance in the therapy of the cerebrovascular diseases. We evaluated the effect of a novel selenium-containing radical scavenger, ebselen (2-phenyl-1,2-benzoisoselenazole-3(2 H)-one, PZ 51, Daiichi Pharmaceutical Co. Ltd.) on ischaemic brain oedema, using a cat middle cerebral artery occlusion (MCAO) model with and without recirculation.

Materials and Methods

In 18 cats (11 controls and 7 drug-treated cats), the left MCA was trans-orbitally exposed under halothane anaesthesia, and was occluded for two hours and recirculated for the subsequent two hours (recirculation: RC groups). In 14 cats (8 controls and 6 drug-treated cats) the left MCA was occluded for 4 hours (prolonged ischaemia: PI groups). Cats were immobilized and mechanically ventilated under 30% O_2 and 70% N_2O. A dose of 10 mg/kg of ebselen was dissolved into 0.5% carboxymethyl cellulose (3 ml/kg) and was administered via nasogastric tube 40 minutes prior to MCA occlusion in the drug treated groups. In RC groups, plasma concentration of selenium was monitored during the experiment. Local cerebral blood flow (*I*CBF) was measured every 30 minutes using a hydrogen clearance method in 6 electrode sites located within the MCA territory. In both the RC and PI groups, *I*CBF values at each electrode during MCA occlusion were averaged and classified according to the severity of ischaemia (0–15, 15–30, 30–100 ml/100 g/min: severe, moderate, and mild ischaemic sites respectively). At the end of the experiment, cats were sacrificed by an injection of saturated potassium chloride. Cortical specific gravity (co-SG) was measured by microgravimetry in the surrounding area at each electrode site, and the co-SG values were analyzed according to the corresponding *I*CBF values during ischaemia. Data are expressed as mean ± SD. The data of physiological parameters and *I*CBF were analyzed using the MANOVA and t-test, and the data of co-SG were analyzed by the Mann-Whitney U-test.

Results

Parameters, such as mean arterial pressure, blood gases, and rectal temperature were kept within physiological ranges and did not differ significantly between the experimental groups. Plasma concentration of selenium in the ebselen-treated RC group was continuously increased $3.0 \pm 0.7 \mu M$ (30 minutes after administration of ebselen), $8.1 \pm 2.3 \mu M$ (1 hour after MCAO), and $11.2 \pm 3.2 \mu M$ (3 hours after MCAO) during the experiment. The change in *I*CBF is shown in Fig. 1. Administration of ebselen exerted no significant influence on *I*CBF before MCA occlusion. In RC groups, no significant difference in *I*CBF was observed during ischaemia. Following recirculation, post-ischaemic hypoperfusion was significantly ameliorated in the severely ischaemic area of the drug-treated group. The ratio of the mean value of *I*CBF in the early re-

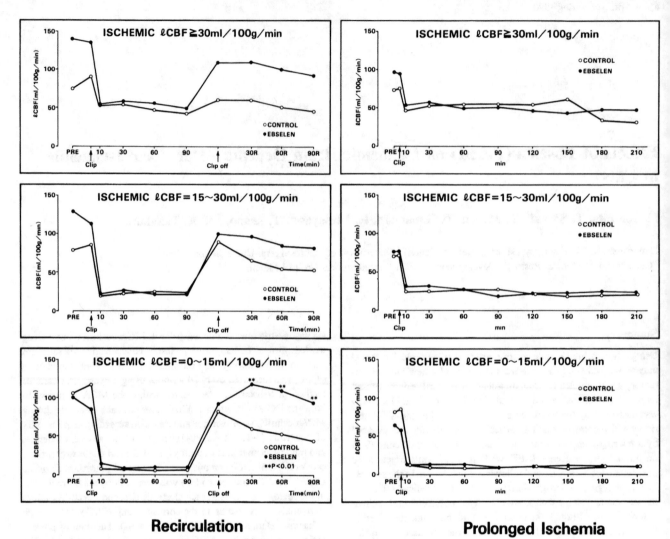

Recirculation   **Prolonged Ischemia**

Fig. 1. Time courses of local cerebral blood flow (*l*CBF) of severely (0–15 ml/100 g/min, bottom), moderately (15–30 ml/100 g/min, middle) and slightly (30–100 ml/100 g/min, upper figures) ischaemic cortical areas in each experimental subgroups

circulated period (mean value of 10 and 30 minutes after recirculation) to the mean value of pre-ischaemic *l*CBF in the drug treated group (1.1 ± 0.08) was significantly higher (p < 0.01) than that of the control group (0.8 ± 0.06). In PI groups, the *l*CBF values did not differ significantly between experimental groups. Data derived from microgravimetry are shown in Fig. 2. In the severely ischaemic areas, ebselen significantly ameliorated ischaemic oedema during prolonged ischaemia (PI group) and following recirculation (RC group).

Discussion

In so much as the pathogenetic mechanism underlying ischaemic brain oedema is not fully elucidated, our previous studies[1, 2] indicate a possible role of reactive

oxygen metabolites and eicosanoid on oedema formation. In those studies, it was suggested that the mechanisms which contribute to formation of oedema differs between focal ischaemia and recirculation. Particularly, indomethacine, a cyclooxygenase inhibitor, significantly ameliorated recirculation oedema and post-ischaemic hypoperfusion, while the same drug had an adverse effect on the oedema of cytotoxic type caused by prolonged ischaemia[2]. In the present study, we tested the effects of a novel synthetic organoselenium, ebselen (PZ 51), which possesses several sites of action as an anti-inflammatory compound[3, 4, 5]: the drug acts like glutathione peroxidase (GSH-Px), catalyses H_2O_2 degradation, inhibits NADH oxidase, inhibits cyclooxygenase and lipoxygenases, and inhibits LTB_4 formation. The present results show that this drug exerts beneficial effects on the ischaemic oedema in the se-

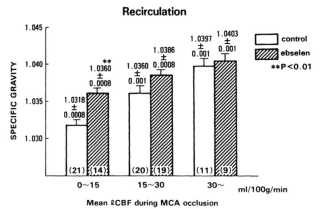

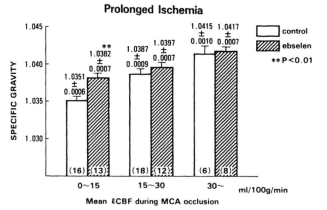

Fig. 2. Cortical specific gravities (co-SG) of severely, moderately, and slightly ischaemic regions of cats with prolonged ischaemia and recirculation

verely ischaemic regions caused by recirculation and by prolonged ischaemia as well. Although the exact site of action is undetermined, the observed anti-oedema effects of ebselen in both prolonged ischaemia and recirculation may be ascribed to the agent's broad-spectrum of activities as stated above. Ebselen may be useful in the treatment of ischaemic brain oedema ensuing from a variety of cerebrovascular diseases.

References

1. Asano T, Koide T, Gotoh O, Johshita H, Hanamura T, Shigeno T, Takakura K (1989) The role of free radicals and eicosanoids in the pathogenetic mechanism underlying ischaemic brain oedema. Mol Chem Neuropathol 10: 101–133
2. Johshita H, Asano T, Hanamura T, Takakura K (1989) Effect of indomethacin and a free radical scavenger on cerebral blood flow and oedema after cerebral artery occlusion in cats. Stroke 20: 788–794
3. Muller A, Cadenas E, Graf P, Sies H (1984) A novel biologically active selenoorganic compound-I. Glutathione peroxidase activity *in vitro* and antioxidant capacity of PZ 51 (ebselen). Biochem Pharmacol 33: 3235–3240
4. Cotgreave IA, Duddy SK, Kass GN, Thompson D, Moldeus P (1989) Studies on the anti-inflammatory activity of ebselen. Biochem Pharmacol 38: 649–656
5. Kuhl P, Borbe HO, Romer A, Fischer H, Parnham MJ (1985) Selective inhibition of leukotriene B_4 formation by ebselen: a novel approach to antiinflammatory therapy. Agents and Actions 17: 366–367

Correspondence: Department of Neurosurgery, Saitama Medical Center, Saitama Medical School, 1981, Kamoda, Kawagoe, 350, Saitama, Japan.

Acta Neurochirurgica, Suppl. 51, 242–244 (1990)
© by Springer-Verlag 1990

Post-ischaemic Treatment with the Prostacycline Analogue TTC-909 Reduces Ischaemic Brain Injury

K. Shima, K. Ohashi, H. Umezawa, H. Chigasaki, Y. Karasawa[1], S. Okuyama[1], H. Araki[1], and S. Otomo[1]

Department of Neurosurgery, National Defense Medical College and [1]Research Center, Taisho Pharmaceutical Co., Ltd., Saitama, Japan

Summary

The effects of stable PGI analogue TTC-909 on CBF and glucose metabolism was studied in the chronic stage of cerebral ischaemia produced by occluding the distal MCA in SHRSP. Administration of TTC-909 (100 ng/kg/day during 7 days) prevented the development of ischaemic oedema and improved secondary metabolic derangement coupled to flow in postischaemic tissues, particularly in the ischaemic rim.

Introduction

Prostacyclin (PGI_2), a prostaglandin synthesized by endothelial cells lining the cardiovascular system, is a potent vasodilator and inhibitor of platelet aggregation[7]. TTC-909 is a chemically stable PGI_2 analogue incoporated into lipid microspheres. We examined the efficacy of TTC-909 on local cerebral blood flow (LCBF) and glucose utilization (LCGU) following focal cerebral ischaemia, and the correlation between their events at the chronic stage of ischaemia was also given attention.

Materials and Methods

Male stroke-prone spontaneously hypertensive rats (SHRSP) at the age of 11 weeks were anaesthetized, and focal ischaemia was produced by occluding the middle cerebral artery (MCA). The left MCA was coagulated distal to striate branches and 0.7–1 mm dorsal to the rhinal fissure[1, 9]. The animals were grouped into three: (a) sham-operated control group, (b) sham(saline)-treated MCA-occlusion group, and (c) TTC-909-treated MCA-occlusion group. TTC-909, 100 ng/kg daily, was given intravenously starting at 30 min following MCA occlusion, and on day 7 following the occlusion, the animals were anaesthetized with halothane, both femoral arteries and veins were catheterized, and blood pressure, rectal temperature and blood gases were monitored. Ten min after the last administration of TTC-909, autoradiographic studies were done.

Local cerebral blood flow (LCBF)

LCBF was measured autoradiographically according to Sakurada et al.[10]. After confirming the acceptable physiological state of each animal, 100 μCi/kg of ^{14}C-iodoantipyrine(^{14}C-IAP) was infused for 30 sec at a constant rate. Arterial blood samples were taken every 5 sec to assess ^{14}C activity. At 30 sec the rat was decapitated, the brain rapidly removed, frozen and cut into 20 μm-thick sections for autoradiographic studies.

Local cerebral glucose utilization (LCGU)

The measurement of LCGU was initiated by the i.v. injection of ^{14}C-2-deoxyglucose(^{14}C-DG, 100 μCi/kg). Timed blood samples were taken for analysis of ^{14}C activity and glucose concentration. The animals were decapitated 45 min after the injection of ^{14}C-DG. Autoradiographs were prepared in the same manner as for LCBF, and LCGU was calculated according to Sokoloff et al.[11]. Since the cerebral glucose content influencing the lumped constant (LC) is presumably close to normal[8], the LC for the normal rat determined by Sokoloff et al.[11] was used.

Brain oedema

Brain water content was determined by the gravimetric method[6] using a bromobenzene-kerosene linear gradient column (r > 0.99). Samples 1 mm in size were dissected from each brain area and put into the column.

Results

There was no significant difference in mean arterial blood pressure at the start of the experiments among all the groups (Table 1). Location of ischaemic brain damage in each animal was limited to the ipsilateral cerebral cortex and did not extend to the striatum. The average infarct size was $12.8 \pm 3.4\%$ (mean ± SD) in the untreated MCA-occlusion group. Treatment with TTC-909 resulted in a insignificant reduction (17%) in size.

Table 1. *Mean Arterial Blood Pressure (MABP) in Each Experimental Group*

Experimental group	MABP (mm Hg)	
	LCBF	LCGU
Sham	188 ± 9 (n = 5)	183 ± 8 (n = 5)
MCAO	180 ± 12 (n = 8)	188 ± 5 (n = 6)
TTC-909	181 ± 10 (n = 5)	190 ± 10 (n = 7)

Values are means S.D. sham, sham-operated control; MCAO, sham-treated MCA occlusion; TTC-909, TTC-909-treated MCA occlusion.

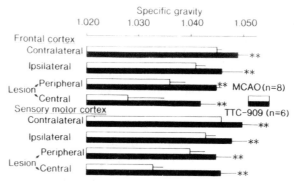

Fig. 2. Effect of TTC-909 on the specific gravity in SHRSP 7 days after MCA occlusion

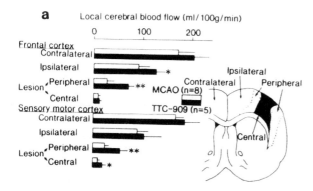

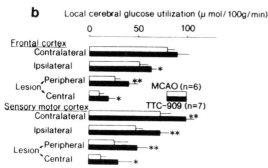

Fig. 1. Effects of TTC-909 on (a) LCBF and (b) LCGU in SHRSP 7 days after MCA occlusion

Marked reduction of LCBF and LCGU was seen unevenly distributed in ischaemic brain regions of the untreated MCA occluded animals as compared to those in sham controls (Fig. 1). A significant reduction in LCGU coupled to CBF was also observed in the following 3 of 13 selected regions ipsilateral and contralateral to the MCA occlusion: caudate-putamen, thalamus (ventromedial nucleus) and amygdala on the ipsilateral side.

With TTC-909 treatment, LCBF and LCGU increased in most of selected regions including contralateral to the MCA occlusion, particularly in the peripheral and surrounding regions of the ischaemic area

that have shown a marked reduction. After treatment with TTC-909, the decreased specific gravity observed in the ischaemic region reverted closely to the values obtained for the non-ischaemic regions (Fig. 2).

Discussion

We obtained evidence that the stable PGI_2 analogue TTC-909 improves postischaemic low blood flow and glucose metabolism, and reduces ischaemic brain oedema. The beneficial effects of TTC-909 were observed in the ischaemic rim and the surrounding areas. We used the less invasive MCA-occlusion model and SHRSP to assess not only the brain changes in chronic ischaemia but also in chronic ischaemia in the presence of hypertension, the main risk factor for stroke[9]. Occlusion of the MCA distal to striate branches results in infarction located in the cortex in SHRSP, but not in normotensive rats. The larger and more reproducible infarcts after MCA occlusion in SHRSP results in an inadequate collateral circulation. Morphological studies have revealed that the lumen diameter of the anterior cerebral artery (ACA)-MCA anastomoses in SHRSP is smaller than that in normotensive rats, even in 5-week-old animals[2]. The vascular resistance of collaterals in SHRSP is greater than in the normotensive rats, possibly as a result of an autoregulatory vasoconstriction of the hypertensive vasculature. We assume that the vasodilating effect of TTC-909 maintains the blood supply to the ischaemic area and prevents development of oedema.

At the level of the mean arterial blood pressure in the SHRSP used in this study, the LCBF did not differ from normotensive rats. This agrees with earlier experimental and clinical studies on the CBF in SHRSP[3,5]. We also noted a correlation between LCBF and LCGU, *i.e.*, perfusion-metabolic coupling, in the areas with a chronic ischaemia. However, the high uptake of deoxyglucose in the cortical rim of the ischaemic area, as is commonly seen in cases of acute stroke, was not observed in any animal. This finding is consistent with the report by Nedergaard *et al.*[8] that the deoxyglucose uptake no longer increased at 20 hours after MCA occlusion.

In addition, the coupled perfusion-metabolic pattern correlated well with the degree of oedema formation. Fredriksson *et al.*[4] found a higher incidence of multifocal BBB opening in SHRSP than in normotensive rats. Therefore, the SHRSP strain has a more severe cerebral ischaemia, oedema formation and metabolic derangement following MCA occlusion.

A double-blind clinical study of TTC-909 is now underway in Japan.

References

1. Coyle P, Jokelainen PT (1983) Differential outcome to middle cerebral artery occlusion in spontaneously hypertensive stroke-prone rats (SHRSP) and Wistar Kyoto (WKY) rats. Stroke 14: 605–611
2. Coyle P (1987) Dorsal cerebral collaterals of stroke-prone spontaneously hypertensive rats (SHRSP) and Wistar Kyoto rats (WKY). Anat Rec 218: 40–44
3. Fredriksson K, Ingvar M, Johansson BB (1984) Regional cerebral blood flow in conscious stroke-prone spontaneously hypertensive rats. J Cereb Blood Flow Metab 4: 103–106
4. Fredriksson K, Kalimo H, Westergren I, Kahrstrom J, Johansson BB (1987) Blood-brain barrier leakage and brain oedema in stroke-prone spontaneously hypertensive rats. Acta Neuropathol (Berl) 74: 259–268
5. Kety SS, Hafkenschiel JH, Jeffers WA, Leopold IH, Shenkin HA (1948) The blood flow, vascular resistance and energy consumption of the brain in essential hypertension. J Clin Invest 27: 511–514
6. Marmarou A, Poll W, Shulman K, Bhagavan H (1978) A simple gravimetric technique for measurement of cerebral oedema. J Neurosurg 49: 530–537
7. Moncada S (1983) Biology and therapeutic potential of prostacyclin. Stroke 14: 157–168
8. Nedergaad M, Gjedde A, Diemer NH (1986) Focal ischaemia of the rat brain: Autoradiographic determination of cerebral glucose utilization, glucose content, and blood flow. J Cereb Blood Flow Metab 6: 414–424
9. Okuyama S, Shimamura H, Kawashima K, Araki H, Kimura M, Aihara H (1990) Protective effects of minaprine in infarction produced by occluding of the middle cerebral artery in stroke-prone spontaneously hypertensive rats. Gen Pharm 21 (in press)
10. Sakurada O, Kennedy C, Jehle J, Brown JD, Carbin GL, Sokoloff L (1978) Measurement of local cerebral blood flow with iodo-^{14}C-antipyrine. Am J Physiol 234: H 59–H 66
11. Sokoloff L, Reivich M, Kennedy C, Des Rosiers MH, Patlak C, Pettigrew K, Sakurada O, Shinohara M (1977) The [^{14}C]deoxyglucose method for the measurement of local cerebral glucose utilization: Theory, procedure and normal values in the conscious and anaesthetized albino rat. J Neurochem 28: 897–916

Correspondence: K. Shima, M.D., D.M.Sc., Department of Neurosurgery, National Defense Medical College, 3-2 Namiki, Tokorozawa, Saitama 359, Japan.

Acta Neurochirurgica, Suppl. 51, 245–247 (1990)

Attenuation of Glutamate-induced Neuronal Swelling and Toxicity in Transgenic Mice Overexpressing Human CuZn-Superoxide Dismutase

P. H. Chan[1,2], **L. Chu**[2], **S. F. Chen**[2], **E. J. Carlson**[3], and **C. J. Epstein**[3,4]

Departments of [1] Neurosurgery, [2] Neurology, [3] Pediatrics, and [4] Biochemistry and Biophysics, University of California, San Francisco, School of Medicine, San Francisco, California, U.S.A.

Summary

The role of oxygen-derived free radicals, superoxide in particular, in the pathogenesis of neuronal cell death induced by glutamate was studied using primary culture cortical neurons from transgenic mice overexpressing human CuZn-superoxide dismutase. Primary cortical neuron cultures were developed form 15-day-old fetuses of both transgenic mice and their normal littermates. Both human CuZn-superoxide dismutase and host mouse CuZn-superoxide dismutase activities in cultured neurons were identified by native gel electrophoresis followed by nitroblue tetrazolium staining. Cultured neurons grown for 10–12 days *in vitro* were exposed briefly to 0.5 mM glutamate for 5 minutes, followed by biochemical and morphological examinations at 2 and 4 hours. Our data have demonstrated that glutamate neurotoxicity is significantly reduced in transgenic neurons at 2 and 4 hours following exposure to glutamate, as measured by the intracellular 3-0-methyl glucose space, the efflux of lactate dehydrogenase, and by phase-contrast and bright-field trypan blue staining. These data indicate that transgenic neurons containing two- to threefold the normal amount of CnZn-superoxide dismutase activity are protected against glutamate neurotoxicity *in vitro*. Our results suggest that oxidative stress play an important role in glutamate-induced neuronal swelling and toxicity.

Introduction

Excitatory neurotransmitter amino acids, glutamate (Glu) in particular, have been implicated in various neurological diseases and neurodegeneration (Choi *et al.* 1987, 1988). Recent studies also suggest that Glu causes neurotoxicity initially by the activation of post-synaptic NMDA receptor, followed by the influx of Ca^{2+}, and then by the amplification of Ca^{2+}-associated biochemical events (Choi 1988). Oxidative stress has been postulated to be one of the possible biochemical events associated with Ca^{2+} mobilization through the activation of phospholipases A 2 and C and the

subsequent metabolic cascade of archidonic acid (Chan and Fishman 1985, Chan *et al.* 1990). Neuronal survival due to substrate deprivation was enhanced by prior treatment that induced cells to take up superoxide dismutase (Saez *et al.* 1987). This increase in survival in cultured neurons was concomitant with the reduction of oxidative stresses, as measured by both the reduction of nitroblue tetrazolium and the oxidation of 2′,7′-dichlorofluorescein diacetate. Furthermore, Glu-induced cytotoxicity in neuronal cell line N 18-RE-105 by the inhibition of cysteine uptake, resulting in lowered glutathione levels leading to oxidative stress and cell death (Murphy *et al.* 1989). These studies, when taken together, suggest that oxidative stress may be one of the common denominators underlying Glu-mediated neuronal cell injury.

We have recently shown that supplementation of CuZn-SOD (SOD-1) through liposomes ameliorates oxygen radical-induced vasogenic oedema, blood-brain barrier permeability changes *in vivo*, and cellular swelling and dysfunction in primary cell cultures of astrocytes (Chan *et al.* 1987, 1988). Epstein *et al.* have successfully developed transgenic mice containing 1.6–6.0-fold increase in human SOD-1 (SOD-1) activity in whole brain (Epstein *et al.* 1987, 1990), and have shown that additional SOD-1 activity is expressed in cultured cortical neurons derived from these transgenic mice. It is our aim to employ these cells to investigate the role of oxidative stress in Glu-induced cell swelling and injury. We anticipate that increases in SOD-1 activity in neurons can affect Glu-mediated neurotoxicity by altering the level of oxidative stress on the cells.

Materials and Methods

Transgenic mice of strain TgHS/SF-218 carrying the human CuZn-SOD gene were produced as described in Epstein *et al.* (1987). A linear 14.5-kilobase Eco RI-Bam HI fragment of human genomic DNA was excised from the recombinant plasmid pH GSOD-SVneo and separated from plasmid sequences prior to microinjection. The Eco-RI-Bam HI DNA fragment contained the entire human CuZn-SOD gene, including the sequences required for expression in transfected cells. Approximately 500 copies of the purified fragment were microinjected into the male pronuclei of B6 SJL zygotes. The founder mice had been bred to produce transgenic offspring caryying the h-SOD-1 gene (Epstein *et al.* 1987, 1990).

Prior to the setting up of primary cultures of cortical neurons, 15-day-old fetuses were quickly removed. The cerebral cortices of the fetuses exhibiting the SOD-1 activity then were pooled and used for primary neuronal cultures as described previously. Cultured neurons, both transgenic and nontransgenic, were grown for 10–12 days and were then exposed to 0.5 mM Glu for 5 minutes. The experiments were terminated with a fresh medium. The efflux of lactate dehydrogenase (LDH) and trypan blue exclusion were studied to determine the degree of cell injury and death at 2 and 4 hours after exposure to Glu. The uptake of 3-0-[^{14}C] methyl glucose into cells at these times also was studied as a measurement of intracellular swelling (Chan *et al.* 1990).

Results

Primary cultures of cortical neurons from transgenic mice and their non-transgenic littermates have been successfully developed. Immunocytochemical staining for the neuron-specific, enolase, and trypan blue ex-clusion staining indicated that more than 95% of the cell population consisted of viable neurons. The 3-0-methyl glucose space of both types of neurons was increased after exposure to 0.5 mM Glu for 5 minutes (Table 1). However, the swelling of the neurons derived from h-SOD-1 transgenic mice was significantly less than the swelling of control neurons at 2 and 4 hours. The efflux of LDH increased with time in both types of neurons. However, the LDH efflux from transgenic neurons increased to 256% and 500% as compared to 383% and 706% respectively, to the efflux from non-transgenic neurons at 2 and 4 hours postincubation, indicating that transgenic neurons overexpressing SOD-1 activity are more resistant to Glu-induced injury. These observations were further confirmed by studies using trypan blue exclusion. At 2 hours after exposure to Glu, only 40% viable cells remained in the culture dish of nontransgenic neurons, whereas 61% viable cells were observed in transgenic neurons. There were 21% more viable cells in transgenic neurons than in nontransgenic controls at 4 hours.

Discussion

Our data clearly demonstrate that primary cortical neurons containing increased functional SOD-1 enzyme activity as the result of the presence of a h-SOD-1 transgene are highly resistant to Glu-induced neurotoxicity. These studies support the hypothesis that oxidative stress may be an important factor in amplifying NMDA receptor-mediated Ca^{2+}-dependent biochemical events that lead to delayed neuronal cell death (Choi 1988). Recent studies have demonstrated that pharmacological blockade of the NMDA receptor with competitive antagonists is the most effective way to reduce Glu-mediated neuron cell death due to ischaemia, hypoglycaemia, and epileptic seizures (Choi *et al.* 1987, Simon *et al.* 1984, Wielock 1985). However, most, if not all, of these studies have employed pharmacological treatment prior to pathological insults, a situation not compatible with treatment regimens in clinical situations. Another strategy is to modify the Ca^{2+}-associated biochemical events (*i.e.*, protease activation), and many experimental approaches have been directed toward this goal with varying degrees of success in ameliorating Glu-induced neuronal cell death. Our present experimental approach is to modify the oxidative state of neurons using genetic manipulation. Our data indicate that oxidative stress is associated with Glu-neurotoxicity, because neuronal swelling (measured by 3-0-methyl glucose space), neuronal cell

Table 1. *Glutamate Neurotoxicity in Transgenic Mice Overexpressing CuZn-SOD Activity*

Determination	Time (hr) post-Glu incubation (% of control)		
	0	2	4
LDH release			
Normal	100	383	706
Transgenic	100	256*	500*
Trypan blue exclusion			
Normal	100	40	26
Transgenic	100	61*	47*
3-OMG space			
Normal	100	284	342
Transgenic	100	159*	256*

Primary cortical neurons (10 days *in vitro*) were incubated with Glu (0.5 mM) for 5 minutes followed by various assays at 2 and 4 hours. * p > 0.05 compared to normal control at same time point. Results are averaged from two to three experiments. Control values for LDH (U/L) are 2.0 ± 0.7 (normal), 1.8 ± 0.2 (transgenic); for 3-0-methyl glucose (3-OMG) space (μl/mg protein) are 5.59 ± 0.74 (normal), transgenic (5.20 ± 0.99); and for trypan blue exclusion (% of viable cells) are 97 ± 5 (normal), 98 ± 6 (transgenic). Mean ± S.E.

injury (measured by LDH efflux), and cell death (measured by trypan blue exclusion) are significantly reduced in cells overexpressing SOD-1 activity. Thus, neuronal cell cultures developed from SOD-1 transgenic mice are useful models for elucidating the cellular mechanisms by which oxidative stress may be associated with neuronal swelling and death due to stroke, trauma, or neurodegeneration.

Acknowledgement

This work is supported in part by NIH grants NS-14543, NS-25372, HD-17001, and AG-08938. We thank Teodosia Zamora for her assistance in breeding the transgenic mice, and Marilyn Stubblebine for preparing the manuscript.

References

1. Chan PH, Fishman RA (1985) Free fatty acids, oxygen free radicals and membrane alterations in brain ischaemia and injury. In: Plum F, Pulsinelli WA (eds) Cerebrovascular diseases. Raven Press, New York, pp 161–169

2. Chan PH, Chen SF, Yu ACH (1988) Induction of intracellular superoxide radical formation by arachidonic acid and by polyunsaturated fatty acids in primary astrocytic cultures. J Neurochem 50: 1185–1193

3. Chan PH, Longar S, Fishman RA (1987) Protective effects of liposome-entrapped superoxide dismutase on post-traumatic brain oedema. Ann Neurol 21: 540–547

4. Chan PH, Chu L, Chen S (1990) Effects of MK-801 on glutamate-induced swelling of astrocytes in primary cell culture. J Neurosci Res 25: 87–93

5. Chan PH, Chen S, Imaizumi S, Chu L, Chan T (1990) New insights into the role of oxygen radicals in cerebral ischaemia. In: Bazan NG, Braquet P, Ginsberg M (eds) Neurochemical correlates of cerebral ischaemia. Plenum Publishing Co., New York (in press)

6. Choi DW, Marlucci-Gedde M, Kriegstein AR (1987) Glutamate neurotoxicity in cortical cell culture. J Neurosci 1: 357–368

7. Choi DW (1988) Glutamate neurotoxicity and diseases of the nervous system. Neuron 1: 623–634

8. Epstein CJ, Avraham KB, Lovett M, Smith S, Elroy-Stein O, Rotman G, Bry C, Groner Y (1987) Transgenic mice with increased CuZn-superoxide dismutase activity: animal model of dosage effects in Down syndrome. Proc Natl Acad Sci USA 84: 8044–8048

9. Epstein CJ, Berger CN, Carlson EJ, Chan PH, Huang TT (1990) Models for Down syndrome: chromosome 21-specific genes in mice. In: Patterson D, Epstein CJ (eds) Molecular genetics of chromosome 21 and Down syndrome. AR Liss, New York (in press)

10. Murphy TH, Miyamoto M, Sastre A, Schnaar RL, Coyle JT (1989) Glutamate toxicity in a neuronal cell line involves inhibition of cysteine transport leading to oxidative stress. Neuron 2: 1547–1558

11. Saez JC, Kesder JA, Bennett MVL, Spray DC (1987) Superoxide dismutase protects cultured neurons against death by starvation. Proc Natl Acad Sci USA 84: 3056–3059

12. Simon RP, Swan JH, Griffiths T, Meldrum BS (1984) Blockade of N-methyl-D-aspartate receptors may protect against ischaemic damage in brain. Science 226: 850–852

13. Wielock T (1985) Hypoglycaemia-induced neuronal damage was prevented by an N-methyl-D-aspartate antagonist. Science 230: 681–683

14. Yu ACH, Chan PH, Fishman RA (1986) Effects of arachidonic acid on glutamate and γ-aminobutyric acid uptake in primary cultures of rat cerebral cortical astrocytes and neurons. J Neurochem 47: 1181–1189

Correspondence: Dr. P. H. Chan, Departments of Neurosurgery and Neurology, Box 0114, University of California, School of Medicine, San Francisco, CA 94143, U.S.A.

Acta Neurochirurgica, Suppl. 51, 248–250 (1990)
© by Springer-Verlag 1990

Effects of Atrial Natriuretic Peptide on Ischaemic Brain Oedema Evaluated by the Proton Magnetic Resonance Method

S. Naruse[1], **Y. Aoki**[1], **Y. Horikawa**[1], **C. Tanaka**[1], **T. Higuchi**[1], **T. Ebisu**[1], **S. Ueda**[1], **S. Kondo**[2], **T. Kiyota**[2], and **H. Hayashi**[2]

[1] Department of Neurosurgery, Kyoto Prefectural University of Medicine, Kyoto, Japan, [2] Pharmaceuticals Lab., Asahi Chemical Industry, Co., Ltd., Kyoto, Japan

Summary

The effect of atrial natriuretic peptide (ANP) on cerebral oedema in rats was examined by magnetic resonance (MR). After occlusion of the left middle cerebral artery (MCA) to induce cerebral ischaemia, rats received continuous infusion of ANP for 24 h at a total dose of 120 μg/kg or 150 μg/kg. Proton relaxation times (T_1 and T_2) of excised oedematous tissue were measured in vitro and the area of the oedematous region was determined in vivo by the use of magnetic resonance imaging (MRI). The administration of ANP was found to decrease the lengthening of both T_1 and T_2 in the oedematous tissues and shown by MRI to decrease the area of the oedematous region, compared with group receiving saline. The topographic observations in vivo suggest that ANP suppress the development of the oedematous region.

Introduction

Atrial natriuretic peptide (ANP) has been shown to be a hormone which takes part in regulation of fluid homeostasis[1,2,4]. We have suggested, further, that ANP has a pharmacological effect on brain oedema, based on our observation that its administration results in suppressed elevation of water and sodium contents in oedematous brain tissue and that this antioedematous effect is not caused by the known systemic effects of ANP such as decrease in blood pressure and plasma volume[6]. Here we apply proton magnetic resonance to further examine the effect of ANP on ischaemic (middle cerebral artery occlusion) brain oedema in rats. The effects of intravenous infusion of ANP on proton relaxation times were examined in vitro with excised brain tissue, as a means of determining the state of the water molecules in the oedematous region, and changes in the size of the region were observed topographically in vivo by magnetic resonance imaging (MRI).

Materials and Methods

Wistar rats (male) aged 8–11 weeks were used. After anaesthetization by intraperitoneal injection of sodium pentobarbital at 50 mg/kg, the main trunk of the left middle cerebral artery (MCA) was occluded at the level of the olfactory tract with a bipolar electrocoagulator according to the method described by Tamura et al.[7]. In the sham-operation group, dura mater and arachnoid membrane were left open without MCA occlusion. Immediately after the operation, infusion of saline containing ANP (120 or 150 μg/kg) through the tail vein was begun and then continued for 24 h at a rate of 0.4 ml/h. Control groups received saline only by a similar intravenous infusion.

For the in vitro measurement of relaxation times, rats were sacrificed after 24 h MCA occlusion and the brain was removed. The cortex of the cerebral hemisphere containing the ischaemic area was excised and used for measurement of proton relaxation times (T_1 and T_2) with a Minispec PC-20 (Bruker, 0.5 tesla). T_1 was measured by the inversion recovery method with eight different interpulse intervals (τ) between the 180° and 90° pulses. T_2 was measured by the Carr-Purcell-Meiboom-Gill method with 10 echoes at 10 msec intervals. Both T_1 and T_2 were calculated as single-component values.

For the in vivo MRI observations, rats were anaesthetized by intraperitoneal injection of sodium pentobarbital at 50 mg/kg after the 24 h MCA occlusion and subjected to MRI with a SCM-200 (JEOL, 4.7 tesla). At intervals of 1.25 mm three coronal slices by T_2-weighted spin-echo (SE) imaging were obtained with 3,000 msec Tr and 96 msec Te, with 2 mm slice thickness. The theoretical imaging resolution was $0.2 \times 0.2 \times 2$ mm. For standardization of signal intensity, a ∅ 4 mm plastic tube containing pure water was held next to the rat head. For each rat, the oedematous area was calculated as the average of the three slices. Following MRI, the rat was sacrificed and the brain was excised for determination of water content (dry-wet method). Statistical comparisons were performed by Student's t-test.

Results

Effects of ANP on T_1 and T_2 values of oedematous tissue

The in vitro MR relaxation times, as the T_1 and T_2 values obtained for the left hemisphere in the untreated

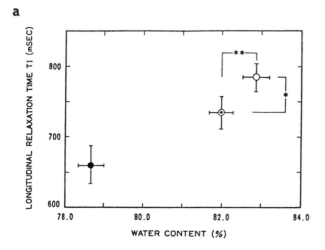

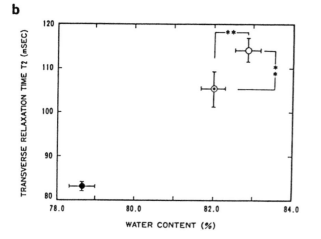

Fig. 1. Effect of ANP on T_1 (a) and T_2 (b) in hemisphere subjected to MCA occlusion. ANP (120 µg/kg) or saline only was intravenously infused during 24 h into rats immediately after left MCA occlusion. Mean ± standard error for: ○ 4 rats receiving saline only; ⊙ 6 rats receiving ANP; ● 5 untreated rats. ** p < 0.01, * p < 0.05

group and in groups subjected to left MCA occlusion and receiving ANP at 120 µg/kg or saline only for 24 h are shown in Fig. 1 a, b. In the group receiving saline only (control), the T_1 and T_2 values were much longer than those of the untreated group. In the group receiving ANP, both T_1 and T_2 were significantly shorter than those of the control (p < 0.05 for T_1 and p < 0.01 for T_2). These differences were consistent with the change of the water content.

Effects of ANP on the size of brain oedematous region

Typical T_2-weighted MRIs in 3 slices for a rat brain following left MCA occlusion and 24 h of saline infusion are shown in Fig. 2 a. The oedematous region caused by the ischaemia is visible as a high intensity area in each image. The mean calculated oedematous area and measured water content obtained for the sham operation, saline treated (control) and ANP treated groups are shown in Fig. 2 b. The group infused with ANP at 150 µg/kg exhibited an area of high intensity significantly smaller than that of the control group (p < 0.01). This difference was consistent with the change of water content.

Discussion

Recent advances in magnetic resonance have led to its utilization as an effective diagnostic method for various disorders. The data obtained through MRI study are highly useful for the diagnosis and elucidation of the pathophysiology of brain oedema[3, 5]. For the investigation of brain oedema, proton relaxation times are

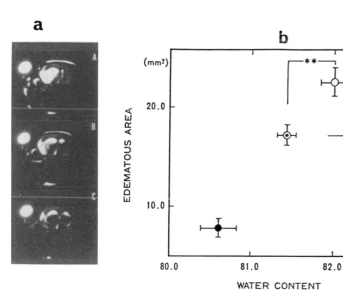

Fig. 2. Topographical evaluation of anti-oedematous effect of ANP by MRI method. (a) T_2-weighted MRIs of 3 coronal slices from 1 rat receiving saline 24 h after occlusion of left MCA. A, B, and C represent the anterior, middle and posterior sections of brain, respectively. The highest signal intensity (↑) represents pure water as the reference to standardize the signal intensity. (b) Effects of ANP on the oedematous area and brain water content. ANP (150 µg/kg) or saline only was intravenously infused during 24 h into rats immediately after left MCA occlusion. Mean ± standard error for: ○ 13 rats receiving saline only; ⊙ 7 rats receiving ANP; ● 4 rats sham operated. ** p < 0.01

sensitive indicators of the state of water molecules in the oedematous tissue. The prolongation of T_1 reflects the increased water content in the oedematous tissue, while the prolongation of T_2 reflects not only the increased water content but also the mobility of abnormally accumulated water molecule in brain tissue[3, 5]. In the present study, the effect of ANP on the state of water molecules in oedematous brain tissue was investigated.

The proton relaxation times T_1 and T_2 of excised brain tissue were measured *in vitro*. The lengthening of the T_1 and T_2 values obtained from the MCA occluded hemisphere was significantly suppressed by the ANP treatment, thus showing that the increase of the intra- and/or extracellular freely-mobile water molecule fraction, which is the main component of oedema fluid, was suppressed by the administration of ANP.

In the ensuring topographical investigation on this anti-oedematous effect of ANP, the known difficulty to quantitatively determine and compare regions of brain oedema by MRI was effectively overcome by measuring the area in the MRI having a signal intensity higher than a given proportion of the signal intensity observed for a pure water sample placed next to the rat head. It has become possible to observe topographically the suppression of the growth of oedematous area by the administration of ANP. On the other hand, in the method of brain water content determination and *in vitro* MR relaxation time measurement, it is only possible to observe the anti-oedematous effect of ANP in respect to the whole brain tissue. The application of MRI, therefore, permitted visual observation of the suppressive effect of ANP on the development of the oedematous region.

This MRI observation of the suppression of oedematous region growth by ANP, particularly in view of the fact that the anti-oedematous effect of ANP can not be attributed to systemic effects such as hypotension or decrease of circulating plasma volume[6], strongly suggests that ANP acts on the oedematous region or its peripheral region through suppression of the abnormal water molecule accumulation in the process of oedema development.

References

1. de Bold AJ (1985) Atrial natriuretic factor: a hormone produced by the heart. Science 230: 767–770
2. Dóczi T, Joó F, Szerdahelyi P, Bodosi M (1987) Regulation of brain water and electrolyte contents: The possible involvement of central atrial natriuretic factor. Neurosurgery 21: 454–458
3. Horikawa Y, Naruse S, Tanaka C, Hirakawa K, Nishikawa H (1986) Proton NMR relaxation times in ischaemic brain oedema. Stroke 17: 1149–1152
4. Israel A, Barbella Y (1986) Diuretic and natriuretic action of rat atrial natriuretic peptide (6-33) administered intracerebroventricularly in rats. Brain Res Bull 17: 141–144
5. Naruse S, Horikawa Y, Tanaka C, Hirakawa K, Nishikawa H, Yoshizaki K (1982) Proton nuclear magnetic resonance studies on brain oedema. J Neurosurg 56: 747–752
6. Naruse S, Takei R, Horikawa Y, Tanaka C, Higuchi T, Ebisu T, Ueda S, Sugahara S, Kondo S, Kiyota T, Hayashi H (1990) Effects of atrial natriuretic peptide on brain oedema: The change of water, sodium and potassium contents in the brain. In this volume
7. Tamura A, Graham DI, McCulloch J, Teasdale GM (1981) Focal cerebral ischaemia in the rat: 1. Description of technique and early neuropathological consequences following middle cerebral artery occlusion. J Cereb Blood Flow Metab 1: 53–60

Correspondence: S. Naruse, M.D., Department of Neurosurgery, Kyoto Prefectural University of Medicine, Kawaramachi-Hirokoji, Kamigyo-Ku, Kyoto 602, Japan.

Acta Neurochirurgica, Suppl. 51, 251–253 (1990)
© by Springer-Verlag 1990

Nimodipine Attenuates Both Ischaemia-induced Brain Oedema and Mortality in a Rat Novel Transient Middle Cerebral Artery Occlusion Model

H. Hara, H. Nagasawa, and **K. Kogure**

Department of Neurology, Institute of Brain Diseases, Tohoku University School of Medicine, Sendai, Japan

Summary

A novel transient middle cerebral artery (MCA) occlusion model in rats was used to evaluate the effects of nimodipine on ^{45}Ca accumulation, brain oedema and mortality. Nimodipine (1 μg/kg/min, IV, 30 min) administered immediately after 3 hr of transient unilateral MCA occlusion significantly attenuated the post-ischaemic increase of tissue water content, partly ^{45}Ca accumulation in the parieto-temporal and frontal cortices ipsilateral to the left MCA occlusion 3 hr after reperfusion. Nimodipine decreased the mortality rate at 6 and 9 hr time points after recirculation, although the survival rate at the 24 hr point after recirculation was not different from the control group. These results suggest that nimodipine has beneficial effects in the early phase of reperfusion period.

Introduction

Focal cerebral ischaemia models of the rats induced by the perpetuated MCA occlusion have been commonly used to evaluate the effects of various drugs on the ischaemic brain damage[8, 12]. In most of the stroke patients, however, recanalization after MCA occlusion is often observed, particularly in the case of cerebral embolism. Thus, a recirculation model after MCA occlusion may be of great value to simulate focal ischaemia in humans. We have developed a transient MCA occlusion model in the rat, which was produced by inserting a piece of silicone-coated nylon thread into the internal carotid artery[2, 7]. The aim of the present report was to investigate the effect of nimodipine on post-ischaemic ^{45}Ca accumulation, brain oedema and mortality after reperfusion using a novel transient MCA occlusion model.

Material and Methods

Male Wistar rats weighing 230–250 g were used. Transient focal ischaemia was induced in rats by means of left MCA occlusion[2, 7]. A piece of nylon thread, one end of which was pre-coated with silicone resin, was inserted into the left common carotid artery in order to occlude the origin of MCA, under anaesthesia (a mixture of 70% N_2O, 30% O_2, and 1% halothane). Three hr after reperfusion (6 hr after induction of ischaemia), the animals were decapitated and tissue samples of both hemispheres were quickly dissected in a humidified chamber. The water content of the samples was determined by the dry-weight method. The doses of test compounds were as follows: Nimodipine (30 μg/kg, 6 ml/kg, Bayer Co. Ltd., FR Germany), nimodipine vehicle (6 ml/kg, consisted of 200 g/l ethanol, 170 g/l macrogol 400, 2 g/l sodium citrate, and 0.3 g/l citrate), 60% glycerol (4 g/kg, 5.3 ml/kg, Wako Pure Chemical Industries Ltd., Japan), glycerol vehicle (5.3 ml/kg, saline). The intravenous administration of these compounds was started immediately after reperfusion using a pump at the rate of 0.2 ml/min for 30 min. ^{45}Ca autoradiography was performed according to the method of Kato et al.[5]. The survival of animals treated with nimodipine or vehicle alone was compared at the time points of 3, 6, 9, 24 hr and 7 days after reperfusion.

Results and Discussion

Nimodipine attenuated partly the abnormal ^{45}Ca accumulation in the MCA and ACA cortex (Fig. 1). Nimodipine significantly reduced the post-ischaemic increase of water content in the frontal and parieto-temporal cortex ipsilateral to the left MCA occlusion 3 hr after reperfusion (Fig. 2 a). Glycerol significantly reduced the increase of water content in the striatum, frontal cortex, and parieto-temporal cortex ipsilateral to the left MCA occlusion 3 hr after reperfusion (Fig. 2 b). Interestingly, glycerol significantly induced a reduction of water content in the striatum and parieto-temporal cortex contralateral to left MCA occlusion. At 3, 6, 9, 24 hr and 7 days after reperfusion, the mortality of the nimodipine-vehicle group (n = 15) was 7, 47, 87, 100 and 100%, respectively. A time-dependent increase in mortality was observed following reperfusion. On the other hand the mortality of the nimodipine

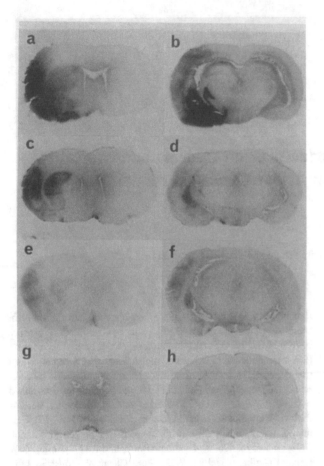

Fig. 1. Representative ⁴⁵Ca autoradiograms in the parieto-temporal cortex ipsilateral to operation in the MCA occlusion model². (a) and (b) control group, (c) and (d) nimodipine-treated group, (e) and (f) glycerol treated group, (g) and (h) sham-operated group

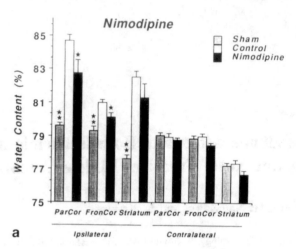

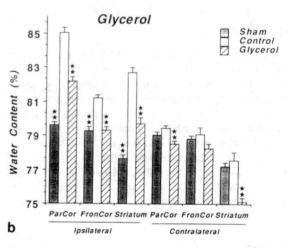

Fig. 2. Effect of nimodipine (a) and glycerol (b) on brain oedema after left MCA occlusion in rats². N = 8–12; control; vehicle-treated ischaemic group, *ParCor*: parieto-temporal cortex, *FronCor*: frontal cortex. * p < 0.05, ** p < 0.01 vs control, Dunnett's multiple range test

group (n = 11) was 0, 0, 27, 82 and 82%, respectively, with statistical significance at 6 (p < 0.05, Fisher's exact probability test) and 9 hr (p < 0.01) after reperfusion.

Calcium antagonists have been reported to be beneficial for the treatment of permanent MCA occlusion in rats[4,8]. In these models, the beneficial effects of drugs were obtained by influencing only the peripheral areas surrounding the ischaemic core. We observed that nimodipine administered immediately after recirculation attenuated the formation of brain oedema in the areas perfused not only by MCA (parieto-temporal cortex) but also by ACA (frontal cortex), although water content in the striatum was not reduced by nimodipine treatment. The failure of nimodipine to attenuate brain oedema in the striatum may be explained by more marked histopathological and metabolic damage in the striatum than in cortex[1]. In contrast, glycerol, a hyperosmotic agent, was potent enough to ameliorate

brain oedema and water content in the striatum ipsilateral to the MCA occlusion. The beneficial effect of nimodipine may partly be explained by an improvement of post-ischaemic hypoperfusion[3,6,10,11]. Another possibility is a reduction of Ca^{2+} overload in neurons[9,13]. Uematsu *et al.* reported that nimodipine prevents an increase in cytosolic free calcium concentrations following cerebral ischaemia *in vivo*[14].

In conclusion, nimodipine treatment is beneficial in early phases of the recirculation period.

References

1. Abe K, Araki T, Kogure K (1988) Recovery from oedema and of protein synthesis differs between the cortex and caudate following transient focal cerebral ischaemia in rats. J Neurochem 51: 1470–1476

2. Hara H, Onodera H, Nagasawa H, Kogure K (1990) Effect of nimodipine on ischaemia-induced brain oedema and mortality in a novel transient middle cerebral artery occlusion model. Japan J Pharmacol 53: 247–253

3. Hoffmeister F, Benz U, Heise A, Krause HP, Neuser V (1982) Behavioural effects of nimodipine in animals. Arzneimittelforsch 32: 347–360

4. Jacewicz M, Brint S, Tanabe J, Pulsinelli W (1989) Continuous nimodipine treatment attenuates cortical infarction in rats subjected to 24 hours of focal cerebral ischaemia. J Cereb Blood Flow Metab 10: 89–96

5. Kato H, Kogure K, Sakamoto N, Watanabe T (1987) Greater disturbance of water and ion homeostasis in the periphery of experimental focal ischaemia. Exp Neurol 96: 118–126

6. Kazda S, Garthoff B, Krause HP, Schlossmann K (1982) Cerebrovascular effects of the calcium antagonistic dihydropyridine derivative nimodipine in animal experiments. Arzneimittelforsch 32: 331–338

7. Nagasawa H, Kogure K (1989) Correlation between cerebral blood flow and histologic changes in a new rat model of middle cerebral artery occlusion. Stroke 20: 1037–1043

8. Nakayama H, Ginsberg MD, Dietrich WD (1988) (s)-Emopamil, a novel calcium channel blocker and serotonin S$_2$ antagonist, markedly reduces infarct size following middle cerebral artery occlusion in the rat. Neurology 38: 1667–1673

9. Siesjo BK (1981) Cell damage in the brain: A speculative synthesis. J Cereb Blood Flow Metab 1: 155–185

10. Steen PA, Newberg LA, Milde JH, Michenfelder JD (1983) Nimodipine improves cerebral blood flow and neurologic recovery after complete cerebral ischaemia in the dog. J Cereb Blood Flow Metab 3: 38–48

11. Steen PA, Newberg LA, Milde JH, Michenfelder JD (1984) Cerebral blood flow and neurologic outcome when nimodipine is given after complete cerebral ischaemia in the dog. J Cereb Blood Flow Metab 4: 82–87

12. Tamura A, Graham DI, McCulloch J, Teasdale GM (1981) Focal cerebral ischaemia in the rat. 1. Description of technique and early neuropathological consequences following middle cerebral artery occlusion. J Cereb Blood Flow Metab 1: 53–60

13. Uematsu D, Greenberg JH, Revich M, Karp A (1988) *In vivo* measurement of cytosolic free calcium during cerebral ischaemia and reperfusion. Ann Neurol 24: 420–428

14. Uematsu D, Greenberg JH, Hickey WF, Reivich M (1989) Nimodipine attenuates both increase in cytosolic free calcium and histological damage following cerebral ischaemia and reperfusion in cats. Stroke 20: 1531–1537

Correspondence: H. Hara, M.D., Department of Neurology, Institute of Brain Diseases, Tohoku University School of Medicine, 1-1 Seiryo-machi, Aoba-ku, Sendai 980, Japan.

Acta Neurochirurgica, Suppl. 51, 254–255 (1990)
© by Springer-Verlag 1990

MR Evaluation of Calcium Entry Blockers with Putative Cerebroprotective Effects in Acute Cerebral Ischaemia

J. Kucharczyk, J. Mintorovich, R. Sevick, H. Asgari, and **M. Moseley**

Neuroradiology Section, University of California Medical Center, San Francisco, California, U.S.A.

Summary

MR imaging and spectroscopy were used to investigate whether two calcium channel entry-blockers, nicardipine and RS-87476 (Syntex), would reduce ischaemic brain damage in barbiturate-anaesthetized cats subjected to permanent unilateral occlusion of the middle cerebral artery (MCA). The evolution of cerebral injury was assessed *in vivo* in a total of 38 cats using a combination of diffusion-weighted and T 2-weighted spin-echo proton MR imaging and phosphorus 31 (P-31) and proton (H-1) MR spectroscopy for up to 12 h following arterial occlusion. Immediately thereafter, the volume of histochemically ischaemic brain tissue was determined planimetrically. In untreated control animals, diffusion-weighted MR images obtained with strong gradient strengths (5.5 gauss/cm) displayed increased signal intensity (oedema) in the ischaemic MCA territory less than 45 min after stroke. These changes were closely correlated with the appearance of abnormal P-31 and H-1 metabolite levels evaluated with surface coil MR spectroscopy. Cats injected with i.v. nicardipine (10 μg/kg bolus, 8 μg/kg/h maintenance) or RS-87476 (2–50 μg/kg bolus, 0.7–17.5 μg/kg/h maintenance) showed a significant reduction in ischaemic injury in the ipsilateral cerebral cortex, internal capsule and basal ganglia. The results of this study suggest that these calcium entry blockers protect against brain damage induced by acute stroke by stabilizing cellular metabolic processes, reducing lactate formation in ischaemic tissues, and attenuating cytotoxic and vasogenic oedema.

Introduction

The role of calcium antagonists in protecting against ischaemia-induced brain injury is controversial[1–5]. The present study was designed to test whether the 1,4-dihydropyridine calcium antagonist, nicardipine, and a new calcium and sodium ion channel modulator RS-87476 (Syntex Research), would reduce cerebral injury induced by permanent middle cerebral artery (MCA) occlusion in cats. Previous work has demonstrated that RS-87476 combines selective inhibition of cerebrovasospasm with a selective action against activators of calcium channels[6]. A neurocytoprotective action of ni-

cardipine in a 20-minute four-vessel rat model of transient forebrain ischaemia has also been reported[3].

MR imaging was used *in vivo* to evaluate whether these calcium entry blockers would prevent or minimize intracerebral accumulation of water if drug administration was delayed until 15 min after MCA occlusion. The anatomical extent of ischaemic brain injury on MR imaging was compared with tissue perfusion deficits demonstrated by a non-ionic intravascular contrast agent, Dysprosium-DTPA-BMA (DyDTPA-BMA). MR images in drug-treated and MCA-occluded control cats were also evaluated in relation to assessments of cerebral metabolic injury obtained via P-31 and H-1 MR spectroscopy, as well as histopathology.

Materials and Methods

The right MCA was isolated via the transorbital approach in 38 anaesthetized cats and occluded just proximal to the origin of the lateral striate arteries. The muscles overlying the ipsilateral parietal fossa were excised unilaterally to optimize the placement of the MR spectroscopy surface coil in a standard location over the MCA territory.

A General Electric CSI (2 Tesla) unit, equipped with self-shielded gradient coils (± 20 gauss/cm, 15 cm bore size) was used. P-3 and H-1 MR spectroscopy was performed with a 2 × 3 cm double-tuned balanced-matched surface coil positioned over the MCA territory. MR imaging was performed with an 8.5 cm inner-diameter proton imaging coil. Successive multi-slice diffusion- (TR 1800/1000 msec, TE 80 msec, gradient strength 5.5 gauss/cm) and T 2-weighted images (TR/TE 2800/80/160 msec), and H 1 and P-31 MR spectra were obtained for 5–12 hours following occlusion. Region-of-interest (ROI) image analyses were carried out in the ischaemic middle temporal cortex, ectosylvian gyrus, internal capsule, and caudate nucleus, and compared with the corresponding uninjured contralateral regions.

To compare anatomic regions of perfusion deficiency with areas of high signal intensity on diffusion-weighted MR imaging, some cats were also given i.v. injections of a non-ionic T 2*-shortening contrast agent, DyDTPA-BMA (0.25, 0.5, 1.0 mmol/kg) (Salutar, Inc.). After injection, the magnetic susceptibility effect was monitored for 15–60 minutes in both ischaemic and normal hemispheres by comparing ROI intensity to pre-contrast T2-weighted and diffusion-weighted ROI intensities.

MR spectra were phased and analyzed with a line-fitting computer simulation program. The inorganic phosphate (Pii) phosphocreatine (PCr) ratio was calculated to quantify the bioenergetic status of the tissues. The peak areas of lactate and N-acetylaspartate (NAA) were resolved and the lactate NAA ratios calculated. At the conclusion of the MR protocol, 10 ml/kg of a 2% solution of TTC was infused transcardially. 24–36 hours later the brain was sectioned coronally at 2–3 mm and immediately examined for histologic evidence of ischaemic damage.

Results

Compared with control cats, infarct size was significantly decreased in animals treated with RS-87476 or nicardipine. In cats receiving nicardipine, there was a clear attenuation of cerebral oedema after MCA occlusion, as evidenced by the reduction in the area of hyperintensity on T2-weighted spin-echo MRI. Subsequent histopathology indicated a 61 ± 23% reduction in infarct size compared with untreated controls at 5 h after arterial occlusion. Animals treated with nicardipine were also more able than controls to maintain preischaemic concentrations of high-energy phosphates and maintaining lower levels of inorganic phosphate.

In animals treated with RS-87476, the infarct size 12 h after occlusion was reduced by an average 70% at the lowest dose (2 µg/kg i.v. bolus followed by continuous infusion at 0.7 µg/kg/h, n = 6), and by 88% at the highest dose (50 µg/kg priming dose, 17.5 µg/kg/h maintenance dose, n = 8). Tissue oedema, observed as areas of signal hyperintensity on diffusion- and T2-weighted spin-echo MR images, was usually confined to small regions of the parietal cortex and basal ganglia. Cats injected with RS-87476 were also better able to stabilize metabolic energy reserves and reduce lactate formation in ischaemic tissues than control animals.

Discussion

The results of this investigation demonstrate that postischaemic treatment with RS-87476 and nicardipine can significantly attenuate ischaemic brain injury induced by permanent unilateral occlusion of the MCA. Administration of these drugs 15 min after arterial occlusion preserved close to normal cerebral metabolism and significantly reduced cerebral oedema and histologically measured infarct size for 5–12 h. Similar degrees of cerebroprotective effect were found for RS-

87476 and nicardipine, and both drugs had only slight systemic hypotensive effects in the dose range tested.

RS-87476 has been shown to combine selective inhibition of cerebrovasospasm, with a novel antagonism of calcium channels at lower concentrations and sodium channels at higher concentrations[6]. The drug also appears to protect against the delayed neuronal cell death which follows 3 days after cerebral ischaemia in the rat and gerbil[6]. Nicardipine has been reported to exert a neurocytoprotective effect when given before or after occlusion in a 10-minute four-vessel rat model of transient forebrain ischaemia[3]. The effects of nicardipine were particularly evident in cerebral tissues like the hippocampus, which are known to be "selectively vulnerable" to acute ischaemia[5]. Nicardipine has also recently been shown to protect hippocampal CA 1 neurons in gerbils surviving 72 hours after 5 minutes of bilateral carotid occlusion[2].

In summary, the sodium/calcium channel modulator RS-87476 and the dihydropyridine calcium channel blocker nicardipine both significantly reduced ischaemic brain damage in cats subjected to permanent unilateral occlusion of the MCA. The protective action of the drugs in this model of acute cerebral ischaemia may be related to their effects on cellular metabolic processes which preserve ATP and reduce oedema formation during the ischaemic period.

References

1. Alps BJ, Calder C, Hass WK, Wilson AD (1988) Comparative protective effects of nicardipine, flunarizine, lidoflazine and nimodipine against ischaemic injury in the hippocampus of the mongolian gerbil. Br J Pharmacol 98: 877–883
2. Alps BJ, Calder C, Wilson AD (1986) The effects of nicardipine on "delayed neuronal death" in the ischaemic gerbil hippocampus. Br J Pharmacol 88: 250–254
3. Alps BJ, Hass WK (1987) The potential beneficial effect of nicardipine in a rat model of transient forebrain ischaemia. Neurology 37: 809–814
4. Kucharczyk J, Chew W, Derugin N, Rollin C, Moseley M, Berry I, Norman D (1989) Nicardipine reduces ischaemic brain injury: An *in vivo* magnetic resonance imaging/spectroscopy study in cats. Stroke 20: 268–274
5. Mumekata K, Hossmann K-A (1987) Effect of 5-minute ischaemia on regional pH and energy state of the gerbil brain: Relation to selective vulnerability of the hippocampus. Stroke 18: 412–417
6. Spedding M, Alps BJ, Patmore L, Kilpatrick AT. Personal Communication 1989
7. Spedding M, Kilpatrick AT, Alps BJ (1989) Activators and inactivators of calcium channels: Effects in the central nervous system. Fundam Clin Pharmacol 3: 35–295·

Correspondence: Dr. J Kucharczyk, Neuroradiology Section, University of California Medical Center, San Francisco, CA 94143, U.S.A.

Acta Neurochirurgica, Suppl. 51, 256–258 (1990)

Effect of Steroid Therapy on Ischaemic Brain Oedema and Blood to Brain Sodium Transport

A. L. Betz[1,2] and **H. C. Coester**[1]

Departments of [1] Surgery (Neurosurgery), [2] Pediatrics, and [2] Neurology, University of Michigan, Ann Arbor, Michigan, U.S.A.

Summary

Dexamethasone has been shown in some studies to reduce ischaemic brain oedema, however, the mechanism is unknown. One possible mechanism is through inhibition of active transport of sodium across the blood-brain barrier (BBB) since some steroids, especially progesterone, inhibit sodium transport in isolated brain capillaries. Therefore, we measured brain oedema and BBB permeability to sodium and a passive permeability tracer, α-aminoisobutyric acid (AIB), 4 hr after middle cerebral artery occlusion (MCAO) in rats that had been treated 1 hr before MCAO with vehicle (control) or 2 mg/kg of either dexamethasone or progesterone. In controls, the water content of tissue in the center of the ischaemic zone was $82.4 \pm 0.2\%$. Brain oedema was significantly reduced following pretreatment with either dexamethasone (80.6 ± 0.1, $p < 0.001$) or progesterone (81.5 ± 0.3, $p < 0.05$). Both steroids also reduced BBB permeability to AIB by about 40% in normal brain but to a lesser extent in ischaemic brain. In contrast, steroid treatment had no effect on BBB permeability to sodium in either normal or ischaemic brain. We conclude that pretreatment with dexamethasone and progesterone reduces brain oedema accumulation during the early stages of ischaemia, however, this effect does not result from a reduction in BBB permeability to sodium.

Introduction

The usefulness of steroid therapy in the treatment of ischaemic brain oedema is controversial[8]. Since the mechanism of action of dexamethasone in reducing brain oedema is not known, it is possible that a different type of steroid would be more consistently effective.

One mechanism by which steroids could reduce brain oedema when the blood-brain barrier (BBB) is intact is by reducing blood to brain sodium transport. We have recently shown that BBB sodium transport is rate limiting for the development of ischaemic oedema[4]. The movement of sodium from blood to brain appears to involve Na,K-ATPase in the brain capillary endothelial cell[2], a transporter whose activity can be reduced *in vitro* by certain steroids[6]. Of particular interest is the

fact that dexamethasone was relatively ineffective in reducing brain capillary Na,K-ATPase activity while progesterone and deoxycorticosterone were approximately 10 times more potent[6]. The study summarized here, was designed to compare the effects of dexamethasone and progesterone on BBB sodium transport and brain oedema formation during ischaemia[3].

Materials and Methods

Focal cerebral ischaemia was produced by occlusion of the middle cerebral artery (MCAO) in adult male Sprague-Dawley rats that were anaesthetized with ketamine (50 mg/kg) and xylazine (10 mg/kg). The method used was similar to that described by Bederson *et al.*[1]. Animals were pre-treated 1 hr before MCAO with 2 mg/kg of either dexamethasone or progesterone given i.p. as a solution in sesame oil containing 2% ethanol and 0.5 mg steroid/ml. Controls received the vehicle alone. Four hours after MCAO, animals were killed by decapitation, the brains were quickly removed and a 7 mm cork borer was used to obtain tissue samples from the center of the ischaemic cerebral cortex and from the corresponding area of the contralateral non-ischaemic cortex[3].

One group of 20 animals was used for measurement of the water content as determined from the wet and dry weights. The second group of 26 animals was used for measurement of BBB permeability to ^{22}Na and ^{3}H-α-aminoisobutyric acid (AIB). The latter compound crosses the BBB by simple diffusion and, therefore, serves as a tracer for passive permeability of the BBB[5]. Animals in this group, received an intravenous injection of ^{22}Na and ^{3}H-AIB 10 min before termination of the experiment and the PS products for these compounds were determined as described previously[3]. Differences between control and steroid-treated groups were tested for significance using ANOVA and Student's t-test with Bonferroni's correction for multiple comparisons. Paired t-tests were used for comparison of ischaemic and non-ischaemic samples.

Results

The cerebral cortex of the ischaemic hemisphere showed a marked increase in water content 4 hr after

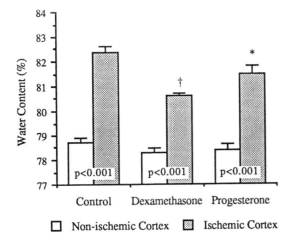

Fig. 1. Effect of treatment with dexamethasone or progesterone on ischaemic brain oedema. Rats were treated 1 hr before MCAO with 2 mg/kg of either steroid or vehicle. Brain water content was determined 4 hr after MCAO in samples taken from the center of the ischaemic zone and from the same area of the contralateral, non-ischaemic cerebral cortex. Values are the means ± SE for determinations on 6–7 animals. The p values shown are for comparison between ischaemic and non-ischaemic samples while * (p < 0.05) and † (p < 0.001) indicate the level of significance between treated and control groups

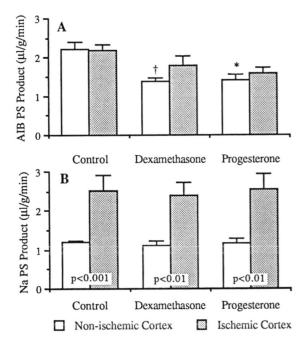

Fig. 2. Effect of treatment with dexamethasone and progesterone on BBB permeability. Rats were treated 1 hr before MCAO with 2 mg/ kg of either steroid or vehicle. The BBB permeabilities (PS product) to AIB (A) and sodium (B) were determined 4 hr after MCAO in samples taken from the center of the ischaemic zone and from the same area of the contralateral, non-ischaemic cerebral cortex. Values are the means ± SE for determinations on 8–10 animals. The p values shown are for comparison between ischaemic and non-ischaemic samples while * (p < 0.05) and † (p < 0.001) indicate the level of significance between treated and control groups

MCAO (Fig. 1). Treatment with either steroid significantly reduced this brain oedema accumulation, although the effectiveness of dexamethasone was greater than that of progesterone.

BBB permeability to AIB was the same in the ischaemic and non-ischaemic cortex of the control group indicating that the BBB was probably still intact (Fig. 2 A). Steroid treatment significantly reduced the BBB permeability to AIB in the normal brain tissue consistent with previous reports[12]. A similar, but less marked reduction was seen in the ischaemic tissue, although, due to the greater variability, this difference was not significant.

There was a significant increase in the PS product for sodium in the ischaemic as compared to the non-ischaemic cortex (Fig. 2 B). However, in contrast to the effect of steroids on BBB permeability to AIB, they had no effect on BBB permeability to sodium in either normal or ischaemic tissue.

Discussion

Our results show that treatment with a high dose of either dexamethasone or progesterone prior to the onset of focal ischaemia significantly attenuates early brain oedema formation. This effect, however, does not appear to be mediated through an inhibition of BBB sodium transport as we had hypothesized. Dexamethasone is known to reduce the passive permeability of the normal BBB[12]. We were able to confirm this by measuring the BBB permeability to AIB and we further demonstrated that progesterone had a similar effect.

In all treatment groups, blood to brain sodium flux was significantly increased in the ischaemic tissue. We have observed a similar selective stimulation of BBB sodium transport during and immediately following focal ischaemia in the gerbil[7], as have Shigeno et al. during focal ischaemia in the cat[11]. We believe that this may result from stimulation of brain capillary Na,K-ATPase by the high extracellular potassium concentration in the ischaemic tissue. Na,K-ATPase in brain capillaries is likely to provide the driving force for active sodium transport across the BBB[2] and its activity is markedly affected by increases in extracellular potassium[10].

The mechanism for the beneficial effect of steroids on early ischaemic oedema formation remains unknown. Since progesterone was nearly as effective as dexamethasone, the mechanism probably does not involve a process mediated by corticosteroid receptors.

Perhaps the steroids act as free radical scavengers and, thereby, permit in the brain cells to better maintain their ion homeostasis. We have proposed this mechanism to explain the effect of other types of free radical scavengers on ischaemic oedema formation[9]. Alternatively, the effect of steroids on ischaemic oedema may be mediated through their effect on passive permeability of the BBB. It is possible that steroids reduce brain oedema by reducing BBB permeability to a solute other than sodium that crosses the BBB by simple diffusion. Future experiments will be directed at determining the role of chloride in brain oedema accumulation and the effect of steroids on chloride flux from blood to brain.

References

1. Bederson JB, Pitts LH, Tsuji M, Nishimura MC, Davis RL, Bartkowski H (1986) Rat middle cerebral artery occlusion: evaluation of the model and development of a neurologic examination. Stroke 17: 472–476
2. Betz AL (1986) Transport of ions across the blood-brain barrier. Fed Proc 45: 2050–2054
3. Betz AL, Coester HC (1990) Effect of steroids on oedema and sodium uptake of the brain during focal ischaemia in rats. Stroke 21: in press
4. Betz AL, Ennis SR, Schielke GP (1989) Blood-brain barrier sodium transport limits development of brain oedema during partial ischaemia in gerbils. Stroke 20: 1253–1259
5. Blasberg RG, Fenstermacher JD, Patlak CS (1983) Transport of α-aminoisobutyric acid across brain capillary and cellular membranes. J Cereb Blood Flow Metab 3: 8–32
6. Chaplin ER, Free RG, Goldstein GW (1981) Inhibition by steroids of the uptake of potassium by capillaries isolated from rat brain. Biochem Pharmacol 30: 241–245
7. Ennis SR, Keep RF, Schielke GP, Betz AL (1990) Decrease in perfusion of cerebral capillaries during incomplete ischaemia and reperfusion. J Cereb Blood Flow Metab 10: 213–220
8. Katzman R, Clasen R, Klatzo I, Meyer JS, Pappius HM, Waltz AG (1977) IV. Brain oedema in stroke. Stroke 8: 512–540
9. Martz D, Beer M, Betz AL (1990) Dimethylthiourea reduces ischaemic brain oedema without affecting cerebral blood flow. J Cereb Blood Flow Metab 10: 352–357
10. Schielke GP, Moises HC, Betz AL (1990) Potassium activation of the Na,K-pump in isolated brain microvessels and synaptosomes. Brain Res, in press
11. Shigeno T, Asano T, Mima T, Takakura K (1989) Effect of enhanced capillary activity on the blood-brain barrier during focal cerebral ischaemia in cats. Stroke 20: 1260–1266
12. Ziylan YZ, LeFauconnier JM, Bernard G, Bourre JM (1988) Effect of dexamethasone on transport of α-aminoisobutyric acid and sucrose across the blood-brain barrier. J Neurochem 51: 1338–1342

Correspondence: A. Lorris Betz, M.D., Ph.D., D3227 Medical Professional Bldg., University of Michigan, Ann Arbor, MI 48109-0718, U.S.A.

Acta Neurochirurgica, Suppl. 51, 259–260 (1990)

Dexamethasone Does not Influence Infarct Size but Causes Haemorrhage in Baboons with Cerebral Infarcts

A. Hartmann, G. Ebhardt[1], Th. Rommel[1], and **A. Meuer**

Neurologische Universitätsklinik Bonn und [1] Städtisches Krankenhaus Köln-Merheim, Federal Republic of Germany

Summary

15 baboons with permanent occlusion of the left MCA were randomly treated with dexamethasone (n = 8; 1 mg/kg day) or 0.9% NaCl (n = 5). Dexamethasone did not reduce the resulting infarct size. However, steroid treated animals had more brain tissue affected by haemorrhage, although this did not seem to affect survival.

Introduction

In previous studies we could show that early given high doses of dexamethasone might improve cerebral blood flow (CBF), autoregulation and CO_2-reactivity in the neighbourhood of the infarcted middle cerebral artery[1]. This, as shown in other animals, goes along with a significant reduction of brain oedema but not with alteration of the infarct size[2].

However, since dexamethasone does not only act on the blood-brain barrier but also on the vascular system itself with possible changes of the morphological integrity, we undertook a third study to investigate morphological effects of the steroid therapy.

Methods and Protocol

15 baboons were included into the protocol and treated randomly with daily intraperitoneal injection of either 1 mg/kg dexamethasone or 1 ml/kg 0.9% NaCl for 7 days. Before start of therapy the left middle cerebral artery was permanently occluded by the transorbital approach. Autopsy was performed on day 7, brains were fixed and evaluated for both, infarct size with assessment of infarcted areas in brain tissue photographs and complications, such as massive and small haemorrhages in both hemispheres.

Results

a) Infarct size: There was no significant difference in infarct volume in both groups (5 control animals, which spontaneously survived till day 7, and 8 treated animals). In the control group, infarct volume was $5.54\% \pm 3.42$ and in the dexamethasone group $4.80\% \pm 2.22$ (Table 1).

Indirect signs of brain oedema (brain swelling, ventricular compression, and midline shift) were present in both groups with no indication of any relevant difference (Table 2).

b) Haemorrhage: Haemorrhages occurred in both groups. In the control group 0.46% of the infarcted hemisphere presented with haemorrhages while bleeding complications were absent in the contralateral side. 0.70% of the infarcted side in the dexamethasone group was haemorrhagic and this was also present in 3 out of 8 contralateral hemispheres occupying 0.12% of the contralateral tissue.

Table 1. *Infarct Volume in Baboons with Middle Cerebral Artery Occlusion*

Animal	Control	Animal	Dexameth.
I	2.74	II	6.73
III	8.94	IV	7.11
VII	9.15	V	1.93
XIII	5.10	VI	5.29
XIV	1.77	VIII	2.58
		IX	5.41
		X	2.27
		XII	7.11
Mean	5.54		4.80
SD	3.42	ns	2.22

Each animal is indicated by a roman number. The infarct volume is given in percent of the total hemispheric volume. 2 control animals died before day 7.

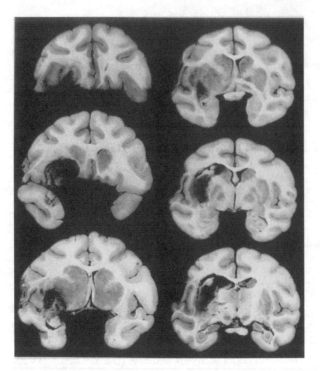

Table 2. *Effect of Dexamethasone on Cerebral Infarct.* Indirect signs of brain oedema in baboons with middle cerebral artery occlusion

	Controls	Dexamethasone
Brain swelling	3/5 (60%)	3/8 (37%)
Ventricular compression	5/5 (100%)	5/8 (62%)
Midline shift	3/5 (60%)	7/8 (87%)

Evaluation on day 7 after permanent clip positioning. The actual number of animals presenting with the individual signs out of the total number of animals is given.

Fig. 1. Massive haemorrhage in middle cerebral artery infarct during treatment with dexamethasone

In the infarcted side the internal capsule, if part of the infarct core, tended to be more affected by bleeding in the control group. The putamen, if part of the infarct core, was more affected in the dexamethasone group. Altogether, no area was significantly more often involved in bleeding complications in any of both groups, however.

Discussion

It was shown that 7 days after acute and permanent occlusion of the middle cerebral artery in the baboon there is no reduction of infarct size by dexamethasone. However, as previously observed, spontaneous death before termination of the protocol (day 7) was observed only in the control group (n = 2). The increased bleeding complication in both, the infarcted and the contralateral hemisphere in the steroid treated group obviously had no serious implication for survival.

Yet, since infarct size was not reduced it might be questioned, whether steroid therapy of acute cerebral infarct actually is effective.

References

1. Hartmann A, Menzel J, Buttinger C, Lange D (1979) Preservation of regional CBF in experimental ischaemic infarction by dexamethasone. Acta Neurol Scand [Suppl] 72: 306–307
2. Barbosa-Coutinho LM, Hartmann A, Hossmann K-A, Rommel T (1985) Effect of serum protein extravasation in experimental brain infarcts in monkeys. Acta Neuropathol (Berl) 65: 255–260

Correspondence: Dr. A. Hartmann, Neurologische Universitätsklinik, D-5300 Bonn, Federal Republic of Germany.

Acta Neurochirurgica, Suppl. 51, 261–262 (1990)
© by Springer-Verlag 1990

Trauma and Brain Oedema (I)

Brain Oedema in Experimental Closed Head Injury in the Rat

W. A. van den Brink, A. Marmarou, and **C. J. J. Avezaat**[1]

Medical College of Virginia, Richmond, Virginia, U.S.A. and the [1] Dijkzigt University Hospital of Rotterdam, The Netherlands

Summary

Development of brain oedema was studied in a new closed head injury (CHI) model of the rat. This acceleration impact models does not produce the dramatic blood pressure surge seen with fluid percussion injury. Sixteen Sprague Dawley rats were separated into 4 groups; 8 survivors sacrified at 4 and 24 hours post injury; and 8 Sham treated animals sacrified at the same time intervals. Brains were analyzed using gravimetric technique[1]. Despite absence of the high post traumatic blood pressure surge, mild oedema was observed in 4 of 5 slices at 4 hours post injury. At 24 hours post injury, significant oedema was observed throughout the brain tissue. The study demonstrates that traumatic oedema develops following acceleration impact within a 24 hour period of CHI. The oedema occurs in the absence of significant brain stem damage and blood pressure rise characteristic of this new CHI model.

Introduction

Cerebral oedema is one of the major factors governing the outcome from severe head injury. The development of oedema following experimental head injury utilizing the fluid percussion model has been reported[2]. However, more recently, we have shown that high level fluid percussion (> 3.5 Atm) is predominantly a model of severe brain stem injury[3]. Moreover, the blood pressure surge exceeding 250 mmHg commonly seen in this model soon after impact[4] clearly exacerbates the breakdown of the blood brain barrier and contributes to the oedema formation. The objective of this research was to investigate the temporal course of oedema formation in a new model of closed head injury where brain stem damage is minimum.

Methods

Sixteen male Sprague Dawley rats were separated in 4 groups: 4 animals were injured and survived for 4 hours, 4 animals were injured and survived for 24 hours, 4 animals were sham operated and survived for 4 hours, 4 animals were sham operated and survived for 24 hours. The brains of non surviving animals were excluded in this study. Preparation existed of inducing anaesthesia in a methophane saturated chamber. When unconscious the rats were endotracheal intubated and anaesthesia was maintained using halothane 1.5% in a NO_2/O_2 mixture of 66%/33%. The preparation of the skull was performed under sterile conditions: a midline scalp incision was made, the skin and periosteum were reflected and the skull was carefully dried. A round stainless steel disc was mounted on the skull using dental acrylic. When the cement was dry, the rat was removed from anaesthesia, positioned under a hollow plexiglas tubing and a sectioned brass weight of 450 gram was dropped from a height of 2 meter on the center of the metal disc. The "helmet" prevented the skull from fractures and the animals were subjected to sudden impact acceleration of the skull. The nature of this injury and biomechanics be described in a future report. The animals were placed on anaesthesia again and after removing the helmet, application of topical analgesics and antibiotics, the skin was sutured and the animal recovered. The sham animals were teated in exactly the same manner with exception of the injury. After the time of the protocol, *i.e.*, 4 or 24 hours, the rats were deeply anaesthetized and sacrificed using 2 cc of saturated KCl intracardiac. Brains were removed quickly and stored in a airtight container in the freezer. Within 3 hours the frosted brains were sliced in a standard way in 5 slices from frontal to occipital pole and kept on ice for 30 minutes. After defrosting microtechniques were used to obtain standard sized pieces of brain that were dropped in the gravimetrical columns. The analysis was performed as described earlier by Marmarou *et al.*[1].

* This work was supported in part by the grants NS 19235 and NS 12587 from the National Institutes of Health. Additional facilities and support were provided by the Richard Roland Reynolds Neurosurgical Research Laboratories.

Table 1

Loss of body weight after 24 hours due to:		gain per 24 hours
operation 2 gram	2 meter trauma 25 gram	no manipulation 5 gram

Table 2. *Brain Water Content (%) per Group.* 3 mm slices of brain tissue from rostral to caudal pole

slice #	Brain water content (%)				Student T-test
	Injury	4 hr	Sham	4 hr	p-value
slice 1	78.91		78.74		N.S.
slice 2	79.21		78.92		<0.01
slice 3	79.00		79.08		N.S.
slice 4	79.16		79.02		N.S.
slice 5	79.60		79.27		N.S.
slice #	Injury	24 hr	Sham	24 hr	p-value
slice 1	78.67		77.85		<0.01
slice 2	79.06		78.77		=0.012
slice 3	79.07		78.67		<0.01
slice 4	79.14		78.64		<0.01
slice 5	79.21		78.98		<0.01

Statistical Analysis

A T-test for unpaired variables was performed to compare the sham groups with the injury groups of corresponding survival time and the sham groups among each other. P-values <0.05 were considered significant.

Results

Brain water contents were averaged among each group. Table 1 demonstrates the mean brain water contents in the four groups. As can be seen the 24 hour sham group, is mildly dehydrated when compared to the 4 hour sham group. We attributed this dehydration to the weight loss often observed after anaesthesia and surgery (Table 2).

The 4 hour survival group shows a trend in oedema development with significant difference only in slice 2. After 24 hours however, the trend in oedema development is significant in all slices. This is more remarkable when the severe weight loss and concomitant dehydration is considered.

Conclusion

Oedema develops within 24 hours in this model of closed head injury. This occurs in the absence of marked brain stem damage or pressure surge that is typical of fluid percussion. Information about the temporal development or clearance of the oedema is lacking and further studies are necessary to document the peak and start of the resolution process. The dynamics of rat closed head injury with accompanying impact and acceleration is an additional experimental model which can be used to elucidate the mechanisms of oedema formation and resolution.

References

1. Marmarou A, Tanaka K, Schulman K (1982) An improved gravimetric measure of cerebral oedema. J Neurosurg 56: 246–253
2. Ellis EF, Chao J, Heizer ML (1989) Brain kininogen following experimental brain injury: evidence for a secondary event. J Neurosurg 71: 437–442
3. Shima K, Marmarou A (in press) Evaluation of brain stem dysfunction following severe fluid percussion head injury to the cat. J Neurosurg (in press)
4. Dixon CE, Lighthall JW, Anderson TE (1988) Physiologic, histopathologic, and cineradiographic characterization of a new Fluid Percussion model of experimental brain injury in the rat. J Neurotrauma 5: 99–104

Correspondence: A. Marmarou, Ph.D., Medical College of Virginia, Division of Neurosurgery, MCV station, Box 508, Richmond, VA 23298, U.S.A.

Acta Neurochirurgica, Suppl. 51, 263–264 (1990)
© by Springer-Verlag 1990

Development of Regional Cerebral Oedema
After Lateral Fluid-Percussion Brain Injury in the Rat

T. K. McIntosh, H. Soares, M. Thomas, and **K. Cloherty**

CNS Injury Laboratory, Surgical Research Center Department of Surgery, University of Connecticut Health Center, Farmington, Connecticut, U.S.A.

Summary

Most studies attempting to characterize post-traumatic oedema formation have focused on the acute postinjury period. We have recently developed a new model of lateral (parasagittal) fluidpercussion (FP) brain injury in the rat. The purpose of the present study was to characterize the temporal course of oedema formation and resolution in this experimental model of brain injury. Male Sprague-Dawley rats (n = 67) were anaesthetized and subjected to FP brain injury of moderate severity. Animals were sacrificed at 1 hour, 6 hours, 24 hours, 2 days, 3 days, 5 days and 7 days after brain injury, brains removed and assayed for water content using either specific gravitimetric or wet weight/dry weight techniques. In the injured left parietal cortex, a significant increase in water content was observed by 6 hours postinjury ($p < 0.05$) that persisted up to 5 days postinjury. A prolonged and significant increase in water content was also observed in the left (ipsilateral) hippocampus which began at 1 hour postinjury ($p < 0.05$) and continued up to 3 days. Other regions examined showed no significant regional oedema after brain injury. These results suggest that lateral FP brain injury produces an early focus oedema that persists for a prolonged period after trauma. This model may be useful in the evaluation of novel pharmacological therapies designed to reduce cerebral oedema after brain injury.

Introduction

The secondary or delayed pathophysiological events that follow traumatic brain injury often include the development of cerebral oedema[1]. The importance of characterizing oedema formation in a relevant clinical model of brain injury is underscored by the fact that 40–50% of head injured patients die from unresolved increased intracranial pressure due to oedema and/or acute lesion development[2, 3]. Midline FP brain injury, a well-characterized experimental model of brain trauma, reproduces many secondary pathophysiological post-traumatic events including oedema formation[4]. Recently, we have developed a new model of lateral (parasagittal) FP brain injury in the rat, centered over the left parietal cortex[5]. Since previous studies attempting to characterize oedema formation in experimental brain injury have focused only on the acute postinjury period, the present study was designed to characterize the temporal course of oedema formation after lateral FP brain injury in the rat.

Materials and Methods

Male Sprague-Dawley rats (300–350 g, n = 67) were anaesthetized with sodium pentobarbital (50 mg/kg i.p.). A craniotomy was made over the left parietal cortex, midway between lambda and bregma for the induction of FP brain injury. All animals were subjected to lateral FP brain injury of moderate severity (2.3–2.6 atmospheres) as previously described[5]. Briefly, a bolus of saline is injected through the craniotomy at high pressure, causing a rapid but brief mechanical deformation of brain tissue.

Animals were reanaesthetized (sodium pentobarbital, 50 mg/kg i.p.) and sacrificed at 1 hour (n = 6), 6 hours (n = 5), 24 hours (n = 5), 2 days (n = 11), 3 days (n = 11), 5 days (n = 11) or 7 days (n = 11). Uninjured animals served as controls (n = 9). Brains were rapidly removed and 2–3 mm (50–100 mg wet weight) whole specimens were dissected on an ice cold glass plate according to the following scheme: injured left parietal cortex, tissue adjacent to the injured left parietal cortex, contralateral cortex, and left and right hippocampi. In an attempt to minimize tissue drying, sections were kept in a tightly sealed humid box until assay. Tissue sections were weighed and baked at 100 °C overnight. Dried pieces were reweighed and water content was calculated as:

wet weight - dry weight/wet weight $\times 100$.

Results and Discussion

A significant increase in tissue water content was observed beginning at 6 hours post-injury in the injured left parietal cortex (Fig. 1). Tissue water content continued to increase throughout the first 24 hours to become maximal at 48 hours post-injury.

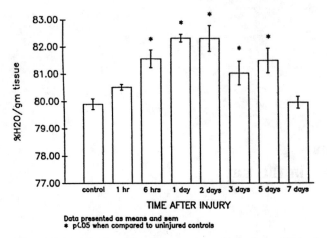

Fig. 1. Changes in percentage water per gram brain tissue obtained from the maximal site of injury (left parietal cortex) from one to seven days following fluid-percussion traumatic brain injury in the rat. * = p < 0.05 when compared to uninjured control animals

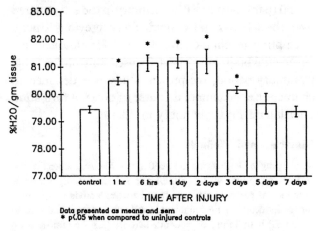

Fig. 2. Changes in percentage water per gram brain tissue obtained from the hippocampus adjacent to the maximal site of injury from one to seven days following fluid-percussion traumatic brain injury in the rat. * = p < 0.05 when compared to uninjured control animals

Significant oedema was observed in the injured cortex up to five days postinjury (Fig. 1). A significant increase in tissue water content was also observed in the left hippocampus, adjacent to the site of injury, beginning at 1 hour post-injury. Like the injured cortex, hippocampal swelling was maximal at 48 hours post-injury (Fig. 2), and resolved by 3 days postinjury.

No significant changes in tissue water concentration were observed following lateral FP injury in the contralateral cortex or hippocampus (results not shown).

The above results suggest that lateral FP brain injury produces an early focus oedema in the injured cortex and ipsilateral hippocampus that persists for a prolonged period after trauma. We feel, therefore, that this model may be useful in the evaluation of novel pharmacological therapies designed to reduce cerebral oedema after traumatic brain injury.

This study was supported, in part, by NIH RO1 NS26818, a Veterans Administration Merit Review award 74R and a grant from the Sunny von Bulow Coma and Head Injury Foundation.

References

1. Becker D (1984) Temporal genesis of primary and secondary brain oedema. In: Baethmann A, Go K, Unterberg A (eds) Mechanisms of secondary brain damage. Plenum Press, New York
2. Becker D, Miller J, Ward J, Greenberg R, Young H, Sakalas R (1977) The outcome from severe head injury with early diagnosis and intensive management. J Neurosurg 47: 491–497
3. Lobato R, Sarabia R, Cordobes F, Rivas J, Adrados A, Carbrera A, Bomez P, Madera A, Lamas E (1988) Posttraumatic cerebral hemispheric swelling. J Neurosurg 68: 417–423
4. McIntosh T, Vink R, Noble L, Yamakami I, Fernyak S, Soares H, Faden A (1989) Traumatic brain injury in the rat: Characterization of a lateral fluid-percussion model. Neuroscience 28: 233–244
5. Tornheim PA (1985) Traumatic oedema in head injury. In: Becker D, Povlishock J (eds) Central nervous system trauma status report. William Byrd, Washington, pp 431–442

Correspondence: K. McIntosh, Ph.D., Department of Surgery, University of Connecticut Health Center, Farmington, CT 06032, U.S.A.

Acta Neurochirurgica, Suppl. 51, 265–267 (1990)
© by Springer-Verlag 1990

Survival and Fibre Outgrowth of Neuronal Cells Transplanted into Brain Areas Associated with Interstitial Oedema*

T. Tsubokawa, Y. Katayama, S. Miyazaki, H. Ogawa, M. Koshinaga, and **K. Ishikawa**[1]

Departments of Neurological Surgery and [1]Pharmacology, Nihon University School of Medicine, Tokyo, Japan

Summary

The influence of interstitial oedema on the survival of fetal raphe cells transplanted into serotonin (5-HT)-denervated rats and the fibre outgrowth from these cells was investigated. Fetal raphe cells were transplanted into the corpus callosum in which long-lasting interstitial oedema had been induced by intracisternal kaolin injection. The 5-HT and 5HIAA levels in the corpus callosum were restored to their maximum within 5–6 weeks post-transplantation regardless of whether interstitial oedema was induced or not. Furthermore, it was appeared that the presence of interstitial oedema even facilitated fibre growth as demonstrated by the 5-HT immunohistochemistry and the restoration of the 5-HT and 5-HIAA levels in brain areas distant from the transplantation sites. These results imply favourable effects of interstitial oedema on the survival of transplanted raphe cells and their fibre outgrowth.

Introduction

Several clinical and experimental studies have indicated that interstitial brain oedema, accumulation of fluid within the extracellular pace, does not primarily cause changes in the electrical activity of neuronal cells. It is not completely clear therefore why interstitial brain oedema ever causes symptoms. It is not even understood whether interstitial brain oedema is a condition which interferes with the recovery of neuronal cells from damage or a condition which facilitates the repair process of damaged neuronal cells.

In the present study, we examine the survival and fibre outgrowth of neuronal cells transplanted into brain areas associated with interstitial oedema in order to determine whether the extracellular environment induced by interstitial oedema is favourable or unfavourable to neuronal cells. Employing the brain-cell

suspension technique[1, 2], we have confirmed that mesencephalic raphe cells of the rat fetus transplanted into the serotonin (5-HT)-denervated adult rat brain can survive, release the specific neurotransmitter, 5-HT, and its metabolite, 5-hydroxy-indolacetic acid (5-HIAA), form functional synaptic connections with the host brain, and modify the behavior of the host animals[6]. The present study was designed to establish whether mesencephalic raphe cells transplanted into brain areas associated with interstitial oedema could survive and give off fibres similarly to those transplanted into non-oedematous brain areas.

Materials and Methods

Male Wistar rats (220–270 g) received desmethylimipramine (10 mg/kg, i.p.) and were then injected with 5,7-dehydroxytriptamine (5,7,-DHT) creatinine sulfate into the cisterna magna percutaneously (5-HT denervation). A separate group of sham animals was injected with vehicle only (sham-denervation). In one group if animals, at 1 day after the intracisternal injection of 5,7-DHT, 0.1 ml of kaolin solution (200 µg/ml saline) was injected again into the cisterna magna[3, 4, 7]. The kaolin-injected animals were retrospectively divided into two groups, one of which revealed ventricular dilatation (>2SD), whereas the other showed no significant dilatation of the ventricle (<2SD). Only the former group animals was used in the present study. In the separate group of animals, 0.1 ml of saline vehicle was injected into the cisterna magna (sham-denervation).

Two weeks after the kaolin injection, transplantation of raphe cells was carried out[1, 2]. A staged pregnant female rat was anaesthetized by intraperitoneal injection of pentobarbital sodium, and prepared for Caesarean section. Fetuses were removed one at a time from the distal end of the uterine horn, at 14 days of gestation. A mesencephalic brain-cell suspension was prepared, bringing the volume equivalent to 10 µl per dissected tissue piece (*i.e.* 100 µl for 10 pieces), and transplanted using sterilized Hamilton syringes and 23-gauge needles into the anterior part of the corpus callosum. The suspension was deposited at a rate of 0.75 µl/min, and a further 3 min were allowed for diffusion prior to moving the needle. Six separate

* This work was supported by a Japan Education Ministry Grant (6048033, 1986–1989) and a Grant from the Ministry of Health and Welfare of Japan (1983–1987).

deposits were made, separated by 0.2 mm, along the same line of needle penetration. The needle was withdrawn 7 min after the final infusion in order to avoid back-diffusion. Rats that developed severe hydrocephalus rarely survived for than 5 weeks after the kaolin injection (*i.e.* 3 weeks post-transplantation). These rats were excluded from the analysis. A separate group of animals underwent transplantation of telencephalic suspension by identical procedures to those for raphe-cell transplantation (sham-transplantation).

All groups of rats were sacrificed at 1–2, 5–6 or 7–8 weeks after the transplantation of measurement of the 5-HT and 5-HIAA levels by high performance liquid chromatography (HPCL). The corpus callosum was dissected out including the white matter up to a distance of 0.3 mm lateral to the midsagittal line, and divided into the anterior and posterior corpus callosum at the coronal plane 4.3 mm posterior to the bregma. Thus, the anterior corpus callosum contained the site of transplantation and the posterior corpus callosum was at least 5.0 mm distant from the transplantation site. Some animals of each group were subjected to 5-HT immunohistochemical examination.

Results

The 5-HT and 5-HIAA levels were markedly decreased by 5-HT denervation and remained at depressed levels at least for 7–8 weeks post-denervation and, consequently, for 5–6 weeks post-sham-transplantation, regardless of whether interstitial oedema was induced or not. The transplantation of fetal raphe cells following 5-HT denervation restored the levels of 5-HT and 5-HIAA in the anterior corpus callosum to their maximum within 5–6 weeks (percent restoration with respect to the normal values after sham-denervation and sham-transplantation: 119.6% and 76.4% respectively), so that these levels were broadly comparable to the 5-HT

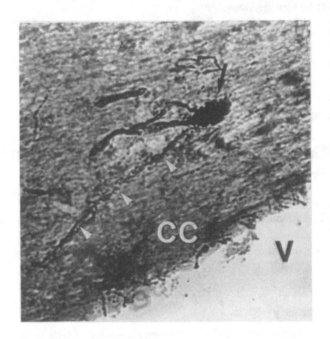

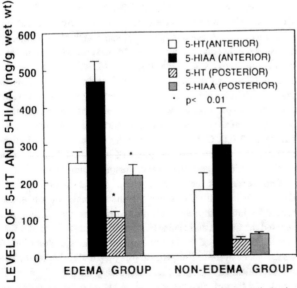

Fig. 1. 5-HT and 5-HIAA levels at 5–6 weeks post-transplantation (7–8 weeks post-denervation) in the anterior and posterior corpus callosum

and 5-HIAA levels in normal rats. No further increases were seen thereafter in the anterior corpus callosum. No significant effect of interstitial oedema on the 5-HT and 5-HIAA levels in the anterior corpus callosum was demonstrated at either 5–6 or 7–8 weeks post-transplantation (Fig. 1).

In contrast, the 5-HT and 5-HIAA levels in the posterior corpus callosum failed to reach the levels of normal rats by 5–6 weeks post-transplantation (percent restoration with respect to the normal values after sham-denervation and sham-transplantation: 76.4% and 83.6%, respectively). The 5-HT and 5-HIAA levels in the posterior corpus callosum attained the levels in the anterior corpus callosum by 7–8 weeks post-transplantation. The delay in the restoration of the 5-HT and 5-HIAA levels in the posterior corpus callosum may reflect the time for fibre outgrowth from the transplantation site into a distant area.

Significantly higher 5-HT and 5-HIAA levels in the posterior corpus callosum were demonstrated in association with the presence of interstitial oedema at 5–6 weeks post-transplantation (Fig. 1). The higher 5-HT and 5-HIAA levels appeared to reflect a greater fibre outgrowth from the site of transplantation to distant areas in the presence of interstitial oedema. No significant effect of interstitial oedema on the 5-HT and 5-HIAA levels of the posterior corpus callosum was demonstrated at 7–8 weeks post-transplantation.

The 5-HT immunohistochemistry revealed that a number of 5-HT-immunoreactive cells survived at the transplantation site even when extensive interstitial oedema was induced. No significant effect of interstitial oedema on the count of surviving 5-HT-immunoreactive cells was demonstrated. There were numerous 5-HT-immunoreactive fibres extending from the transplantation site into distant brain areas (Fig. 2). Consistent with the neurochemical data indicating a greater fibre outgrowth in the presence of interstitial oedema, the presence of more abundant 5-HT-immunoreactive fibres in association with interstitial oedema was suggested at 5–6 weeks post-transplantation (Fig. 2).

Discussion

The results of the present study demonstrate that fetal raphe cells transplanted into brain areas associated with interstitial oedema can survive in a similar manner to raphe cells transplanted into non-oedematous brain areas. Therefore, insofar as interstitial oedema is concerned, a moderate increase in intracranial pressure, an increase in extracellular fluid volume, and moderate disturbance of the microcirculation may not greatly influence the survival of transplanted raphe cells.

The present findings further suggest that interstitial oedema represents a condition which may even be favourable for the survival of transplanted raphe cells and fibre outgrowth into remote brain areas. It has been reported that contrast of transplanted brain cells with CSF facilitates their growth[4]. The interstitial oedema in hydrocephalus may have similar effects on transplanted brain cells. Such oedema results in an enlarged extracellular space. Since this condition is similar to that seen in the immature brain in which active fibre growth occurs, it is tempting to speculate that the enlarged extracellular space may represent another factor facilitating fibre growth.

It has been demonstrated that fetal brain cells are able to survive anoxia better than mature cells[2]. The effects of disturbances at the microcirculation were not examined in the present study. Our data appear to suggest that unfavourable effects of interstitial oedema, is they exist, might be produced through disturbances of the microcirculation.

References

1. Auerbach S, Zhou F, Jacobs BL, Azmitia E (1985) Serotonin turnover in raphe neurons transplanted into rat hippocampus. Neurosci Lett 61: 147–152
2. Bjorklund A, Stenevi U, Schmidt RH, Dunnet SB, Gage FH (1983) Intracerebral grafting of neuronal cell suspension. I. Introduction and general methods of preparation. Acta Physiol Scand 522 [Suppl]: 1–7
3. Gonzalez-Darder J, Barbera J, Cerda-Nicolas M, Segura D, Broseta J, Barcia-Salorio JL (1984) Sequential morphological and functional changes in kaolin-induced hydrocephalus. J Neurosurg 61: 918–924
4. Hochwald G (1984) Animal models of hydrocephalus. In: Shapiro K, Marmarou A, Portnoy H (eds) Hydrocephylus. Raven Press, New York, pp 199–213
5. Nishino H, Ono T, Takahashi J (1986) Transplants in the peri and intraventricular region grow better than those in the central parenchyma of the caudate. Neurosci Lett 64: 184–190
6. Tsubokawa T, Katayama Y, Miyazaki S, Ogawa H, Iwasaki M, Shibanoki S, Ishikawa K (1988) Supranormal levels of serotonin and its metabolite after raphe cell transplantation in serotonin-denervated rat hippocampus. Brain Res Bull 20: 303–306
7. Tsubokawa T, Katayama Y, Kawamata T (1988) Impaired hippocampal plasticity in experimental chronic hydrocephalus. Brain Injury 2: 19–30

Correspondence: T. Tsubokawa, M.D., Department of Neurological Surgery, Nihon University School of Medicine, Tokyo 173, Japan.

Acta Neurochirurgica, Suppl. 51, 268–270 (1990)

Effect of Murine Recombinant Interleukin-1 on Brain Oedema in the Rat*

C. R. Gordon, R. S. Merchant, A. Marmarou, C. D. Rice, J. T. Marsh, and **H. F. Young**

Division of Neurosurgery, Medical College of Virginia, Richmond, Virginia, U.S.A.

Summary

We investigated the effects of murine recombinant interleukin-1 (rIL-1, Du Pont) *in vivo* in the normal rat brain and here report both local and systemic effects of centrally administered rIL-1. Normal rats were given single or multiple atraumatic doses of either rIL-1 or and equal volume (5 µl) of vehicle for control comparison. All dosages of intraparenchymal rIL-1 produced a uniform a hyperthermic response and concomitant lethargy. There was a related anorexia beyond fever duration. Histologic examination of intraparenchymal injection tracts revealed fibrillary whorls of oedema and a cellular infiltrate surrounding the rIL-1 tract, while similar changes were less prominent in control injection tracts. Repeated high doses of rIL-1 produced significantly higher concentrations of brain water as measured by the gravimetric technique. We conclude that rIL-1 is not only a potent chemoattractant, but is also an edigematic agent when adminstered in high doses.

Introduction

Interleukin-1 beta (IL-1) is a naturally occurring cytokine which is centrally important in the acute phase response. Produced chiefly by the macrophage, it modulates a wide variety of physiologic activities. Beyond its pyrogenic effect on the supraoptic nucleus in the hypothalamus in response to bacterial endotoxin, multiple systemic effects are now more completely understood. Both systemic and central nervous system effects have been documented. Of interest to us is the cytokine's role in inflammation and tissue insult. IL-1 acts on both cellular and humoral components of the immune system to affect leukocyte chemotaxis and antibody production. Importantly, IL-1 appears to exert a positive feedback effect on its own production while at the same time inducing the release of a variety of other biologic response modifiers and immunologic hormones.

The present study was predicated on evidence that IL-1 is present in the ventricular fluid of patients with traumatic brain injury, leading to speculation that rIL-1 might be responsible at least in part to the formation of post-traumatic brain oedema. We first studied any histopathologic changes following a single intracerebral injection of murine rIL-1 in normal rat brain in an effort to established doese-dependent central and systemic responses. The second phase of our study examined the effects of chronic intraparenchymal rIL-1 exposure. The murine rIL-1 used in these studies was generously provided by Doctor Mary E. Neville of the E. I. Du Pont Company (Glenolden, PA).

Materials and Methods

Infusion Technique: Central to all experiments was the atraumatic delivery of rIL-1 into brain parenchyma which was accomplished by injection through a preimplanted canula. Canulae (Small Parts inc., Jacksonville, Fla.) were placed under stereotaxic guidance over 1 mm burr hole and were directed to the parietal lobe. Jeweler's screws were placed in the outer table of the cranium in a triangular distribution around the canula and secured with dental epoxy. Injection needles fit precisely into canulae and extended 1.5 mm into parenchyma. To avoid misinterpretation of possible post-traumatic changes as opposed to primary rIL-1 mediated changes, 5–7 days were allotted between canula placement and rIL-1 injection.

Brain water quantification: Using the gravimetric technique as described by Marmarou et al.[2], we determined brain water content after intracerebral injection of either rIL-1 or control excipient. Specific gravity of brain was determined sequentially in a gradient column of bromobenzene and kerosene. After standardization of 100 ml column with potassium chloride suspension drops, individual 1 mm squared brain sections were suspended and brain water content was calculated from specific gravity.

Results

Temperature Changes: Following an initial post-anaesthesia hypothermia, an intense fever spike occurred in rIL-1 treated animals which was not observed in

* This work was supported in part by The Brain Tumor Research Fund, with additional facilities and support provided by the Richard Roland Reynolds Neurosurgical Research Laboratories.

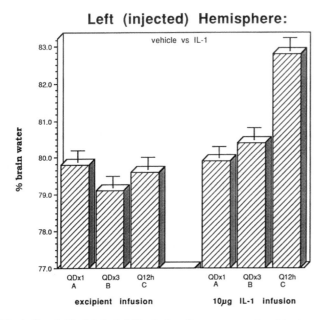

Fig. 1. Graph "Left Injected Hemisphere" compares regional brain water content (BWC) in the area of substance infusion. Multiple daily injections of rIL-1 caused significant brain oedema as compared to controls whereas single or daily injections caused only a moderate rise in BWC

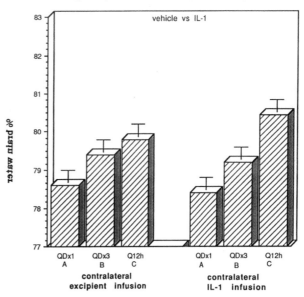

Fig. 2. Graph "Right Injected Hemisphere" compares regional brain water content (BWC) contralateral to the area of substance infusion. Multiple daily injections of rIL-1 caused significant brain oedema as compared to controls even in the contralateral side, suggesting a global oedema and effect. Daily dosing of rIL-1 was limited by systemic toxicity rather than any measurable brain oedema

excipient controls. The hyperthermia was not dose dependent and peaked 10–12 h following injection. After reaching maximal temperature, there was a rapid return to baseline. During the period of hyperthermia, animals were observed to be lethargic and normal behavior was seen once temperature returns to baseline. Daily injection of excipient under short acting methfluorane anaesthesia had no apparent untoward effects on body temperature. Daily injection of high dose rIL-1, was associated with a persistent hyperthermia for 2–3 days followed by hypothermia and death. Low-dose rIL-1 infusions (0.1 µg) produced only a chronic, low-grade fever.

Weight changes: Single injection of excipient was associated with a 2–3% loss in body weight. This weight change was transient and animals return to baseline weight over the following 2–3 days. Animals injected with from 0.1 µg to 5.0 µg rIL-1 showed similar transient changes while 10 µg rIL-1 produced a more profound weight loss of between 10 and 17% of total body weight. This cachexia resolved over the ensuing 2 weeks with weight eventually returning to baseline. Daily injections of excipient did not significantly affect animal weight. Daily injections of low doses of rIL-1 were also not accompanied by significant deviations from baseline. High-dose (10 µg) injections, however, were as-

sociated a significant cachexia and approximately 50% of the animals did not survive the third injection. These animals were lethargic, febrile, and refused to consume food or water.

Injection tract histology: Injection tracts from excipient and rIL-1 injected brains were examined and compared. Both injection tracts were filled with red blood cells and debris. Circular whorls of fibrillar character, indicative of brain oedema, were seen surrounding the rIL-1 injection tract. Similar findings were not seen in the area of excipient injection. While lymphocytes and granulocytes surrounded both sets of injection sites, the rIL-1 tracts demonstrated comparatively greater cellularity.

Brain water content: Brain water content (BWC) was quantified by the gravimetric technique. To avoid any possible inter-hemispheric contamination of injected materials, animals used in this study bore a single, left-sided canula only. The composite Figs. 1 and 2 are data gathered from 36 rats and express percent brain water found 1 mm on either side of the injection area. Three groups of animals were used for BWC measurements. The dose of rIL-1 was held constant (10 µg/5 µl), while the administration frequency varied. Control groups received an equal volume of excipient (1% normal rat serum). Group A was given a single dose of rIL-1 or excipient. Group B was given daily doses for three days, while group C was dosed twice within a 24-hour-period. A trend to oedema was noted in group A comparing the intracerebral injection site of single dose of rIL-1 compared with vehicle-injected hemisphere. This trend persists in group B when the same dose rIL-1 was injected daily for 3 days and compared with similarly injected vehicle controls. Group C, injected twice daily, demonstrated significantly greater BWC around the rIL-1 injected site compared to the same area of excipient injection. Higher amounts of water also appeared in the contralateral hemisphere of animals receiving rIL-1 in group C compared with their excipient controls.

Discussion

This study shows that intracerebral rIL-1 injections of all concentrations examined produce systemic effects. Hyperthermia was dosage-independent and remarkably uniform in both degree and duration. Moreover, a systemic cachexia was noted and weight loss was correlated positively with rIL-1 dosages. An intracerebral rIL-1 injection appears to compromise an intact BBB. This compromise was significant at high doses and short lived in these experiments. The compromise of the BBB was demonstrated histologically by observation of rIL-1 injection tracts surrounding oedematous changes with proximal cellular infiltrate.

We found repeated high doses of intraparenchymal rIL-1 to be edigematic, employing the gravimetric technique to quantify BWC. Given its multiplicity of actions, it is conceivable that rIL-1 stimulates a cascade of events which ultimately promotes brain oedema.

References

1. McClain CM, Cohen D, Ott L, Dinarello CA, Young B (1987) Ventricular fluid interleukin-1 activity in patients with head injury. J Lab Clin Med 110: 48–54
2. Marmarou A, Tanaka K, Shulman K (1982) An improved gravimetric measure of cerebral oedema. J Neurosurg 56: 246–253

Correspondence: A. Marmarou, Ph.D., Medical College of Virginia, Division of Neurosurgery, MCV Station, Box 508, Richmond, VA 23298, U.S.A.

Acta Neurochirurgica, Suppl. 51, 271–273 (1990)

Early Cellular Swelling in Experimental Traumatic Brain Injury: A Phenomenon Mediated by Excitatory Amino Acids

Y. Katayama[1, 2], **D. P. Becker**[1], **T. Tamura**[1], and **K. Ikezaki**[1]

[1] Division of Neurosurgery, University of California at Los Angeles, California, U.S.A., and [2] Department of Neurological Surgery, Nihon University School of Medicine, Tokyo, Japan

Summary

Early cellular swelling following fluid-percussion brain injury was demonstrated *in vivo* by means of microdialysis in the rat. When rapid cellular swelling occurs, water moves from the extracellular space (ECS) into the cells and the extracellular concentration of ECS marker which does not move into the cells increases. Cellular swelling was therefore demonstrated *in vivo* as an increase in the dialysate concentration of ^{14}C-sucrose ([^{14}C-sucrose]$_d$) pre-perfused as an ECS marker. The increase in [^{14}C-sucrose]$_d$ occurred concomitantly with an increase in the dialysate concentration of K^+ ([K^+]$_d$) representing large ionic fluxes. The increases in [K^+]$_d$ and [^{14}C-sucrose]$_d$ were both inhibited by kynurenic acid, a broad spectrum antagonist of excitatory amino acids (EAAs), which was administered through the dialysis probe. These findings suggest that early cellular swelling following traumatic brain injury is a result of ionic fluxes mediated by EAAs.

Introduction

Our previous studies[4, 7] have revealed that massive ionic fluxes represent a major event taking place in neuronal cells following traumatic brain injury. We have also demonstrated that indiscriminate release of excitatory amino acids (EAAs)[4, 7], and EAA-coupled ion channels may play a vital role in producing these ionic events[4, 5, 7]. Changes in ionic permeability in response to neurotransmitter release are characteristic of neurons as excitable cells. Thus, these phenomena are unique to neuronal cells as compared to any other cell within the body.

Such ionic events usually cause water movement from the extracellular space (ECS) into the cells and cellular swelling, since the osmotic pressure of intracellular impermeable anions may no longer be counterbalanced. We have previously demonstrated early cellular swelling following fluid-percussion brain (FP) injury *in vivo* as an increase in ECS markers[6]. When rapid cellular swelling occurs, water moves from the ECS into the cells and the concentration of ECS markers which do not move into the cells increases. Microdialysis was employed for both the introduction and concentration measurement of ECS marker.

Since the major component of the ionic events following FP injury has been demonstrated to be mediated by EAAs[4, 6, 7], we postulated that there may also be a causal relationship between EAA release and early cellular swelling. EAAs have been shown *in vitro* to cause cellular swelling as well as large ionic fluxes across the neuronal plasma membrane[2, 3]. The present study was undertaken to test the hypothesis that early cellular swelling following traumatic brain injury represents a phenomenon mediated by EAAs *in vivo*.

Materials and Methods

Young female Sprague-Dawley rats weighing 180–240 g were used. They were anesthetized with a mixture of O_2 (33%), N_2O (66%) and enflurane (1.5–2.0 ml/min), and placed in a stereotaxic frame. The rectal temperature was maintained at 37.0–38.0 °C with a heating pad. A pair of dialysis probes (O.D., 300 u; effective length, 2 or 3 mm; cut off, 20,000 MW) were positioned vertically in the hippocampus (CA1/dentate area) bilaterally.

The method employed for detecting cellular swelling involved the perfusion of focal brain regions with ^{14}C-sucrose, as an ECS marker, and subsequent measurement of the extracellular concentration of ^{14}C-sucrose, both performed by microdialysis[5]. The probes were initially perfused with temperature-maintained, Ringer solution containing ^{14}C-sucrose (10 mM) at a rate of 5.0 µl/min. Sodium kynurenate (KYN; 0.1–10 mM) was administered through one of the two probes throughout the experiment (test probe). This dialysis with the other probe served as the control (control probe). The osmolarity of each perfusate was kept equivalent to that of normal Ringer solution by changing the concentration of sodium chloride.

At 20 min after the start of ^{14}C-sucrose perfusion, the perfusate for the dialysis was switched to temperature-maintained, K^+-free Ringer solution and measurements of the dialysate concentrations of K^+ and ^{14}C-sucrose ($[K^+]_d$ and $[^{14}$C-sucrose$]_d$) were initiated. The dialysate fractions were collected at 1-min intervals for measurement of their K^+ concentration and radioactivity. The length of the outlet tube was shortened in order to make a dead space of 5.0 µl, so that the second dialysate fraction collected after FP injury corresponded to the initial post-trauma fraction.

FP injury (2–4 atm) was inflicted at the vertex at 13–18 min after the termination of ^{14}C-sucrose perfusion. The probes were temporarily withdrawn at the moment of injury in order to avoid tissue damage due to brain movement, and were re-inserted immediately thereafter. Following the experiment, autoradiograms were obtained on Kodak X-ray films.

Results

Ionic Fluxes and Cellular Swelling

As reported previously[4, 5, 7], the large, tetrodotoxin (TTX)-insensitive increase in $[K^+]_d$ which follows moderate FP injury (unconsciousness > 250 s) corresponded to $73 \pm 6\%$ of the increase in $[K^+]_d$ induced by complete ischaemia in the same animals at the end of the experiment. It is well established that the maximum level of extracellular potassium attained by complete ischemia is 80–90 mM. Therefore, the maximum level of extracellular potassium reached following moderate FP injury was far above the physiological ceiling level. An increase in $[^{14}$C-sucrose$]_d$ occurred in parallel with this $[K^+]_d$ increase (Fig. 1). In contrast, no clear increase was noted in association with the small TTX-sensitive increase in $[K^+]_d$ which followed mild FP injury (unconsciousness < 250 s)[4, 7]. Only a slight increase in $[K^+]_d$ and substantially no increase in $[^{14}$C-sucrose$]_d$ were induced by re-positioning of the probe, indicating that the increases in $[K^+]_d$ and $[^{14}$C-sucrose$]_d$ following FP injury were not merely the result of probe insertion.

Effects of Excitatory Amino Acid Antagonist

KYN, a broad spectrum antagonist of EAAs[2, 3], administered through the dialysis probe effectively attenuated the large $[K^+]_d$ increase after moderate FP injury. Fifty % of the maximum effect was obtained at a perfusate concentration of 5 mM. The maximal effect of KYN was 55–65% inhibition. The remaining component of the $[K^+]_d$ increase was comparable to the small TTX-sensitive increase in $[K^+]_d$ induced by mild FP injury. The administration of KYN also inhibited the increase in $[^{14}$C-sucrose$]_d$ at a perfusate concentration of more than 5 mM, as compared to the increases observed with the other probe perfused without KYN (Fig. 1, Table 1).

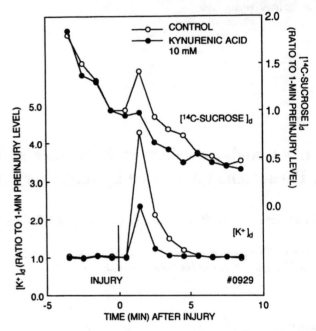

Fig. 1. A representative example of changes in $[^{14}$C-sucrose$]_d$ and $[K^+]_d$ following fluid-percussion brain injury. Open circles, control probe perfused without kynurenic acid; solid circles, test probe perfused with kynurenic acid (perfusate concentration, 10 mM)

Table 1. *Changes* in $[K^+]_{da}$ and $[^{14}$C-sucrose$]_d$ Following Fluid-percussion injury*

	Increase in $[K^+]_d$	Increase in $[^{14}$C-sucrose$]_d$
Control probe	4.4 ± 0.3	1.3 ± 0.1
Test probe**	2.7 ± 0.4***	1.0 ± 0.1***

* Ratio to 1-min preinjury level; ** kynurenic acid 10–20 mM; *** (n = 10), p < 0.01.

Discussion

The present results demonstrated that $[^{14}$C-sucrose$]_d$ increased concomitantly with the increase in $[K^+]_d$ employed as an indicator of the ionic flux. Sucrose is an ECS marker, which does not enter either the cells or capillaries, and is frequently used in *in vitro* experiments. Since the absolute amount of ^{14}C-sucrose within the ECS is unlikely to be altered, the only explanation for the observed increase in $[^{14}$C-sucrose$]_d$ appeared to be a decrease in water in the ECS[5]. The increase in $[^{14}$C-sucrose$]_d$ is likely to reflect water movement from the ECS into the cells, *i.e.*, cellular swelling, which is expected to occur as a consequence of the ionic fluxes.

The magnitude of the increase in $[^{14}$C-sucrose$]_d$ following FP injury was at most 1.3-fold baseline which is comparable to the change in $[^{14}C\text{-sucrose}]_d$ during cerebral ischaemia[5]. While this represents a value discounted by a natural decrease in $[^{14}C\text{-sucrose}]_d$ due to the continuous diffusion of ^{14}C-sucrose, the magnitude of the increase is obviously too small to explain the large increase in extracellular K^+ or EAAs[4, 7].

Assuming that the probe efficiency for drug delivery was approximately 10%, a 5 µM perfusate concentration of KYN would be expected to produce a 500 mM tissue concentration. Thus concentration of KYN, in hippocampal cells *in vitro*, blocks 50% of the depolarization mediated by kainate-preferring subreceptor and 90% of the depolarization mediated by NMDA-preferring subreceptor[3]. This, the effect of KYN is observed within the order of concentration that blocks the EAA-coupled ion channels. EAAs have been shown *in vitro* to cause cellular swelling as a direct consequence of large ionic fluxes[1, 2]. The results of the present study indicate that the early cellular swelling observed *in vivo* following traumatic brain injury is a result of ionic fluxes mediated by EAAs.

References

1. Chan PH, Fishman RA, Lee JL (1979) Effects of excitatory neurotransmitter amino acids on swelling of rat brain cortical slices. J Neurochem 33: 1309–1315

2. Choi DW, Koh J-Y, Peters S (1988) Pharmacology of glutamate neurotoxicity in cortical cell culture: Attenuation by NMDA antagonist. J Neurosci 8: 185–196

3. Ganong AH, Cotman CW (1986) Kynurenic acid and quinolinic acid act at N-methyl-D-aspartate receptors in the rat hippocampus. J Pharmacol Exp Therap 236: 293–299

4. Katayama Y, Becker DP, Tamura T, Hovda D (1990) Massive increase in extracellular potassium and indiscriminative glutamte release after concussive brain injury. J Neurosurg in press

5. Katayama Y, Becker DP, Tamura T, Tsubokawa T (1990) Cellular swelling during cerebral ischaemia demonstrated by microdialysis *in vivo*: Preliminary data indicating the role of excitatory amino acids (this volume)

6. Katayama Y, Cheung MK, Alves A, Becker DP (1989) Ion fluxes and cell swelling after experimental traumatic brain injury: The role of excitatory amino acids. In: Hoff JT, Betz AL (eds) Intracranial pressure, Vol 7. Springer, Berlin Heidelberg New York, pp 584–588

7. Katayama Y, Cheung MK, Gorman L, Tamura T, Becker DP (1988) Increase in extracellular glutamate and associated massive ionic fluxes following concussive brain injury. Neurosci Abstr 14: 1154

Correspondence: Y. Katayama, M.D. Department of Neurological Surgery, Nihon University School of Medicine, Tokyo 173, Japan (Visiting Professor in continnum, Division of Neurosurgery, University of California at Los Angeles, CA 90024, U.S.A.)

Acta Neurochirurgica, Suppl. 51, 274–276 (1990)

Changes in Extracellular Glutamate Concentration After Acute Subdural Haematoma in the Rat – Evidence for an "Excitotoxic" Mechanism?

R. Bullock[1], **S. Butcher,** and **J. McCulloch**

[1] University Department of Neurosurgery, Institute of Neurological Sciences, Southern General Hospital, Glasgow, Scotland

Summary

Using a rat model of subdural haematoma which is associated with ischaemic damage in the ipsilateral hemisphere, we have measured cerebral blood flow and release of excitatory amino acids after the haematoma. A more than sevenfold rise in glutamate and aspartate, persisting for forty minutes occurred in the severely ischaemic cortex ($CBF < 5$ ml 100 gm^{-1} min^{-1}) and a threefold, sustained rise was seen in hippocampus, although CBF was preserved (85 ml/100 g/ $^{-1}$ min^{-1}). Excitotoxic mechanisms may, therefore, be involved in the ischaemic damage associated with subdural haematoma.

Introduction

Intracranial haematomas are the main cause of preventable brain damage after head injuy, and acute subdural haematoma (ASDH) is the commonest of these[2, 4]. In spite of its importance, little is known of the mechanisms by which ASDHs produce brain damage[4]. We have devised a rat model of ASDH which produces reproducible ischaemic damage in up to 16% of the volume of the hemisphere beneath the haematoma[6]. To study the mechanisms causing this ischaemia, we have measured extracellular excitatory amino acids in the ischaemic zone of microdialysis, and related these measurements to changes in regional cerebral blood flow (rCBF), two hours after the haematoma.

Materials and Methods

We studied eleven adult Sprague-Dawley rats, anaesthetized throughout the experiment using halothane and a nitrous oxide: oxygen mixture (70:30%). Tracheostomy and cannulation of femoral vessels was first carried out and animals were ventilated to normocarbia and normoxia.

Microdialysis

The animals were immobilized in a stereotactic frame, and the skull exposed. Using stereotactic co-ordinates, burr holes were made overlying the left parietal cortex and hippocampus, and the dura incized. After prior calibration, microdialysis probes (2 mm active surface) were implanted into cortex and hippocampus, cemented into position and perfused at 2.5 microlitres/min Krebs bicarbonate solution[5]. After stabilisation for one hour, 20 minute dialysate samples were taken prior to, and for two hours after the haematoma or sham operation. Aspartate, glutamate, glutamine, taurine, alanine and valine were determined in the dialysate using high performance liquid chromatography with fluorescence detection.

Induction of Haematoma

A second burr hole was made in the left parietal cortex at the coronal suture, and a curved cannula was placed in the subdural space and cemented into position. In the seven ASDH animals, 0.4 mls of autologous blood was slowly injected into the subdural space. In the control animals, no injection was made.

CHF Measurement

Two hours after the induction of the haematoma of sham procedure, CBF was measured using the 14 C iodo-antipyrene method of Sakurada et al.[8]. Semi-serial twenty micron sections through the forebrain were cut and autoradiograms made. The exact position of the microdialysis probes in the cortex and hippocampus was identified, and CBF was measured in the tissue adjacent to the probe and in other representative structures (Table 1).

Results

Adequate oxygenation, normocapnia and normothermia were maintained throughout the experiment. Animals with ASDH demonstrated a marked Cushing's response of induction of the haematoma, and blood pressure was significantly greater at 15 minutes after the haematoma than the level in the control group.

Changes in Extracellular Amino Acids

Figure 1 demonstrates the changes in dialysate glutamate and aspartate after ASDH in the cortex and hip-

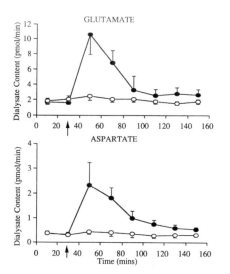
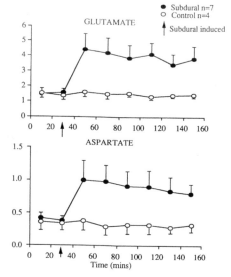

Fig. 1. Release of glutamate and aspartate in ischaemic cortex (left) and hippocampus (right) after subdural haematoma

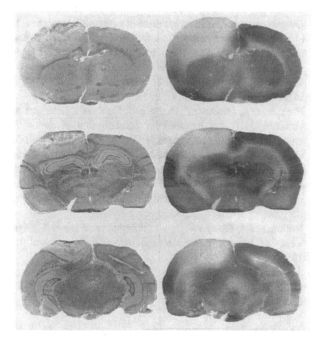

Fig. 2. Regional cerebral blood flow after subdural haematoma: Left column – H and E stained brain sections to show subdural haematoma overlying cortex. Right column – ^{14}C iodoantipyrene autoradiograms made from the same section

Table 1. *Regional CBF 2 Hours After Subdural Haematoma or Sham Operation. Mean + SEM ml/100 g/min*

Region	Subdural haematoma n = 7	Sham operation n = 4
Cortex beneath haematoma	<5±1**	81±11
Ipsilateral frontal cortex	33±8*	133±23
Contralateral cortex	132±20	154±46
Ipsilateral cingulate cortex	126±41	188±44
Ipsilateral caudate	86±14	151±44
Hypothalamus	73±9	89±18
Cortical zone (3 mm Diameter) around dialysis probe	16±6***	60±9
Ipsilateral hippocampus		
C$_{A1}$ Oriens	63±7*	96±9
Molecular layer	82±7	104±12
Hippocampal zone (3 mm diameter) around dialysis probe	85±5***	114±15
Contralateral hippocampus		
C$_{A1}$ Oriens	74±6	99±13
Molecular layer	79±5	110±18

* p<0.05 Unpaired "t"-test.
** p<0.01.
*** p<0.05 (Mann-Whitney) (not normally distributed data).

pocampus. Glutamate and aspartate increased markedly in cortical extracellular fluid after ASDH-glutamate by 775% and aspartate by 780% of basal levels. For all amino acids measured, the peak increase over basal levels was significant. (P<0.05 Mann-Whitney). The magnitude and temporal pattern of extracellular glutamate levels in the hippocampus differed from that in the neocortex. Glutamate rose by 339% after in-

duction of the haematoma, and aspartate by 280% and this rise was more sustained-levels remained at three times basal for the duration of the experiment (P<0.05 Mann-Whitney). Increases were seen in metabolic amino acids, but were less marked.

Cerebral Blood Flow (CBF)

Induction of ASDH produced a large zone of profoundly reduced CBF in the frontal and parietal cortex under the haematoma but CBF was well preserved in the ipsilateral hippocampus (Fig. 2, Table 1). In the controls, a small zone of reduced CBF was seen around the cortical microdialysis probes but flow was well preserved around the hippocampal probe. Wide variations in CBF from <5 mls to 96 ml/100 g/min were seen around dialysis probes in individual animals. The relationship between peak glutamate release in each animal and CBF in a 3 mm zone of tissue surrounding the probe was tested, but no threshold relationship was found.

Discussion

Regional CBF measurements two hours after an ASDH shown profound flow reductions affecting about 20–30% of the ipsilateral hemisphere, in this model. The most marked reductions occurred in cortex below the haematoma. This findings of cortical CBF values subthreshold for ischaemic damage, accords well with neuropathological findings, both in this model, and in patients who die after an ASDH[3, 6] (Fig. 2, Table 1).

Several studies have recently shown that the consequences of an episode of cerebral ischaemia maybe greatly influenced by the effects of excitatory amino acids (EAAs), such as glutamate and aspartate[3, 7]. In particular, blockade of glutamatergic receptors using NMDA antagonist drugs, has reduced the size of an ischaemic lesion by up to 65%, in some models[1]. Evidence from parallel studies in this model (Inglis, *et al.* – this volume) has shown a marked increase in glucose metabolism, both in the ischaemic cortex, and in both hippocampi, suggestive of and excitatory process. The sevenfold increase in EAAs in ischaemic cortex after ASDH accords with a transient "excitotoxic" process, and may have therapeutic implications, in this clinical setting.

The threefold, sustained increase in EAAs in the hippocampus and the increase in glucose metabolism suggests a more prolonged excitatory process, and may give clues to a mechanism for the widespread hippocampal damage seen in humans who die after ASDH[4].

References

1. Bullock R, McCulloch J, Graham DI *et al* (1990) Focal ischaemic damage is reduced by CPP-ene – Studies in two animal models. Stroke (in press)
2. Bullock R, Teasdale G (1989) Surgical management of traumatic intracranial haematomas. In: Braakman (ed) Handbook of clinical neurology. Head injury, Vol 24 (in press)
3. Butcher SP, Bullock R, Graham DI *et al* (1990) Occlusion of the middle cerebral artery induces release of neuro-active amino acids in rat striatum and cortex: Correlation with neuropathological outcome. In: Proc. of 3rd International Symposium on the pharmacology of cerebral ischaemia. Marburg, Germany (in press)
4. Graham DI, Adams JH, Doyle D (1968) Ischaemic brain damage in fatal non-missile head injuries. J Neurol Sci 39: 213–234
5. Hamberger A, Butcher SP, Hagberg H, Jacobsen I, Lehmann A, Sandberg M (1986) Extracellular concentration of excitatory amino acids: Effects of hyperexcitation, hypoglycaemia and ischaemia. In: Roberts PJ, Storm-Mathison J, Bradford HF (eds) Excitatory amino acids. MacMillan Press, London, pp 409–422
6. Miller JD, Bullock R, Graham DI *et al* (1990) Ischaemic brain damage in a model of acute subdural haematoma. Neurosurgery (in press)
7. Rothman SM, Olney JW (1986) Glutamate and the pathophysiology of hypoxic ischaemic brain damage. Ann Neurol 19: 105–111
8. Sakurada O, Kennedy C, Jehle JW *et al* (1978) Measurements of local cerebral blood flow with [14]C Iodo-antipyrene. Am J Pyhsiol 234: H59–H66

Correspondence: R. Bullock, Ph.D., FRCS (SN) Ed, University Department of Neurosurgery, Institute of Neurological Sciences, Southern General Hospital, Glasgow G51 4TF, Scotland.

Acta Neurochirurgica, Suppl. 51, 277–279 (1990)
© by Springer-Verlag 1990

Ischaemic Brain Damage Associated with Tissue Hypermetabolism in Acute Subdural Haematoma: Reduction by a Glutamate Antagonist

F. M. Inglis, R. Bullock[1], M. H. Chen, D. I. Graham[2], J. D. Miller, and J. McCulloch

University Departments of [1] Neurosurgery and [2] Neuropathology, Institute of Neurological Sciences, Southern General Hospital, and The Wellcome Surgical Institute, and Hugh Fraser Neuroscience Laboratories, Glasgow

Summary

Ischaemia results in elevated extracellular glutamate concentrations, and drugs which act at the N-methyl-D-aspartate sub-type of glutamate receptor have been shown to decrease ischaemic brain damage. Because almost all patients who die after severe head injury demonstrate ischaemic brain damage, and acute subdural haematoma (ASDH) is one of the commonest complications of severe head injury, we have studied this condition in a rat model. Using double-label autoradiography, we have measured the effects of ASDH on cerebral glucose utilization and cerebral blood flow (RCBF). Following ASDH, increased glucose utilisation was observed in some cortical and hippocampal structures, without concomitant increases in blood flow. Directly below the ASDH, blood flow and glucose utilization were profoundly reduced. Pre-treatment with D-CPP-ene, a competitive NMDA antagonist, resulted in amelioration of hypermetabolism induced by the ASDH. The results suggest that NMDA antagonists may prevent ischaemic brain damage after ASDH by reducing hypermetabolism, induced by glutamatergic mechanisms.

Introduction

Acute subdural haematoma (ASDH) is the most common mass lesion which complicates severe head injury: following ASDH 60% of patients die or remain severely disabled[2, 3, 5]. Ischaemic damage is nearly always present in the hippocampus of those who die, but the etiology of this damage is not understood[3]. The hippocampus contains the highest concentration of glutamatergic synapses in the mammalian CNS, and the hierarchy of selective vulnerability of neurons within the CNS to ischaemic damage corresponds well with the distribution of glutamate receptors. The hippocampal CA 1 pyramidal cells, which contain the highest density of NMDA-sensitive glutamate binding sites, are particularly susceptible to ischaemic damage, in animal models and humans[6].

The excitatory amino acid glutamate has been implicated in the pathophysiological alterations which accompany cerebral ischaemia[2, 6]. Experimental evidence has shown that glutamate is "excitotoxic" in high extracellular concentrations, and raised extracellular concentrations of glutamate have been demonstrated following both global and focal ischaemia[6] (Bullock et al., this volume).

The purpose of this study was to examine the metabolic events, in terms of cerebral glucose consumption (LCGU) and blood flow (RCBF), following ASDH in the rat, and to assess the effect of a competitive NMDA antagonist (D-CPP-ene) upon LCGU and RCBF after ASDH.

Materials and Methods

Experiments were performed on adult male Sprague-Dawley rats under light halothane anaesthesia. The femoral vessels were cannulated for measurement of physiological parameters and autoradiography. An ASDH was produced by injection of 0.4 ml autologous blood into the subdural space[5]. Sham-operated rats underwent the same surgical procedure, without an injection of blood. Drug-treated rats received 15 mg/kg D-CPP-ene, fifteen minutes prior to production of ASDH. The animals were then restrained lightly by a loose plaster cast around the hindlimbs and abdomen, and allowed to regain consciousness.

Using double-label autoradiography, cerebral metabolism and blood flow distribution was studied in anatomically defined structures in each animal. Local cerebral glucose utilization (LGCU) was estimated using the quantitative 14 C 2-deoxyglucose technique using the rate and lumped constants for normal rats[8]. Regional cerebral blood flow (RCBF) was simultaneously measured using 99 mTc-HMPAO[1]. After in vivo labelling, animals were sacrificed, the brain was removed, frozen rapidly and sectioned. The high specific activity of 99 mTc-HMPAO allows autoradiograms of RCBF to be obtained with short exposures: (tritium-sensitive film for four hours). Sections were set aside for 72 hours to allow technetium decay and deoxy-glucose autoradiograms were then produced from the same sections by using 14 C-sensitive film. RCBF was quantified by standardization of optical densities (OD) of each structure measured against the OD of the caudate contralateral to the ASDH. These ratios, expressed as fractions of caudate OD, were compared to similarly standardized LCGU values in order to examine the relationship between RCBF and metabolism.

Results

Effect of ASDH upon Cerebral Metabolism and Flow

De-oxyglucose autoradiographs revealed a profound reduction of glucose uptake in the ischaemic area, directly beneath the haemtome. In contrast, marked hypermetabolism was persent in the area bordering the ischaemic tissue ("ischaemia penumbra") and in both hippocampi (Fig. 1). While RCBF was also profoundly reduced in the area of ischaemia, there was not evidence of simultaneously increased blood flow in the "penumbral" region, or hippocampal structures.

In sham-operated animals, there was a strong correlation between local glucose uptake and RCBF for all structures measured. In animals with ASDH, however, regions with glucose "hypermetabolism", particularly the hippocampus, deviated from this correlation, suggesting alteration in the coupling of LCGU and RCBF in those areas.

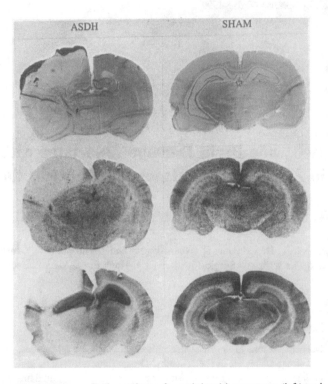

Fig. 1. **Top row:** Brain sections after subdural haematoma (left) and sham operation (right) stained with haematoxylin and eosin (×4) note the zone of pallor of staining under the haematoma ischaemic damage). **Middle row:** 99 mTc HMPAO RCBF autoradiograms – note zone of very low RCBF under ASDH. **Bottom row:** 14C 2-deoxyglucose autoradiograms corresponding to the same brain section. Note the "hypermetabolism" in hippocampus, and periphery of the ischaemic zone after ASDH

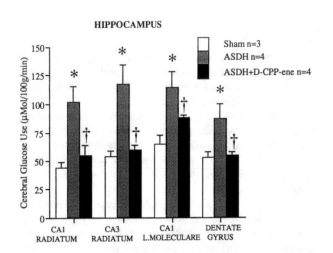

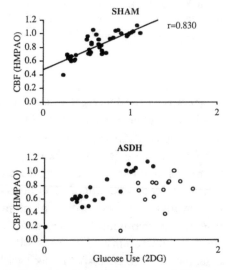

Fig. 2. Above, histogram comparisons of LCGU in the hippocampus; *indicates significant difference ($p < 0.05$) from sham-treated rats; *indicates significant difference ($p < 0.05$) from ASDH rats. Below, comparisons of local glucose utilization and bloodflow following sham or ASDH in individual animals. Cerebral bloodflow is quantified in each animal by standardization of the optical density (OD) of each structure against the OD of the caudate contralateral to ASDH. Ratios of caudate OD were compared to similary standardized glucose utilization values. Open circles indicate structures for which mean values of glucose use are significantly increased. The broken line in the ASDH figure represents the regression line for the sham animal. Note the reduction in cerebral bloodflow

Effect of D-CPP-ene upon cerebral metabolism and flow after ASDH

In the majority of the 42 brain regions studied, increases in LCGU induced by ASDH were significantly ameliorated by D-CPP-ene pretreatment (Fig. 2). In posterior cingulate, however, LCGU was significantly increased by D-CPP-ene pretreatment.

Discussion

These studies show that an ASDH causes marked "hypermetabolism" in the ischaemic periphery of the associated cortical infarct, and bilaterally in the hippocampus, at two hours, unaccompanied by a concomitant increase in RCBF, in the same tissues (Fig. 2 below) (see also Bullock, this volume). This pattern of glucose hypermetabolism has been shown to persist for four hours after middle cerebral artery occlusion (MCAO) and in both this subdural model, and MCAO, a marked increase in extracellular glutamate (sevenfold and 20-fold respectively) has been demonstrated. This is strong evidence that an excitotoxic process is implicated in the ischaemic damage which follows ASDH.

The apparent failure of RCBF to increase in the areas of increased metabolism maybe a feature of this ASDH model, and maybe due to raised intracranial pressure[5]. This RCBF metabolism mismatch may delay clearance of EAAs from ischaemic of "hypermetabolic" tissue thus potentially prolonging their excitotoxic action.

In parallel studies with this model, we have shown that pretreatment with the competitive NMDA antagonist D-CPP-ene significantly reduces the volume of cortical infarction under the ASDH[2]. The present studies have shown that D-CPP-ene pretreatment reduces the "penumbral" and hippocampal hypermetabolism, seen two hours after ASDH (Fig. 2). This accords with its putative blockade of glutamatergic hyperactivation.

D-CPP-ene pretreatment also significantly increased LCGU in posterior cingulate. This accords with the effect of non-competitive NMDA antagonists on parts of the limbic system, and the propensity of this class of drugs to induce psychomotor effects[4].

These studies suggest that competitive glutamate antagonists may prevent ischaemic damage after ASDH by reducing hypermetabolism, induced by glutamatergic mechanisms.

References

1. Andersen AR (1989) 99 mTc hexamethylene – propylene amine oxine (99 mTc HMPAO): Basic kinetic studies of a tracer of cerebral blood flow. Cerebrovasc Brain Metabol Reviews 1: 288–318
2. Bullock R, McCulloch J, Graham DI *et al* (1990) Focal ischaemic damage is reduced by CPP-ene – Studies in two animals models. Stroke (in press)
3. Graham DI, Ford I, Adams JH, Doyle D, Teasdale GM, Lawrence AE, McLellan DR (1989) Ischaemic brain damage is still fatal non-missile head injury. J Neurol Neurosurg Psychiatry 52: 346–350
4. McCulloch J; Kurumaji A, Park CK *et al* (1989) Cerebrovascular and metabolic consequences of N-methyl D-aspartate (NMDA) receptor blockade. In: Seylaz J, Mackenzie ET (eds) Neurotransmission and cerebrovascular function, Vol I. Excerpta Medica
5. Miller JD, Bullock R, Graham DI *et al* (1990) Ischaemic brain damage in a model of acute subdural haematoma. Neurosurgery (in press)
6. Rothman SM, Olney JW (1986) Glutamate and the pathophysiology of hypoxic ischaemic brain damage. Ann Neurol 19: 105–111
7. Shirashi K, Sharp FR, Simon RP (1989) Sequential metabolic changes in rat brain following middle cerebral artery occlusion: A de-oxyglucose study. J Cereb Blood Flow Metabol 9: 765–773
8. Sokoloff L, Reivitch M, Kennedy C *et al* (1977) The 14 C deoxyglucose method for measurement of local cerebral glucose utilization: theory, procedures and normal values in the conscious and anaesthetized albino rat. J Neurochem 28: 897–916

Correspondence: F. M. Inglis, B.Sc., University Department of Neurosurgery, Institute of Neurological Sciences, Southern General Hospital, Glasgow G51 4TF, Scotland.

Acta Neurochirurgica, Suppl. 51, 280–282 (1990)
© by Springer-Verlag 1990

Autoradiographic Patterns of Brain Interstitial Fluid Flow After Collagenase-induced Haemorrhage in Rat

G. A. Rosenberg, E. Estrada, M. Wesley, and **W. T. Kyner**

Departments of Neurology, Physiology, and Mathematics and Statistics, University of New Mexico,
and the Neurology and Research Services, Veterans Affairs Medical Center, Albuquerque, New Mexico, U.S.A.

Summary

Cerebral oedema accompanies intracerebral haemorrhage. We induced intracranial bleeding by the intracerebral injection of bacterial collagenase. There was oedema observed both at the haematoma site in the caudate/putamen and bilaterally in the hippocampal regions. To determine the role of vasogenic oedema spread from the site of injury, we studied by autoradiography the distribution of extracellular markers injected along with the collagenase. Both ^{14}C-dextran (m.w. 70,000) and ^{14}C-sucrose (m.w. 341) spread away from the injection site into both hippocampal regions in a similar pattern, suggesting bulk flow. Vasogenic oedema secondary to a haemorrhagic lesion in the caudate/putamen is an important cause of the oedema observed in both hippocampal regions in our model.

Introduction

Intracranial bleeding occurs in stroke, trauma, and tumours[12]. Cerebral oedema and mass effect has been shown by the direct injection of blood into the brains of animals to compromise cerebral blood flow[6, 11]. We developed a reproducible method to produce intracerebral haemorrhage in rat by the infusion of bacterial collagenase into brain[9]. Earlier, we found that collagenase-induced haemorrhage in the caudate/putamen of rat caused oedema both at the site of bleeding and in both posterior regions. We hypothesized that the oedema seen at a distance from the injury site was due to spread of interstitial fluid and oedema from the injury area. To test that hypothesis we injected intracerebrally radiolabeled extracellular markers along with the lesion-producing enzyme and used autoradiography to study the patterns of the tracers' distribution[7].

Methods

Twenty-one Sprague-Dawley rats were studied. They were anaesthetized with 50 mg/kg pentobarbital i.p. A 23 gauge infusion needle was stereotactically implanted in the caudate/putamen. Two µl of saline containing 0.5U of bacterial collagenase (Type VII, Sigma Corp.) was infused over 9 minutes with a Harvard microinfusion pump. The infusate contained either ^{14}C-sucrose (m.w. 341) or ^{14}C-dextran (m.w. 70,000). Twelve experimental animals had the isotopes infused along with the collagenase, while 9 control animals had only the isotope infused. The skin was sutured closed and the animals recovered and resumed normal activities. Animals infused with labeled sucrose were sacrificed at either 2 or 4 hrs after infusion (3 experimental and 3 controls). Dextran-infused animals were studied at 2, 4, 24, and 48 hours (9 experimental and 6 controls). The animals were sacrificed with an overdose of pentobarbital. The brains were rapidly removed and frozen. Frozen sections were made at 20 µm intervals in a cryostat and prepared for autoradiography. The slides were stained with threonine or cresyl violet. The X-ray images were analyzed with a Megavision computer image analysis system. The anatomical regions were traced on the computer images, and the traced images were superimposed on the autoradiograms.

Results

At 4 hours the ^{14}C-sucrose was localized to the caudate/putamen in all of the control animals. In contrast all of the animals injected with collagenase and ^{14}C-sucrose had spread of the tracer at 4 hours along white matter fiber tracts into the dorsal hippocampi. By 24 hours both the control and collagenase injected animals had only a faint image from the sucrose tracer.

Because the concentration of sucrose was so low at 24 hours, a heavier molecule, ^{14}C-dextran, was also employed as an extracellular marker. A similar pattern of spread of tracer was observed with the ^{14}C-dextran-injected animals. All of the control animals had the tracer confined to the site of injection (Fig. 1). Animals injected with both the tracer and the collagenase developed a pattern of spread at 4 hours that resembled that seen in the sucrose-injected animals. Unlike sucrose, dextran was observed at the site of injury in the

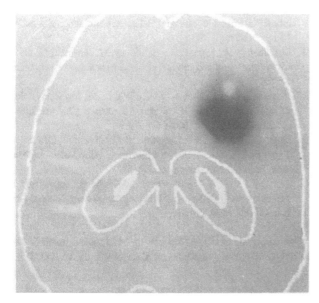

Fig. 1. Autoradiogram 4 hours after injection of ^{14}C-dextran into a control animal. Tracer (dark region) is confined to the caudate/putamen. The white lines are over the pyramidal cells of the hippocampus and the granule cells of the dentate gyrus

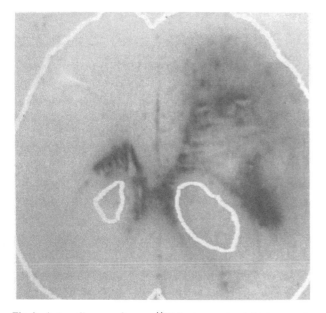

Fig. 2. Autoradiogram from a ^{14}C-dextran animal 24 hours after collagenase injection. The dextran is seen around both hippocampal regions (white lines) and in the injury site

posterior regions at 24 hours (Fig. 2). The control dextran animals at 24 hours showed tracer confined to the region of the injection. By 48 hours the tracer was mainly seen in the ipsilateral hippocampal regions in a pattern that was consistent with persistent tracer in the perivascular spaces.

Discussion

We used a recently developed model of collagenase-induced haemorrhage to study with autoradiography the movement of extracellular radiolabeled tracers away from a site of injury. In our model of collagenase-induced haemorrhage, we observed that the vasogenic oedema was spread away from the injury in a characteristic pattern that was similar for tracers of different molecular weights. This suggests that the oedema in the posterior hippocampal regions is due to bulk of oedema and entrained tracers along white matter fiber tracts and perivascular spaces into the tissue surrounding the dorsal hippocampi.

In our earlier studies of this model we found that the distant oedema was present at 4 and 24 hours, but had begun to resolve by 48 hours. This time-course was consistent with the patterns observed for the isotope studies. The time-course of the vasogenic oedema formation around the hippocampi paralleled that of the oedema measured in the earlier study in that region.

Transport of oedema in white matter is observed in cold-injury oedema where it is enhanced by hypertension and raised CSF pressure[3, 10]. There is bulk flow of interstitial fluid through brain tissue[1, 8]. Larger molecules diffuse more slowly through the brain than smaller ones, while bulk flow rates are similar for all substances[2]. The patterns observed in the autoradiograms for both sucrose and dextran are compatible with bulk flow in the white matter and along perivascular spaces[4, 5].

Bacterial collagenase may be acting on the collagen in the basal lamina of capillaries. In our model bleeding is observed by 20 minutes after injection of the enzyme and continues in multiple sites to coalesce into a haemorrhagic mass[9]. Our results indicate that one of the causes of the oedema in the region of the hippocampus is that the vasogenic oedema formed at the site of injury moves via fiber tracts and perivascular spaces into that region. Other factors that may be involved remain to be identified.

Acknowledgements

The study was supported by grants from the NIH and the Veterans Administration. Milo Navratil provided excellent technical assistance. Alisa Sherwood helped prepare the manuscript.

References

1. Cserr HF, Ostrach LH (1974) Bulk flow of interstitial fluid after intracranial injection of blue dextran 2000. Exp Neurol 45: 50–60
2. Fenstermacher JD, Patlak CS (1976) The movements of water and solutes in the brains of mammals. In: Pappius HM, Feindel W (eds) The dynamics of brain oedema. Springer, New York

3. Klatzo I, Wisniewski H, Smith DE (1965) Observations on penetration of serum proteins into the central nervous system. Prog Brain Res 15: 73–88

4. Rennels ML, Gregory TF, Blaumanis OR, Fujimoto K, Grady PA (1985) Evidence for a "paravascular" fluid circulation in the mammalian central nervous system, provided by the rapid distribution of tracer protein throughout the brain from the subarachnoid space. Brain Res 326: 47–63

5. Reulen HJ, Tsuyumu M, Tack A, Fenske AR, Prioleau GR (1978) Clearance of oedema fluid into cerebrospinal fluid. A mechanism for resolution of vasogenic brain oedema. J Neurosurg 48: 754–764

6. Ropper AH, Zervas NT (1982) Cerebral blood flow after experimental basal ganglia haemorrhage. Ann Neurol 11: 266–271

7. Rosenberg GA, Barrett J, Estrada E, Brayer J, Kyner WT (1988) Selective effect of mannitol-induced hyperosmolality on brain interstitial fluid and water content in white matter. Metab Brain Dis 3: 217–227

8. Rosenberg GA, Kyner WT, Estrada E (1980) Bulk flow of brain interstitial fluid under normal and hyperosmolar conditions. Am J Physiol 238: 42–49

9. Rosenberg GA, Mun-Bryce S, Wesley M, Kornfeld M (1990) Collagenase-induced intracerebral haemorrhage in rat. Stroke 21: 801–807

10. Rosenberg GA, Saland L, Kyner WT (1983) Pathophysiology of periventricular tissue changes with raised CSF pressure in cats. J Neurosurg 59: 606–611

11. Sinar EJ, Mendelow AD, Graham DI, Teasdale GM (1987) Experimental intracerebral haemorrhage: effects of a temporary mass lesion. J Neurosurg 66: 568–576

12. Zulch KJ (1971) Haemorrhage, thrombosis, embolism. In: Minckler J (ed) Pathology of the nervous system. McGraw-Hill Book Corp, New York

Correspondence: G. A. Rosenberg, M.D., Department of Neurology, UNM School of Medicine, Albuquerque, NM 87131, U.S.A.

Acta Neurochirurgica, Suppl. 51, 283–285 (1990)
© by Springer-Verlag 1990

Analysis of MRI and SPECT in Patients with Acute Head Injury

H. Fumeya, K. Ito, O. Yamagiwa, N. Funatsu, T. Okada, S. Asahi, H. Ogura, M. Kubo, and **T. Oba**

Yokohama Shintoshi Neurosurgical Hospital, Jokohama, Japan

Summary

Traumatic lesions defined by magnetic resonance (MR) imaging were divided into two groups according to findings on computed tomography (CT). This classification reflected difference in the regional cerebral blood flow (rCBF). In the contusional lesions which CT could demonstrate, rCBF varied from hyperperfusion to hypoperfusion, while it was almost always decreased in the lesions which CT could not detect. These results suggest that the former may include a mixture of brain oedema and hyperemia and the latter may imply brain oedema. MR imaging can reveal the minor oedema which CT fails to show in patients with acute head injury.

Introduction

Since MR imaging was widely used for evaluation of neurotrauma, many authors have reported that MR imaging, especially T_2-weighted MR imaging, is more sensitive in detection of the cerebral contusion than CT and that it can disclose the traumatic lesions which CT cannot demonstrate in some patients[2,3,4,8]. Significance of these lesions is poorly understood, however. Although MR imaging indicates increased water content in the brain tissue, it may not be possible to make a specific diagnosis. Areas of high intensity shown by T_2-weighted MR imaging may be interpreted as brain oedema, nonhaemorrhagic contusion, ischaemic infarct, and so on.

This study was designed to examine rCBF in the lesions by single photon emission computed tomography (SPECT) and to determine the pathophysiology of them.

Clinical Materials and Methods

From April 1987 to July 1989, we studied CT and T_2-weighted MR images (0.15 Tesla, spin echo method: TR 2000 TE 100 msec) in 269 patients with head injury within 48 hours after trauma and found MR imaging to be superior to CT in visualizing traumatic lesions in 83 patients. In 24 of 83 patients, we accomplished SPECT study within ten days after trauma.

rCBF was assessed using a Shimadzu SET-031 (Tokyo, Japan) and intravenous injection of [123I] iodoamphetamine (IMP) or inhalation of 133Xe gas. IMP images were routinely obtained 10–30 minutes and then five hours after injection. Doses injected were 185–259 MBq (5–7 mCi) in 10 mg IMP. IMP uptake results were not corrected for isotope decay and were expressed relative to the contralateral symmetrical value. rCBF was measured using the 133Xe inhalation method described by Obrist[7] and calculated according to Kanno and Lassen[5]. Cerebral blood volume (CBV) was assessed using intravenous injection of 99mTc-DTPA-human serum albumin. Doses injected were 740 MBq (20 mCi).

Results and Discussion

rCBF in a total of 51 contusional lesions was assessed by SPECT study obtained from the 24 patients. These lesions were divided into two groups. Group 1 comprised the 38 lesions seen on CT and MR imaging. Group 2 comprised the 13 lesions seen on MR imaging with improved visualization on MR imaging, not seen on CT in spite of using the windowing techniques. Table 1 shows rCBF in the lesions. Difference between the groups is significant ($p < 0.05$, chi-square test).

In Group 1, rCBF was revealed to be increased in 14 lesions (37%) and decreased of normal in 24. These lesions appeared as areas of various density in CT and as areas of high intensity occasionally accompanied by

Table 1. *rCBF in the Traumatic Lesions Defined by MR Imaging*

	Group 1	Group 2
Increased	14	0
Normal	4	3
Decreased	20	10

Group 1: seen on CT and MR with improved visualization on MR.
Group 2: seen on MR, not seen on CT in spite of using the windowing techniques.

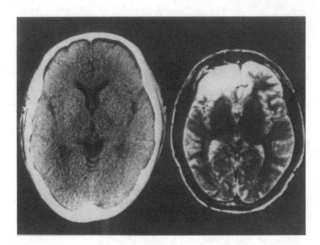

Fig. 1. CT and MR of a 41-year-old man who fell from a bicycle and was hit on his occiput. He was agitated and combative. Left: CT revealed no abnormality. Right: T_2-weighted MR imaging disclosed high intensity areas in the frontal lobes

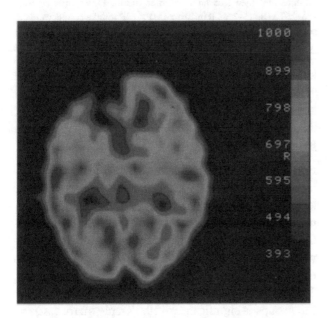

Fig. 2. SPECT of the same patients as Fig. 1. Decreased rCBF in the left frontal lobe was considered to be responsible for the posttraumatic psychosis. This lesion seemed to be brain oedema

low intensity in T_2-weighted MR imaging. These findings suggest that hyperemic lesions also appear as areas of high intensity in T_2-weighted MR imaging, as well as nonhaemorrhagic contusion and brain oedema.

In four patients, CBV was assessed by SPECT. It showed that local CBV was increased in the lesions where rCBF was high. Accordingly, hyperemia was thought to be brought by local vasodilatation. It is conceivable that the high local CBV may increase the hydrogen content and prolong the T_2 relaxation time of the tissue.

In Group 2, hyperemia was not found and rCBF was proved to be decreased in ten lesions and normal in three. CT failed to show these lesions. All of them appeared as areas of high intensity in T_2-weighted MR imaging (Fig. 1).

Many previous studies report that rCBF in the cerebral contusion defined by CT is variable. They find that rCBF is often increased in the contused brain and reduced in the brain tissue that is declared to be oedematous[1, 6].

The lesions which only MR imaging could demonstrate and CT failed are presumed to be brain oedema because rCBF proved to be decreased in most patients (Fig. 2). If oedema is accompanied by enough tissue water, it results in an area of hypodensity of CT. MR imaging can reveal subtle change of the relaxation time and demonstrate the lesion which makes little changes of the X-ray attenuation values. Therefore it seems probable that MR imaging can detect the minor brain oedema which appears as an area of normal brain density in CT.

These lesions often decreased in size on follow-up MR imaging and disappeared completely in some patients. This supports the concept that these lesions represent brain oedema which resolves in a while. Several clinical symptoms caused by the lesions only MR imaging can detect are reported, such as acute posttraumatic psychosis[3] and transient traumatic cerebellar dysfuntion[2]. Because they are temporary syndromes and no neurological deficit persists, they are distinguished from the cerebral contusion. This also supports that the traumatic lesions only MR imaging can detect may represent brain oedema rather than contusion.

Acknowledgements

We thank Mr. Hiromi Takahashi, Mr. Takashi Kanai, and Mr. Hideaki Sudo for technical assistance.

References

1. Enevoldsen EM, Jensen FT (1977) Compartmental analysis of regional cerebral blood flow in patients with acute severe head injuries. J Neurosurg 47: 699–712
2. Fumeya H, Ito K, Okiyama K, Funatsu N, Shiwaku T, Ogura H, Oba T (1990) MR imaging of traumatic cerebellar dysfunction. Neurol Surg 18: 279–283
3. Gandy SE, Snow RB, Zimmerman RD, Deck MDF (1984) Cranial nuclear magnetic resonance imaging in head trauma. Ann Neurol 16: 254–257
4. Hesselink JR, Dowd CF, Healy ME, Hajek P, Baker LL, Luerssen TG (1988) MR imaging of brain contusions: a comparative study with CT. AJNR 9: 269–278

5. Kanno I, Lassen NA (1979) Two methods for calculating regional cerebral blood flow from emission computed tomography of inert gas concentration. J Comput Assist Tomogr 3: 71–76

6. Langfitt TW, Obrist WD (1981) Cerebral blood flow and metabolism after intracranial trauma. Prog Neurol Surg, Vol 10. Karger, Basel, pp 14–48

7. Obrist WD, Thompson HK, King CH, Wang HS (1967) Determination of regional cerebral blood flow by inhalation of [133]-xenon. Circ Res 20: 124–135

8. Snow RB, Zimmerman RD, Gandy SE, Deck MDF (1986) Comparison of magnetic resonance imaging and computed tomography in the evaluation of head injury. Neurosurgery 18: 45–52

Correspondence: H. Fumeya, M.D., Yokohama Shintoshi Neurosurgical Hospital, Eda 433, Midoriku, Yokohama 227, Japan.

Acta Neurochirurgica, Suppl. 51, 286–288 (1990)
© by Springer-Verlag 1990

The Time Course of Vasogenic Oedema After Focal Human Head Injury – Evidence from SPECT Mapping of Blood Brain Barrier Defects

R. Bullock[1], **P. Statham, J. Patterson, D. Wyper, D. Hadley,** and **E. Teasdale**

[1] University Department of Neurosurgery, Institute of Neurological Sciences, Southern General Hospital, Glasgow, Scotland

Summary

We have tomographically mapped changes in the blood brain barrier (BBB) (99 mTc Pertechnetate) in 20 patients with acute contusions, and four with acute subdural haematomas *in situ*. The changes were related to regional CBF, (99 mTc HMPAO SPECT) T_2 weighted MRI scans, CT abnormalities and the clinical features. Seventy-five percent of contusions were accompanied by a BBB abnormality, usually a "halo" around the lesion, which was more common in scans made after the second day. All contusions demonstrated "oedema" as a zone of "T_2" signal on MRI or a zone of lucency on CT, and all were accompanied by a focal zone of low CBF on SPECT. Early contusional oedema appears to be cytotoxic but in certain cases, delayed blood brain barrier lesions develop, suggesting a vasogenic component.

Introduction

Focal cerebral contusions can be found in almost all severe head injuries. The results of post-mortem, of computed tomography (CT) scanning, magnetic resonance (MRI) scanning, and by direct microgravimetric measurement of brain water content show that almost all contusions are accompanied by local oedema[1,3]. Serial CT scanning studies have shown that peri-contusional oedema is progressive, and in a third of patients causes increasing mass effect[4]. Neurosurgeons need to know which lesions need to be resected and which are safe to be treated conservatively. This knowledge depends upon increased understanding of the pathophysiology of contusions in man and very few such studies have hitherto been performed.

We have, therefore, applied MRI and single photon emission computed tomographic (SPECT) mapping of cerebral blood flow (CBF) and the "blood brain barrier" (BBB) to a group of patients with severe head injury and focal cerebral contusion or acute subdural haematoma. The aim of these studies was:

1. To determine the incidence and time course of abnormalities in the "blood brain barrier" after both cerebral contusion and acute subdural haematoma.

2. To establish the relationship between rCBF and abnormalities of the "blood brain barrier" and the extent of oedema was demonstrated by T_2 weighted MRI scanning.

Patients and Methods

Twenty-four patients (age 22 to 69 years) were studied from 1–21 days after a severe head injury. SPECT imaging (NOVO 810 scanner) was carried out using 99 mTc labelled HMPAO (Ceretec) for blood flow mapping and 99 mTc labelled Pertechnetate for "blood brain barrier" mapping. Up to eight SPECT slices 10 mm apart were made and abnormalities were graded visually on a 0–3 scale, in comparison with the contralateral hemisphere, and normal volunteers. Grading was based upon the number of slices upon which abnormalities were seen, and the size and intensity of the area of abnormal uptake. MRI scanning was performed with a 0.15 TESLA Picker resistive unit and efforts were made to align MRI and SPECT slices. In ten patients, a single study was performed and in six, two BBB studies were carried out 1–18 days apart. In seven patients, CBF studies were performed within 48 hours of a blood brain barrier mapping study, allowing comparison of the two techniques. In a further 34 patients with contusions, HMPAO CBF mapping studies were performed without BBB studies.

Results

Incidence of "Blood Brain Barrier" Abnormalities – Contusion

Of the 20 patients with contusions, in 5 (25%) the pertechnetate studies did not show abnormality of the BBB, to pertechnetate. Nevertheless, in each of these,

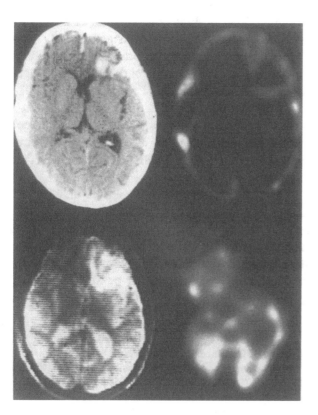

Fig. 1. Abnormalities of the blood brain barrier, and regional blood flow, 10 days after cerebral contusion. **Top left:** CT scan showing right frontal contusion. **Top right:** ^{99m}Tc pertechnetate SPECT BBB mapping study-note the "halo" of BBB abnormality around the contusion. **Bottom left:** "T_2" weighted MR scan, note the zone of oedema around the contusion. **Bottom right:** ^{99m}Tc HMPAO CBF mapping study-note the zone of reduced CBF at the contusion site

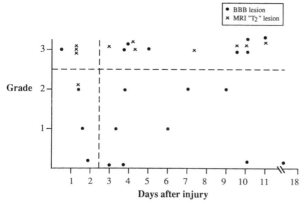

Fig. 2. Time course of blood brain barrier abnormality and MRI "T_2 oedema" after focal contusion

altered signals indicative of oedema were present around the contusion, on CT or "T_2" weighted MRI, or both. In the remaining 15 patients, BBB lesions were present (75%). There was no difference in the severity

or nature of the signal change in either MRI, or CT between these patients and those without a BBB abnormality.

Regional CBF and Blood Brain Barrier Studies – Contusions

In every patient with a focal cerebral contusion there was a surrounding zone of reduced CBF. The zone of low flow on SPECT closely matched the area of contusion itself, but was usually smaller than the area in which "T_2" oedema was present on MRI scanning. Precise assessment of the topographical relationships in the results of the different scans (SPECT, CT, MRI) was difficult because of the limitations of partial volume averaging and spatial resolution of the SPECT system (Fig. 1). In all patients in whom a BBB abnormality was seen, low flow areas on the HMPAO mapping study were also present.

Blood brain barrier lesions were usually maximal at the periphery of focal cerebral contusions and frequently a "halo" appearance was produced where a barrier lesion developed surrounding an area of solid blood clot within the contusion.

Time Course of BBB Abnormalities – Contusions

Patients studied within 2 days of injury had fewer severe BBB abormalities (1 out 4) than those studied in the subsequent period (8 out of 16). Yet three of those studied early had severe contusions with "grade 3" oedema on "T_2" MRI, or CT scans (Fig. 2).

Among the six patients who underwent serial BBB mapping, no clear pattern of evolution was see: in two, the lesions remained unchanged, in one, the lesion was less after surgical removal, while in a second it progressed after surgery.

Acute Subdural Haematomas

Four patients had an acute subdural haematoma that had not been evacuated at the time of SPECT. Two of the patients werre conscious, the subdural clot was not causing mass effect on CT, and both were treated conservatively. In the other two patients, the subdural haematomas had recurred after craniotomy. BBB mapping studies were performed at three, nine and ten days (two patients) after injury. In the one patient scanned at three days after injury, no BBB lesion was seen, while in the three patients scanned later, a band of BBB breakdown was seen underlying the haematoma, corresponding to the position of the displaced cortex.

Discussion

With the increasing use of CT scanning in less severe head injuries, more contusions are being demonstrated, often in conscious patients. Some authors advocate removal of all focal contusions >2 cms, because some cause worsening of mass effect, and raised intracranial pressure[2]. In the majority, however, contusions will resolve spontaneously leaving an area of cortical atrophy. Surgery for all such lesions is thus not necessary. Equally, however, the policy of using neurological observation to detect clinical deterioration in patients with focal lesions, relying on this as an indication for surgical resection, is dangerous because some patients will deteriorate suddenly without warning and this may result in death or disability.

These studies have shown that oedema, as shown by "T_2" MRI and CT scanning, is a universal finding, even early after a contusion, and that a focal zone of low CBF in relation to the contusion was always present. In certain patients, BBB breakdown to pertechnetate (a polar molecule of MW 187 d) is present, indicating a BBB defect. BBB lesions were absent in 25% of contusions, and were more frequent after the first 48 hours. These findings suggest that particularly early after injury, "contusion" oedema is predominantly cytotoxic in nature. No clear pattern of injury could be identified in the group with BBB lesions and not all the contusions were large. In four patients, surgery became necessary because the contusion and swelling caused high ICP; in three of these, prior SPECT BBB studies had shown a "grad 3" BBB abnormality.

Pertechnetate scanning has been used successfully as a diagnostic technique for chronic subdural haematoma, so that our finding of a band of BBB breakdown present under a SDH, during the "subacute" period (7–14 days) is not surprising, and may represent early neovascularisation of a developing subdural membrane, or an "open" BBB due to the effect of ischaemia.

Treatment of early contusion oedema may be best achieved by improving CBF to the area or preventing neuronal damage by drugs. The delayed mass effect which occurs in some contusions may be due to late BBB opening, and further studies are needed to establish whether BBB scanning either by SPECT or MRI can identify patients at risk of this complication.

References

1. Adams JH, Graham DI, Scott G *et al* (1980) Brain damage in fatal non-missile head injury. J Clin Pathol 33: 1132–1145
2. Becker DP, Gudeman SP (1989) Textbook of head injury. Saunders, Baltimore
3. Bullock R, Smith R, Favier J *et al* (1985) Brain specific gravity and CT density measurements after human head injury. J Neurosurg 63: 64–68
4. Kobayashi S, Nakazawa S, Otsuk T (1983) Clinical value of serial computer tomography with severe head injury. Surg Neurol 20: 25–29

Correspondence: R. Bullock, Ph.D., FRCS (SN) Ed., University Department of Neurosurgery, Insitute of Neurological Sciences, Southern General Hospital, Glasgow G51 4TF, Scotland.

Acta Neurochirurgica, Suppl. 51, 289–291 (1990)

Intracerebral Haematoma: Aetiology and Haematoma Volume Determine the Amount and Progression of Brain Oedema

P. F. X. Statham[1] and **N. V. Todd**[2]

[1] Department of Clinical Neurosciences, Western General Hospital, Edinburgh and [2] Institute of Neurological Sciences (INS), Southern General Hospital, Glasgow, Scotland

Summary

In a study of 182 patients with a traumatic, 'spontaneous', or aneurysmal intracerebral haematoma (ICH) a significant correlation was found between the amount of focal brain oedema seen on computed tomogram (CT) and both the aetiology and the size of the haematoma. Traumatic haematomas were associated with twice the oedema per unit volume of haematoma, and a doubling of median oedema volume on second CT (performed in 18 patients), compared to spontaneous or aneurysmal haematomas.

Introduction

Intracerebral haematomas are associated with local brain oedema, which may increase the mass effect of a haematoma and adversely affect clinical outcome[1]. The amount of oedema seen on CT is variable. We looked at the effect of three clinical and radiological features:

1. The aetiology of the haematoma (traumatic, spontaneous, aneurysmal)

2. The volume of the haematoma in CT,

3. The time after ictus/injury, and progression of oedema with time.

Method

A consecutive series of patients with a CT diagnosis of ICH (greater than 4 mls in volume) were identified from a diagnostic database in the INS, Glasgow. Only patients with traumatic, spontaneous, or aneurysmal haematomas were recruited.

The aetiology of the patient's haematoma was determined from case records. The volume of ICH, and the volume of surrounding oedema were determined using CT planimetry. Haematoma was defined as the region of increased density, similar to that of blood, within the brain, including reduced density regions within it. Brain oedema was defined as the region of reduced density around the ICH, which has been shown to correlate with reduced specific gravity and increased brain water content per unit volume of brain[2, 3, 4].

Where a second or subsequent CT had been performed the change in oedema volume was measured, but those patients who had undergone an operation for removal of haematoma, or who had re-bled from an aneurysm were excluded. Statistical analysis was performed using non-parametric tests (Kendall's rank correlation and Wilcoxon rank sum tests).

Results

Sixty six patients had a traumatic haematoma, 59 a spontaneous haematoma, which included 8 with an arteriovenous malformation (AVM), and 57 and aneurysmal haematoma.

Aetiology and Oedema Volumae

Figure 1a shows the relationship between the aetiology of the haematoma and volume of oedema surrounding it. Open bars represent median haematoma volume, and shaded bars represent median oedema volume for each aetiological group (Tr, traumatic Sp, spontaneous An, aneurysmal). Traumatic and spontaneous haematomas were associated with a similar median oedema volume, although spontaneous haematomas were bigger ($p < 0.01$). Aneurysmal haematomas were associated with half the volume of oedema compared to the other two groups ($p < 0.01$).

When standardised for unit volume of haematoma (Fig. 1b), traumatic haematomas are associated with twice the oedema of the spontaneous group, and nearly three times that of the aneurysmal group ($p < 0.01$)

Haematoma Volume and Oedema Volume

There was a significant correlation between haematoma and oedema volume for the group as a whole,

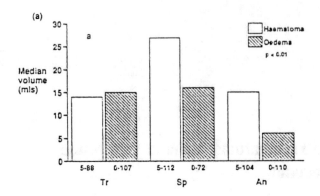

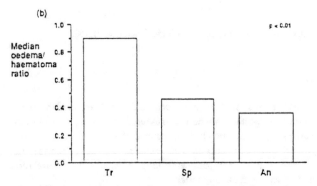

Fig. 1. Histograms showing (a) haematoma volume, oedema volume and aetiology of the haematoma (b) median of oedema/haematoma ratios and aetiology of the haematoma

Time After Ictus/Injury and Progression of Oedema

There was no clear relationship between time from ictus or injury and amount of oedema. In 18 patients a subsequent CT was performed, a median of 5–6 days after admission for each group. Table 1 demonstrates the difference in volume of oedema on the second CT compared with the first. In this subseries of patients those with traumatic haematomas showed an overall doubling in oedema volume, which was not seen in patients with spontaneous or aneurysmal haematomas.

Table 1

	Traumatic	Spontaneous	Aneurysmal
Change in oedema volume	$n = 6$	$n = 6$	$n = 6$
Increase	5	2	3
Unchanged	0	1	0
Decrease	1	3	3
Net change	+1.17	+0.03	+0.03

Discussion

In a study of the management of patients with an intracerebral haematoma we found significant differences in management and outcome based on the aetiology of the haematoma. It was suggested that differences in the amount of oedema around the ICH could account for this. This study has demonstrated that the amount of oedema is related to both the aetiology of the ICH, and it's size.

and for each aetiological group, showing that a large haematoma is more likely to be associated with more oedema than a small haematoma. Kendall's correlation coefficients were 0.31 for traumatic ICH ($p < 0.001$), 0.22 for spontaneous ICH ($p < .0.03$) and 0.45 for aneurysmal ICH ($p < 0.001$).

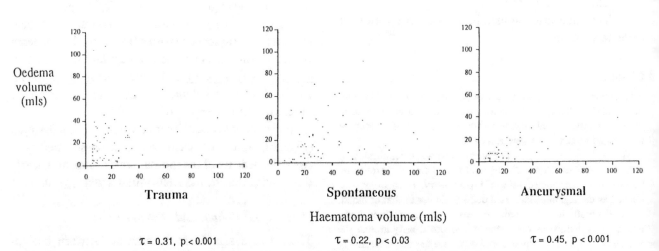

$\tau = 0.31, \, p < 0.001$ $\tau = 0.22, \, p < 0.03$ $\tau = 0.45, \, p < 0.001$

Fig. 2. Scatter diagrams of volume of haematoma vs volume of oedema per patient, for each aetiological group

Brain oedema was defined by CT appearances alone, and although there are both biological and physical limitations in enterpreting this measurement it corresponds well with clinical features[1, 4]. The striking differences demonstrated in the amount of oedema around traumatic haematomas compared with spontaneous or aneurysmal ICHs suggests that the injury, or factors relating to it, as well as the haematoma, determines the amount of oedema. The correlation between volume of haematoma and volume of oedema could be explained by the increased surface area of a larger haematoma giving rise to more oedema.

Because relatively few patients had a second CT it was not surprising that no clear relationship was demonstrated between time after injury and amount of oedema. In patients with traumatic haematomas who had a second CT, the mass effect caused by progression of oedema may partly explain delayed clinical deterioration. This change in the volume of oedema on the second CT was greater than could be accounted for by the reduction in haematoma size or density.

References

1. Penn R (1980) Cerebral oedema and neurological function: CT, evoked responses, and clinical examination. In: Cervos-Navarro J, Ferszt R (eds) Advances in neurology, Vol 28: Brain oedema. Raven Press, New York, pp 383–394
2. Bullock R, Smith R, Favier J, du Trevou M, Blake G (1985) Brain specific gravity measurements after human head injury. J Neurosurg 63: 64–68
3. Lansch W, Baethmann A, Kazner E (1981) Computed tomography of brain oedema. In de Vleiger M, de Lange SA, Beks JWF (eds) Brain oedema. Wiley, New York, pp 67–98
4. Clasen RA, Huckman MS, von Roenn KA, Pandolfi S, Laing I, Lobick JL (1981) A correlative study in human experimental vasogenic cerebral oedema. Comput Assist Tomogr 5(3): 313–327

Correspondence: P. F. X. Statham, M.D., Department of Clinical Neurosciences, Western General Hospital, Grewe Road, Edinburgh EH4 2XU, Scotland.

Acta Neurochirurgica, Suppl. 51, 293–295 (1990)

Trauma and Brain Oedema (II)

Gadolinium-DTPA Enhanced Magnetic Resonance Imaging in Human Head Injury

D. A. Lang, D. M. Hadley, G. M. Teasdale, P. Macpherson, and **E. Teasdale**

Institute of Neurological Sciences, Southern General Hospital, Glasgow, U.K.

Summary

We used low field magnetic resonance imaging and Gadolinium-DTPA to study 10 recently head injured patients. On the pre-contrast images we identified 36 abnormalities in 9 of the 10 patients. After contrast enhancement was seen in 9 of these lesions in 3 patients imaged 5–6 days after injury. Altered blood brain barrier permeability to gadolinium was not seen in 6 patients studied within 4 days of trauma. We did not identify any lesions with gadolinium that were not present on the pre-contrast images.

We do not think that the routine use of contrast in head injury is justified since clinically important lesions will be present on routine sequences. However the use of gadolinium may provide information about the nature and timing of the underlying pathophysiological changes. In human head injury cytotoxic oedema occurs early and vasogenic oedema does not occur until 5–6 days after injury.

Introduction

Paramagnetic contrast agents produce MR enhancement in a variety of acute and chronic intracranial disorders as a result of leakage of the contrast from the intravascular compartment to the parenchyma in areas where blood brain barrier permeability is increased[2, 4, 8]. Such agents have not been used in patients with an acute head injury.

We have investigated a series of recently head injured patients using magnetic resonance imaging (MRI) and Gadolinium-DTPA. Our aims were to determine if more information about primary brain damage could be obtained with contrast assisted imaging; we also investigated if lesions shown by standard sequences had evidence of altered blood-brain barrier permeability to gadolinium.

Patients and Methods

We selected 10 patients with an acute head injury. The patients were initially admitted to a general hospital and then transferred to the regional neurosurgical unit at the Institute of Neurological Sciences Glasgow; the indication for transfer was persisting impairment of conscious level in 9 patients and the development of focal signs in one. The patient's age, conscious level[9] and neurological signs were recorded. We selected patients with a variety of patterns and severity of head injury. All patients had CT scans. MRI – We used a Picker "VISTA" 1100 0.15 Tesla resistive imager which operates at 6.38 MH to obtain our images. We used our standard head sequences (TR 2000/TE 80 and TR 2000/TI 400/TE 40) which give us T2 and T1-weighted images respectively. After obtaining the standard sequences we administered Gadolinium intravenously and repeatedly scanned the patients using the T1-weighted sequence for at least 1 hr. 5 patients received gadolinium in a low dose (0.1 mmols/kg) and the other 5 received 0.2 mmols/kg. The time interval from injury ranged from 36 hours to 6 days.

Abnormalities were noted in each case before and after contrast. We derived an index of enhancement which allowed us to study in each patient the time course of enhancement.

Results

Patients

There were 10 patients in the study. Their ages ranged from 19–60 and there were 9 males. 4 patients had been involved in road traffic accidents 1 had fallen from a horse and 4 had simple falls. In 1 patient the circumstances of the injury were unknown. 7 patients had impaired conscious level on admission. At the time of imaging 4 patients were in coma (GCS $\leqslant$ 8), 3 patients had impaired consciousness (GCS 9–14) and 3 were alert and orientated.

Table 1. *MRI Findings*

Time from injury to imaging	Gd-DTPA dose mmol/kg	Lesions on 1st MRI*			Number	Lesions enhancing after Gd-DTPA
		c/scw	dw	e/c		
36 hours	0.2	+	−	−	5	−
39 hours	0.1	+	−	−	4	−
2 days	0.1	+	−	−	5	−
3 days	0.1	+	−	−	6	−
3 days	0.1	+	+	−	6	−
4 days	0.1	+	−	−	1	−
4 days	0.2	−	−	−	0	−
6 days	0.2	+	−	−	2	+
6 days	0.2	+	−	−	4	+
6 days	0.2	+	−	+	3	+

* *c* cortex, *scw* subcortical white matter, *dw* deep white matter, *ec* extracerebral haematoma.

Radiology

7 patients had a skull fracture.

CT scanning. 3 patients had a normal CT scan. 6 patients had contusions in a number of sites and 1 patient had small extradural and intracerebral haematomas.

MR imaging. Of the 10 patients, 9 had abnormalities due to the recent head injury. 8 patients had cortical contusions which extended into the subcortical white matter and 1 patient had a purely cortical lesion. 2 of these patients had a second lesion. One had "shearing" lesions in the deep white matter and the other had a small extracerebral haematoma. 36 lesions were identified in the 9 patients.

Effect of Gadolinium. Each patient showed the normal distinct enhancement of the pituitary gland, nasal mucosa and cavernous sinuses. 6 patients had focal enhancement of the dura underlying a skull fracture. Despite this, no new lesions were seen after contrast. In 6 patients with 27 lesions, there was no change in the signal intensity of the lesions after gadolinium. The patient with the normal images pre-contrast had normal post-contrast images. Three patients' images showed intraparenchymal lesions (9 lesions) which enhanced (Table 1). These three patients were all studied 5–6 days after injury. One of these patients had cortical and "gliding contusions"[1]. After contrast all of the lesions enhanced. The effect was seen within 6 minutes of contrast injection. The entire lesion was hyperintense by 40 minutes and enhancement persisted for several hours thereafter. The second patient had a "burst" temporal lobe, a frontal haemorrhagic contusion and a small extradural haematoma. All of the intracerebral lesions showed enhancement, either in the hypointense oedematous brain around the contusion haematomas or in the periphery of lesions not containing central haemorrhage. This effect was seen within 9 minutes of contrast injection. The third patient had multiple contusions; enhancement was seen peripherally and was most marked by 35 minutes.

Discussion

MR enhancement did not identify a new lesion in any patient; clinically relevant lesions will be shown therefore on standard sequences. However, the pattern and timing of enhancement provides information about the nature of traumatic brain lesions.

Gadolinium "leaks" from the intravascular space into the parenchyma when blood brain barrier permeability is increased[2, 4, 8]. We did not see intraparenchymal enhancement in patients studied within the first few days after head injury. This pattern is in accord with studies using CT contrast[7] and with preliminary observations with SPECT imaging[3]. Thus lesions shown by MRI and CT are not accompanied by a gross increase in blood-brain barrier permeability at this stage.

The implications of our findings are that traumatic oedema in man initially has the characteristics of cytotoxic swelling and that vasogenic oedema may not appear until relatively late[6]. Thus head injury models with an early vasogenic component, such as the cortical freeze lesion[5], may not be relevant to human head injury.

The findings of the present study re-emphasize the relevance of cerebral ischaemia in the mechanisms of acute traumatic brain damage.

References

1. Adams JH, Gennarelli TA, Graham DI (1982) Brain damage in non missile head injury: observations in man and subhuman primates. In: Smith WT, Cavanagh JB (eds) Recent advances in neuropathology, Vol 2 Chap 7. Churchill Livingstone, Edinburgh London
2. Brasch RC (1983) Work in progress: methods of contrast enhancement for NMR imaging and potential applications. Radiology 147: 781–788
3. Bullock R, Statham P, Patterson J *et al* (1989) Tomographic mapping of CBF, CBV and blood-brain barrier changes in humans after focal head injury using SPECT-mechanisms for late deterioration. In: Hoff JT, Betz AL (eds) Intracranial pressure VII. Springer, Berlin Heidelberg New York, pp 637–639
4. Caille JM, Lemanceau B, Bonnemain B (1981) Gadolinium as a contast agent for NMR. AJNR 2: 517–526
5. Klatzo I, Wisniewski H, Steinwall O *et al* (1967) Dynamics of cold injury oedema. In: Klatzo I, Seitelberger F (eds) Brain oedema. Springer, New York, p 554
6. Klatzo I (1976) Neuropathological aspects of brain oedema. J Neuropathol Exp Neurol 26: 1
7. Mauser HW, Van Nieuwenhuizen O, Veiga Pires JA (1984) Is contrast enhanced CT indicated in acute head injury? Neuroadiology 26: 31–32
8. McNamara MT (1987) Paramagnetic contrast media for magnetic resonance imaging of the central nervous system. In: Brant-Zawadzki M, Norman D (eds) Magnetic resonance imaging of the central nervous system. Raven Press, New York, pp 97–107
9. Teasdale G, Murray G, Parker L *et al* (1979) Adding up the Glasgow Coma Score. Acta Neurochir (Wien) [Suppl] 28: 13–16

Correspondence: D. A. Lang, M.D., Institute of Neurological Sciences, Southern General Hospital, Glasgow, U.K.

Acta Neurochirurgica, Suppl. 51, 296–299 (1990)

Blood-Brain Barrier Damage in Traumatic Brain Contusions

N. V. Todd and D. I. Graham

Institute of Neurological Sciences, Southern General Hospital, Glasgow, Scotland

Summary

Plasma proteins were used as an endogenous marker of blood-brain barrier damage in 19 patients dying with traumatic cortical contusions. Patients survived for a few hours to 31 days after head injury. Eight proteins (M WT 61-2,500 $\times$ 10^3) were demonstrated with standard immunohistochemical techniques. Proteins were not found in "control" brains or in macroscopically normal parasagittal cortex in the head injury patients. Proteins were found in all of the macroscopic contusion in all brains. Protein leakage appeared to be from the contusion itself. Protein staining around histologically normal vessels was unusual. There was a gradient of staining from the macroscopic contusion into the surrounding brain. There was a trend for staining to be most marked between 3 and 8 days survival after head injury. There was no gradient of leakage by molecular size of the protein.

Introduction

Mechanical brain injury is associated with a vascular response which includes changes in systemic haemodynamics and also in cerebral microvessels[6]. Following experimental concussive head injury in animals extravasation of plasma proteins into brain parenchyma has been shown using protein tracers such as Evan's blue and fluorescein[3, 7]. We report a study of fatal human head injury in man in which plasma proteins immunohistochemically stained within brain parenchyma have been used as an endogenous tracer to determine protein leakage in and around traumatic brain contusions.

Materials and Methods

The brains of 19 adult male patients dying from head injury were fixed in 4% buffered formaldehyde. The time from head injury to death ranged from a few hours to 31 days. Macroscopic contusions of the cortical surface were sampled and embedded in paraffin wax blocks. Samples were also taken from macroscopically normal parasagittal cortex bilaterally in 8 of these 19 brains. Additional control samples were taken from 2 further brains of patients dying from causes not associated with central nervous system damage.

Six u serial sections were cut from each block and immunohistochemically stained for 8 different serum proteins using the peroxidase-antiperoxidase (PAP) method[8]. The paraffin sections were dried onto poly-l-lysine coated slides at 60 °C for 48 hours pretreated with 0.3% H$_2$O$_2$ and 0.0125% trypsin (Sigma) at room temperature for 40 minutes, and preincubated for 20 minutes with normal mouse IgG in 5% bovine serum albumin/PBSA at pH 7.6 to block non-specific binding by Fc-receptors. Sections were incubated overnight at 4 °C with the following antibodies (and dilutions): pre-albumin (1 : 500), albumin (1 : 10,000), ceruloplasmin (1 : 500), IgG (1 : 10,000), IgA (1 : 5,000), fibronectin (1 : 2,000), IgM (1 : 5,000) and b-lipoprotein (1 : 500) (polyclonal antibodies, Dakopatts). Sections were then incubated for 30 minutes at room temperature with secondary swine antirabbit antibody (1 : 50 Dakopatts) and then with rabbit PAP complex (1 : 200 Dakopatts) and stained with 3,3' diaminobenzidine (Sigma). Sections were counterstained with haemotoxylin and eosin. All samples also had one section stained for glial fibrillary acidic protein (1 : 500, Dakopatts). Staining was graded by 2 observers as either absent ($= 0$), present with light staining ($= 1$) or present with heavy staining ($= 2$). Staining controls were performed by omission of the primary antibodies.

Results

Controls: Sections from parasagittal cortex of the 2 brains from patients dying without CNS damage did not show staining of either gray or white matter for any of the proteins. Staining of proteins was found within blood vessels in these sections.

Head Injury Brains: Parasagittal cortex sections taken from macroscopically normal parasagittal cortex had a normal histological appearance. Plasma proteins were not identified in gray or white matter in any of the 8 brains although intravascular staining was present.

Macroscopic Contusion: Samples were taken from cortical regions which had a macroscopic contusion

Table 1. *Number of Brains Showing Protein Leakage*

	Protein present	Number of proteins	
		5–8	1–4
Contusion	19 (100%)	19	
Infarct	19 (100%)	17	2
Parenchyma	18 (95%)	11	7

Number of Proteins Leaking

	Heavy staining	Light staining	Not present	Protein present
Contusion	114	33	2	147 (99%)
Infarct	86	42	15	128 (89%)
Parenchyma	46	43	58	89 (60%)

Nineteen head injury brains.

Eight proteins per brain (total number of sections may be less than 152 [19 × 8] because of a few technical failures).

with surrounding more normal brain. A common morphological pattern was found in these small contusions. There was usually a region of grossly disrupted brain with extravasation of blood which we have called the contusion. This was surrounded by a zone of tissue necrosis without extravasation of erythrocytes which we have called the infarct. Most distant from the contusion was the more or less morphologically normal brain which we have called the parenchyma. Separate contusions were often found within one block.

The presence of one or more plasma proteins was found in all blocks (Table 1). Within the *contusion* one or more proteins were present in all 19 brains and only 2 of 149 sections failed to show protein staining. This staining was heavy in 114 and light in 33 sections. Within the *infarct* one or more proteins were present in all 19 brains but proteins were not found in 15 of 143 sections. Staining was heavy in 86 and light in 42 sections. Within the *distant parenchyma* one or more proteins were present in 18 of the 19 brains. No protein was found in 58 of 147 sections. Staining was heavy in 46 and light in 43 sections. Thus there was a gradient of staining from the contusion through the infarct to the distant parenchyma. Protein extravasation was found in 99% of sections within the contusion, 89% in the infarct and only 60% in the parenchyma. Staining (where present) was heavy in 83% of sections in the contusion, 67% in the infarct and 52% within the parenchyma.

Protein was commonly confined to the white matter and a clear distinction between gray and white matter could be found for one or more proteins in 18 of 19 brains. Perivascular protein leakage from histologically normal vessels was uncommon although intravascular staining was usually seen. Perivascular staining for one or more proteins was found in 3 brains and in a total

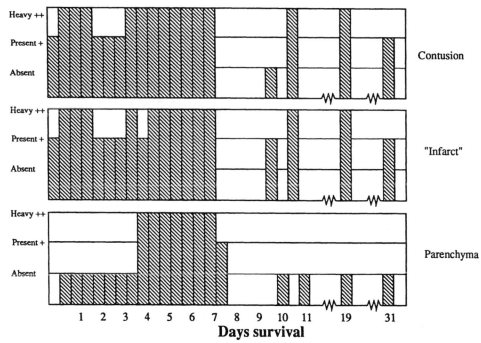

Fig. 1. Protein staining and survival – Pre-albumin

of 13 sections (8%). We did not find a gradient by molecular size. At all time points following head injury larger proteins were as likely to be found as the smaller proteins. There was also no clear temporal profile to protein leakage (Fig. 1). Within the contusion and the infarct proteins were very commonly present and mainly heavily stained. Within the distant parenchyma there was a tendency for less protein to be present within the first 3 days following head injury and for staining to be less intense after 10 days. From day 0 to day 2 proteins were found in 15 of 48 sections (31%) and staining was heavy in 5 (33%). Between day 3 and day 9 proteins were found in 53 of 69 sections (77%) and in 34 staining was heavy (64%). From day 10 to day 31 protein staining was found in 21 of 30 sections (70%) with heavy staining in 7 (33%).

Discussion

Plasma proteins are normally not transferred into the brain and can therefore be used as an endogenous marker of blood-brain barrier integrity. The presence of plasma proteins within the brain implies blood-brain barrier dysfunction at some point following head injury. The rate of plasma protein breakdown within brain is not known so the time at which protein entry occurred cannot be determined in an individual. Equally the route of protein entry and the quantitative amount of protein movement into brain cannot be determined from these methods. Nevertheless they do allow us to determine whether protein leakage is a feature of brain contusions in fatal human head injury. We used antibodies to 8 different plasma proteins which have a range of molecular weight (MW) and hydrodynamic radius (HR) from the smallest protein prealbumin MW = 61×10^3 HR = 32.5 to the largest b-lipoprotein MW = $2,500 \times 10^3$ HR = 124.0. The absence of staining in gray and white matter in the 2 control brains confirmed that antigens recognized by the polyclonal antibodies used are not expressed by the brain nor do plasma proteins leak into the brain as a postmortem event. These methods can therefore be used to mark blood-brain barrier damage.

Protein was found universally within macroscopic contusions which indeed should be the case given the presence of frank haemorrhage. Protein was also found within the surrounding region of tissue necrosis in the absence of erythrocytes. The source of the protein in the infarct and the distant parenchyma appeared to be the contusion itself. The gradient of staining both in the number of proteins leaking and in the intensity of staining was from the contusion to the infarct to the

parenchyma. In no brain was this gradient reversed. There was little evidence to suggest that intact microvessels within the parenchyma were the source of protein leakage. Whether protein staining in the infarct and parenchyma represented continued protein leakage from the contusion/infarct or whether it simply represented diffusion of protein from the extravasated blood within the macroscopic contusion is not known.

We do not know at what time point this protein extravasation occurred. In experimental head injury there is immediate blood-brain barrier damage[4]. This is secondary to the short-lasting surge of arterial hypertension that follows head injury with forced cerebral vasodilatation and barrier damage[1]. Our results showed a trend toward maximal protein extravasation occurring between day 3 and day 10. This would suggest that the blood-brain barrier dysfunction occurring around traumatic contusions does not occur solely at the time of head injury. This is supported by the absence of staining in macroscopically normal parasagittal cortex in these brains. However, it is possible that the source of the protein is simply the macroscopic contusion with diffusion into surrounding brain producing increased levels between days 3 and 10 then gradual metabolism of protein and reduced intensity of staining beyond day 10. Whether there is continued protein leakage around traumatic contusions in man as well as immediate post-traumatic barrier damage remains unresolved from our data.

Leakage was not related to the molecular size of protein: large proteins were as frequently found as smaller proteins. If protein leakage around traumatic contusions was from intact blood vessels, rather than from extravasated protein, this would suggest that the route of protein leakage is either via a large interendothelial pore, or protein was transferred non-selectively via vesicles. Increased vesicular transfer has been demonstrated following concussive head injury in a number of species (*e.g.* 5) although endothelial changes without increased numbers of vesicle has been found in one of the best models of diffuse axonal injury[2]. In our material there were only a few sections in which protein was found leaking from vessels and we cannot comment upon the route of protein leakage from this material.

Conclusion

Protein leakage was found in and around small contusions in fatal human head injury. This protein leakage was highly localised. Protein leakage was found at all durations of survival. Protein leakage was not selective by molecular size and tended to be confined to white matter. There was no evidence for generalised blood-brain barrier damage in patients dying with small brain contusions.

References

1. DeWitt DS, Jenkins LW, Wei EP, Lutz H, Becker DP, Kontos HA (1986) Effects of fluid percussion brain injury on regional cerebral blood flow and pial arteriole diameter. J Neurosurg 64: 787–794

2. Maxwell WL, Irvine A, Adams JH, Graham DI, Gennarelli TA (1988) Response of the cerebral microvasculature to brain injury. J Pathology 155: 327–335

3. Ommaya AK, Rokoff SD, Baldwin M (1964) Experimental concussion, a first report. J Neurosurg 21: 249–265

4. Povlishock JT, Becker DP, Sullivan HG, Miller JD (1978) Vascular permeability alterations to horseradish peroxidase in experimental head injury. Brain Res 153: 223–239

5. Povlishock JT, Kontos HA (1982) The pathophysiology of pial and intraparenchymal vascular dysfunction. In: Grossman RG, Gildenberg PL (eds) Head injury: Basic and clinical aspects. Raven Press, New York, pp 15–29

6. Povlishock JT (1986) The morphogenic responses to experimental head injuries of varying severity. In: Becker DP, Povlishock JT (eds) Central nervous system status report. Nat Inst Health, pp 443–452

7. Rinder L, Olsson Y (1968) Studies on vascular permeability changes in experimental brain concussion. Acta Neuropathol (Berl) 11: 183–200

8. Sternberger LA, Hardy PH, Cuculis JJ, Meyer HG (1970) The unlabelled antibody enzyme method of immunohistochemistry. J Histochem Cytochem 118: 315–333

Correspondence: N. V. Todd, M.D., Institute of Neurological Sciences, Southern General Hospital, Glasgow G51 4TF, Scotland.

Acta Neurochirurgica, Suppl. 51, 300–301 (1990)

Large and Small "Holes" in the Brain: Reversible or Irreversible Changes in Head Injury

L. F. Marshall and **Th. Gautille**

Division of Neurosurgery, School of Medicine, University of California, San Diego, California, U.S.A.

Summary

We prospectively studied the frequency and course of large ($> 25\,cc$) and small ($< 25\,cc$) areas of decreased density in 491 patients with severe closed head injury entered into the Traumatic Coma Data Bank (TCDB). The frequency of such areas and of subarachnoid haemorrhage on initial and subsequent CT scans were recorded. The frequency of large "holes" increased from 8 on the initial CT scan to 24 on scans done from day 4 to day 10. Half of these lesions either completely or almost completely resolved 14 days or more following injury. In patients with small "holes" the frequency increased from 24 to 77, but on scans performed 14 days or more following injury, 47% had completely disappeared. The presence of subarachnoid haemorrhage on the initial scan predicted the development of large areas of decreased density, but it did not predict the development of small areas of decreased density.

The disappearance of a substantial number of these areas of decreased density ("holes") indicate that these areas do not necessarily represent areas of cerebral infarction. Patients with closed head injury are at risk for the development of what appear to be regional areas of cerebral ischaemia, but subarachnoid haemorrhage only predicts the development of large areas with these changes. Pharmacologic trials with calcium channel blocking agents or NMDA receptor antagonists in head injured patients appear warranted.

Introduction

Cerebral ischaemia, from a variety of causes, is a well known complication of severe head injury. Vasospasm has been identified in up to 20% of patients with severe head injury, but its long term consequences in head injury have not been systematically studied[1, 3, 4]. In an attempt to determine the role of subarachnoid haemorrhage with vasospasm in the development of regions of cerebral ischaemia, we prospectively studied 491 patients who had survived at least 96 hours and in whom an initial CT scan and subsequent CT scans were available for study within the Traumatic Coma Data Bank (TCDB). This study was designed to de-

termine the frequency of the development of areas of decreased density within small and large arterial territories and to learn, over time, whether these densities resulted in infarction or simply reflected, at least in some instances, areas of decreased perfusion or oedema which did not go on to infarction[2].

Materials and Methods

All patients entered into the Traumatic Coma Data Bank of the United States and who survived for more than 96 hours were included in the study. These scans were sequentially studied to determine the frequency of (a) subarachnoid haemorrhage and (b) areas of decreased density which were further classified into large or small. Large areas were those $> 25\,ccs$ in size, and small those $< 25\,ccs$ in size although most were $< 10\,cc$. Such lesions could not include any areas of increased density suggestive of contusion or haemorrhage in an attempt to limit the study only to pure areas of subsequent development of decreased density. The frequency of subarachnoid haemorrhage and its relationship to the development of "holes" both large and small was also studied within this cohort.

Results

Subarachnoid haemorrhage was present in 1/3 of the 491 patients. The number of patients with large areas of decreased density increased from 8 on the initial CT to 24 on scans performed between 96 and 240 hours and in slightly more than half of these patients subarachnoid haemorrhage was present on the initial scan ($p < 0.05$).

The frequency of small areas of decreased density increased from 17 to 77, but in this group there was absolutely no association between subarachnoid haemorrhage on the initial CT scan and the development of small "holes".

In 91% of patients a CT scan done at 14 days from injury or longer was available for further study. In 50%

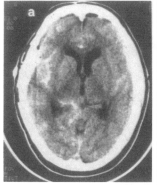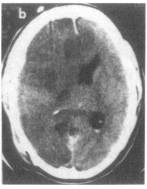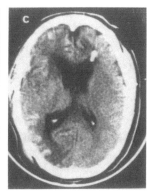

Fig. 1. (a) Day 0 – Initial CT scan demonstrates subarachnoid haemorrhage; no shift. Pt GCS 6 on admission. (b) Day 8 – Massive decreased density vascular distribution right hemisphere. (c) Day 28 – Further improvement right frontal area. Area in occipital region sole area of "infarction"

Table 1. *Profile of Development of CT Evidence and Resolution of Delayed Cerebral Ischaemia in 491 Severely Head Injured Patients*

	Day 0	Days 4–10	Days > 14
Small "holes"	17	77	41
Large "holes"	8	24	12

of the patients with large "holes", there was almost complete or complete resolution of the area of decreased density. In 47% of the patients with small "holes", there was complete disappearance indicating that these areas do not necessarily represent areas of cerebral infarction. A series of CT scans in one patient with severe head injury and associated subarachnoid haemorrhage is shown in Fig. 1 a through c. Note the development and subsequent resolution of large areas of decreased density with only very little CT evidence of infarction in a patient who ultimately had no detectable neurological deficit involving the right hemisphere.

Discussion

CT areas of decreased density have been typically interpreted as representing areas of infarction in patients with subarachnoid haemorrhage from any cause. Here we present information to suggest that in fact infarction is not the only outcome of such lesions, but rather that if tissue perfusion improves this may, in fact, be a completely or partially reversible tissue injury reflecting areas of damaged tissue which can recover. The high frequency of these lesions occurring in approximately 1/5 of patients in the Data Bank indicate that trials in

head injured patients which incorporate agents which prevent or reduce cerebral ischaemia in focal and global models of brain injury are in order. Calcium channel blocking agents such as Nimodipine and/or NMDA receptor antagonists would seem to be appropriate agents for trial in targeted patients. The fact that subarachnoid haemorrhage was only predictive of large areas of decreased density indicates that either all patients with closed head injury should be considered as candidates or that other information, such as that which is available from transcranial doppler studies, needs to be employed to identify patients at particular risk.

This study further indicates that attention to brain perfusion during the acute phase of head injury is essential. Policies which result in dehydration, or reduction in cardiac output or mean arterial pressure are unsound and are likely to result in the exacerbation of ischaemic injury in these particularly vulnerable patients.

References

1. Marshall LF, Bruce DA, Bruno L, Langfitt TW (1978) Vertebrobasilar spasm: a significant cause of neurological deficit in head injury. J Neurosurg 48: 560–564
2. Miller JD, Corales RL (1981) Brain oedema as a result of head injury: Fact or fallacy? In: de Vlieger M, deLange S, Beks JWF (eds) Brain oedema. John Wiley & Sons, New York, pp 99–115
3. Wilkins RH, Alexander JA, Odom GL (1968) Intracranial arterial spasm: a clinical analysis. J Neurosurg 29: 121–134
4. Wilkins RH, Odom GL (1970) Intracranial arterial spasm associated with craniocerebral trauma. J Neurosurg 32: 626–633

Correspondence: L. F. Marshall, M.D., Division of Neurosurgery, School of Medicine, University of California, San Diego, California, U.S.A.

Acta Neurochirurgica, Suppl. 51, 302–304 (1990)

Hemispheric CBF-Alterations in the Time Course of Focal and Diffuse Brain Injury

J. Meixensberger, A. Brawanski, M. Holzschuh, I. Danhauser-Leistner[1], and W. Ullrich

Department of Neurosurgery and [1] Anaesthesiology, University of Würzburg, Federal Republic of Germany

Summary

Hemispheric CBF-alterations were studied in the time course of focal and diffuse brain injury in a series of 25 head injured patients. Repeated 133 Xe CBF measurements with a mobile 10 detector system were performed in order to evaluate hemispheric CBF and cerebral vasoreactivity after change of $PaCO_2$. Focal brain injury influenced hemispheric CBF varying in the time course: A hyperperfusion could be found within the first seven days after injury in 55 percent, whereas a hypoperfusion could be detected during the whole examination period in 18 percent. In diffuse injury we never saw such CBF abnormalities. On the other hand hemispheric CO_2-reactivity was disturbed in focal and diffuse lesions, whereas a correlation between altered CBF at rest and CO_2-reactivity could be only detected in 40% in focal injury. This investigations demonstrates no significant general, however, a partial influence of morphological damage upon cerebral microcirculation.

Introduction

Disturbances of cerebral circulation are important in the development of secondary brain injury and are one of the major determinants of outcome after head injury[3]. However, there are only a few clinical examinations studying the influences of focal and diffuse brain injury upon cerebral microcirculation. On the other hand experimental data revealed progressive CBF changes which may be responsible for secondary brain damage[1]. Therefore we were interested in the following questions:

1. How is hemispheric CBF influenced by focal and diffuse brain injury?

2. What is the time course of hemispheric CBF in such conditions?

3. Is there a correlation between hemispheric CBF, cerebrovascular reactivity and focal and diffuse brain lesions?

Patients and Methods

A total of 25 patients (17 male, 8 female) were studied during the first and 14th day after traumatic brain injury in the Neurointensive Care Unit in the Department of Neurosurgery University Würzburg. Their ages ranged from 7–79 years, with an average of 32.5 years. On admission all patients had a GCS-score less than 8. All patients were sedated, intubated and ventilated. Beside epidural ICP monitoring we performed serial CAT scans to evaluate form and extent of the brain injury. Nineteen patients (76%) revealed focal lesions (mass lesion and/or mass oedema) on one hemisphere, whereas the remaining six patients (24%) had multiple, diffuse lesions on both hemispheres or a diffuse brain oedema.

A mobile unit for the bedside measurements (Novo Cerebrograph 10 a) with 10 detectors, five placed over each hemisphere, was utilized for the CBF measurements. Regional CBF was measured after intravenous administration of 10 to 15 mCi of Xenon-133. As CBF-parameter we used the Initial slope index (ISI) according to Risberg et al.[5] by the Obrist model[4]. Whenever possible we performed two CBF measurements with different $PaCO_2$ levels in order to determine rest CBF under therapeutic conditions and CBF-reactivity to $PaCO_2$ changes at various time intervals 1–14 days after head injury. Specific subgroups depending on the time interval (1.–3., 4.–7. and 8.–14. day after injury) were formed. We calculated hemispheric rest CBF and determined hemispheric CO_2-reactivity as CBF change calculated per 1 mmHg $PaCO_2$ change. An interhemispheric CBF difference of 8% was assumed as significant CBF abnormality. Additionally we monitored continuously arterial blood pressure, determined arterial $PaCO_2$ blood gas content and measured body temperature. Statistical analysis was done using Student's t-test for dependent and independent samples.

Results

Correlation of Hemispheric CBF to Focal and Diffuse Brain Injury

Fifty CBF measurements were analyzed at a mean $PaCO_2$ of 31.2 ± 3.2 mmHg. Table 1 summarizes the absolute hemispheric CBF values on side of lesion and

Table 1. *Hemispheric CBF in the time course of focal and diffuse brain injury*

	1.–3.	4.–7.	8.–14.	Day after injury
Focal injury (n = 40)	*42.4 ± 8.7 **38.3 ± 6.5	37 ± 13.3 33.2 ± 10.7	40.2 ± 14 40.7 ± 15.5	
Diffuse injury (n = 10)	33.3 ± 9.7 32.9 ± 9.5	25.2 ± 3.7 26 ± 3.7	– –	

n number of measurements.
Values given as mean hemispheric CBF ± SD/ISI units.
* Mean hemispheric CBF on the side of lesion, ** mean hemispheric CBF contralaterally.

FOCAL LESION

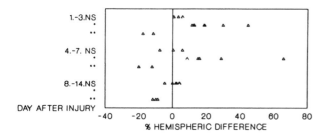

DIFFUSE LESION

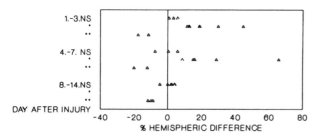

Fig. 1. % interhemispheric CBF difference in the time course of focal and diffuse brain injury. Significant interhemispheric CBF difference > 8%. *NS* not significant, * hyperperfusion on the side of lesion, ** hypoperfusion on the side of lesion

contralaterally in the time course after injury. Focal injury influenced hemispheric CBF varying in the time course after lesion: Between the first and seventh day after injury we observed a hyperperfusion in 55% of the cases of focal injury. In diffuse injury we never saw such a hyperperfusion, but we found low absolute mean CBF values on both hemispheres. Between the different subgroups no statistically significant difference between $PaCO_2$, MABP, body temperature, haemoglobin and age could be found.

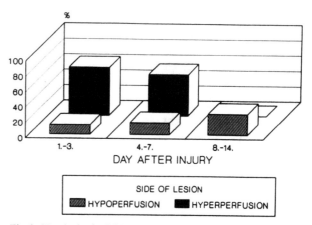

Fig. 2. Hemispheric CBF and $PaCO_2$-reactivity in the time course of focal brain injury. 1: % of the cases with significant interhemispheric difference (> 8%) of CBF and $PaCO_2$-reactivity on the side of lesion 2: % of the cases without interhemispheric difference of CBF and significant disturbances of hemispheric $PaCO_2$-reactivity

The % interhemispheric differences for CBF emphasized the time depending influence of focal injury on cerebral blood flow which could not be detected in diffuse injury (Fig. 1). Additionally we observed a focal hypoperfusion corresponding to lesion in CAT scan throughout the whole investigation period.

Correlation of Hemispheric CO_2-Reactivity and Hemispheric CBF to Focal and Diffuse Brain Injury

Forty activation measurements were analyzed with an induced mean $PaCO_2$ change of 6.7 mmHg. CO_2-reactivity is only partially influenced by focal injuries as well as by diffuse lesions in the time course after injury. A reduction of CO_2-reactivity correlated with a disturbance in rest hemispheric CBF in focal lesion could be detected in 40% of the cases during the first seven days after injury (Fig. 2). However, we observed a reduction of CO_2-reactivity without altered rest hemispheric CBF in focal as well as in diffuse injury.

Conclusion

Our results confirm knowledge from SPECT measurements[2] pointing out the fact, that morphological brain damage visualized by CAT scan is partially accompanied by disturbances of cerebral circulation. We observed hyperperfusion on side of focal injury as the more common phenomenon in the early stage after brain trauma. Thereby hyperperfusion pointed out the possible role of haemodynamics for brain swelling and oedema and consecutively for secondary raised intracranial hypertension. On the other hand hemispheric hypoperfusion in focal brain injury may also increase the risk of ischaemic damage. At least both, hypoperfusion and hyperperfusion may play a possible role in prognosis of damaged brain. However, further clinical studies are necessary to confirm experimental data and improve our knowledge of altered cerebral circulation and metabolism. This might give us a better therapeutic approach of head injured patients.

References

1. Bullock R, Mendelow AD, Teasdale GM *et al* (1984) Intracranial haemorrhage induced at arterial pressure in the rat. Part 1. Neurol Res 6: 184–188
2. Bullock R, Statham P, Patterson J *et al* (1989) Tomographic mapping of CBF, CBV, and blood brain barrier changes in humans after focal head injury using SPECT-mechanisms for late deterioration. In: Hoff JT, Betz AL (eds) Intracranial pressure VII. Springer, Berlin Heidelberg New York
3. Miller JD (1985) Head injury and brain ischaemia – implications for therapy. Br J Anaest 57: 120–129
4. Obrist WD, Thomson HK Jr, Wang HS *et al* (1975) Regional cerebral blood flow estimated by 133 Xenon inhalation. Stroke 6: 145–256
5. Risberg J, Ali Z, Wilson EM *et al* (1975) Regional cerebral blood flow by 133 Xenon inhalation. Preliminary evaluation of an initial slope index in patients with unstable flow compartments. Stroke 6: 142–148

Correspondence: J. Meixensberger, M.D., Neurochirurgische Klinik und Poliklinik, Universität Würzburg, Josef-Schneider-Strasse 11, D-8700 Würzburg, Federal Republic of Germany.

Acta Neurochirurgica, Suppl. 51, 305–307 (1990)

Correlations Between Brain Oedema Volume on CT and CSF Dynamics in Severely Head Injured Patients

J. Tjuvajev, J. Eelmäe, M. Kuklane, T. Tomberg, and **A. Tikk**

Tartu University, Estonia, USSR

Summary

The CSF dynamics were studied in 18 patients with severe head injury who remained comatose over 6 hours after trauma (GCS < 8). Amount of brain oedema was estimated by CT tomodensitometry. In addition, CSF parameters of PVI, Elastance (E), compliance (C) and resorption resistance (R) were calculated from serial bolus infusion tests. We observed a decrease in viscoelastic parameters as indexed by PVI, however, no increase in resistance to CSF outflow. It was noted that patients with lower PVI developed more severe brain oedema during the 3–5 day post traumatic period. From these data, we conclude that buffering capacity in severe head injury is mainly affected by the volume of brain oedema and not by the haematoma volume.

Introduction

In several recent studies[4] it was shown that there is a clear relationship between pressure-volume index (PVI) decrease measured soon after injury and subsequent development of intracranial pressure (ICP) elevation and outcome. It was also supposed[3] that the rise of ICP is mediated primary through the vasculature in response to intracranial swelling, but other works[5] do not support this contention. Elevation of ICP as the routine criterium of intensive treatment in head injury reflects only decompensation in volume-pressure relationships in cranio-spinal compartment and thus is not a sufficient criterium of brain oedema dynamical changes during vigorous treatment for ICP control.

Materials and Methods

In this study we investigated the relationships between parameters of CSF dynamics and volume of brain oedema on CT in 18 patients with severe head injury who remained comatose over 6 hours after trauma (GCS < 8). The study was performed during the first 36 hours, a 3–5 day period and 1 month after trauma. Volume of brain oedema was estimated on CT by tomodensitometry (16–22 HU) and volume of oedema per slice of 10 mm (volumetric oedema index "VEI") was calculated from the various slices in each patient. Parameters of CSF dynamics such as PVI, elastance (E), compliance (C), CSF resorption resistance (R), Ayala index (AI) were calculated from data obtained during repetitive 2 ml bolus infusion test described earlier[7]. In other series of 12 patients the level of CSF acidosis was measured additionally.

Results

Average age of patients was 39.5 ± 19.6 years and average GCS score was $5.88 + 01.8$. Preventively dehydrative therapy was supplied before the operation and ICP remained at $10.4 + 6.2$ mmHg. We have found that in acute stage the craniospinal system (CSS) viscoelastic parameters showed a decrease of $PVI = 16.21 + 14.31$ ml, and $C = 0.44 + 0.43$ ml/mmHg; increase of $E = 2.1 + 1.87$ mmHg/ml and no increase of $R = 1.7 + 1.6$ mmHg/ml/min. On CT the summary volume of haematomas (VSum) was $75.2 + 16.37$ ml, midline shift index (MI) $0.87 + 0.12$ and volumetric brain oedema index (VEI = oedema vol./number of CT scans) was $48.1 + 18.9$ ml.

In acute severe head injury the increase of VSum was accompanied with increase of E but there was no significant decrease of PVI or increase of ICP. Brain oedema appeared already 24 hours after injury and grew in size and severity during the 3 5 days post injury. Using the multiplicative model of regression analysis it was found that the volume of brain oedema quantitatively expressed as VEI correlates with E (r = 0.59, p < 0.05) (Fig. 1), with C(inf) (r = − 0.57, p < 0.05), with C(with) (r = − 0.63), with PVI (r = − 0.36, p < 0.01) (Fig. 2) and AI (r = − 0.64, p < 0.05), but there was no correlation with ICP and pulsality amplitude. It was noted, that patients with lower PVI or C and higher E values develop more severe brain oe-

E mmHg/ml

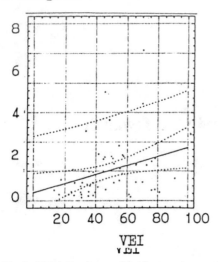

VEI

Fig. 1. Multiplicative model of regression analysis between brain elastance (E) and volumetric oedema index (VEI) r = 0.59, p < 0.05

PVI ml

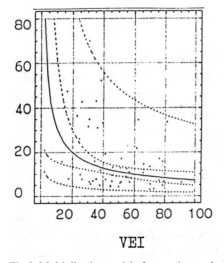

VEI

Fig. 2. Multiplicative model of regression analysis between pressure-volume index (PVI) and volumetric oedema index (VEI) r = − 0.36, p < 0.01

dema during days 3–5 postinjury and need more aggressive treatment. The increase of VEI was reflected as an increase of E and decrease of PVI. In other series of 12 patients a close correlation was found between CSF acidosis and ICP (r = − 0.54, p < 0.05), especially in patients with ICP over 7.0 mmHg (r = − 0.95), and with E (r = − 0.62, p < 0.05). However, there was no significant correlation between CSF acidosis and PVI. Coma was more severe in cases with larger midline

shift, larger haematoma and higher E and lower PVI. There was no significant correlation between midline shift and ICP that shows the compensatory role of midline shift in intracranial volume/ICP system.

Discussion

The alterations in the intracranial volume/pressure system after severe head injury due to brain oedema and intracranial lesions causes the ICP elevation after the exhaustion of the compensatory mechanisms. According to several studies[1, 2, 6] the early aggressive treatment is recommended to control the ICP. However, it is obvious that the management of severe head injury must not be only concentrated on keeping the ICP within normal ranges, but also on treating brain oedema. The ICP monitoring in the compensated state of the intracranial volume/pressure-system does not reveal the dynamic changes of brain oedema.

From the described pathophysiological correlations between the volume of brain oedema (VEI) and the visco-elastic parameters of the craniospinal compartment we conclude that buffering capacity of CSS in severe head injury is mainly affected by the volume of brain oedema and not by the haematoma volume.

The increase of CSF acidosis in traumatic brain oedema was found to correlate with E and correspondently with VEI, a finding that also supports our conclusions. This correlations are especially valid in cases with ICP kept within the normal ranges. Recently it was shown that therapy intensity level rises with the elevations of ICP[1] and that therapies be directed to increasing PVI and reducing ICP. Our data also support this concept. Thus PVI, E, C, and AI must be used as the criteria in treatment of brain oedema not only in head injury.

References

1. Becker DP, Miller JD, Ward JD *et al* (1977) The outcome from severe head injury with early diagnosis and intensive management. J Neurosurg 47: 491–502
2. Frieden HG, Ekstedt J (1983) Volume/pressure relationship of the cerebrospinal space in humans. Neurosurgery 13: 351–366
3. Marmarou A, Maset AL, Ward JD, Sung Choi, Brooks D, Lutz HA, Moulton RJ, Muizellaar JP, DeSalles A, Young HF *et al* (1987) Contribution of CSF and vascular factors to elevation of ICP in severely injured patients. J Neurosurg 66: 883–890
4. Maset A, Marmarou A, Ward JD, Sung Choi, Lutz HA, Brooks D, Moulton RJ, DeSalles A, Muizellaar JP, Turner H, Young HF *et al* (1987) Pressure-volume index in head injury. J Neurosurg 67: 832–840

5. Muizellaar JP, Marmarou A, DeSalles A, Ward JD, Zimmerman RS, Zhongchiao Li, Choi SC, Young HF *et al* (1989) Cerebral blood flow and metabolism in severely head-injured children. J Neurosurg 71: 63–67

6. Saul TG, Ducker TB (1982) Effect of intracranial pressure monitoring and aggressive treatment on mortality in severe head injury. J Neurosurg 56: 498–503

7. Tikk A, Eelmae J (1985) CSF Hydrodynamics studied by means of a simple constant volume injection technique. In: Abstracts of the 6th International Symposium on Intracranial Pressure. ICP and Mechanism of brain damage. June 13th, 1985. Glasgow, Scotland. Glasgow, pp 11–12

Correspondence: J. Tjuvajev, M.D., Tartu University, Estonia, USSR.

Acta Neurochirurgica, Suppl. 51, 308–310 (1990)

Oedema Fluid Formation Within Contused Brain Tissue as a Cause of Medically Uncontrollable Elevation of Intracranial Pressure: The Role of Surgical Therapy

Y. Katayama, T. Tsubokawa, S. Miyazaki, T. Kawamata, and **A. Yoshino**

Department of Neurological Surgery, Nihon University School of Medicine, Tokyo, Japan

Summary

In patients with focal cerebral contusions, medical therapies have generally been advocated unless haemorrhages significantly contributing to the elevated intracranial pressure (ICP) exist. We report here several lines of clinical evidence which indicate that (1) enormous amount of extracellular oedema fluid is formed within contused brain tissue, (2) the formation of extracellular oedema fluid within contused brain tissue alone can be a cause of medically uncontrollable elevation of ICP and (3) surgical excision of the contused brain tissue provides excellent control of the elevated ICP in such patients. The excision of contused brain tissue appears to be the only therapy currently available to alleviate the formation of extracellular oedema fluid in cerebral contusions. We believe that, if ICP is elevated primarily by extracellular oedema due to cerebral contusions and the elevated ICP is medically uncontrollable, surgical excision of contused brain tissue should be carried out without delay regardless of the size of associated haemorrhages.

Introduction

Elevated intracranial pressure (ICP) is a frequent cause of death after head trauma[4]. Medical therapies have generally been advocated for elevated ICP unless haemorrhages significantly contributing to the elevation of ICP exist. Despite intensive medical therapy, however, elevated ICP after head trauma in many patients is uncontrollable. We report here clinical evidence which indicates that elevated ICP in a certain group of patients should be treated surgically rather than medically regardless of the size of haemorrhage. Thus, when extensive extracellular oedema surrounding cerebral contusion is the cause of elevated ICP, surgical excision of the contused brain tissue, which appears to represent removal of the source of extracellular oedema fluid formation, is the only therapy which provides satisfactory control of the elevated ICP.

Patients and Methods

During the last 4 years, we experienced 21 adult patients with frontal lobe contusions who showed medically uncontrollable elevation of ICP (> 40 mm Hg). Patients who developed haemorrhages sufficiently large to consider surgical evacuation (diameter > 3 cm) were excluded from the present study. Characteristics of cerebral contusion were investigated with Glasgow coma scale (GCS), computed tomography (CT) scans, magnetic resonance imagings (MRI), and ICP monitorings. Formation rate of the cerebrospinal fluid (CSF) was determined in 11 patients by CSF volume manipulation and ICP measurements via ventricular catheter employing a method described by Marmarou et al.[3]. The patients were separated into 2 groups; group I patients, treated before 1987 (n = 8) were maintained under intensive, standardized medical therapy and group II patients, treated after 1987 (n = 13) were subjected to surgical excision of contused brain tissue. Outcome was evaluated at 3 months after head trauma. Age, sex, GCS and ICP levels after trauma were comparable between these 2 groups.

Results

Characteristics of Contusion Oedema

All patients investigated in the present study showed delayed and progressive neurological deterioration during the first 24–36 hours post-trauma. None of the patients presented with a GCS score less than 8 on admission. Eighteen of the 21 patients deteriorated to a GCS score less than 8 by 36 hours post-trauma. ICP monitoring was initiated at 8–16 hours post-trauma which generally corresponded to the time when the progressive neurological deterioration became apparent. ICP levels higher than 40 mm Hg were reached within 24–36 hours post-trauma. Following characteristics were observed in these patients.

(1) All of the patients developed contusion haemorrhages which were often not detected by initial CT

scans but were revealed by subsequent CT scans taken 6–12 hours post-trauma. The progressive increase in ICP and neurological deterioration, however, did not coincide with the enlargement of intracerebral high density lesions in many cases investigated in the present study. Thus, the haemorrhage often reached the maximum size within 6–12 hours post-trauma while the ICP continued to rise and the GCS further deteriorated.

(2) CT scans and MRI demonstrated typical characteristics of extracellular oedema. Careful examination of CT scans revealed a progressive increase in oedema surrounding the high density lesions and an increase in the volume of the affected brain areas, although the high density lesions remained unchanged. On MRI scans large amounts of oedema fluid are seen extending along the white matter fiber tracts and reaching the ventricular lining. A frequent finding was the formation of a niveau within the high density lesions suggesting that the blood constituents are floating in a voluminous interstitial oedema fluid. Elevated ICP levels in these cases are therefore most likely to be due to accumulation of extracellular oedema.

(3) The production CSF in this particular group of patients was often markedly elevated. It therefore may be suggested that an enormous amount of extracellular oedema fluid is formed in the contused brain tissue (Table 1).

Effects of Surgical Therapy for Contusion Oedema

Medical therapies chosen in group I patients did not provide satisfactory control of elevated ICP. Osmotic diuretics such as mannitol have a dramatic effect on ICP but this effect is only transient. Ventricular drainage was performed in 6 of the patients and reduced ICP at best by 10 mm Hg. ICP progressively increased often for more than 48 hours despite intensive medical therapies and caused death in 4 patients. Three other

Table 1

	n	CSF formation rate* (ml/min)	
		Mean ± SD	Range
Group I	5	0.52 ± 0.13	0.4–0.7
Group II	6	0.52 ± 0.14	0.4–0.8
Control*	5	0.37 ± 0.12	0.2–0.5

* 36 hours post-trauma. ** Head trauma patients with pathologies other than focal cerebral contusions. Group I + II vs. Control, $p < 0.05$ (unpaired t test).

Table 2

	n	ICP (mmHg)*	Mortality rate (%)
Group I	8	40–72	88
Group II	13	40–60	22

* 36 hours post-trauma. Group I vs. Group II, $p < 0.01$ (χ^2 test).

patients remained comatose for a prolonged period and eventually died from medical complications.

In group II patients, surgical excision of contused brain tissue was performed within 36 hours post-trauma. Excellent control of ICP and recovery in GCS were quickly achieved by surgery in these patients. The mortality rate was 88% in group I patients and 23% in group II patients. Thus, significantly better outcome was obtained by surgical therapy than by medical therapies (χ^2 test, $p < 0.01$, Table 1).

Discussion

Experimental studies have shown that cerebral contusions are associated with increased cerebrovascular permeability which results in extracellular oedema. CT scans and MRI in the present study demonstrated that the patients with cerebral contusions developed oedema which is analogous to extracellular oedema seen experimentally. Marmarou et al.[3] have demonstrated that CSF formation rate exceeding 0.74 ml/min is observed in some of their severely injured head trauma patients. They assumed that the high rate of CSF formation in such patients represents oedema clearance through the CSF system, as observed experimentally[5]. The present study confirms these observations and further indicates that the high rate of CSF formation is frequently seen in patients with cerebral contusion. Thus, the increase in CSF formation rate in these patients is consistent with findings in CT scans and MRI which suggest production of enormous amount of extracellular oedema fluid within the contused brain tissue.

In earlier CT studies, besides epidural or subdural haematomas, contusion haemorrhages with delayed progressive enlargement after the trauma were emphasized as a major cause of delayed neurological deterioration[2]. The present study in addition has demonstrated that the increase in ICP and the neurological deterioration may even continue after the intracerebral high density lesions have reached their maximum size. This suggests that formation of considerable amounts of oedema fluid within the contused brain tissue may play a decisive role in producing medically uncontrollable elevation of ICP in such patients. The high density lesions in these patients may partly be a result of diapedesis of blood cells from cerebral vessels with increased permeability and therefore represent an epiphenomenon of the accumulation of extracellular oedema fluid which is also caused by increased cerebrovascular permeability.

There is no established medical treatment which effectively inhibits increased cerebrovascular permeability within contused brain tissue. The present study demonstrates that an effective therapy to ameliorate this potentially fatal oedema is surgical excision of the contused brain tissue. Part of this effect probably can be explained by the reduction in intracranial volume. The main effect of the excision of contused brain tissue, however, is the removal of the source of extracellular oedema fluid formation. Arabi and Long[1] have demonstrated that removal of a cold lesion prevents production and propagation of extracellular oedema in experimental animals. These data indicate that, if ICP in this group of patients becomes medically uncontrollable, surgical excision of contused brain tissue should be considered.

An elevated ICP is the most frequent cause of ischaemia following brain trauma. Traumatized brain is more vulnerable to secondary ischaemic insults than normal brain. In order to control increased ICP, the most appropriate therapy for each pathology should be chosen and uniform medical therapies disregarding the underlying mechanisms must be avoided. Based on the results of the present study, we believe that surgical excision of contused brain tissue should be carried out without delay, if ICP is elevated primarily by extracellular oedema due to cerebral contusion and becomes medically uncontrollable.

References

1. Arabi B, Long DM (1979) Dynamics of cerebral oedema: The role of intact vascular bed in the production and propagation of vasogenic brain oedema. J Neurosurg 51: 779–784
2. Katayama Y, Tsubokawa T, Miyazaki S (1989) Two types of delayed traumatic intracerebral haematomas: Differential forms of treatment. Neurosurg Rev 12 [Suppl]: 231–236
3. Marmarou A, Maset AL, Ward JD, Choi S, Brooks D, Lutz H, Moulton RJ, Muizalaar JP, DeSalles A, Young HF (1987) Contribution of CSF and vascular factors to elevation of ICP in severe head-injured patients. J Neurosurg 66: 883–890
4. Miller JD, Becker DP, Ward JD, Sullivan HG, Adams WE, Rosner MJ (1977) Significance of intracerebral hypertension in severe head injury. J Neurosurg 47: 503–516
5. Reulen HJ, Tsuyumu M, Tack A, Fenske AR, Prioleau GR (1978) Clearance of oedema fluid into cerebrospinal fluid. J Neurosurg 48: 754–764

Correspondence: Y. Katayama, M.D., Department of Neurological Surgery, Nihon University School of Medicine, Tokyo 173, Japan.

Acta Neurochirurgica, Suppl. 51, 311–314 (1990)

Magnetic Resonance Imaging and Neurobehavioural Outcome in Traumatic Brain Injury

J. C. Godersky, L. R. Gentry, D. Tranel, G. N. Dyste, and **K. R. Danks**

Division of Neurosurgery, Department of Surgery, University of Iowa College of Medicine, C42 GH, University of Iowa Hospitals, Iowa City, Iowa, U.S.A.

Summary

Magnetic resonance (MR) imaging is a sensitive means of detecting haemorrhagic and nonhaemorrhagic forms of brain injury. This study correlates the neurobehavioural (NB) deficits in 49 adult patients with lesions detected by MR imaging. MR imaging was performed 2–19 days following trauma, analyzed for the injury type and graded for severity. A battery of NB tests was performed prior to hospital discharge or at the time of initial follow-up visit (31 patients). 15 patients were so severely impaired that testing could not be done and 3 died prior to discharge. The NB test scores were grouped into 3 levels of impairment.

The overall NB scores were compared with MR lesion severity ratings and a positive correlation found ($r = 0.47$). In addition, lesion severity, type and location resulted in specific NB deficits. We conclude that the lesion location and severity can be accurately identified by acute phase MR are associated with specific types of neurobehavioural deficits in a high percentage of testable patients.

Introduction

Outcome following head injury is dependent upon multiple factors. These include the nature and severity of the trauma, the existence of secondary insults (hypoxia, hypotension, seizures, and elevated intracranial pressure) and patient age[6]. Lesion appearance on Computed Tomographic scans (CT) has been used to predict neuropsychological outcome from brain injury[7]. Although CT is invaluable in head injury management by its ability to delineate sites of intracranial haemorrhage[5], the presence of mass effect and midline shift, it is relatively insensitive in identifying areas of non-haemorrhagic brain injury[2]. Autopsy data indicate that axonal damage (diffuse axonal injury – DAI) is a frequent accompaniment of head injury[1]. DAI is often nonhaemorrhagic and difficult to visualize by CT[3]. Magnetic resonance imaging (MR) is superior to CT in its ability to recognize DAI, nonhaemorrhagic areas

of cortical contusion (CC) and subcortical gray matter injury (SCG). In this study we correlated neurobehavioural outcome with lesions identified by MR in 49 patients with acute brain injuries.

Methods

From July 1984 to January 1988, adult patients with head injury were selected for study if they had MR imaging performed during the acute phase of their illness and underwent neurobehavioural evaluation or were too severely affected to undergo testing prior to discharge. The specifics of the MR imaging sequences have been described[3]. Data were recorded from chart review, including initial Glasgow coma scale (GCS) and Glasgow outcome scale (GOS) at last follow-up. MR studies were reviewed and lesions analyzed for the presence of diffuse axonal injury (DAI), cortical contusion (CC), and subcortical gray matter injury (SCG). These were each graded on a scale 1–3 (Table 1). Neurobehaviour evaluation included assessment of intellect, memory, orientation, language, speech, visuospatial perception and construction, executive control and personality. These were also graded on a 0–3 scale and an overall score calculated. Those individuals who had died or that were so severely affected as to not be testable were placed into Grade 3 for data analysis.

Results

The study population consisted of 40 males and 9 females ranging in age from 14–74 years, mean 28. Admission GCS ranged from 4–15 with a mean of 9. MR was obtained when the patients were clinically stable, with a mean delay of 7 days (range 2–19 days) following injury.

In order to determine if there was a relationship between the injury severity and NB test scores, the patients were divided into three categories based on the initial GCS. Group 1 consisted of those with GCS of 3–8, group 2 – GCS of 9–11, and group 3, GCS of

Table 1. *Magnetic Resonance Grading Scale*

DAI

1. 1 or 2 lobes, subcortical white matter (SCW) only, 1 to 4 lesions
2. 3 or 4 lobes, SCW, corpus callosum (CC), >5 lesions or >5 mm in size to 10 mm
3. 5 or more lobes, SCW, CC, brain stem (BS), lesions >10 mm mass effect, cistern compressions

Cortical Contusion

1. 1 or 2 lobes, cortex only, <2 cm greatest dimension, 1–4 lesions
2. 3–4 lobes, cortex + SCW, 2–4 cm greatest dimension, >5 lesions
3. 5 or more lobes, cortex + SCW, >4 cm greatest dimension, mass effect cistern compression

Subcortical Gray Matter Lesions

2. 0–9 mm, no mass effect, unilateral
3. >10 mm, mass effect, bilateral

Neurobehaviour Evaluation and Scoring

Intellectual Functioning	Verbal
	Non-verbal
Memory	Verbal
	Non-verbal
Speech	
Language	
Visuo-spatial	Perception
	Construction
Orientation	
Awareness of deficits	
Executive control	
Personality	

Scoring for each neurobehaviour category:

  0 = Normal
  1 = Mild deficit
  2 = Moderate deficit
  3 = Severe deficit

Overall neurobehaviour score obtained by adding each category:

  Grade 1 = 0–10
  Grade 2 = 11–20
  Grade 3 = >20

12–15. The mean values of the overall NB score and GOS for each GCS group were then compared by analysis of variance. Group 1 differed significantly from groups 2 and 3 ($p = 0.001$). There was no difference when comparing group 2 to group 3. We also compared the additive MRI score with the overall NB score for each patient by regression analysis and found a positive correlation ($r = 0.47$). This correlation was felt to support the validity of the MRI severity scale.

The primary aim of the study was to determine if the lesions identified on MR caused specific deficits in neurological function as assessed by neurobehavioural testing. Of the 49 patients, 31 were testable with a variety of instruments assessing the functions listed in Table 2. The lesions seen on MR were classified as to type and location within the brain and severity (Table 1). Lesions consistent with DAI involved the frontal or temporal (F-TP) lobes, parietal-occipital (P-O) regions or corpus callosum (CC). Cortical contusions were studied as a whole and subdivided by lobe, while the subcortical gray (SCG) matter lesions were evaluated as a single group. Table 2 lists the frequency of neurobehavioural deficits (moderate or severe in degree) for each lesion type and location. All associations above 50% are highlighted. Speech and language defects were found in only 26 to 33% of patients. This lower frequency probably reflects the lack of separation of lesions into those in the right or left hemisphere.

Table 2. *Percent of Moderate to Severe NB Deficits in Specific MRI Lesions*

	F-TP DAI		Corpus callosum DAI		P-O DAI		All cortical cont		Frontal CC		Temporal CC		P-O CC		SCG	
Intellectual functioning (Verbal)	9/19	47%	8/17	47%			13/25	52%							3/3	100%
Intellectual functioning (Nonverbal)	13/19	68%	13/17	76%			15/25	60%							3/3	100%
Memory (Verbal)	14/19	74%	12/17	71%					13/17	76%	14/18	78%			3/3	100%
Memory (Non-verbal)	15/19	79%	13/17	76%					11/17	65%	14/18	78%			3/3	100%
Visuo-spatial Perception					6/8	75%							4/5	80%		
Visuo-spatial Construction					5/8	63%							4/5	80%		
Orientation	9/19	47%	8/17	47%			19/25	76%								
Awareness of Deficits			6/17	35%					9/17	52%						
Personality	13/19	68%	11/17	65%												
Executive Control	14/19	74%	13/17	76%					16/17	94%			3/5	60%		

Discussion

Magnetic resonance provides a sensitive method of detecting areas of brain injury. MR is superior to computed tomography (CT) in identifying nonhaemorrhagic lesion and equal to CT in detecting lesions containing haemorrhage[2]. A number of investigations have been done comparing CT lesion type, location and severity with outcome[5, 7]. More recently, additional studies have demonstrated neurological deficits in association with trauma evident on MR[4, 8]. This study provides further information in this area by correlation of lesion type (DAI, CC, SCG injuries), location and severity with deficits in neurological functions that are mediated by specific brain regions. Examples include impairments in verbal and nonverbal memory in 74% and 79% of those patients with frontotemporal DAI and 65–78% of those with frontal or temporal contusions. Visuo-spatial defects were present in 63–80% of those with parietal-occipital lesions attributed to DAI or cortical contusion. The results are displayed in percentages because the small number of patients does not allow satisfactory statistical analysis. We chose to assess neurobehaviour impairments that were moderate or severe and therefore more likely to interfere with the individuals function.

This study involved MR abnormalities detected during the acute phase of trauma. Our findings that these lesions are associated with neurological impairment in a large percentage of patients disagrees with Wilson *et al.*[8] who concluded that only abnormalities identified in the chronic phase correlate with NB impairments.

Our classification of MR lesion severity was based on increasing lesion size, depth of brain involvement and multiplicity of abnormalities. A positive correlation was found between this scale and NB dysfunction. Wilson *et al.*[8] found a correlation between depth of lesions within the brain on MR and persistence and severity of NB defects. We concur with this observation and our MR grading scale reflects this observation. All individuals with subcortical gray matter lesions, on example of deep brain involvement, had impairment in memory and intellectual function as did a high percentage (71–76%) of those with corpus callosum injury.

References

1. Adams JH, Graham DI, Parker LS, Doyle D (1980) Brain damage in fatal non-missile head injury. J Clin Pathol 33: 1132–1145

2. Gentry LR, Godersky JC, Thompson B, Dunn VD (1988) Prospective comparative study of intermediate field MR and CT in the evaluation of closed head trauma. AJNR 9: 91–100

3. Gentry LR, Thompson B, Godersky JC (1988) Trauma to the corpus callosum: MR features. AJNR 9: 1129–1138

4. Levin HS, Handel SF, Goldman AM, Eisenberg HM, Guinto FC (1985) Magnetic resonance imaging after "diffuse" nonmissile head injury. Arch Neurol 42: 963–968

5. Lobato RD, Cordobes F, Rivas JJ *et al* (1983) Outcome from severe head injury related to the type of intracranial lesion. J Neurosurg 59: 762–774

6. Saul TG, Ducker TB (1982) Effect of intracranial pressure monitoring and aggressive treatment on mortality in severe head injury. J Neurosurg 56: 498–503

7. Uzzell BP, Dolinskas CA, Wiser RF, Langfitt TW (1987) Influence of lesions detected by computed tomography on outcome and neuropsychological recovery after severe head injury. Neurosurg 20: 396–402

8. Wilson JTL, Wiedemann KD, Hadley DM, Condon B, Teasdale G, Brooks DN (1988) Early and late magnetic resonance imaging and neuropsychological outcome after head injury. J Neurol Neurosurg Psychiatry 51: 391–396

Correspondence: J. C. Godersky, M.D., Division of Neurosurgery, Department of Surgery, University of Iowa, College of Medicine, C42 GH, Iowa City, IA 52242, U.S.A.

Acta Neurochirurgica, Suppl. 51, 315–316 (1990)
© by Springer-Verlag 1990

The Effect of Nimodipine on Outcome After Head Injury: A Prospective Randomised Control Trial

The British/Finnish Co-operative Head Injury Trial Group*: G. Teasdale[1], I. Bailey[2], A. Bell[3], J. Gray[2], R. Gullan[4], U. Heiskanan[5], P. V. Marks[6], H. Marsh[3], A. D. Mendelow[7], G. Murray[8], J. Ohman[5], G. Quaghebeur[4], J. Sinar[7], A. Skene[9], and A. Waters[6]

[1] Department of Neurosurgery, Southern General Hospital, [2] Department of Neurosurgery, Royal Victoria Hospital, Belfast, [3] Department of Neurosurgery, Atkinson Morley's Hospital, London, [4] Regional Neurosurgical Unit, Brook General Hospital, London, [5] University of Helsinki, [6] Department of Neurosurgery, Addenbrooke's Hospital, Cambridge, [7] Department of Neurosurgery, Regional Neurological Centre, Newcastle Upon Tyne, [8] Department of Surgery, Western Infirmary, Glasgow, [9] Clinical Trial Data Centre, University of Nottingham

Summary

To study the effect of nimodipine on the outcome of head injury, three hundred and fifty-two patients who were not obeying commands were randomised to placebo or nimodipine (2 mg per hour intravenously for 7 days). The 2 groups were well matched for important prognostic features. Six months after injury, more of the patients who were given nimodipine had a favourable outcome (moderate/good recovery) than in the control group, but the increase in favourable outcome (8%) was not significant statistically.

Introduction

The calcium antagonist nimodipine has been shown to reduce brain damage in experimental models of cerebral ischaemia[4, 7] and haemorrhage[6]. Clinically it is effective in the prevention of ischaemia after subarachnoid haemorrhage[15] and improves outcome from occlusive stroke[1]. Ischaemic brain damage is found in the majority of fatal head injuries[2] and is the most common mechanism for secondary brain damage. A preliminary report suggested that nimodipine might benefit head injured patients through relieving traumatic vasospasm[3]. We have therefore carried out a randomised prospective study to discover the value of nimodipine treatment of severely head injured patients.

Methods

The study was performed in neurosurgical departments in Belfast, Cambridge, Helsinki, London (Atkinson Morley Hospital, Brook Hospital), and Newcastle. Adult patients with a non-missile head injury were entered into the study if they did not obey simple commands within the first 24 hours of injury. Patients were excluded if they were haemodynamically unstable, clinically brain dead, pregnant or had renal, hepatic, pulmonary or cardiac decompensation. Informed consent was obtained from a relative. Outcome was assessed 6 months after injury using the Glasgow Outcome Scale.

Treatment

Nimodipine or matching placebo were given intravenously. The initial dose was 1 mg per hour increased to 2 mg if the blood pressure did not decline. Treatment was continued for 7 days.

Results

The groups randomised to nimodipine or placebo were well balanced for major prognostic factors (Table 1). Eighty three per cent of patients were admitted to the Neurosurgical Units within 6 hours of injury and 68% were entered into the study within 12 hours.

Six months after injury a favourable outcome (moderate/good recovery) had occurred in 52% of patients given nimodipine compared with 49% in the placebo group. This 8% relative increase in a favourable outcome (95% confidence intervals + 33% − 12%) was not statistically significant. The number of deaths were very close in the two groups, and in 94% death was judged to be due to a direct effect of the injury.

Discussion

The results of this first randomised prospective controlled trial of nimodipine in head injured patients show

* The study was performed in co-operation with Bayer UK Ltd. and Bayer Sverige AB.

Table 1. *Features of Patients Analyzed*

	Patients given nimodipine	Patients given placebo
Total	176	175
Age mean (years)	35.4	35.5
Male	155 (88%)	140 (80%)
Cause of injury		
Road traffic accident	99 (56%)	102 (58%)
Fall under influence of alcohol	28 (16%)	28 (16%)
Skull fracture	98 (55%)	98 (55%)
CT scan findings		
Contusions	102 (58%)	110 (63%)
Intracranial haematoma	88 (50%)	87 (50%)
Raised ICP without haematoma	20 (11%)	25 (14%)

a modest, statistically insignificant benefit. Our data, therefore, do not provide an indication to use nimodipine in head injured patients. Instead, further trials are needed, with a larger sample size, in order to determine if the modest benefit observed in this stage is consistent and clinically useful.

References

1. Gelmers HJ, Gorter K, de Weerdt JD *et al* (1988) A controlled trial of nimodipine in acute ischaemic stroke. New Engl J Med 318: 203–207
2. Graham DI, Ford I, Hume-Adams J *et al* (1989) Ischaemic brain damage is still common in fatal non-missile head injury. J Neurol Neurosurg Psychiatry 62: 346–350
3. Kostron H, Twerdy K, Stampfl G *et al* (1984) Treatment of the traumatic cerebral vasospasm with the calcium channel blocker nimodipine: A preliminary report. Neurol Res 6: 29–31
4. Mohamed AA, Gotoh O, Graham DI *et al* (1985) Effect of pretreatment with the calcium antagonist nimodipine on local cerebral blood flow and histopathology after middle cerebral artery occlusion. Ann Neurol 18 (No 6): 705–711
5. Pickard JD, Murray GD, Illingworth R *et al* (1989) Effect of oral nimodipine on the incidence of cerebral infarction and outcome at three months following subarachnoid haemorrhage: British Aneurysm Nimodipine Trial (BRANT). J Neurol Neurosurg Psychiatry 52: 140
6. Sinar EJ, Mendelow AD, Graham DI *et al* (1988) Intracerebral haemorrhage: the effect of pretreatment with nimodipine. J Neurol Neurosurg Psychiatry 51: 651–662
7. Steen PA, Gisvold SE, Milde JH *et al* (1985) Nimodipine improves outcome when given after complete ischaemia in primates. Anaesthesiology 62: 406–414

Correspondence: G. Teasdale, M.D., Department of Neurosurgery, Institute of Neurological Sciences, The Southern General Hospital, Glasgow G51 4TF, Scotland.

Acta Neurochirurgica, Suppl. 51, 317–319 (1990)

Effect of THAM on Brain Oedema in Experimental Brain Injury*

K. Yoshida, F. Corwin, and **A. Marmarou**

Division of Neurosurgery and Radiation Physics, Medical College of Virginia, Richmond, Virginia, U.S.A.

Summary

The metabolic brain acidosis after trauma has been thought to increase brain oedema and contribute to neurologic deterioration. Amelioration of the brain acidosis either by systemic buffering agents or by hyperventilation has been proposed as a method of treatment. The objective of this study was to explore brain oedema and the metabolic changes in brain that occur with the use of hyperventilation, Tromethamine and combination (THAM and hyperventilation) therapy in experimental fluid-percussion brain injury. Brain lactate, brain pH, inorganic phosphate (Pi) and ATP were measured by ^1H and ^{31}P magnetic resonance spectroscopy. Also Water content in brain tissue using the specific gravimetric technique were determined in 32 cats. Prolonged hyperventilation provided relative ischaemia in brain tissue and promoted more production of brain lactate, no recovery of PCr/Pi ratio, and no decrease in brain oedema. On the other hand the administration of THAM served to decrease production of brain lactate and brain oedema and promoted the recovery of cerebral energy dysfunction. THAM ameliorates the deleterious effects of hyperventilation by minimizing energy disturbance and also decreases brain oedema. We conclude that THAM may be effective in reducing brain tissue acidosis and helpful as a metabolic stabilizing agent following severe head injury.

Introduction

Clinical studies of severely head injured patients have revealed disturbances in energy metabolism as evidenced by elevated levels of lactate in cerebrospinal fluid (CSF). Many investigators have interpreted these increased lactate levels to represent a condition of brain tissue acidosis which may be deleterious to neurologic recovery. Seitz and Ocker[1] related the degree of CSF lactic acidosis to brain oedema and believed that this acidosis was at least partly causal in development of brain swelling. Amelioration of the brain acidosis either by systemic buffering agents or hyperventilation has

been proposed as a method of treatment. The objective of this study was to explore the metabolic changes and oedema in brain that occur with the use of hyperventilation, THAM (tromethamine; tris [hydroxymethyl] aminomethane) and combination (THAM and hyperventilation) therapy in experimental fluid-percussion injury.

Materials and Methods

Thirty two cats were anaesthetized with intravenous methohexical sodium (1–2 mg/kg) and maintained with intravenous a-chloralose (40 mg/kg initially then 20 mg/kg after 8 hours). They were tracheotomized and mechanically ventilated. Muscle paralysis was then obtained with intravenous pancronium bromide (1 mg/kg). Catheters were placed in the femoral artery for pressure monitoring and blood sample withdrawal and in the femoral vein for drug infusion. Two coils placed for magnetic resonance studies in both hemispheres, one for hydrogen and one for phosphate spectroscopy. After surgical preparation, control MRS measurements of brain lactate and brain pH, phosphocreatine (PCr) and inorganic phosphate (Pi) were taken. After control measurement of energy metabolism, the animals were subjected to a left lateral fluid-percussion trauma via an opening in the skull. The animals were then randomized to the one of four groups; 1) THAM treatment group (n = 11), 2) Hyperventilation treatment group (n = 7), and 3) THAM and Hyperventilation treatment group (n = 7). 4) Untreated group (n = 4). 15 minutes after trauma, the initial dose of THAM calculated using arterial pH, PCO_2, and HCO_{3-}, was injected from the femoral vein. Following this, a maintenance dose (1 cc/kg) was injected every hour until seven hours after trauma. Hyperventilation started 15 minutes after head trauma. The rate and volume of the respirator were adjusted to give an end-expiratory pCO_2 of 20 mm Hg during hyperventilation. After injury, the animals returned to the magnet and followed for 8 hours after trauma. At sacrifice brains were removed for measurement of oedema. The measurement of water content in brain was obtained from sham operated animals.

Calculation of THAM Dose

THAM was administered intravenously as a 0.3 molar solution. Calculations of dosage are based on the amount of base required to

---

* This work was supported in part by grants NS 12587 and NS 19235 from the National Institute of Health. Additional facilities and support were provided by the Richard Roland Reynolds Neurosurgical Research Laboratories.

Table 1. *The Water Content in Cat Brain 8 Hours After Trauma*

	Left			Right		
	Brain			Brain		
	white	stem	gray	white	stem	gray
Sham	68.927** ±0.073	70.169 ±0.112	80.322** ±0.135	69.067** ±0.040	70.241 ±0.103	80.416** ±0.083
Untreated	69.836 ±0.194	70.656 ±0.231	81.223 ±0.044	69.736 ±0.191	70.505 ±0.219	81.254 ±0.062
THAM	69.065** ±0.104	70.368 ±0.151	80.643** ±0.074	69.084* ±0.144	70.017 ±0.112	80.624** ±0.059
Hypervent	69.860 ±0.224	70.214 ±0.127	81.190 ±0.136	69.873 ±0.175	70.507 ±0.111	81.158 ±0.073
Combination	69.626 ±0.190	70.238 ±0.171	80.624** ±0.154	69.013** ±0.080	70.266 ±0.203	80.560** ±0.125

* $p < 0.05$, ** $p < 0.01$ compared with untreated.
All data are mean ± SEM % water.

raise pH to 7.6 (using a Henderson-Hasselbach normogram). An empirical formula used for calculation of the dosage tromethamine is given by:

(1) THAM ml = body weight (kg) × base deficit

Results

Magnetic Resonance Spectroscopy

Following injury, the PCr/Pi ratio, which is an index of cerebral energy depletion decreased to 76% in untreated animals, 79% with THAM, 68% with hyperventilation, and 66% with combination THAM and hyperventilation. The PCr/Pi returned to a normal level in 8 hours in animals treated with THAM and THAM in combination with hyperventilation. The brain lactate index increased to 157% in the hyperventilation group after trauma. THAM plus hyperventilation reduced BLI to 142%, while the minimum rise of 126% was associated with treatment by THAM alone. Despite this lactate production, brain pH varied within a narrow range in each group during the experiment period.

Water Content in the Brain

The water content of the cat brain measured by the gravimetric technique is shown in Table 1. In the THAM treatment and combination treatment groups, the water content of white matter and gray matter was significantly decreased compared with untreated injured cat brains ($p < 0.01$). Also, water content in the brain stem region tended to decrease with THAM or

combination therapy. However, the reduction in brain stem did not reach statistical levels of significance.

Discussion

THAM is an amine alkalinizing agent, and since 30% of THAM is in the non-ionized form, it can cross the plasmatic membrane and directly effect intracellular acidosis[2]. This study shows that lactate production in the experimentally traumatized cat brain is ameliorated when animals are treated immediately post injury with THAM. Controlled hyperventilation has been recommended as a potentially useful therapeutic method in head injured patients and is used routinely. However, our clinical data indicated that sustained hyperventilation to $PaCO_2$ of less than 24 for 5 days is deleterious and results in poorer outcome when compared with standard therapy[3]. Also our results indicate that maximal lactate production and minimal recovery of oxidative stores (PCr/Pi) was associated with those animals treated with sustained hyperventilation following their injury because hyperventilation provided relative ischaemia. On the other hand, PCr/Pi ratios of hyperventilated animals treated with THAM return to control levels.

THAM is eliminated in its ionized form with an equimolar amount of bicarbonate, thus acting at least partially as an osmotic diuretic[4]. This offers one explanation of our finding that THAM therapy alone or in combination with hyperventilation resulted in a small but significant reduction in brain oedema following traumatic injury. It is also possible that THAM prevented the development of brain oedema by improving the extracellular milieu. Our experiments cannot differentiate between the mechanisms. Nevertheless, this reduction occurred in the cortex and white matter of both left and right hemispheres and is consistent with work by Gaab who also found that THAM was effective in reducing brain oedema[5]. These data provide one explanation for the deleterious effect of sustained hyperventilation and the amelioration of these effects when treated with THAM. Hyperventilation should be used sparingly and if sustained hyperventilation is necessary for ICP control, THAM should be administered.

References

1. Seitz HD, Ocker K (1977) The prognostic and therapeutic importance of changes in the CSF during the acute stage of brain injury. Acta Neurologica 38: 211–231

2. Rosner MJ, Becker DP (1984) Experimental brain injury: Successful therapy with the weak base, tromethamine with an overview of CNS acidosis. J Neurosurg 60: 961–971

3. Ward JD, Choi S, Marmarou A *et al* (1989) Effect of prophylactic hyperventilation on outcome in patients with severe head injury. In: Hoff JT (ed) Intracranial pressure VII. Springer, Berlin Heidelberg New York, pp 630–633

4. Goetz RH, Selmonosky A, State D (1961) Anuria of hypovolemic shock relieved by tris (hydroxymethyl) aminomethane (THAM). Surg Gynecol Obstet 112: 724–728

5. Gaab M, Knoblich OE, Spohr A *et al* (1980) Effect of THAM on ICP, EEG and tissue oedema parameters in experimental and clinical brain oedema. In: Shulman K, Marmarou A, Miller JD, Becker DP, Hochwald GM, Brock M (eds) Intracranial pressure IV. Springer, Berlin Heidelberg New York, pp 85–87

Correspondence: A. Marmarou, Ph.D., Medical College of Virginia, Division of Neurosurgery, MCV Station, Box 508, Richmond, VA 23298, U.S.A.

Acta Neurochirurgica, Suppl. 51, 320–323 (1990)

A Comparative Analysis of THAM (Tris-buffer) in Traumatic Brain Oedema

M. R. Gaab, K. Seegers, R. J. Smedema, H. E. Heissler, and **Ch. Goetz**

Neurosurgical Department, Hannover Medical School, Hannover, Federal Republic of Germany

Summary

The effect of THAM on brain oedema parameters was initially investigated in animals with cold brain lesion; THAM was then used in head injury patients, ICP, SAP and CPP were analyzed. In the experiments with rats after freezing lesion, THAM was compared to equivalent doses of Na-bicarbonate. The animals were artificially respirated and sacrificed 6 h after trauma. THAM did significantly reduce water (wet-dry weight technique) and sodium contents in both hemispheres, whereas bicarbonate was ineffective. The potassium contents were even preserved at almost normal levels.

In 80 patients receiving alternatively THAM (18–36 g/ 100–200 ml/1–2 h), mannitol (20%, 125–250 ml/20–40 min) or sorbitol (40%, 70–140 ml/20–40 min), the ICP rapidly decreased following THAM infusion. The maximal fall in ICP (33%) was equal to that with mannitol and sorbitol. The slope of ICP decrease was equal with THAM and Mannitol but steeper with sorbitol. With THAM, however, the effect on ICP lasts longer than with osmotherapy. The EEG improved more rapidly after THAM. As shown by blood plasma values, the action of THAM is not based on osmotic effects. The increases in pH and especially in base excess suggest an intracerebral buffering. The encouraging results with THAM require a randomized clinical trial after severe head injury which is presently prepared.

Introduction

Already in 1962 Dos and Nahas[3] described a decrease in ICP following administration of THAM (Tris-buffer, generic name trometamol[11]) in animal experiments. In 1976, Akioka and coworkers described a decrease in ICP with THAM infusion in patients after severe head injury[1]. Our group found a rapid decrease in ICP and a stabilization of cerebral perfusion pressure (CPP) in cats after cold lesion[5, 9, 10]; the effect on the EEG in these animals was better than with osmotherapy[5, 9, 10], and also the survival time[6] was significantly longer than with osmotherapy. In first clinical observations, a decrease in ICP and an EEG improvement was also seen in patients with traumatic brain oedema[6, 9]. However, these reports remained sporadic, and besides encouraging experiences THAM has not yet systematically been investigated.

Material, Patients and Methods

Animal Experiments

Male SP-rats with 250–300 g b.wt. were anaesthetized with 6–7 mg ketamine (Ketanest®) i.p., and a cold lesion was induced in the right hemisphere by placing a 5×6 mm freezing probe on the intact skull with a stereotactic apparatus[7, 8]. Freezing time was 3 min with $-73\,°C$[7, 8]. Due to the breath-depressing effect of THAM[11], all animals were continuously respirated with 70% N_2O and 30% O_2 using the Brackebusch respirator (simultaneous ventilation of 4 rats[2]). In a randomized manner, animals remained untreated (oedema, no fluid), received infusion mixture (0.45%, NaCl 2.5% glucose) as vehicle (oedema, vehicle) or were treated with 0.33 mmol Na-bicarbonate/ml or with 0.66 mmol THAM/ml i.v. (equivalent buffering capacity[10]). In addition, animals without trauma with respiration ("respir., no oedema") or breathing spontaneously ("controls") were investigated. In animals receiving infusion, 1 ml bolus was injected 5 min after injury, followed by a constant-rate infusion (Perfusor®) with 5 ml/6 h. Six hours after injury, the animals were decapitated, and water, sodium and potassium were measured separately in each hemisphere wet-dry weight technique; flame photometry[7, 8].

Clinical Investigation

21 patients with intracranial hypertension (ICP $\geqslant 25$ mm Hg) due to traumatic brain swelling were alternately treated with THAM, Mannitol or sorbitol. ICP was recorded with epidural GAELTEC sensor, systemic blood pressure (SAP) with temporal artery catheter[5]. Data were registered with a computerized bedside Neuromonitor[5], which in addition allowed a continuous record of the EEG which was analyzed via FFT (CSA, chronospectra[5]). Blood osmolality, pH, pCO_2 and base excess were measured at 15 min intervals (Fig. 2).

By alternating therapy 202 drug dosages were evaluated: 80 with THAM i.v., 18–36 g/100–200 ml/1–2 h; 82 with 20% mannitol (125–250 ml, 20–40 min) and 40 infusions of 40% sorbitol (70–140 ml, 20–40 min). Then we calculated (Fig. 1):

- maximal rel. decrease in ICP (dICP)
- maximal rel. decrease in SAP (dSAP)
- time of maximal ICP decrease after start of therapy (tICP, min)

Table 1. *Edema Parameters in Rats with Cold Lesion, Therapy with THAM versus Bicarbonate.* Water and sodium uptake significantly less with THAM, potassium almost normalized

	Water content (%)		sodium (mmol/kg dry weight)		potassium	
	right	left	right	left	right	left
Controls	76.97 ±0.92	77.01 ±0.82	158 ±13	156 ±12	355 ±20	359 ±15
Respirated, no oedema	76.81 ±1.14	76.84 ±0.98	156 ±27	167 ±18	354 ±35	348 ±10
Oedema, no fluid	79.18 ±0.62	77.71 ±0.74	261 ±36	196 ±33	331 ±26	346 ±23
Oedema, vehicle	79.56 ±1.00	77.92 ±0.83	292 ±46	239 ±24	320 ±29	370 ±24
Sodiumbicarbonat	79.14 ±1.04	77.66 ±1.14	258 ±52	173 ±40	331 ±28	357 ±28
THAM	78.18 ±0.74	77.27 ±0.70	201 ±35	164 ±20	345 ±31	353 ±18

   – slope of ICP decrease (vICP, mm Hg/min)

   – duration of ICP decrease until reaching again 50% of pre-infusion value (tdec, min)

   The statistical analysis was done with SPSS-X, Vers. 3.3.

Results

Animal Experiments

6 h after cold injury, a significant increase in water and sodium content is seen in both hemispheres (Table 1), much more pronounced on the traumatized side. Po-

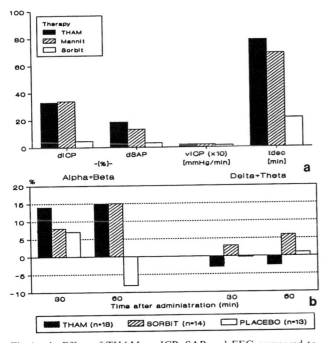

Fig. 1 a, b. Effect of THAM on ICP, SAP and EEG compared to osmotherapy. THAM at least as effective on ICP and SAP-dICP max. rel. fall in ICP, dSAP max. rel. decrease in SAP, vICP slope of ICP decrease (mmHg/min), tdec duration of ICP decrease (min) (a). EEG improvement better with THAM (b)

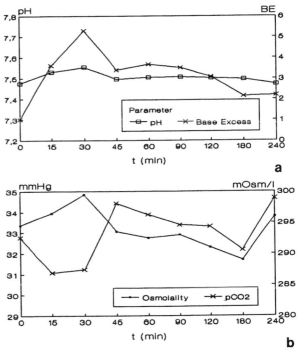

Fig. 2 a, b. Effect of THAM on blood parameters: Significant increase in pH and base excess

tassium levels are significantly decreased ipsilaterally (Table 1). All values are worse in animals receiving vehicle than in rats without any fluid administration after trauma (Table 1).

Bicarbonate infusion does not significantly improve water and sodium contents compared to these two control groups; only the potassium content of the traumatized hemisphere remains higher than with vehicle infusion alone but is not different to that in animals with fluid restriction.

With THAM treatment, however, a considerable and highly significant improvement of all oedema parameters is seen: Water and sodium contents remain much lower, and the potassium content is even preserved at an almost normal level even in the traumatized hemisphere (Table 1).

Clinical Investigation

In all patients with the exception of one (male, 18 years) with very high ICP and a CPP below 30 mm Hg THAM did rapidly lower the ICP. In this patient also mannitol and sorbitol remained ineffective, suggesting a disturbed cerebral circulation. In the other patients, the mean maximal decrease in ICP was 33.2% (Fig. 1), it was achieved 59 min after start of therapy and had a mean duration of 79 min. The SAP was lowered to some extent (18.6%, often reduction of a hypertensive Cushing response); the increase in CPP was therefore less dramatic than the fall in ICP. The effects of mannitol on all parameters were almost identical to that of THAM (Fig. 1); only the mean duration of ICP decrease was 10 min shorter with 69 min. The effect of sorbitol was even shorter with 57 min; however, the slope in ICP decrease was more than twofold steeper (0.52 mm Hg/min) than with THAM (0.21 mm Hg/min) or with mannitol (0.17 mm Hg/min). In spite of the more rapid fall in ICP with sorbitol, the effect on the EEG is seen earlier with THAM and is more pronounced (Fig. 1): The alpha- and beta-power rise immediately after the onset of the THAM infusion, and the slow wave activity decreases in contrast to sorbitol therapy.

The effect of THAM is not based on an osmotic dehydration as shown by blood osmolality (Fig. 2); also the blood pCO_2 is not affected as a possible factor for the ICP. Besides a well known hypoglycemic effect[11], the only changes are seen in blood pH and base excess (Fig. 2). The significant increase in these two parameters goes parallel to the effect on ICP.

Discussion

Experimental and clinical results show that THAM might be a potent drug in therapy and prophylaxis of brain oedema after head injury. Not only in the cold lesion model of our experiments, which are not completely comparable to brain contusion, but also in a fluid percussion model (Marmarou and coworkers, this volume) THAM improves significantly the oedema parameters like water and sodium uptake of the brain tissue. In addition, the preservation of an almost normal potassium content may indicate a protection of the neurons after trauma.

The effect on ICP and the improvement in CPP in patients is at least as good as with mannitol and sorbitol. Degree and duration of ICP improvement are almost identical with THAM compared to mannitol; the effect even lasts longer. The decrease in ICP with sorbitol is faster but lasts shorter than with THAM. In summarizing, THAM is at least as potent as osmotherapy; the effect on the EEG seems to be better. The rapid EEG improvement seen before a significant decrease in ICP suggests a direct effect on the tissue. According to the parameters measured in the blood, a direct buffering action of THAM in traumatized brain tissue is assumed. THAM rapidly crosses the blood-brain barrier[11] in contrast to bicarbonate which may explain the inefficacy of bicarbonate in our experiments. A direct buffering activity is shown by the data from Marmarou (this volume), and also explains the breath depressing effect of THAM[11].

In a first pilot study, Rosner *et al.*[12] found not only a significant ICP decrease in patients with head injury after THAM, but also an increased survival rate. However, the number of patients was only small (n = 37). The encouraging results up to now suggest a larger clinical trial with THAM in patients with head injury. We are presently preparing such a multicenter THAM trauma study.

References

1. Akioka T, Ota K, Matsumoto A *et al* (1976) The effect of THAM on acute intracranial hypertension. In: Beks JWF, Bosch DA, Brock M (eds) Intracranial pressure III. Springer, Berlin Heidelberg New York, pp 219–233
2. Brackebusch HD, Cunitz G, Lidl H, Weis KH (1974) Simultaneous ventilation of small animals by way of a time-controlled pressure generator. Pharmacol 11: 241–246
3. Dos SJ, Nahas GG, Pappen EM (1962) Experimental correction of hypercapnic intracranial hypertension. Anaesthesiology 23: 46–50

4. Gaab MR, Knoblich OE, Spohr A *et al* (1980) Effect of THAM on ICP, EEG and tissue oedema parameters in experimental and clinical brain oedema. In: Shulman K, Marmarou A, Miller JD, Becker DP, Hochwald GM, Brock M (eds) Intracranial pressure IV. Springer, Berlin Heidelberg New York, pp 664–668

5. Gaab MR (1986) Routine computerized neuromonitoring. In: Miller MD, Teasdale GM, Rowan JO, Galbraith SL, Mendelow AD (eds) Intracranial pressure VI. Springer, Berlin Heidelberg New York, pp 240–247

6. Gaab MR, Seegers K, Goetz C (1989) THAM (Tromethamine, "Tris-Buffer"): Effective therapy of traumatic brain swelling? In: Hoff JT, Betz AL (eds) Intracranial pressure VII. Springer, Berlin Heidelberg New York, pp 616–619

7. Kerscher-Habbaba I (1984) Die Therapie des Hirnödems mit Phenobarbital und Dexamethason im Experiment. M.D. Thesis, Medical Faculty, University of Würzburg, Germany

8. Kesting U (1984) Therapie des experimentellen Hirnödems mit Natriumbikarbonat und THAM. M.D. Thesis, Medical Faculty, University of Würzburg, Germany

9. Knoblich OE, Gaab MR (1978) Comparison of the effects of osmotherapy, hyperventilation and THAM on brain pressure and EEG in experimental and clinical brain oedema. In: Frowein RA (ed) Advances in neurosurgery, Vol 5. Springer, Berlin Heidelberg New York, pp 336–345

10. Knoblich OE, Gaab MR (1978) Wirkung von Hyperventilation und THAM-Behandlung auf Hirndruck und elektrische Hirnaktivität im experimentellen Hirnödem. Neurochirurgia (Stuttgart) 21: 109–119

11. Nahas GG (1966) The clinical pharmacology of THAM. Clin Pharm Ther 4: 784–803

12. Rosner MJ, Elias KG, Coley I (1989) Prospective, randomized trial of THAM therapy in severe brain injury: Preliminary results. In: Hoff JT, Betz AL (eds) Intracranial pressure VII. Springer, Berlin Heidelberg New York, pp 611–616

Correspondence: Prof. Dr. M. R. Gaab, Neurochirurgische Klinik, Medizinische Hochschule Hannover, Konstanty-Gutschow-Strasse 8, D-3000 Hannover 61, Federal Republic of Germany.

Acta Neurochirurgica, Suppl. 51, 324–325 (1990)
© by Springer-Verlag 1990

A Comparison of the Cerebral and Haemodynamic Effects of Mannitol and Hypertonic Saline in an Animal Model of Brain Injury

M. H. Zornow, Y. S. Oh, and **M. S. Scheller**

Neuroanaesthesia Research, University of California, San Diego, La Jolla, California, U.S.A.

Summary

There has recently been an increased interest in the use of hypertonic saline solutions in the fluid resuscitation of trauma victims and to control intracranial hypertension. In this study, the cerebral and haemodynamic effects of a 3.2% hypertonic saline solution were compared with those of either a 0.9% saline or 20% mannitol solution in a rabbit model of brain injury. Forty-five minutes following the creation of a left hemispheric cryogenic brain lesion, equal volumes of hypertonic saline, 0.9% saline, or mannitol were infused over a 5 minute period. Monitored variables over the ensuing 120 minutes included mean arterial pressure, central venous pressure, intracranial pressure, hematocrit, serum osmolality and oncotic pressure. Upon conclusion of the two hour study period, hemispheric water contents were determined by the wet/dry weight method.

There were no significant differences in mean arterial pressure between the three groups at any point during the experiment. Plasma osmolality was significantly increased by 10–11 mOsm/kg in both the mannitol and hypertonic groups. The infusion of either mannitol or hypertonic saline produced a transient decrease in intracranial pressure lasting approximately 60 minutes whereas animals in the saline group demonstrated a continual increase in intracranial pressure. The lesioned hemisphere demonstrated a significantly greater water content than the non-lesioned hemisphere.

Introduction

Many trauma victims sustain serious head injuries in addition to their peripheral injuries. In these patients, uncontrollable intracranial hypertension and cerebral oedema are the most common causes of mortality for patients that reach a hospital alive[1, 2]. The intracranial hypertension that develops following brain injury has been commonly treated by the intravenous infusion of various hypertonic solutions. Mannitol is currently the most commonly used hypertonic agent for the treatment of cerebral oedema. Recently, there has been a case report[3] of intracranial hypertension which was resistant to treatment with large doses of mannitol, but

was successfully controlled following the infusion of a hypertonic saline solution.

This study was designed to compare the cerebral and haemodynamic effects of the intravenous administration of equi-osmolar solutions of mannitol (20% solution, calculated osmolarity = 1098 mOsm/L) and hypertonic saline (3.2% solution, calculated osmolarity = 1098 mOsm/L) in an animal model of brain injury and intracranial hypertension. The cerebral and haemodynamic effects of these hypertonic solutions were contrasted with a third group of animals that received an equal volume of normal saline (0.9% solution, calculated osmolarity = 310 mOsm/L).

Methods

Following approval by the Institutional Animal Care Committee, 21 New Zealand white rabbits were anaesthetized with 4% halothane in oxygen, intubated, and then mechanically ventilated with 0.7% halothane in 70% nitrous oxide in oxygen so as to achieve an end-tidal CO_2 of 35–40 mm Hg. Esophageal temperature was servo-controlled with infrared heat lamps to 37 °C. Following infiltration with 0.25% bupivicaine, femoral arterial and central venous catheters were inserted via a groin incision. The scalp was incised in the midline and reflected laterally to expose the skull. A 2 mm diameter burr hole located 8 mm lateral to the sagittal and 9 mm posterior to the coronal sutures was created over the right hemisphere for the insertion of an intracranial pressure monitor (Camino Laboratories, San Diego, CA). A metal funnel with a neck diameter of 1 cm was positioned and cemented to the skull overlying the left cerebral hemisphere.

Baseline values of the monitored variables (mean arterial pressure, central venous pressure, intracranial pressure, osmolality, oncotic pressure, hematocrit and arterial blood gases) were recorded and a cryogenic lesion created by pouring liquid nitrogen into the funnel positioned over the left hemisphere for a period of 90 seconds. The animals were then left undisturbed for 45 minutes to allow for the development of cerebral oedema and intracranial hypertension.

At the end of this 45 minutes, monitored variables were again recorded and the animals were randomly allocated to receive a 5 minute infusion of one of the following solutions: 1. $-10\,\text{ml/kg}$ of 20% mannitol (equivalent of 2 grams/kg, calculated osmolarity = 1098 mOsm/L), 2. $-10\,\text{ml/kg}$ of 3.2% hypertonic saline (calculated osmolarity = 1098 mOsm/L), or 3. $-10\,\text{ml/kg}$ of 0.9% saline (calculated osmolarity = 308 mOsm/L). Monitored variables were recorded upon completion of the infusion, and at 15, 30, 60, 90, and 120 minutes thereafter. Upon completion of the study period, the animals were killed with an intravenous bolus to T-61. The brains were rapidly removed and samples obtained from each hemisphere for the wet/dry weight measurement of water content.

Data for the three groups were examined using analysis of variance followed by multiple comparison testing.

Results

There were no significant differences in mean arterial pressure between the three groups at any point during the experiment. $PaCO_2$ was maintained at 35–40 mm Hg. ICP increased by $9.5 \pm 5.0\,\text{mm Hg}$ (mean ± SD) over the 45 minute period following the application of the liquid nitrogen. Central venous pressure increased transiently by $2.7 \pm 1.3\,\text{mm Hg}$ during the infusion of either mannitol or hypertonic saline. The infusion of either hypertonic saline or mannitol caused similar decreases in ICP lasting approximately 60–90 minutes. Animals receiving saline demonstrated a continuous increase in ICP (see Fig. 1). Plasma osmolality showed similar increases in both the mannitol and hypertonic saline groups (see Fig. 2). Brain water content in the lesioned hemisphere (80.56%) was significantly greater than that of the non-lesioned hemisphere (79.30%, $p < 0.0001$).

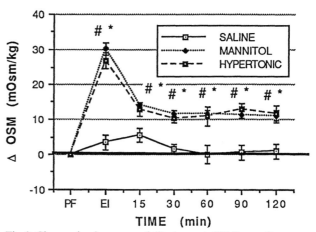

Fig. 2. Changes in plasma osmolality (mean ± SEM) over the course of the experiment. *PF* = 45 minutes post-freeze, *EI* = end infusion of solution. $\triangle OSM$ = change in OSM from PF value. # = hypertonic saline group > saline group, $p < 0.05$. * = manntiol group > saline group, $p < 0.05$. There were no differences between the hypertonic saline and mannitol groups at any time point

Discussion

This study demonstrates that hypertonic saline is as effective as mannitol in the acute reduction of intracranial hypertension. There was no evidence, however, that hypertonic saline was superior to mannitol in terms of either the duration or degree of decrease in ICP. Infusion of either hypertonic solution resulted in a transient increase in CVP due to movement of water from the intracellular to extracellular compartments. The observation that plasma osmolality was not different between the two hypertonic groups over the course of the experiment suggests that mannitol, Na^+, and Cl^- are confined to a similar degree to the extracellular space. Although plasma sodium concentrations were not measured, based upon the increase in plasma osmolality it can be estimated that peak increase in serum sodium concentrations in the hypertonic group were 15 meq/L and that these decreased to approximately 6 meq/L 15 minutes after completion of the infusion. In summary, equi-osmolar solutions of either mannitol or hypertonic saline produce similar acute decreases of ICP in this animal model of brain injury.

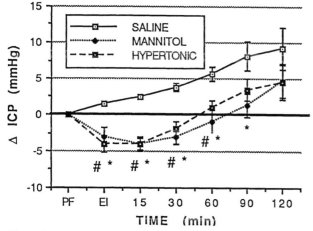

Fig. 1. Changes in ICP (mean ± SEM) over the course of the experiment. *PF* = 45 minutes post-freeze, *EI* = end infusion of solution. $\triangle ICP$ = change in ICP from PF value. # = hypertonic saline group < saline group, $p < 0.05$. * = manntiol group < saline group, $p < 0.05$. There were no differences between the hypertonic saline and mannitol groups at any time point

References

1. Baker CC, Oppenheimer L, Stephens B, Lewis FR, Trunkey DD (1980) Epidemiology of trauma deaths. Am J Surg 140: 144–150
2. Baxt WG, Moody P (1987) The differential survival of trauma patients. J Trauma 27: 602–606
3. Worthley LIG, Cooper DJ, Jones N (1988) Treatment of resistant intracranial hypertension with hypertonic saline. J Neurosurg 68: 478–481

Correspondence: M. H. Zornow, M.D., Neuroanaesthesia Research, M-029, University of California at San Diego, La Jolla, CA 92093, U.S.A.

Acta Neurochirurgica, Suppl. 51, 326–328 (1990)

Traumatic Brain Swelling and Operative Decompression: A Prospective Investigation

M. R. Gaab, M. Rittierodt, M. Lorenz, and **H. E. Heissler**

Neurosurgical Department, Hannover Medical School, Hannover, Federal Republic of Germany

Summary

Since 1978, *decompressive craniotomy* was performed according to a standardized protocol. Exclusion criteria were age $\geqslant 40$ years, deleterious primary brain damage, operable space occupying lesions, larger infarctions in CT scan or irreversible brain stem incarceration/ischaemic damage as shown by bulbar syndrome, loss in BAEP or oscillating flow in TCD. Indication was given by progressive intracranial hypertension not controllable by conservative methods, if ICP decompensation was correlated with clinical (GCS, extension spasms, mydriasis) and electrophysiological (EEG, SEP, CCT) deteriorations. 18 patients were decompressed by unilateral, 19 by bilateral craniotomy with large fronto-parieto-temporal bone flap and a dura enlargement by use of temporal muscle/fascia. 37 patients at an age of $18 \pm 7 (4–34)$ years were operated 5 h–10 d after trauma. Recovery was surprisingly good: only 5 died, 2 due to an ARDS; 3 remained vegetative, all others achieved full social rehabilitation[14] or remained moderately disabled[12]*. The best predictor of a favourable outcome was an initial posttraumatic GCS $\geqslant 7$. These encouraging results suggest a routine use of operative decompression in younger patients with delayed posttraumatic decompensation before irreversible ischaemic damage occurs.

Introduction

A *decompressive craniotomy, i.e.* a removal of an uni- or bilateral bone flap of the skull, has already been described by Von Bergmann as a treatment of traumatic brain swelling. The value of operative decompression in traumatic brain oedema, however, remains controversial. Especially after introduction of CT scanning, of intensive care with respiration and of ICP guided conservative treatment in patients with severe head injury, the indication for this operative procedure is often denied[2]. It is usually only accepted as a last resort in uncontrollable intracranial hypertension. No clear criteria for indication and operative technique

have been defined, and the data on outcome are not convincing due to small numbers, unclear pathophysiology and/or insufficient documentation. However, published results[3, 4] and own preliminary experiences[5, 6] are encouraging, if decompression is achieved before irreversible cerebral vasoparalysis occurs. In our animal experiments with cold lesion, the operative decompression was more effective on survival rates, ICP and EEG than any other therapy[7]. Since 1978 decompressive craniectomy is therefore performed in head injury patients according to a standardized protocol (Table 1) based on clinical symptoms, CT and ICP.

Patients and Methods

Initially, patients up to an age of $\leqslant 30$ y were included; this age limit is now increased to $\leqslant 40$ y. Decompressive craniotomy was performed if the criteria on Table 1 were fulfilled. No indication is seen in patients with a bulbar syndrome (GCS 3) initially after trauma (extreme primary injury) or with signs of cerebral circulatory arrest (GCS 3 with bilateral fixed mydriasis, ICP at SAP level for $\geqslant 10$ min, oscillating flow in transcranial doppler sonography, TCD).

The ICP was continuously registered with ventricle catheter or with epidural GAELTEC sensor in every patient. Since 1985 EEG and SEP-monitoring were included for assessing the extent of primary brain injury (BAEP, CCT) and for defining the time of operation (decrease in SEP-amplitude and increase in CCT, Fig. 1). Since 1987 also a TCD of the MCA flow velocity is used; decompensating ICP is associated with an extreme increase in MCA pulsatility index.

A large fronto-parieto-temporal bone flap was removed on both sides (19 patients) or unilaterally with clearly lateralized brain oedema in CT scan (18 patients). The age of the 37 patients (26 male, 11 female) ranged between 4–34 years with a mean of 18 ± 7 years. The interval between trauma and operation was 5 h–10 days.

The decompressive craniectomy includes a wide dura opening which is essential for an ICP normalization. The dura is then elongated by using the temporal muscle and its fascia as a graft which allows a watertight suture. After 4 weeks up to 12 months according

---

* In 3 patients status not determined yet – follow-up interval $\geqslant 1$ year.

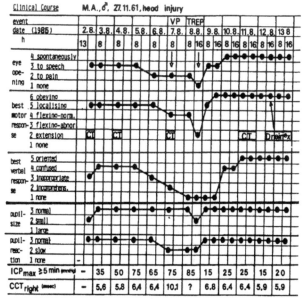

Fig. 1. Clinical course, max. ICP ≥ 5 min and central conduction time of SEP (CCT): Uncontrollable intracranial hypertension with clinical and electrophysiological deterioration – rapid improvement after bilateral decompression. Full rehabilitation

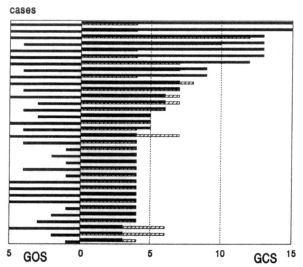

Fig. 2. Prognosis best predicted by initial GCS after trauma (solid) – GCS first day (hatched) – no bad outcome with initial GCS ≥ 7

to recovery, the skull is closed with a Palacos implant or by using the own bone flap which can be preserved by implanting it subcutaneously into the thigh. This autografting is restricted to 6–8 weeks after decompression, afterwards the bone shrinks and does no longer fit into the craniotomy.

In all surviving patients, the neurosurgical therapy was followed by an intensive rehabilitation program with physiotherapy, psychological care, professional training, social rehabilitation etc. The outcome was assessed 1 year after trauma with the Glasgow Outcome Scale (GOS 1 = death,   2 = vegetative,   3 = severly   disabled, 4 = moderately disabled, 5 = rehabilitated).

Results

The outcome was surprisingly good (Fig. 2) if the decompression was performed before irreversible ischaemic brain damage occurs. Only 5 patients died, two of them due to pulmonary complications (ARDS). In the remaining 3 non-survivors, the brain continued to swell and reached ICP values above 70 mm Hg and finally equalled to systemic blood pressures in spite of large decompressions; in one of these patients the necrotic brain swelling even perforated the galea sutures. CT showed large hemispheric infarctions in these patients. It must be noted that these 3 patients and the 3 patients which remained vegetative all underwent decompression in a bad clinical state (GCS 3, uni- or bilateral loss of pupillary reaction, Fig. 2).

In the remaining patients, the ICP was controlled after decompressive craniotomy. During operation, the brain surface usually moves only a few mm out of the skull in contrast to the malignant "erectio cerebri" seen after trepanation of ischaemic brains with peracute subdural haematomas. A watertight dura closure is easily achieved by using the temporal muscle with its fascia for prolongating the dura. The later cranioplasty with Palacos or now preferably with the autologous bone does not give any operative or cosmetical problem.

Complications were one wound infection with secondary healing after implantation of a Septopal® chain (moderate disability after 1 year) and 13 subdural effusions. 8 of these hygromas disappeared spontaneously/after puncture; 5 were associated with hydrocephalic ventricles and were successfully treated by VP-shunting.

As the most important predictor of outcome the severity of the initial (primary) brain injury can be defined (Fig. 2): In all patients with an initial post-traumatic GCS ≥ 8 or a GCS ≥ 6 on day one after trauma a good prognosis can be expected if decompression is achieved before intracranial hypertension results in irreversible brain damage.

Discussion

According to our results, there is good evidence on a beneficial effect of a timely and appropriate operative decompression based on clear pathophysiological criteria according to Table 1. A good outcome is almost guaranteed in young patients with only slight or moderate primary brain injury with delayed development

Table 1. *Decompressive Craniotomy-Indication*

---

* Younger age (initial limit ≤ 30 y, now ≤ 40 y)

* No deleterious primary brain damage

* No space occupying lesion (haematoma, focal contusion, hygroma, ventricle enlargement) which could directly be operated

* No larger infarction areas in CT

* ICP cannot be compensated by conservative means

* Intracranial hypertension associated with clinical (GCS, extension spasm, mydriasis), electrophysiological (EEG, SEP-amplitudes, CCT) and TCD (increase in pulsatility index, fall in diastolic flow) deterioration

* Decompression before irreversible brain stem incarceration or ischaemic brain damage (no terminal bulbar brain syndrome, wave V in BAEP present, no oscillating flow or worse in TCD)

Table 2. *Decompressive Craniotomy-Outcome*

Dead	5
Vegetative	3
Disabled	12
Full	14
Rehabilitated	

of brain swelling as the only cause of clinical deterioration. At least in younger patients (a clear age limit cannot be defined) with decompensating posttraumatic brain swelling, a decompressive craniotomy should be considered routinely if conservative therapy fails. The wide craniotomy must include an opening and prolongation of the dura, and must be done before irreversible ischaemic secondary brain damage is induced by loss in cerebral perfusion pressure. In defining the operative indication, a skillfull clinical observation, repeated CT scanning and continuous reliable ICP monitoring are mandatory. In addition, electrophysiology (EEG, SEP, BAEP) and Transcranial Doppler Sonography (TCD) are helpful in proper (early!) operative timing.

References

1. von Bergmann E (1880) Die Lehre von den Kopfverletzungen. Enke, Stuttgart
2. Cranial decompression (editorial). Lancet I (1988) 1204
3. Kjellberg RM (1971) Bifrontal decompression craniotomy for massive cerebral oedema. J Neurosurg 34: 488–493
4. Karlen J, Stula D (1987) Dekompressive Kraniotomie bei schwerem Schädelhirntrauma nach erfolgloser Behandlung mit Barbituraten. Neurochirurgia (Stuttgart) 30: 35–39
5. Gruss P, Gaab MR, Miltner F, Sörensen N (1979) Zur Reaktionsweise des kindlichen Gehirns am Beispiel sogenannter Dekompressionsoperationen. Z Kinderchir 32: 12–28
6. Sörensen N, Gaab MR, Gruss P, Halves E, Miltner F (1982) Decompressive craniectomy, an ultimate therapy in cranio-cerebral trauma. In: Monographia Paediatrica, Vol 15. Karger, Basel, pp 96–99
7. Gaab MR, Knoblich OE, Fuhrmeister U, Pflughaupt KW, Dietrich K (1979) Comparison of the effects of surgical decompression and resection of local oedema in the therapy of experimental brain trauma. Child's Brain 5: 484–498

Correspondence: Prof. Dr. M. R. Gaab, Neurochirurgische Klinik, Medizinische Hochschule Hannover, Konstanty-Gutschow-Strasse 8, D-3000 Hannover 61, Federal Republic of Germany.

Acta Neurochirurgica, Suppl. 51, 329–330 (1990)

The Novel 21-aminosteroid U-74006F Attenuates Cerebral Oedema and Improves Survival After Brain Injury in the Rat

T. K. McIntosh, M. Banbury, D. Smith, and **M. Thomas**

CNS Injury Laboratory, Surgical Research Center, Department of Surgery,
University of Connecticut Health Center, Farmington, Connecticut, U.S.A.

Summary

The present study evaluated the effect of the non-glucocorticoid 21-aminosteroid U74006F on the development of regional cerebral oedema after lateral fluid-percussion (FP) brain injury in the rat. Male Sprague-Dawley rats (n = 20) were anaesthetized and subjected to lateral FP brain injury of moderate severity (2.5–2.6 atmospheres). Fifteen minutes after brain injury, animals randomly received an i.v. bolus of either U74006F (3 mg/kg, n = 11) followed by a second bolus (3 mg/kg) at 3 hours vs buffered saline vehicle (equal volume, n = 9). At 48 hours postinjury, animals were sacrificed and brains tissue assayed for water content using wet weight/dry weight methodology. Administration of U74006F significantly attenuated the increase in water content observed in control animals in the ipsilateral hippocampus (adjacent to the site of maximal injury, $p < 0.05$). Administration of U74006F also significantly reduced post-injury mortality from 28% in control animals to zero in treated animals ($p < 0.001$). These results suggest that lipid peroxidation may be involved in the pathophysiological sequelae of brain injury and that 21-aminosteroids may be beneficial in the treatment of brain injury.

Introduction

Severe traumatic brain injury is often followed by secondary or delayed pathophysiological events including alterations in cerebral metabolism, dysregulation of cerebral blood flow and the onset of cerebral oedema (see ref[1] for review). Post-traumatic cerebral oedema has been related to the release or activation of a number of endogenous autodestructive "injury" factors including free radicals, leukotrienes, free fatty acids and other breakdown products of the arachidonic acid cascade[2]. A growing body of biochemical, physiological and pharmacological evidence has suggested that lipid peroxidation induced by the above factors may play a key role in the pathophysiology of central nervous system (CNS) injury[3]. A considerable body of experimental evidence has demonstrated that the use of high dose glucocorticoid steroids (30 mg/kg) can promote both short- and long-term neurological recovery after experimental CNS trauma[4, 5]. To a large extent, these effects of high dose glucocorticoids may be due to an inhibition of post-traumatic lipid peroxidation[4, 6].

A novel series of non-glucocorticoid 21-aminosteroids has been developed which lack glucocorticoid activity but are more effective inhibitors of nervous tissue lipid peroxidation than the glucocorticoid steroids[7]. Recently, the 21-aminosteroid U-74006F has been shown to be a potent inhibitor of iron dependent lipid peroxidation and is extremely effective in improving outcome in experimental models of stroke, vasogenic oedema, subarachnoid haemorrhage, spinal cord trauma in the cat and brain injury in the mouse (see[8]). In the present study, we evaluated the effect of U-74006F administration on the development of regional cerebral oedema following lateral (parsaggital) fluid-percussion (FP) brain injury in the rat.

Materials and Methods

Male Sprague-Dawley rats (300–350 g, n = 20) were anaesthetized with sodium pentobarbital (50 mg/kg i.p.). A craniotomy was made over the left parietal cortex, midway between lambda and bregma for the induction of FP brain injury. The femoral vein was cannulated for drug administration. All animals were subjected to lateral FP brain injury of moderate severity (2.5–2.6 atmospheres) as previously described[9]. Briefly, a bolus of saline is injected through the craniotomy at high pressure, causing a rapid but brief mechanical deformation of brain tissue over the left hemisphere.

Fifteen minutes after FP brain injury, animals were assigned to randomly receive an intravenous (i.v.) bolus of either U-74006F (3 mg/kg, n = 11) followed by a second bolus (3 mg/kg) at 3 hours vs. buffered saline vehicle (equal volume, n = 9). At 48 hours postinjury, the timepoint of maximal regional cerebral oedema in this

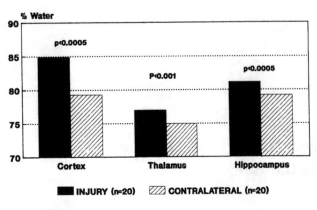

cortex = injury site

Fig. 1. Regional cerebral oedema (increase in tissue water concentration) in the injury site (left parietal cortex), left hippocampus and left thalamus at 48 hours following FP brain injury. Statistical significance is expressed when injured tissue was compared with contralateral (uninjured) homologous brain regions

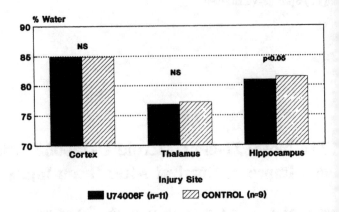

cortex = injury site

Fig. 2. Regional cerebral oedema at 48 hours following FP brain injury and treatment at 3 hours postinjury with either U-74006F (3 mg/kg, n = 11) or buffered saline vehicle (equal volume, n = 9)

model, animals were sacrificed, brains removed and dissected according to the following scheme: injured left parietal cortex, contralateral right parietal cortex, bilateral hippocampi, bilateral thalami. Tissue sections were assayed for water content using wet weight/dry weight methodology.

Results and Discussion

FP brain injury resulted in significant regional oedema at 48 hours postinjury in the injury site (left parietal cortex, $p < 0.001$), hippocampus ipsilateral to the injury site ($p < 0.001$) and thalamus ipsilateral to the injury site ($p < 0.001$) when compared to both (a) contralateral homologous brain tissue or (b) identical brain regions taken from uninjured control animals (see Fig. 1).

Administration of U-74006F at 15 minutes following FP brain injury had no effect on tissue water concentrations in the injury site or thalamus ipsilateral to the injury site (Fig. 2). U-74006F also did not affect tissue water concentration in the hemisphere contralateral to the injury site. However, U-74006F significantly reduced brain oedema observed in the hippocampus ipsilateral to the injury site ($p < 0.06$, Fig. 2).

Fluid-Percussion traumatic brain injury caused a 28 percent mortality in saline-treated control animals. All animals treated with U-74006F at 15 minutes after FP brain injury survived. This difference in mortality was statistically significant (Fisher's Exact Probability Test, $p = 0.01$). These results suggest that lipid peroxidative injury may play a role in the pathophysiological sequelae of traumatic brain injury and that 21-aminosteroids may be therapeutically beneficial in the treatment of brain trauma.

Acknowledgements

This study was supported, in part, by NIH RO1 NS26818 and a University of Connecticut Health Center HCRAC Faculty Grant.

References

1. Anderson D, Saunders R, Demediuk P, Dugan L, Braughler J, Hall E, Means E, Horrocks L (1986) Lipid hydrolysis and peroxidation in injured spinal cord. CNS Trauma 2: 257–268
2. Becker DP, Povlishock J (1985) CNS trauma status report. William Byrd Press, Washington DC
3. Braughler J, Hall E (1988) High dose methylprednisolone and CNS injury. J Neurosurg 64: 985–986
4. Braughler J, Bregenzer J, Chase R, Duncan L, Jacobsen E, McCall J (1987) Novel 21-aminosteroids as potent inhibitors of iron-dependent lipid peroxidation. J Biol Chem 262: 10434–10440
5. Hall E, Braughler J (1982) Glucocorticoid mechanisms in acute spinal cord injury: A review and rationale. Surg Neurol 18: 320–327
6. Hall E, Braughler J (1986) Role of lipid peroxidation in posttraumatic spinal cord degeneration. CNS Trauma 3: 281–294
7. Hall E, Yonkers P, McCall J, Braughler J (1988) Effects of the 21-aminosteroid U74006F on experimental head injury in mice. J Neurosurg 68: 456–461
8. Ellis E, Police R, Rice L, Grabeel M, Holt S (1989) Increased plasma PGE 2, 12-HETE levels following experimental concussive brain injury. J Neurotrauma 6: 31–37
9. McIntosh T, Vink R, Noble L, Yamakami I, Fernyak S, Soares H, Faden A (1989) Traumatic brain injury in the rat: Characterization of a lateral fluid-percussion model. Neuroscience 28: 233–241

Correspondence: T. K. McIntosh, Ph.D., Department of Surgery, University of Connecticut Health Center, Farmington, CT 06032, U.S.A.

Acta Neurochirurgica, Suppl. 51, 331–333 (1990)
© by Springer-Verlag 1990

The Increase in Local Cerebral Glucose Utilization Following Fluid Percussion Brain Injury is Prevented with Kynurenic Acid and is Associated with an Increase in Calcium

D. A. Hovda, A. Yoshino, T. Kawamata, Y. Katayama, I. Fineman, and **D. P. Becker**

Division of Neurosurgery, UCLA School of Medicine, University of California at Los Angeles, California, U.S.A.

Summary

Immediately following a lateral fluid percussion brain injury, the cerebral cortex and hippocampus ipsilateral to the percussion show a marked accumulation of calcium and a pronounced increase in glucose metabolism. To determine if this increase in glucose metabolism was related to the indiscriminate release of the excitatory amino acid (EAA) glutamate, kynurenic acid (an EAA antagonist) was perfused into the cerebral cortex through a microdialysis probe for 30 min prior to injury. The results show that adding kynurenic acid to the extracellular space prior to trauma prevents the injury-induced increase in glucose utilization. These results indicate that calcium contributes to the ionic fluxes that are typically seen following brain injury and supports the concept of an increased energy demand upon cells to drive pumping mechanisms in order to restore membrane ionic balance.

Introduction

Following a concussive brain injury cells are exposed to an increase in extracellular potassium which in part is the result of an injury-induced release of excitatory amino acids (EAA) particularly glutamate[6]. In addition, previous work has suggested that calcium may also contribute to this post concussion ionic flux[1].

As predicted from the concept of functional compartmentalization of energy, ionic perturbation across the cell membrane would selectively increase glucose utilization in order to activate pumping mechanisms[8, 14]. This increase in glucose metabolism as measured with 2-[^{14}C]deoxyglucose (DG) has been reported following cerebral ischaemia[11] and has been suggested to be present in some structures following concussion[3, 4, 10]. However, these DG-concussion studies addressed the metabolic changes beginning at 10, 20 and 60 min following injury. It is well documented that the potassium flux occuring following concussion is restricted to the first 5 min after trauma, therefore,

studies need to be conducted during the actual period of ionic perturbation.

The following study was conducted to determine if calcium accumulation does in fact exist following cerebral concussion contributing to the potassium flux previously reported[6]. Furthermore, these experiments were designed to measure the extent of glucose metabolism immediately following concussion and to determine if, like the ionic flux of potassium, the EAAs play a role in the metabolic alterations.

Materials and Methods

Young male Sprague-Dawley rats (n = 37, wt: 250–300 g) were studied immediately following sham operations or after a fluid percussion brain injury (3.7–4.3 atm) as previously described, using a mixture of 33% oxygen 66% nitrous oxide and enflurane (1.5–2.0 ml/min) as an anaesthetic[6]. For both the calcium and DG studies the percussion was delivered 1.0 mm posterior to bregma and 6.0 mm lateral to the midline. For studies involving the intraparenchymal dialysis of the EAA antagonist kynurenic acid, the percussion was delivered on the midline 2.0 mm anterior to bregma. For the calcium and DG experiments ^{45}Calcium (100 uCi) or DG (150 mCi/kg) was slowly injected (i.v.) 30 sec prior to injury. Autoradiographic analysis of the ^{45}Calcium experiments was conducted using optical densitometry. To control for film development time and concentration differences, optical densities were expressed in terms of relative asymmetry (left − right/left + right). For the DG studies local cerebral metabolic rates for glucose (lCMRgcl; μmol/100 g/min) were calculated using the method described by Sokoloff *et al.* (1977)[12].

To determine if the EAAs were involved in the changes in glucose utilization two dialysis probes (O.D. = 300 μm; effective length = 3.0 mm; cut off = 20,000 MW) were positioned bilaterally 3.5 mm posterior to bregma and 3.5 lateral to the midline. One probe was perfused with Ringer's solution and the other with 10 mM kynurenic acid for 30 min prior to injury. Sham controls consisted of 3 groups. Sham 1 animals underwent the standard anaesthesia and surgical procedures without the microdialysis or fluid-percussion

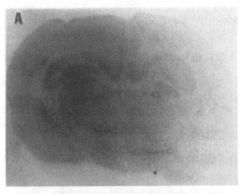

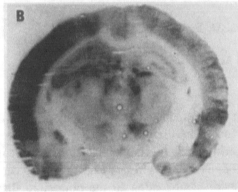

Fig. 1. Autoradiographs depicting the accumulation of calcium (A) and the increase in glucose utilization (B) immediately following a lateral fluid percussion brain injury. Note that the same regions which show the calcium accumulation also exhibit the increase in glucose utilization

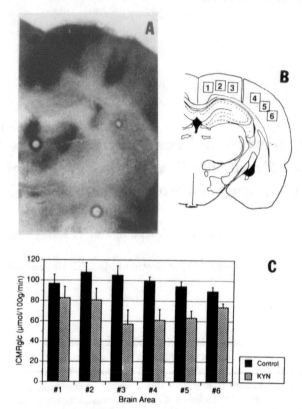

Fig. 2. (A) Autoradiograph of the region surrounding the dialysis probe which was perfused with kynurenic acid (10 mM) 30 min prior to the fluid percussion brain injury. (B) Diagram depicting the regions surrounding the dialysis probe where lCMRglc values were calculated. (C) Mean (± SEM) lCMRglc of the regions depicted in (B). Note the relationship between the distance from the probe and the rate of lCMRglc

injury. Sham 2 animals did not receive microdialysis however they endured the fluid-percussion. Sham 3 animals were administered the test drugs and Ringer's solution but were not injured. Following the dialysis perfusion and the ensuing injury the animals were processed for DG autoradiography and lCMRglc were compared between selected regions surrounding the tip of the dialysis probe.

Results

Calcium Accumulation: Immediately following injury regions of the brain, particularly the cerebral cortex and hippocampus, exhibited a marked accumulation of calcium (Fig. 1). This calcium accumulation was seen throughout the entire anterior-posterior extent reaching levels of 0.075–0.125. This represents a 200–350% increase of calcium accumulation compared to sham controls. However, this accumulation was not as pronounced within other regions including the thalamus or brain stem areas.

Glucose Metabolism: The same regions which showed an increase in calcium accumulation also exhibited high rates of lCMRglc ranging from 84.05–140.5, which were approximately 94% higher than that of sham controls. This increase in lCMRglc

was seen to some extent bilaterally, however the greatest effect was seen in the hemisphere ipsilateral to the percussion site. Just as in the calcium studies this increase of lCMRglc was not restricted to the primary percussion site but instead encompassed virtually the entire cerebral cortex and hippocampus.

Kynurenic Acid Dialysis: The concussion-induced increase in lCMRglc was greatly reduced (if not blocked completely) by pretreatment of kynurenic acid as compared to sham controls. As previously described[7] the dialysis procedure can administer drugs to a spherical area (6 mm in diameter) centered at the tip of the probe. This was similar to the area of near normal lCMRglc in animals who sustained a central percussion injury following 30 min of dialysis of kynurenic acid (Fig. 2).

Discussion

These results indicate that following concussive brain injury calcium contributes to the ionic fluxes seen following fluid percussion brain injury. An increase in calcium accumulation has been previously reported following cerebral ischaemia[2, 9] and spinal cord contusion[13]. In addition evidence of a possible accumulation of intracellular calcium has been shown following a lateral fluid-percussion injury within the hippocampus as determined by histological methods[1].

The same regions of the brain which exhibited a marked accumulation of calcium also exhibited an increase in lCMRglc. An increase in glucose utilization has also been described immediately following cerebral ischaemia[11] and there are a few studies that have addressed glucose metabolism following traumatic brain injury[3, 4, 10]. However, these traumatic brain injury studies have measured lCMRglc in only a few structures and at times after injury when the ionic balance is already reestablished[6]. The injury-induced increase in lCMRglc is most likely related to the massive release of potassium seen following similar injury[6] and the accumulation of calcium demonstrated in the current study.

It appears as if the ionic and metabolic perturbations seen following concussive brain injury are due, at least in part, to the injury induced release of the EAA glutamate[6]. The current study revealed that just as the EAA antagonist kynurenic acid can reduce the release of potassium[6], it also prevents the increase in glucose utilization seen immediately following concussive brain injury.

The fluid percussion injury studied in the current investigation does not result in cell death. Therefore, these ionic and metabolic changes are occuring in cells not mechanically damaged but that are exposed to an altered extracellular milieu produced by the concussion. We propose that these events, which occur during the first few minutes following trauma, lead to long term consequences of cellular metabolic disruption producing a period of vulnerability. It has previously been described that a state of vulnerability exists following concussive brain injury during which if cells are exposed to a second (normally sublethal) insult, they may not survive[5].

Acknowledgements

This work was supported by the Lind Lawrence Foundation, the Sunny Von Bulow Coma and Head Trauma Foundation and by an NIH grant #1 RO1 NS27544-01A1.

References

1. Cortez SC, McIntosh TK, Noble LJ (1989) Experimental fluid percussion brain injury: vascular disruption and neuronal and glial alterations. Brain Res 482: 271–282
2. Dienel GA (1984) Regional accumulation of calcium in postischaemic rat brain. J Neurochem 43: 913–925
3. Hayes RL, Katayama Y, Jenkins LW, Lyeth BG, Clifton GL, Gunter J, Povlishock JT, Young HF (1988) Regional rates of glucose utilization in the cat following concussive head injury. J Neurotrauma 5: 121–137
4. Hayes RL, Pechura CM, Katayama Y, Povlishock JT, Giebel ML, Becker DP (1984) Activation of pontine cholinergic sites implicated in unconsciousness following cerebral concussion in the cat. Science 223: 301–303
5. Jenkins LW, Moszynski K, Lyeth BG, Lewelt W, Dewitt DS, Allen A, Dixon CE, Povlishock JT, Majewski TJ, Clifton GL, Young HF, Becker DP, Hayes RL (1989) Increased vulnerability of the mildly traumatized rat brain to cerebral ischaemia: the use of controlled secondary ischaemia as a research tool to identify common or different mechanisms contributing to mechanical and ischaemic brain injury. Brain Res 477: 211–224
6. Katayama Y, Becker DP, Tamura T, Hovda DA (in press) Massive increases in extracellular potassium and the indiscriminate release of glutamate following concussive brain injury. J Neurosurg
7. Katayama Y, Cheung MK, Alves A, Becker DP (1989) Ion fluxes and cell swelling in experimental traumatic brain injury: The role of excitatory amino acids. In: Hoff JT, Betz AL (eds) Intracranial pressure VII. Springer, Berlin Heidelberg New York
8. Meyer FB, Anderson RE, Sundt Jr TM (1990) The novel dihydronaphthyridine Ca^{2+} channel blocker CI-951 improves CBF, brain pH$_i$, and EEG recovery in focal cerebral ischaemia. J Cereb Blood Flow Metab 10: 97–103
9. Rappaport ZH, Young W, Flamm ES (1987) Regional brain calcium changes in the rat middle cerebral artery occlusion model of ischaemia. Stroke 18: 760–764
10. Shah KR, West M (1983) The effect of concussion on cerebral uptake of 2-deoxy-D-glucose in rat. Neurosci Lett 40: 287–291
11. Shiraishi K, Sharp FR, Simon RP (1989) Sequential metabolic changes in rat brain following middle cerebral artery occlusion: A 2-deoxyglucose study. J Cereb Blood Flow Metab 9: 765–773
12. Sokoloff L, Reivich M, Kennedy C, Des Rosiers MH, Patlak CS, Pettigrew KD, Sakurada O, Shinohara M (1977) The [14 C]deoxyglucose method for the measurement of local cerebral glucose utilization: Theory, procedure, and normal values in the conscious and anaesthetized albino rat. J Neurochem 28: 897–916
13. Young W, Yen V, Blight A (1982) Extracellular calcium ionic activity in experimental spinal cord contusion. Brain Res 253: 105–113
14. Anderson BJ, Marmarou A (1989) Energy compartmentalization in neural tissue. J Cereb Blood Flow Metab 9: 386

Correspondence: D. A. Hovda, Ph.D., Division of Neurosurgery, UCLA School of Medicine, CHS 74-140, University of California at Los Angeles, Los Angeles, CA 90024–6901, U.S.A.

Acta Neurochirurgica, Suppl. 51, 335–337 (1990)

Various Disease Processes and Brain Oedema I (Hypertension, Hyperammonemia, Hydrocephalus)

Presence of Transendothelial Channels in Cerebral Endothelium in Chronic Hypertension

S. Nag

Department of Pathology (Neuropathology), Queen's University and Kingston General Hospital, Kingston, Ontario, Canada

Summary

In this model hypertension developed as early as 1 wk. post-surgery and was associated with reduction in Ca^{2+}-ATPase activity in cerebral vessels indicating that abnormalities in ionic calcium in vessel walls occur early in the evolution of hypertension. This study supports previous observations[1] that cerebral cortical arterioles develop increased permeability to endogenous plasma proteins in chronic hypertension. The principal mechanism resulting in this increased permeability is enhanced pinocytosis. Ca^{2+}-ATPase localisation in endothelial pinocytotic vesicles helped to localise transendothelial channels in occasional vessels of hypertensive rats.

The latter findings reinforce the concept that in pathologic states associated with cerebral oedema, pinocytotic vesicles fuse to form transendothelial channels which transport plasma proteins into brain.

Introduction

Plasma membrane calcium-activated adenosine-triphosphatase (Ca^{2+}-ATPase) plays an important role in maintaining the 10^4 gradient of calcium between cells and medium. Thus, any change in activity of this enzyme should cause an alteration in intracellular calcium levels.

Our previous study[3] has reported reduced Ca^{2+}-ATPase activity in cerebral endothelium of arterioles permeable to horseradish peroxidase in acute hypertension. This ultracytochemical study was undertaken to localise Ca^{2+}-ATPase at weekly intervals in cerebral endothelium of rats following induction of chronic renal hypertension to determine when changes in Ca^{2+}-ATPase occur.

Materials and Methods

Male Wistar-Furth rats weighing 120–140 g were used in this study. Chronic hypertension was induced by constricting the left renal artery with a silver clip having a gap of 0.007", following a right nephrectomy. Controls were sham-operated. The systolic blood pressure of animals was measured at weekly intervals under light methoxyflurane anaesthesia, using the tail cuff sphygmomanometric apparatus.

Hypertensive and normotensive rats were sacrificed at weekly intervals starting with one week post-surgery for the induction of hypertension, up to a period of 5 weeks. At the time of killing intra-aortic pressures were measured via a polyethylene catheter (PE 50) connected to a pressure transducer. The femoral vein was cannulated for administration of Evans blue. Each rat was injected intravenously with 1 ml. of 2% Evans blue which was circulated for 8 minutes. Thereafter a thoracotomy was done and fixative was perfused via a cannula in the ascending aorta at a pressure of 120 mm Hg. The fixative solution consisted of 2% paraformaldehyde and 0.2% glutaraldehyde in 0.1 M cacodylate buffer (pH 7.2) containing 4.35% sucrose.

Brains were removed and placed in the same fixative for 2 hr. at room temperature. They were then chopped at 40 μm intervals using a Smith-Farquhar tissue chopper. Slices from the occipito-temporal lobes were then reacted for demonstration of Ca^{2+}-ATPase as described previously[3]. Brain slices were processed for electron microscopy as described previously[2] and ultrathin sections were examined unstained or following staining with lead citrate using a Hitachi H500 electron microscope.

Results

A rise in blood pressure was observed as early as 1 week following the surgery for induction of hypertension. At the time of sacrifice most rats had blood pressure over 200 mm Hg. The blood gases and pH of all animals was in the normal range.

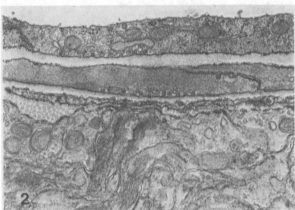

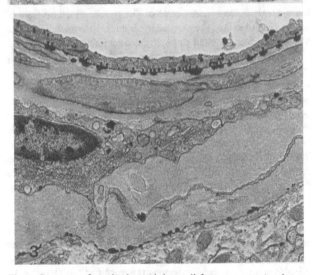

None of the hypertensive rats demonstrated increased permeability to Evans blue at the time of sacrifice.

Ca^{2+}-ATPase: Localisation of Ca^{2+}-ATPase in vessels of normotensive rats was similar to our previous observations[3]. There was discontinuous distribution of Ca^{2+}-ATPase on the outer plasma membranes of endothelial, smooth muscle and adventitial cells of cerebral cortical arterioles (Fig. 1). Localisation of Ca^{2+}-ATPase along plasma membranes of pinocytotic vesicles in endothelial and smooth muscle cells was quite striking.

Hypertensive rats showed reduced ultracytochemical localisation of Ca^{2+}-ATPase along plasma membranes of endothelial and smooth muscle cells of cortical arterioles (Fig. 2). This change was apparent as early as 1 wk. following the surgery for induction of hypertension.

Since vessels were sampled from the occipito-temporal lobes, a region where increased vascular permeability consistently develops in hypertension, ultrastructural evidence of increased arteriolar permeability to endogenous plasma proteins was detected. Six of the 15 arterioles studied showed protein extravasation which was evident as a fine granular electron-dense

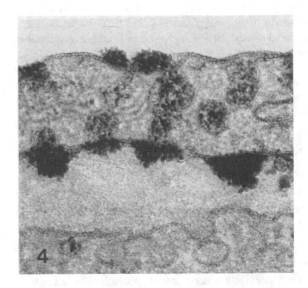

Fig. 1. Segment of cerebral arteriolar wall from a normotensive rat showing Ca^{2+}-ATPase localisation. Reaction product is present in a discontinuous manner on the outer plasma membrane of endothelial and smooth muscle cells and nerve processes. × 26,000

Fig. 2. Segment of arteriolar wall from a hypertensive rat showing reduced Ca^{2+}-ATPase along the plasma membranes of endothelial, smooth muscle and adventitial cells. × 27,200

Fig. 3. Segment of arteriolar wall from a rat with chronic hypertension of 5 weeks duration. Note the extravasation of endogenous plasma proteins which appears as granular electron-dense material in the basement membrane of endothelial and smooth muscle cells. Plasma membrane Ca^{2+}-ATPase is reduced in all components of

Fig. 4. Segment of arteriolar endothelium from a hypertensive animal showing Ca^{2+}-ATPase along the plasma membranes of pinocytotic vesicles which have fused to form a transendothelial channel. At −10 °C tilt. × 119,000

the vessel wall. Ca^{2+}-ATPase localisation highlights the increased numbers of endothelial pinocytotic vesicles. × 13,600

material distending the vessel wall (Fig. 3). The endothelium of such arterioles showed increased numbers of pinocytotic vesicles. Occasional arterioles showed chains of fused vesicles extending from the luminal to the abluminal plasma membrane of endothelium. Such transendothelial channels were readily apparent because of Ca^{2+}-ATPase localisation in the plasma membranes of the pinocytotic vesicles (Fig. 4).

Acknowledgements

The skilled technical assistance of Mrs. Verna Norkum and Mr. Blake Gubbins is gratefully acknowledged. This work was supported by the Heart and Stroke Foundation of Ontario.

References

1. Nag S (1984) Cerebral changes in chronic hypertension: combined permeability and immunohistochemical studies. Acta Neuropathol (Berl) 62: 178–184
2. Nag S (1985) Ultrastructural localisation of monosaccharide residues on cerebral endothelium. Lab Invest 5: 553–558
3. Nag S (1988) Localisation of calcium-activated adenosine-triphosphatase (Ca^{2+}-ATPase) in intracerebral arterioles in acute hypertension. Acta Neuropathol (Berl) 75: 547–553

Correspondence: Dr. S. Nag, Department of Pathology (Neuropathology), Queen's University and Kingston General Hospital, Kingston, Ontario, Canada K7L 3N6.

Acta Neurochirurgica, Suppl. 51, 338–340 (1990)
© by Springer-Verlag 1990

Blood-Brain Barrier Disruption Caused by Impairment of Cerebral Autoregulation During Chronic Cerebral Vasospasm in Primates

Y. Handa[1], **H. Takeuchi**, **M. Kabuto**, **H. Kobayashi**, **H. Kawano**, **K. Hosotani**, and **M. Hayashi**

[1] Department of Neurosurgery, Fukui Medical School, Fukui, Japan

Summary

Impairment of the cerebral autoregulation and its effects on the development of brain oedema during chronic cerebral vasospasm after subarachnoid haemorrhage (SAH) were studied in primates. The unilateral induction of SAH by clot placement around cerebral arteries produced a moderate degree of vasospasm (more than 40% reduction of vessel caliber on angiogram) seven days after SAH. Abolishment of autoregulation was observed in the territories of the cerebral hemisphere supplied by the vasospastic arteries. It was found that in this region an increase in cerebral perfusion pressure easily produced a marked elevation of the cerebral blood flow over the upper limit of autoregulation threshold resulting in disruption of the blood-brain barrier.

Introduction

Delayed cerebral ischaemia due to chronic cerebral vasospasm is the most important factor affecting morbidity and mortality after subarachnoid haemorrhage (SAH). In many therapeutical attempts for preventing the development of ischaemic neurological deficits, augmentation of cerebral blood flow (CBF) by induced hypertension or hypervolemia is widely used in patients with vasospasm. This management method is based on the hypothesis that the cerebral autoregulation is impaired in vasospasm, and that an increase in cerebral perfusion pressure (CPP) may improve the ischaemic state[6, 7]. However, it must be considered that induced hypertension often worsens the clinical condition due to the development of brain oedema[6, 9]. Although several clinical studies have been performed to investigate the impairment of autoregulation during cerebral vasospasm[3, 8], a precise evaluation has not been made because of ethical problems. Presently, we studied to what degree the autoregulation is impaired and how this affects the development of brain oedema during chronic cerebral vasospasm in primates.

Materials and Methods

Twenty female cynomolgous monkeys (macaca fascicularis) weighing 2.2 to 3.0 kg were used in this study. The animals were divided into three groups consisting of SAH group I, SAH group II and a sham-operated group. In all animals, angiography was performed before and seven days (Day 7) after SAH induction to evaluate the degree of vasospasm. For induction of SAH, in the SAH group I and II, an autologous blood clot was placed around the cerebral arteries at the right side of the circle of Willis. After the angiogram on Day 7, regional CBF was measured in animals of the SAH group I and the sham-operated group during altering the mean arterial blood pressure (ABP). An increase or decrease in ABP was made in steps of 20 mm Hg by intravenous infusion of pressor or depressor drug in order to change the mean ABP level between 40 to 180 mm Hg. Regional CBF was measured in the bilateral parietal cortex by using the hydrogen clearance method. At the end of the measurements, the animals were sacrificed by exsanguination. In the SAH group II, the animals were sacrificed after angiography without induced hypertension. All animals were perfused transcardially with physiological saline followed by 4% paraformaldehyde solution. For evaluation of blood-brain barrier (BBB) disruption, the brain was studied microscopically by an immunohistochemical method using anti-albumin antibodies.

Results

In the sham-operated group, the cerebral arteries had no vasospasm on Day 7. In the SAH group, all animals developed moderate to severe vasospasm of the cerebral arteries at the right side (clot side) but not at the left side. The average reduction in vessel caliber of the cerebral arteries (ICA + MCA + ACA) on Day 7 compared to baseline values was significantly ($p < 0.01$) greater in the SAH group I ($-51.2 \pm 9.9\%$) and the SAH group II ($-48.8 \pm 5.8\%$) than in the sham-operated group ($-2.0 \pm 2.2\%$).

The mean ABPs recorded after Day 7 at angiography showed no significant differences. They were

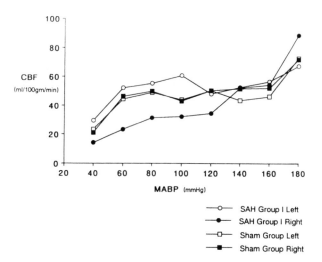

Fig. 1. The relationship between mean cerebral blood flow (CBF) and mean arterial blood pressure (ABP) in the SAH group I and the sham-operated group during induced hypertension and hypotension

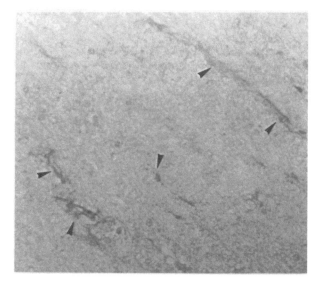

Fig. 2. Microscopical observation of extravasation of serum albumin in the parietal white matter of the spasm side hemisphere. Arrow heads indicate staining of albumin. Immunohistochemical staining with anti-albumin antibody, $\times 270$

111 ± 14.0 mm Hg (mean $\pm$ SD) in the SAH group I, 115 ± 12.6 mm Hg in the SAH group II, and 110 ± 6.4 mm Hg in the sham-operated group, respectively.

The bilaterally assessed regional CBF in the sham-operated group, and non-clot side in the SAH group I showed no significant differences between each other measured at mean ABP levels of 60 to 160 mm Hg, indicating a preservation of autoregulation. In the SAH group I, the regional CBF at the clot side was signif-

icantly ($p < 0.05$) lower than that at the non-clot side at mean ABP levels of 40 to 100 mm Hg, while it was significantly higher at a mean ABP level of 180 mm Hg. The regional CBF at the clot side increased in parallel with the increase in ABP with a significant correlation coefficient, indicating the impairment of the cerebral autoregulation (Fig. 1).

The immunohistochemical study revealed extravasation of serum albumin only in brain tissue of the spasm side in the SAH group I. Extravasation of albumin was observed around small arterioles in the parietal white matter as well as grey matter. Extravasation of serum albumin could not be definitely assessed in brain tissues of both, the sham-operated group or SAH group II.

Discussion

The present results confirm that the cerebral autoregulation is impaired in the territory of cerebral hemispheres supplied by arteries presenting with chronic vasospasm. In this region, an increase in CPP easily produces a marked elevation of CBF over the upper limit of the autoregulation threshold resulting in disruption of the BBB. Although several experimental studies have reported an impairment of autoregulation during vasospasm or after SAH[1,5], none have evaluated a correlation between the impaired autoregulation and an angiographically detected chronic cerebral vasospasm. In these studies, SAH was induced diffusely in the subarachnoid space by cisternal blood injection or arterial puncture. These procedures of induction of SAH may produce an elevation of the intracranial pressure (ICP), which also may elicit an impairment of the cerebral autoregulation. In the present primate model, there was no elevation of ICP during SAH induction. Furthermore, abolishment of autoregulation occurred only in the spasm side hemisphere. For these reasons, it is concluded that in the present study, impairment of autoregulation was caused chiefly by the pathological condition due to chronic vasospasm.

As mentioned above, it is suggested that an increase in ABP often worsens the brain function due to the development of brain oedema in patients with chronic vasospasm. We must determine the safety range of an increased ABP level, which is enough to maintain the cerebral circulation for normal brain function in spite of impaired autoregulation. In the present study, it is confirmed that the upper limit of an increase in ABP should be approximately at 40% of the baseline value to keep CBF normal, and this level is similar to the results described in clinical studies[6, 7]. It is described that in the wall of vasospastic arteries, there is a disappearance of sympathetic nerve fibers[2], which constrict the cerebral arteries during severe hypertension[4]. Furthermore, it is reported that vasospastic arteries showed no pharmacological responses to vasoactive agents[10]. These findings may explain that the vasospastic arteries can not respond to maintain CBF against a marked elevation of ABP. The present results suggest that hemodynamic therapy in patients with vasospasm following SAH should be carefully attempted because of the potent risk of the development of vasogenic brain oedema.

References

1. Boisvert DP, Overton TR, Weir B, Grace MG (1978) Cerebral arterial responses to induced hypertension following subarachnoid haemorrhage in the monkey. J Neurosurg 49: 75–83
2. Edvinsson L, Eglund N, Owman CH, Sahlin CH, Svendgaard N-A (1982) Reduced noradrenaline uptake and retention in cerebrovascular nerves associated with angiographically visible vasoconstriction following experimental subarachnoid hemorrhage in rabbits. Brain Res Bull 9: 799–805
3. Heilbrun MP, Olesen J, Lassen NA (1972) Regional cerebral blood flow studies in subarachnoid hemorrhage. J Neurosurg 37: 36–44
4. Heistad DD, Marcus ML, Abboud FM (1978) Experimental attempts to unmask effects of neural stimuli on cerebral blood flow. In: Purves (ed) Cerebral vascular smooth muscle and its control. Elsevier, Amsterdam, pp 97–111
5. Jakubowski J, Bell BA. Symon L, Zawirsko MB, Francis DM (1982) A primate model of subarachnoid hemorrhage: Change in regional blood flow, autoregulation, carbon dioxide reactivity, and central conduction time. Stroke 13: 601–611
6. Kassell NF, Peerless SJ, Duward QJ, Beck DW, Drake CG, Adams HP (1982) Treatment of ischemic deficits from vasospasm with intravascular volume expansion and induced hypertension. Neurosurgery 11: 337–343
7. Kosnik EJ, Hunt WE (1976) Postoperative hypertension in the management of patients with intracranial arterial aneurysms. J Neurosurg 45: 148–154
8. Nornes H, Knutzen HB, Wikeby P (1977) Cerebral arterial blood flow and aneurysm surgery, Part 2: Induced hypotension and autoregulatory capacity. J Neurosurg 47: 819–827
9. Shigeno T, Fritschka E, Brock M, Schramm J, Shigeno S, Cervós-Navarro J (1982) Cerebral edema following experimental subarachnoid hemorrhage. Stroke 13: 368–379
10. Toda N, Ozaki T, Ohta T (1977) Cerebrovascular sensitivity to vasoconstricting agents induced by subarachnoid hemorrhage and vasospasm in dogs. J Neurosurg 46: 296–303

Correspondence: Y. Handa, M.D., Department of Neurosurgery, Fukui Medical School, Matsuoka-chou, Yoshida-gun, Fukui 910-11, Japan.

Acta Neurochirurgica, Suppl. 51, 341–343 (1990)

Effect of L-arginine on Haemoglobin-induced Inhibition of Endothelium-dependent Relaxation of Isolated Cerebral Arteries

L. Schilling, A. A. Parsons*, J. R. L. Mackert, and **M. Wahl**

Department of Physiology, University of Munich, Federal Republic of Germany

Summary

The effects of Oxy-haemoglobin were investigated in the rat isolated basilar artery and compared to Methylene blue and N^G-Nitro-L-Arginine (NOLAG), as these compounds interfere with the relaxation induced by the endothelium-derived relaxing factor. Acetylcholine induced a concentration-related relaxation of serotonin-precontracted arteries which was further enhanced by the application of L-arginine. Similarly, the spasmolytic activity of L-arginine was enhanced by acetylcholine. Oxy-haemoglobin, Methylene blue and NOLAG induced an increase of resting tension and inhibited acetylcholine-induced relaxation. Furthermore, inhibition of acetylcholine-induced relaxation by Oxy-haemoglobin and NOLAG but not that by Methylene blue was partially reversed by L-arginine.

Introduction

Subarachnoid haemorrhage is often followed by the development of delayed cerebroarterial spasm. The pathogenesis of this complication is not yet clear, although an imbalance between vasodilator and vasoconstrictor mechanisms has been suggested to play an important role. Factors which may promote development of vasospasm are (i) release of blood components or endothelial compounds which may act then as vasoconstrictors (for review see ref.[12]), (ii) a denervation supersensitivity of the smooth muscle cells to constrictor substances[11], (iii) depletion of dilatory transmitters from perivascular nerves[3], and (iv) disruption of endothelial cells inhibiting the release of dilatory compounds. Two distinct factors, prostacyclin (for review see ref.[8]) and the endothelium-derived relaxing factor (EDRF)[1] have been described. Recently, EDRF has been identified as nitric oxide (NO)[9] synthesized from the precursor L-arginine (L-Arg)[12]. The presence of blood products such as Oxy-haemoglobin (Oxy-Hb) in the subarachnoid space would, therefore, inhibit the effects of EDRF due to its NO binding properties[2].

In the present study we have therefore investigated whether L-Arg possesses spasmolytic activity and assessed its effect on ACh-induced relaxation, which is mediated by release of EDRF. We also compared the effects of Oxy-Hb on ACh- and L-Arg-induced relaxation with Methylene blue (MB), an inhibitor of the soluble guanylate cyclase and N^G-Nitro-L-Arginine (NOLAG), a structural analog of L-Arg and inhibitor of NO synthesis.

Materials and Methods

Normotensive rats (male Sprague-Dawley, 180–300 g) were decapitated under ether anesthesia and the basilar artery removed. Each artery was divided into four segments and set up for measurement of isometric contraction[4] in a modified Krebs solution of the following composition (mmol/l): NaCl, 119; KCl, 4.6; NaH_2PO_4, 1.2; $CaCl_2 \cdot 2H_2O$, 1.5; $MgCl$, 1.2; $NaHCO_3$, 15; glucose, 10. The Krebs solution was maintained at 37 °C and bubbled with 90% O_2/10% CO_2 to maintain pH at 7.25–7.30.

The vessels were placed under a resting tone of 2 mN. On sustained tension to serotonin (5-HT) (1 µmol/l) concentration-effect curves to either ACh or L-Arg (100 nmol/l-100µmol/l) were obtained. In the presence of 100 µmol/l ACh a concentration-effect curve to L-Arg was also constructed and vice versa. Only vessels that relaxed to application of ACh, indicating a functionally intact endothelium were used in this study.

After a wash period the vessels were preincubated with Oxy-Hb (10 µmol/l) for 20 to 30 minutes. Once sustained tension was attained concentration-effect curves to either ACh or L-Arg were formed. L-arginine or ACh were applied in the presence of 100 µmol/l ACh or L-Arg, respectively, as described on 5-HT spasm (see above).

In a separate series of experiments, after preincubation with Oxy-Hb (1 or 10 µmol/l), the vessels were further contracted by application of 5-HT (1 µmol/l) and the effects of ACh and of L-Arg applied in the presence of 100 µmol/l ACh were investigated. L-Arg was added

* Royal Society European Exchange Fellow and supported by the Alexander von Humboldt Foundation

in the presence of 100 µmol/l ACh. Similar experiments were also performed after preincubation with either MB (10 µmol/l) or NO-LAG (10 µmol/l).

The degree of contraction was calculated in percent of contraction to isotonic 124 mmol/l KCl-Krebs solution. Relaxation was expressed in percent of precontraction and is given as mean ± standard error (SEM).

All compounds used were obtained from Sigma (FRG). Oxy-haemoglobin was prepared from bovine haemoglobin as described by Martin *et al.*[7].

Results and Discussion

Oxy-haemoglobin induced a very variable but concentration related increase of resting tension. Mean contraction was for 1 µmol/l Oxy-Hb 8.8 ± 3.8% and for 10 µmol/l Oxy-Hb 53.8 ± 8.7% of contraction to 124 mmol/l KCl-Krebs. NOLAG and MB also produced contraction (11.1 ± 6.1% for 10 µmol/l NOLAG and 10.8 ± 4.7% for 10 µmol/l MB) but did not show a marked concentration relationship. When 5-HT (1 µmol/l) was applied to vessels preincubated with 1 or 10 µmol/1 Oxy-Hb, NOLAG or MB the resulting contractions were similar to those found in the presence of 5-HT alone.

On 5-HT (1 µmol/l) spasm both ACh and L-Arg produced a concentration related relaxation with ACh being more potent than L-Arg (Fig. 1). The relaxation induced by 100 µmol/l ACh could be enhanced by cu-

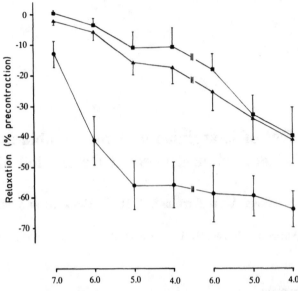

Fig. 2. Effects of acetylcholine (ACh) (left panel) and L-arginine (L-Arg) in the presence of 100 µmol/l ACh (right panel) on rat isolated basilar artery precontracted with 5-HT alone (●–●) and in the presence of Oxy-haemoglobin (▲–▲ 1 µmol/l, ■–■ 10 µmol/l). Results shown are mean ± SEM of ⩾ 5 values

mulative application of L-Arg, and the relaxation induced by L-Arg was enhanced by ACh. The respective concentration-effect curves are shown in Fig. 1. In vessels precontracted with Oxy-Hb (10 µmol/l), ACh or L-Arg alone still induced relaxation. However, this relaxation was markedly inhibited as compared to that on 5-HT spasm (Fig. 1). Addition of L-Arg in the presence of 100 µmol/l ACh, and vice versa, produced concentration-related further relaxations (Fig. 1).

When the vessels were precontracted by Oxy-Hb (1 or 10 µmol/l) and 5-HT (1 µmol/l) together, ACh-induced relaxation was also inhibited as it was in Oxy-Hb spasm alone. Again relaxation to 100 µmol/l ACh could be enhanced by the addition of L-Arg even in the presence of the higher concentration of Oxy-Hb (10 µmol/l) (Fig. 2). When NOLAG (10 µmol/l) instead of Oxy-Hb was present in the bath similar effects were observed for ACh and L-Arg. In contrast, MB (10 µmol/l) induced inhibition of ACh relaxation was not reversed by L-Arg (results not shown).

In the present study ACh induced an approximately 50% reversal of spasm in accordance to the observation by Lai *et al.*[6] and similar to that described for human basilar[5]. The results are consistent with the hypothesis of EDRF being NO and derived from L-Arg, as Oxy-Hb, NOLAG and MB (compounds which interfere with the synthesis of NO or prevent the activation of the soluble guanylate cyclase by NO) inhibited ACh-

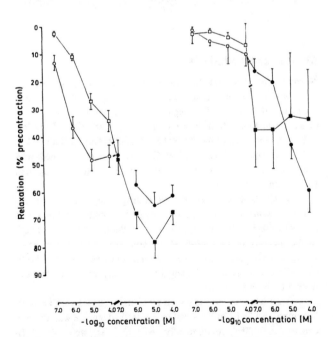

Fig. 1. Acetylcholine (ACh) and L-arginine (L-Arg) induced relaxation of rat isolated basilar artery precontracted with either 5-HT (1 µmol/l) (left panel) or Oxy-haemoglobin (10 µmol/l) (right panel). ○–○ ACh alone, ●–● L-Arg in the presence of 100 µmol/l ACh, □–□ L-Arg alone, ■–■ ACh in the presence of 100 µmol/l L-Arg. Results shown are mean ± SEM of ⩾ 5 values

induced relaxation in a comparable manner. Furthermore the spasmolytic activity of L-Arg alone and its enhancing action on ACh-induced relaxation may indicate the availability of L-Arg being the rate determining step in NO production in rat isolated basilar artery.

The contractile activity of Oxy-Hb and its inhibition of EDRF-mediated relaxation shown here support the hypothesis that Oxy-Hb may play an important role in the pathogenesis of cerebral vasospasm following subarachnoid haemorrhage. An unexpected finding was that the inhibition by Oxy-Hb of ACh-induced relaxation could be at least partially reversed by application of L-Arg. This may indicate that enhancement of the endogenous vasodilatory mechanism which is mediated by the effects of EDRF could provide a new promising approach in the treatment of cerebral vasospasm.

References

1. Furchgott RF, Zawadzki JV (1980) The obligatory role of endothelial cells in the relaxation of arterial smooth muscle by acetylcholine. Nature 288: 373–376
2. Gibson QH, Roughton FJW (1957) The kinetics and equilibria of the reactions of nitric oxide with sheep haemoglobin. J Physiol 136: 507–526
3. Hara H, Nosko M, Weir B (1986) Cerebral perivascular nerves in subarachnoid haemorrhage. A histochemical and immunohistochemical study. J Neurosurg 65: 531–539
4. Hoegestaett ED, Andersson K-E, Edvinsson L (1983) Mechanical properties of rat cerebral arteries as studied by a sensitive device for recording of mechanical activity in isolated small blood vessels. Acta Physiol Scand 117: 49–61
5. Kanamaru K, Waga S, Fujimoto K, Itoh H, Kubo Y (1989) Endothelium-dependent relaxation of human basilar arteries. Stroke 20: 1208–1211
6. Lai FM, Cobuzzi A, Shepherd C, Tanikella T, Hoffman A, Cervoni P (1989) Endothelium-dependent basilar and aortic vascular responses in normotensive and coarctation hypertensive rats. Life Sci 45: 607–614
7. Martin W, Villani GM, Jothianandan D, Furchgott RF (1985) Selective blockade of endothelium-dependent and glyceryl trinitrate-induced relaxation by haemoglobin and by methylene blue in the rabbit aorta. J Pharmacol Exp Ther 232: 708–716
8. Moncada S (1982) Biological importance of prostacyclin. Br J Pharmacol 76: 3–31
9. Palmer RMJ, Ferrige AG, Moncada S (1987) Nitric oxide release accounts for the biological activity of endothelium -derived relaxing factor. Nature 327: 524–526
10. Palmer RMJ, Ashton DS, Moncada S (1988) Vascular and endothelial cells synthesize nitric oxide from L-arginine. Nature 333: 664–666
11. Svendgaard N-A, Edvinsson L, Owman C, Sahlin C (1977) Increased sensitivity of the basilar artery to norepinephrine and 5-hydroxytryptamine following experimental subarachnoid haemorrhage. Surg Neurol 8: 191–195
12. Wahl M (1985) A review of neurotransmitter and hormones implicated in mediating cerebral vasospasm. In: Voth D, Glees P (eds) Cerebral vascular spasm. de Gruyter, Berlin, pp 223–230

Correspondence: Dr. L. Schilling, Department of Physiology, University of Munich, Pettenkoferstrasse 12, D-8000 Munich 2, Federal Republic of Germany.

Acta Neurochirurgica, Suppl. 51, 344–345 (1990)
© by Springer-Verlag 1990

Cerebrovascular Permeability in Acute Hypertension: Effect of Flunarizine

S. Nag and **L. Young**

Department of Pathology (Neuropathology) and Kingston General Hospital, Queen's University, Kingston, Ontario, Canada.

Summary

The pattern of Evans blue extravasation in the brain in norepinephrine-induced acute hypertension is similar to our previous observations using horseradish peroxidase as a tracer. Pretreatment with flunarizine IV resulted in significant reduction of RISA leakage in all regions of the brains of acutely hypertensive rats. The reduction in RISA leakage in the drug-treated hypertensive group is not attributable to differences in the blood pressure elevations which were not significantly different in both groups. These studies suggest a role for calcium in the increased endothelial permeability occurring in cerebral vessels in acute hypertension. Further morphological studies are required to determine whether flunarizine reduces permeability by decreasing pinocytosis.

Introduction

Our previous studies (Nag 1988; 1990) suggest that in acute hypertension influx of calcium into cerebral endothelium could mediate increased permeability by enhanced pinocytosis. This hypothesis was tested by studying the effect of pretreatment with the calcium entry blocker Flunarizine on the increased cerebrovascular permeability to radio-iodinated human serum albumin (RISA) occurring in norepinephrine-induced acute hypertension. Reduction in cerebrovascular permeability in the presence of Flunarizine would support our hypothesis that calcium is involved in mediating increased permeability in acute hypertension.

Materials and Methods

Wistar-Furth rats (200–250 gm.) were anesthetised by an intraperitoneal injection of Sodium Amytal 10 mg/100 gm. Polyethylene cannulas were inserted in both femoral arteries. One cannula was connected to a pressure transducer for continuous monitoring of the blood pressure and the other was used for the withdrawal of blood samples for measurement of blood gases and of total plasma radioactivity. The polyethylene cannula inserted into the femoral vein was used for the administration of test substances.

Rats were injected intravenously with a bolus of 10 µCi of Iodine 125 labelled RISA (ICN Radiochemicals, Irvine, CA). Ten minutes later 1 ml of 2% Evans blue was injected intravenously followed 2 minutes later by infusion of norepinephrine (3 µg/ml/min) over a 2 minute period. Experiments were terminated 8 minutes after the administration of the hypertensive agent by an intravenous injection of 0.75 ml Pentobarbital. Blood samples were withdrawn at 6 min, 12 min and prior to killing the rat for measurement of plasma radioactivity.

Experimental Groups

The experimental groups consisted of: (1) Normotensive saline controls, (2) Norepinephrine-induced hypertensive group, (3) Hypertensive rats pretreated with Flunarizine (Janssen Pharmaceutica Inc., Mississauga, Ontario, Canada) as follows: (a) 1 mg/Kg IP, 30 min prior to the injection of RISA; (b) 1 mg/Kg and (3) 2.5 mg/Kg IV, 5 min. prior to the induction of hypertension.

Following a thoracotomy, rats were perfused with saline for 10 minutes. Brains were rapidly removed and divided into 6 parts designated as frontal lobes, frontal cortex, occipitotemporal cortex, occipital lobes, pons and cerebellum. The white matter was excised from the regions designated as frontal and occipitotemporal cortex. Radioactivity in these regions and the plasma samples were measured using a gamma counter and the amount of leakage of ^{125}I-labelled RISA into brain tissue was expressed as a percentage of plasma radioactivity, as follows:

$$\frac{\text{cpm/mg of brain}}{\text{cpm/µl plasma}} \times 100 = \%\ \text{protein transfer}$$

Values of the mean percent protein transfer in the whole brain as well as the mean percent protein transfer in individual brain regions in the different experimental groups were analysed using the analysis of variance. In addition, results were compared using the Wilkinson repeated measures analysis of variance in which correlation between regions is taken into account.

In the hypertensive non-drug treated and hypertensive drug-treated groups, only rats having Evans blue lesions were used for the statistical analysis.

Results

The blood pressure response to the infused norepinephrine was similar to our previous observations (Nag and Harik, 1987). The mean of the peak blood pressure level in the hypertensive group was 224 ± 8 while hypertensive rats injected with 2.5 mg/Kg flunarizine had a peak, mean pressure of 219 ± 16 mm Hg. Significant differences were not observed in the blood pressure of the hypertensive rats as compared with the drug-treated hypertensive rats.

The pH and blood gases of all rats were in the normal range.

Permeability to Evans blue: Hypertensive rats developed increased permeability to Evans blue in multifocal areas of the cortex, being most prominent in the occipital and occipitotemporal lobes.

Permeability to RISA: There was a significant (p < 0.001) increase in total RISA leakage in brains of hypertensive rats as compared with normotensive controls (Fig. 1). When individual regions of normotensive and hypertensive rats were compared there was some variation in the degree of permeability alterations with more significant increases in permeability in the frontal and occipital lobes and cerebellum (p < 0.001) than in the other regions studied (p < 0.01) (Fig. 2).

The total RISA leakage in brains of hypertensive rats pretreated with flunarizine IP was not significantly different from non-drug treated hypertensive rats. A significant decrease in total RISA leakage in brain was observed following pretreatment of hypertensive rats

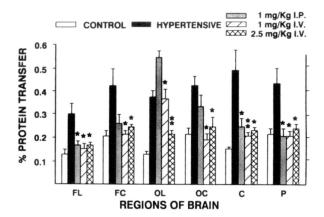

Fig. 2. Leakage of RISA in different regions of the brain in controls rats, norepinephrine-induced acutely hypertensive rats and flunarizine-treated hypertensive rats. The bars denote mean values ± S.D. The regions in which flunarizine significantly reduced permeability is marked with asterisks. *:p < 0.05; **:p < 0.005. *FL* = frontal lobes; *FC* = frontal cortex; *OL* = occipital lobes; *OC* = occipital cortex; *C* = cerebellum; *P* = pons

with IV flunarizine 1 mg/Kg (p < 0.005) and 2.5 mg/Kg (p < 0.001).

The amount of RISA leakage in individual brain regions was significantly reduced in hypertensive animals pretreated with flunarizine. Intraperitoneal administration of flunarizine was associated with a significant decrease in permeability in the frontal and occipital lobes, cerebellum and pons (p < 0.05). Intravenously administered flunarizine was more effective in reducing permeability in most brain regions. Administration of 2.5 mg/Kg Flunarizine was associated with a significant reduction of RISA leakage in the occipital lobes and cerebellum (p < 0.005) and in other brain regions (p < 0.05).

Acknowledgements

The skilled technical assistance of Mrs. Verna Norkum is gratefully acknowledged. This work was supported by the Heart and Stroke Foundation of Ontario.

References

1. Nag S (1988) Localisation of calcium-activated adenosine-triphosphatase (Ca^{2+}-ATPase) in intracerebral arterioles in acute hypertension. Acta Neuropathol (Berl) 75: 547–553
2. Nag S (1990) Ultracytochemical localisation of Na$^+$, K$^+$-ATPase in cerebral endothelium in acute hypertension. Acta Neuropathol (Berl) in press
3. Nag S, Harik SI (1987) Cerebrovascular permeability to horseradish peroxidase in hypertensive rats: effects of unilateral locus ceruleus lesion. Acta Neuropathol (Berl) 73: 247–253

Correspondence: Dr. S. Nag, Department of Pathology (Neuropathology) and Kingston General Hospital, Queen's University, Kingston, Ontario, Canada K7L 3N6.

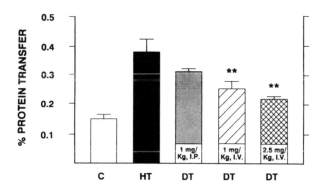

Fig. 1. Total RISA leakage in brain in norepinephrine-induced acute hypertension. The bars denote mean values ± S.D. There is a significant increase in permeability in brains of hypertensive rats as compared with the normotensive controls. Pretreatment with flunarizine IP did not have a significant effect on the permeability alterations occurring in this model. A significant reduction in leakage of RISA was observed following IV administration of 1 mg/Kg and 2.5 mg/Kg of the drug. (*C* = control; *HT* = acutely hypertensive rats; *DT* = flunarizine-treated hypertensive rats; **:p < 0.005)

Acta Neurochirurgica, Suppl. 51, 346–347 (1990)

Glutamine Synthetase Inhibition Prevents Cerebral Oedema During Hyperammonemia

H. Takahashi, R. C. Koehler, S. W. Brusilow, and **R. J. Traystman**

Department of Anesthesiology/Critical Care Medicine, and Pediatrics, The Johns Hopkins Medical Institution, Baltimore, Maryland, U.S.A.

Summary

The relationship between cerebral oedema and cerebral glutamine accumulation was investigated during acute hyperammonemia in anesthetized rats. Six hours of ammonium acetate infusion resulted in an increase in cortical glutamine concentration and a decrease in specific gravity. Pretreatment with methionine sulfoximine inhibited glutamine synthetase, prevented the increase in glutamine during hyperammonemia, and prevented the decrease in specific gravity. We conclude that the increase in brain water content is linked to the glutamine accumulation derived from the detoxification of ammonia by glutamine synthetase.

Introduction

Brain swelling and elevated intracranial pressure frequently develop in hyperammonemia associated with hepatic insufficiency[3], congenital urea cycle enzyme defects[10], and Reye's Syndrome[9]. However, the pathophysiological mechanism for the increase in the brain water content is poorly understood. Because of the lack of the urea cycle enzyme system, the brain detoxifies the increased ammonia load mainly by synthesizing glutamine from glutamate and ammonia by glutamine synthetase, which is localized in astrocytes[4, 6]. Histopathological investigations have revealed swelling of astrocytes with hyperammonemia[7].

We hypothesized that the glutamine accumulation would play an important role in the development of brain oedema[2], and tested if the increase in brain water was linked to this glutamine accumulation by inhibiting glutamine synthetase with methionine sulfoximine (MSO).

Materials and Methods

Male Wistar rats (350–400 g) were anesthetized with pentobarbital intraperitoneally, tracheotomized and ventilated. A tail artery was cannulated for monitoring the blood pressure and blood sampling.

The right external jugular vein was cannulated for continuous infusion of 0.2 M of sodium acetate or 0.2 M of ammonium acetate for 6 hr (55 µmol/kg/min). Three hours before the infusion, rats were pretreated intraperitoneally with either 1 ml/kg of sterile water or 150 mg/kg of MSO.

Brain biopsy was performed at the end of the six hour infusion. In one cohort from each group, a cortical specimen was rapidly freeze-clamped for later analysis of glutamine synthetase activity[8] and glutamine by HPLC. In another cohort, brain water content was measured of a cortical specimen by the specific gravity technique[5]. Plasma ammonia was assayed by the cation exchange-colorimetric technique[1] and plasma osmolarity was measured by freezing point depression. All values were expressed as means ± SE.

Results

With sodium acetate infusion, plasma ammonium levels were $31 \pm 4 \mu M$ and the cortical glutamine concentration was $5.6 \pm 0.4 \mu mole/g$ wet tissue (n = 5). With six hours of ammonium acetate infusion, plasma ammonium levels were $583 \pm 34 \mu M$ and cortical glutamine ($18.8 \pm 1.6 \mu mole/g$; n = 7) was significantly higher. Pretreatment with MSO inhibited cortical glutamine synthehase activity by 64% in both sodium and ammonium acetate groups. With MSO plus sodium acetate administration, plasma ammonium levels were $104 \pm 13 \mu M$ and cortical glutamine concentration was $1.8 \pm 0.4 \mu mole/g$ (n = 6). With MSO plus ammonium acetate, plasma ammonium levels were $692 \pm 40 \mu M$, but cortical glutamine levels ($2.6 \pm 0.4 \mu mole/g$; n = 8) were significantly less than in either group not pretreated with MSO.

Cortical specific gravity was 1.0452 ± 0.0003 (n = 11) in the control group receiving sodium acetate, whereas it was significantly decreased in the ammonium acetate group (1.0424 ± 0.0003; n = 11). This decrease in specific gravity was not observed in the ammonium acetate group pretrated with MSO (1.0446 ± 0.0003; $n = 9$). With MSO plus sodium acetate, specific gravity was significantly elevated (1.0462 ± 0.0003; $n = 12$).

Discussion

This study demonstrates that the increased brain cortical water content resulting from acute hyperammonemia is associated with elevated brain glutamine levels, and that reducing glutamine accumulation by MSO ameliorates brain oedema. Glutamine synthesis from glutamate and ammonia in astrocytes has been known to be the major ammonia detoxication mechanism in brain. However, the consequent $13.2\,\mu mol/g$ increase in glutamine could represent as much as a 3% increase in idiogenic osmoles if there were no other major changes in osmotically active constituents. The decrease in specific gravity represents approximately a 2–3% increase in the concentration of water and thus corresponds approximately to that predicted by the osmotic effect of glutamine accumulation. Moreover, if glutamine accumulates primarily in astrocytes the osmotic load in this compartment would be greater than reflected by the whole tissue measurements, thereby explaining astrocyte swelling seen by others[7]. Thus, our results suggest that glutamine accumulation increases the intracellular osmoles and that this osmotic load might directly cause brain oedema.

References

1. Brusilow SW, Batshaw ML, Waber L (1982) Neonatal hyperammonemic coma. Adv Pediatr 29: 69–103
2. Brusilow SW, Traystman RJ (1986) Letter to the Editor. N Engl J Med 314: 786
3. Butterworth RF, Giguere JF, Michaud J, Lavoie J, Layrargues GP (1987) Ammonia: key factor in the pathogenesis of hepatic encephalopathy. Neurochem Pathol 6: 1–12
4. Cooper AJL, Plum F (1987) Biochemistry and physiology of brain ammonia. Physiol Rev 67: 440–519
5. Marmarou A, Poll W, Shulman K, Bhagavan H (1978) A simple gravimetric technique for measurement of cerebral edema. J Neurosurg 49: 530–537
6. Norenberg MD, Martinez-Hernandez A (1979) Fine structural localization of glutamine synthetase in astrocytes of rat brain. Brain Res 161: 303–310
7. Pilbeam CM, McD Anderson R, Bhathal PS (1983) The brain in experimental portal-systemic encephalopathy. II. Water and electrolyte changes. J Pathol 140: 347–355
8. Rao SLN, Meister A (1972) *In vivo* formation of methionine sulfoximine phosphate, a protein bound metabolite of methonine sulfoximine. Biochemistry 11: 1123–1127
9. Shaywitz AB, Rothstein P, Venes JL (1980) Monitoring and management of increased intracranial pressure in Reye syndrome: Result in 29 children. Pediatrics 66: 198–204
10. Zimmerman S, Bachman C, Columbo J (1981) Ultrastructural pathology in congenital defects of the urea cycle: ornithine transcarbamylase and carbamyl phosphate synthetase deficiency. Virchows Arch [A] 393: 321–331

Correspondence: Dr. R. C. Koehler, Department of Anesthesiology & Critical Care Medicine, The Johns Hopkins Hospital, 600 N. Wolfe Street, Baltimore, MD 21205, U.S.A.

Acta Neurochirurgica, Suppl. 51, 348–350 (1990)

Hydrocephalic Oedema in Normal-Pressure Hydrocephalus

N. Tamaki, T. Nagashima, K. Ehara, T. Shirakuni, and **S. Matsumoto**

Department of Neurosurgery, Kobe University School of Medicine, Kobe, Japan

Summary

The hydrocephalic oedema in normal-pressure hydrocephalus (NPH) was evaluated by measurement of the relaxation time of protons of the water molecules of brain tissue. Patients with NPH were divided into two groups: shunt responders and shunt non-responders. In the group of shunt responders both T1 and T2 of periventricular white matter were significantly prolonged compared to those of controls, and shortened after shunting. Both T1 and T2 of white matter were significantly longer than of gray matter, while a reversed relationship was seen in normal controls. However, in the group of shunt non-responders, T1 of white matter was significantly prolonged, while T2 of the same area not. There was no change in either T1 or T2 of this region after shunting. Both T1 and T2 were almost the same in white and gray matter in shunt non-responders. It is suggested that periventricular abnormalities seen in various diseases may be distinguished on the basis of the relaxation behavior of protons of tissue water.

Introduction

It is generally accepted that the relaxation times of protons in measurements of nuclear magnetic resonance are very sensitive to assess changes in water content of the tissue, and possibly in the state of biological water binding.

Hydrocephalic oedema in normal-pressure hydrocephalus (NPH) was evaluated by measurement of relaxation times of the protons of brain water: a differentiation of periventricular white matter abnormalities was thereby investigated. A regression of periventricular oedema after ventricular shunting in patients with NPH was also documented.

Materials and Methods

Twenty-one patients at a mean age of 64 with NPH underwent ventriculoperitoneal shunting. The diagnosis was made on the basis of the clinical findings. The causes of NPH included idiopathic origin in 13 patients, head injury in 5, and intracranial bleeding in 3. Patients were classified into three groups: the first group consisted of 14

patients with "true" NPH, who responded to shunting, the second of 7 patients with suspected but not confirmed NPH, who did not respond to shunting, and the third of 17 control patients with a mean age of 64, who had no proven intracranial organic diseases.

Magnetic resonance (MR) imaging was performed in a total of 38 patients for obtaining longitudinal (T1) and transverse relaxation time (T2) images of the brain with a 0.15 Tesla, resistive MRI unit (Picker International). The relaxation times were read directly from the calculated T1 and T2 images produced by a two-point method by computer algorithms using a combination of two different pulse sequences. T1 and T2 were measured in regions of interest, such as the periventricular white matter and cortical gray matter. Post-shunt MR imaging was also performed, and T1 and T2 measurements were made in the same fashion at an average of 51 days after ventricular shunting, varying from 15 to 134 days.

Results

While both T1 and T2 were significantly longer in gray matter than in white matter in the control group, they were longer in white matter than in gray matter of the true NPH group. There was, however, no significant difference in either T1 or T2 between gray matter and white matter in the "false", i.e. not confirmed NPH group.

T1 and T2 of the regions of interest were compared between the three groups. The preshunt T1 of the white matter in the true NPH group was significantly prolonged when compared to that of the controls or of the "false" NPH group. T1 of white matter in the "false" NPH group was also significantly longer than in the control group. However, there was no difference in T1 of gray matter among the three groups (Fig. 1, upper panel). As far as the preshunt T2 in white matter is concerned, this was significantly prolonged in the true NPH group when compared to the control and false NPH groups. There was no difference in T1 between the false NPH group and the controls nor in

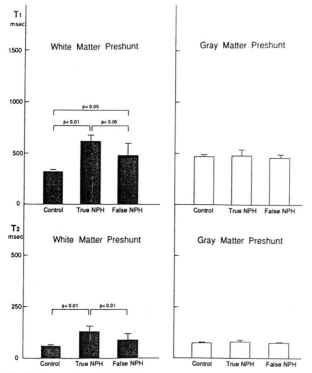

Fig. 1. Comparison of T_1 & T_2 among control, true NPH and false NPH groups

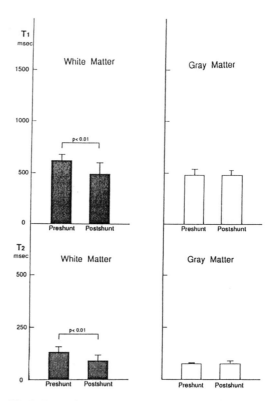

Fig. 2. Comparison between pre- and postshunt T_1 & T_2 in true NPH

preshunt T2 of gray matter among the three groups (Fig. 1, lower panel).

While there was no change in T1 and T2 of gray matter after shunting, both T1 and T2 of white matter in the true NPH group became significantly shorter after shunting (Fig. 2). In the false NPH group, however, T1 and T2 did not change after shunting in both areas. When T1 and T2 after CSF shunting in the true NPH group were compared with those of the control group, both T1 and T2 of the white matter in the true NPH group were still significantly prolonged, even after an average of 51 days. There was no difference in T1 and T2 after shunting in gray matter between the true NPH and control group.

Discussion

There is a growing awareness that only part of the tissue water is freely diffusable in the tissue, ("free" water). A considerable portion is "bound" or "structured". Free water might correspond to bulk water, the properties and relaxation times of which might be similar to water in an aqueous solution. Conversely, bound water is motionally restricted, with much shorter relaxation times than bulk water. This might apply for the water bound to high molecular substance or membrane surfaces in hydration layers[6, 7, 10, 13].

Both T1 and T2 were prolonged in lesions associated with an increased water content[2, 3]. While prolongation of T1 was found to be correlated with an increased water content and decreased protein concentration[4, 14], T2 is influenced not only by the water content, but also by the physical environment of the protons[1, 10]. It is suggested that T2 is related to the free water fraction, characterized by a high molecular mobility and high diffusability[11].

Early experience has indicated that MR imaging is especially useful for investigations of periventricular abnormalities and to distinguish various forms of dementia. But periventricular hyperintensity proved to be non-specific for some diseases. Periventricular hyperintensity in T2-weighted images has been observed in a variety of diseases causing periventricular white matter abnormalities including oedema, ischaemia, gliosis, encephalomalacia, haemorrhage, neoplasm, axonal loss, degeneration, demyelination, dysmyelination, necrosis and lipid changes[5, 9, 12, 17]. Many authors suggested that normal subjects, especially elderly frequently show a periventricular hyperintensity in T2 weighted images[8, 12, 17]. The reason why normal subjects have a periventricular hyperintensity in T2 weighted images may relate with the fact that the regions anterior and lateral to the frontal horns have a loose network of axons with a low myelin content, and with an ependymitis granularis representing as a patchy loss of the ependyma associated with gliosis. Further, the flow of interstitial fluid in this region of the brain amounting to 33% of total CSF formation[13] tends to converge there[16]. These factors contribute to an increased water content in this area, resulting in periventricular hyperintensity in normal T2 weighted images. Thus, the images alone do not provide for a differentiation between various dementia states in the elderly, nor for a prediction of the response to ventricular shunting in patients with NPH.

The results obtained in the present study indicate that the greater prolongation of T2 of the periventricular white matter in the "true" NPH group as compared to the "false" NPH group might be attributable to an increase in the free water fraction. The lower prolongation of T2 compared to T1 in periventricular white matter of the "false" NPH group might be due to differences in molecular properties and motility of the water protons related to degeneration, ischaemia or demyelination of the periventricular white matter. The difference in relaxation behaviour of the periventricular white matter between the "true" and "false" NPH group in the present study, therefore, may be explained by different properties of the two fractions of biological water and the physico-chemical environment.

References

1. Barnes D, McDonald WI, Jonson G *et al* (1987) Quantitative nuclear magnetic resonance imaging: Characterization of experimental cerebral oedema. J Neurol Neurosurg Psychiatry 50: 125–133
2. Bederson JB, Bartkowski HM, Moon K *et al* (1984) Nuclear magnetic resonance imaging and spectroscopy in experimental brain edema in a rat model. J Neurosurg 64: 795–802
3. Belton PS, Packer KJ (1974) Pulsed NMR studies of water in striated muscle. III. The effects of water content. Biochim Biophys Acta 354: 307–314
4. Bottomley PA, Hardy CJ, Argersinger RE *et al* (1987) A review of H-1 nuclear magnetic resonance relaxation in pathology: Are T1 and T2 diagnostic? Med Phys 14: 1–37
5. Bradley WG Jr, Waluch V, Brant-Zawadzki M *et al* (1984) Patchy, periventricular white matter lesions in the elderly: A common observation during NMR imaging. Noninvasive Med Imag 1: 35–41
6. Cooke R, Kuntz ID (1974) The properties of water in biological systems. Ann Rev Biophys Bioeng 3: 95–126
7. Furuse M, Gonda T, Inao S *et al* (1987) Thermal analysis on water compounds in brain tissue. Quantitative determination of free and bound water fractions. Brain Nerve (Tokyo) 39: 761–767
8. George AE, de Leon MJ, Kalnin A *et al* (1986) Leucoencephalopathy in normal and pathologic aging: 2. MRI of brain lucencies. AJNR 7: 567–570
9. Gerard G, Weisberg LA (1986) Magnetic resonance imaging in adult white matter disorders and hydrocephalus. Semin Neurol 6: 17–27
10. Go KG, Edzes HT (1975) Water in brain edema. Observations by the pulsed nuclear magnetic resonance technique. Arch Neurol 32: 462–465
11. Hansen JR (1971) Pulsed NMR study of water mobility in muscle and brain tissue. Biochim Biophys Acta 230: 482–486
12. Kertesz A, Black SE, Tokar G *et al* (1988) Periventricular and subcortical hyperintensities on magnetic resonance imaging. 'Rims, caps, and unidentified bright objects'. Arch Neurol 45: 404–408
13. Pollay M, Curi F (1967) Secretion of cerebrospinal fluid by ventricular ependyma of the rabbit. Am J Physiol 213: 1031–1038
14. Saryan LA, Hollis DP, Economou JS *et al* (1974) Brief communication: Nuclear magnetic resonance studies of cancer. IV. Correlation of water content with tissue relaxation times. J Natl Cancer Inst 52: 599–602
15. Sze G, De Armond SJ, Brant-Zawadzki M *et al* (1986) Foci of MRI signal (Pseudo lesions) anterior to the frontal horns: Histologic correlation of a normal finding. AJNR 7: 381–387
16. Tamaki N, Yamashita H, Kimura M *et al* (1990) Changes in the components and content of biological water in the brain in experimental rabbit hydrocephalus. J Neurosurg (in press)
17. Zimmerman RD, Fleming CA, Lee BCP *et al* (1986) Periventricular hyperintensity as seen by magnetic resonance. Prevalence and significance. AJNR 7: 13–20

Correspondence: N. Tamaki, M.D., Department of Neurosurgery, Kobe University School of Medicine, 5-1 Kusunoki-Cho, 7-Chome, Chuo-Ku, Kobe 650, Japan.

Acta Neurochirurgica, Suppl. 51, 351–353 (1990)

Local Cerebral Blood Flow and Glucose Metabolism in Hydrostatic Brain Oedema

H. Umezawa, K. Shima, H. Chigasaki, K. Sato[1], and S. Ishii[1]

Department of Neurosurgery, National Defence Medical College Saitama and
[1] Department of Neurosurgery, Juntendo University Tokyo, Japan

Summary

Two hydrostatic factors such as acute hypertension and decompressive craniectomy were chosen and assessment was focused on how the hydrostatic pressure gradient altered the cerebrovascular dynamics and metabolism during the process of development of brain oedema. Hydrostatic oedema was induced by bolus injection of autologous blood through the common carotid artery in Sprague-Dawley rats. Rats were divided into two groups, one with craniectomy (Cr+) and the other without craniectomy (Cr−). Animals were sacrificed immediately, 24 and 48 h after the hypertensive insult. Brain water content was determined by the gravimetric method. Regional cerebral blood flow (rCBF) and glucose metabolism (lCGU) were measured by quantitative autoradiographic methods using ^{14}C-iodoantipyrine and ^{14}C-deoxyglucose, respectively. The hypertensive insult produced multifocal lesions stained by Evans blue. In the brains of the Cr− group, there was a transient increase in water content and no significant change of rCBF and lCGU. In the Cr+ group, the increase in water content was pronounced and continued until 48 h later. In addition, misery perfusion was observed at 24 h after the insult and both rCBF and lCGU were significantly decreased after 48 h. These results indicate that the increased hydrostatic pressure gradient enhances tissue damage and causes the reopening of blood-brain barrier.

Introduction

Recent study has revealed that the hydrostatic pressure gradient is closely related to the biomechanical aspect of the formation and spread of brain oedema. This study was designed to identify the effect of the gradient between arterial pressure and intracranial pressure on the hydrostatic factors and how these factors affect the net movement of oedema fluid and resultant changes from the cerebrovascular dynamics and metabolism. For this purpose, two hydrostatic factors, hypertension and decompressive craniectomy, were chosen.

Materials and Methods

Ninety six Sprague-Dawley rats weighing 300–400 g were used. After transoral intubation, the animals were anesthetized with halothane and ventilated with a respirator.

Under aspectic technique, the left common carotid and external carotid artery, femoral artery and vein were cannulated. The animals were divided into two groups, one with craniectomy (Cr+ group) and the other without craniectomy (Cr− group). The craniectomy was performed at the left frontoparietal skull. The hypertensive insult was induced via the left common carotid artery with a bolus injection of heparinized blood of 2 ml. The infusion pressure was measured via the left external carotid artery. The ICP was monitored throughout the experiment using a catheter inserted into the cisterna magna. Blood gases were maintained within normal range. As a tracer of BBB, 2% EB was injected intravenously.

In order to determine the primary and secondary leakage of macromolecules through the BBB, injection of EB was performed at different time intervals.

Group 1: injection prior to hypertensive insult.

Group 2: injection at 24 hours after hypertensive insult.

Rats in group 1 were sacrificed at 0, 24 and 48 hours after the insult and those in group 2 were sacrificed 48 hours later. After sacrifice, the brains were removed and cut in serial coronal sections. A small piece of tissue was taken for the measurement of water content. Water content was determined with a gravimetric method. Regional cerebral blood flow (rCBF) was measured at 56 regions using iodo [^{14}C] antipyrine according to the method of Sakurada et al.[5]. Local cerebral glucose utilization (lCGU) was also determined at the same region using [^{14}C] deoxyglucose[7]. The rats with their left common carotid arteries ligated (sham-operated) were used as controls. The infusion pressure of the Cr+ group was 395 mm Hg and that of the Cr− group was 408 mm Hg. There was no significant difference in the infusion pressure between both groups.

Results

The extravasation of EB was observed in nearly all animals in both the Cr+ and Cr− groups and rec-

Table 1. *Blood Pressure and ICP Response in Unilateral Hypertensive Insult*

	SBP (mmHg)			ICP (mmHg)		
	Resting value	During maneuver	After maneuver	Resting value	During maneuver	After maneuver
Cr (−) group (n=5)	127.4±2.98	145.8±6.24	133.8±4.44	6.4±0.83	22.1±5.38	7.6±0.97
Cr (+) group (n=5)	126.8±6.03	130.4±6.37	129.0±4.28	2.6±0.68 *	9.0±1.65 **	3.0±0.54 *

* P<0.005 VS Cr (−) group
** P<0.025 VS Cr (−) group

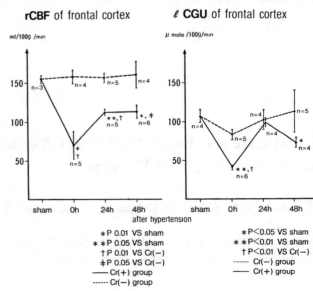

rCBF of frontal cortex   ℓCGU of frontal cortex

* P 0.01 VS sham
** P 0.05 VS sham
† P 0.01 VS Cr(−)
‡ P 0.05 VS Cr(−)
——— Cr(+) group
------- Cr(−) group

* P<0.05 VS sham
** P<0.01 VS sham
† P<0.01 VS Cr(−)
------- Cr(−) group
——— Cr(+) group

Fig. 2. The change of rCBF and lCGU in both Cr+ and Cr− groups immediately, 24 and 48 hours after hypertensive insult

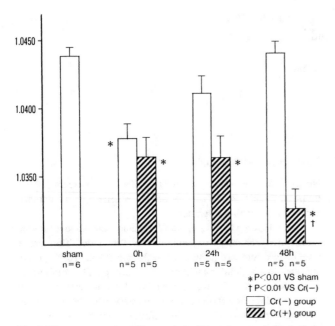

* P<0.01 VS sham
† P<0.01 VS Cr(−)
☐ Cr(−) group
▨ Cr(+) group

Fig. 1. Water content (specific gravity) of frontal cortex in both Cr+ and Cr− groups immediately, 24 and 48 hours after hypertensive insult

ognized as patchy spots over the frontoparietal cortex, predominantly at the watershed areas of the anterior and middle cerebral arteries, and the thalamus, the caudate putamen complex, and hippocampus. Twenty four and 48 hours after the insult, diffuse and homogenous extravasation of EB was observed. Compared with the Cr− group, the extravasated EB spots were usually more dense and widely distributed in the Cr+ group. There was no EB extravasation in the Cr− group when the dye was injected 24 hours after the hypertensive insult. However, in the craniectomized animals in which EB was injected at 24 hours after the insult and sacrifice was at 48 hours, extravasated EB spots were recognized in the parietal cortex, confined to the area behind the craniectomized portion. In the resting state, the ICP in the Cr+ group was significantly lower than that in the Cr− group. During the hypertensive insult, the ICP of the Cr+ group was

9.0 mm Hg, whereas that of the Cr− group reached to 22.1 mm Hg (Table 1). Specific gravities of the Cr− group were transiently decreased immediately after the insult but gradually returned to the control value. On the other hand, the value of the Cr+ group was significantly lower immediately after the insult and this tendency was most pronounced at 48 hours later (Fig. 1). The rCBF of the Cr− group was not significantly different from that of the sham operated controls in all portions examined. On the other side in the Cr+ group, the value of the frontal cortex was most significantly reduced and reached 45% of the sham operated controls immediately after the insult. Then rCBF showed a tendency to increase 24 hours later to 71% of that of the sham operated controls (Fig. 2). There was a transient decrease in lCGU in the Cr− group. In the Cr+ group, a marked decrease in lCGU was recognized immediately after the insult to 38% of that of the sham operated controls. After 24 hours, lCGU returned to the control level. At 48 hours later, however, a secondary decrease of lCGU ocurred to 68% of the control value (Fig. 2).

Discussion

Arterial hypertension is commonly used as a method for reversible opening of BBB. The mechanism of extravasation of EB albumin is thought to be caused by opening of the tight junctions due to the forced dilatation of the vessels and/or by the increase of pinocytic vesicles. From the aspect of pressure difference between vessels and tissue, the hydrostatic pressure gradient seems to be a major factor in determining both the degree of the initial insult and the successive deterioration, especially during and after the hypertensive insult. According to a previous report of our laboratory, hydrostatic brain oedema is initiated by the hydrostatic pressure gradient, either between blood vessels and tissue or between areas with different tissue pressure[2]. In this study, the greater hydrostatic pressure gradient introduced by a combination of craniectomy and hypertension caused more dense and widely distributed EB extravasation than hypertension alone. Furthermore, a significant increase of water content developed immediately after the insult and lasted up to 48 hours later. In contrast, there was only a transient increase of water content in the Cr− group. Kogure *et al.* reported that protein leakage is presumably accompanied by an equal bidirectional water movement, and that the extravasation of macromolecules per se does not necessarily cause brain oedema as long as cerebral metabolism remains normal[1]. The explanation of metabolic derangements related to the edematous process and tissue pressure change derived from the hydrostatic factors is as follows. After craniectomy, the brain tissue pressure of craniectomized portion is reduced to nearly the atmospheric pressure. Thus the tissue pressure gradient, once developed, seems to be amplified and to last longer[2]. The prolonged maintenance of a tissue pressure gradient, either between craniectomized and non-craniectomized hemispheres, or between well perfused and watershed areas, is thought to be a major driving force, propelling extravasation of plasma-like fluid[2].

Kontos *et al.* noted that acute severe hypertension causes both prolonged vasodilation and morphological and metabolic damage of arteriolar endothelium[3]. In our model, with an amplified hydrostatic pressure gradient, the damaged vascular bed is rendered leaky, resulting in profound cerebral oedema. The microvasculature is thought to be secondarily compromised by the accumulated oedema fluid and the microcirculation is impaired due to the increased vascular resistance. The cerebral metabolism is reduced subsequently in the acute phase. At 24 hours after the insult, misery perfusion was observed in the Cr+ group. It is reasonable to assume that the structural changes due to oedema alter the environment of an essentially normal brain leading to a relative ischaemic state with resultant metabolic damage[6]. At 48 hours after the insult, both rCBF and lCGU were reduced in the Cr+ group and a spongy state of cerebral structures was recognized in this group. Under this condition, reopening of BBB was recognized. In this experiment, the initial BBB opening was elicited by the hydrostatic pressure alone.

The hydrostatic oedema resulted in a derangement of the cerebral microcirculation and metabolism leading to anaerobic glycolysis. The oedema was prolonged or aggravated at this stage, accompanied in part by cytotoxic factors. Thus, a vicious circle of oedema development may appear in the brain tissue when both cerebrovascular dynamics and metabolism are impaired. With morphological and metabolic damage of the vascular wall, the second BBB opening was elicited which is similar to that in vasogenic oedema.

References

1. Kogure K, Busto R, Scheinberg P (1981) The role of hydrostatic pressure in ischemic brain edema. Ann Neurol 9: 273–282
2. Koike J (1983) Permeability change of cerebral vessels in acute hypertension and external decompression. Neurol Med Chir (Tokyo) 23: 325–335
3. Kontos H, Wei E *et al* (1981) Mechanism of cerebral arteriolar abnormalities after acute hypertension. Am J Physiol 240 (Heart Circ Physiol 9): H511–H527
4. Kuroiwa T, Ting P, Martinez H, Klatzo I (1985) The biphasic opening of the blood-brain barrier to proteins following temporary middle cerebral artery occlusion. Acta Neuropathol (Berl) 68: 122–129
5. Sakurada O, Kennedy C, Sokoloff L (1978) Measurement of local cerebral blood flow with iodo (^{14}C) antipyrine. Am J Physiol 234: H59–H66
6. Sokrab T, Johansson B (1988) A transient hypertensive opening of the blood-brain barrier can lead to brain damage. Acta Neuropathol (Berl) 75: 557–565
7. Sokoloff L, Sakurada O, Shinohara IN (1977) The ^{14}C deoxyglucose method for the measurement of local cerebral glucose utilization: theory, procedure and normal values in the conscious and anesthetized albino rat. J Neurochem 28: 899–916

Correspondence: H. Umezawa, M.D., Department of Neurosurgery, National Defence Medical College Saitama, Japan.

Acta Neurochirurgica, Suppl. 51, 354–356 (1990)
© by Springer-Verlag 1990

Changes in Free Water Content and Energy Metabolism of the Brain in Experimental Hydrocephalus

N. Tamaki, T. Nagashima, K. Ehara, M. Kimura, S. Matsumoto, and **N. Iriguchi**

Department of Neurosurgery, Kobe University School of Medicine, Kobe and Asahi Chemical Industry Inc.,
System Engineering Laboratory, Atsugi, Japan

Summary

In acute and subacute hydrocephalus periventricular oedema is most prominent. At these stages of hydrocephalus, the free water content is increased and the bound water content, to the contrary, significantly decreased in the periventricular white matter. The bioenergetic state is also altered. In the chronic stage the ratio of free-to-bound water content returns to a level near the control value, leading to a decrease of periventricular oedema by formation of alternative pathways of CSF absorption. The bioenergetic state was slightly altered at this stage.

Introduction

The aim of this work is to investigate changes in free water content of the brain, the evolution of periventricular oedema, and the cerebral energy metabolism in experimental hydrocephalus.

Materials and Methods

Thirty four Japanese white rabbits (mean body weight, 3.2 kg) were used for measurement of free water content of hydrocephalic brain, and 6 mongrel dogs (mean body weight, 11.5 kg) were used for NMR-measurement of P-31 spectra.

Experimental hydrocephalus was induced in 28 rabbits and 5 dogs by percutaneous injection of 0.1.–0.2 ml/kg of 50% kaolin into the cisterna magna. The remaining 6 rabbits and 1 dog had no treatment and were used as controls.

Animals were divided into three groups according to the interval between the cisternal injection of kaolin and measurement of the free water content or study of magnetic resonance spectroscopy. The first group consisting of 6 rabbits and 1 dog had acute hydrocephalus, with an interval of 1 week or less; the second group consisting of 14 rabbits and 2 dogs had subacute hydrocephalus, with an interval of 1–3 weeks; and the third group consisting of 8 rabbits and 1 dog had chronic hydrocephalus; with an interval of 3 weeks or more.

Measurement of free water content was made as follows. The animals were sacrificed by an overdose of pentobarbital on the day of observation. A coronal section was made at the level of the re-trochiasmal region after removal of the brain by calvarial craniectomy. The percentage of total water content of the periventricular white matter and the cortical gray matter was measured by the conventional drying-weighing method using the following equation:

$$\% \text{ Total water content} = \frac{\text{wet weight} - \text{dry weight}}{\text{wet weight}} \times 100$$

The percentage of free and bound water was measured by thermal analysis using differential scanning calorimetry.

The amount of free water was calculated from the area under the thermal curve, which corresponded to the calorific value in melting the ice formed by free water in the sample frozen at $-60\,°C$. Calculation of free water content was determined by the thermal curve, and the percentage of free and bound water content, was made by using a computer on the basis of the following equations:

$$\text{Free water content (mg)} = \frac{\text{Calorific value (cal)}}{79.7 \times 10^{-3} (\text{cal/mg})}$$

$$\text{Total water content (mg)} = \text{wet weight} \times \frac{\% \text{ total water content}}{100}$$

$$\% \text{ Free water content} = \frac{\text{free water content (mg)}}{\text{total water content (mg)}} \times 100$$

$$\% \text{ Bound water content} = \frac{\text{bound water content (mg)}}{\text{total water content (mg)}} \times 100$$

MR imaging and measurement of P-31 MR spectra were performed as follows. An Asahi Mark J 200 whole-body MR imaging system with a 2.0 T, 100-cm-bore superconducting magnet was used at 34.35 MHz for P-31 spectroscopy and at 85.2 MHz for proton imaging. The calculated T1- and T2 images were obtained by combination of the two kinds of pulse sequences. The extent of the periventricular oedema was assessed on the basis of the periventricular high signal intensity zone on T1 and T2 images. P-31 MR spectroscopy was then carried out using an $8 \times 5\,cm$ square, two-turn surface coil as a radio-frequency coil, which was positioned on the head of the dog and tightly secured. During MR imaging of the brain and measurement of P-31 MR spectra, the animals were kept under anaesthesia by intravenous administration of 20 mg/kg pentobarbital, followed by additional intraperitoneal injections.

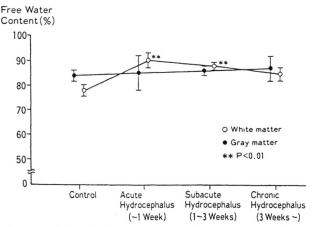

Fig. 1. Graph showing the percentage of free water content of periventricular white matter and cortical gray matter in control animals and in animals with acute, subacute, and chronic hydrocephalus. Values are expressed by means ± standard deviation

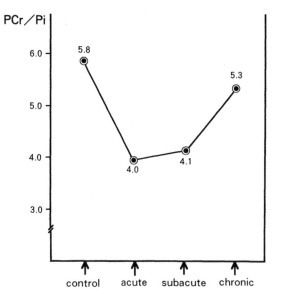

Fig. 2. Time course of the PCr Pi ratio in acutely, subacutely and chronically hydrocephalic brains

Results

There was neither a statistically significant difference in total water content of the periventricular white matter nor the cortical gray matter between control animals and the animals in any stage of hydrocephalus.

However, the free water content increased most significantly in acute hydrocephalus, while gradually in subacute hydrocephalus with a statistically significant difference as compared to that of the control animals. The free water content in chronic hydrocephalus decreased then to the normal level. There was no statistically significant difference in free water content of the cortical gray matter between control animals and hydrocephalic animals in any stage of hydrocephalus (Fig. 1).

The bound water content of the periventricular white matter decreased significantly in acute and subacute hydrocephalus. The bound water content of the periventricular white matter was lowest in acute hydrocephalus, and increased gradually in subacute and chronic hydrocephalus. It returned to the normal range in chronic hydrocephalus. However, there was no statistically significant difference in the bound water content of the gray matter between control animals and hydrocephalic animals in any stage.

The periventricular oedema was most prominent in acute and subacute stages, when the free water content of the periventricular white matter was significantly increased. There was less prominent periventricular oedema in the chronic stage, when the free water content returned to the control level.

The P-31 spectra obtained from the intact brain of the control animal (lower) showed the characteristic resonances of the α, β, and γ P-31 nuclei of adenosine triphosphate (ATP), phosphocreatine (PCr), phosphodiesters (PD), inorganic phosphate (Pi), and phosphomonoesters (PM).

The P-31 spectra obtained from an acutely hydrocephalic animal showed no significant change in ATP signals, but the PCr relative to Pi was reduced. The spectra obtained from the brain of a chronically hydrocephalic animal at 50 days after induction showed a peak pattern almost identical to that seen in the control.

The PCr/Pi ratio was 4.0 in acute and 4.1 in subacute and 5.3 in the chronic stage, while that observed in the control animal was 5.8. There was a decline in PCr/Pi ratio in the acute and subacute stages, when the periventricular oedema was prominent. The PCr/Pi ratio in the chronically hydrocephalic animal, which showed mild periventricular oedema on a T1 image, returned to a level near the control value. Thus, the bioenergetic state was altered in the acute and subacute stages of hydrocephalus but only little in chronic hydrocephalus (Fig. 2).

Discussion

It is generally agreed that periventricular oedema occurs in the development of hydrocephalus[3, 9, 10, 15], although there are disagreements as to the changes in water content of the hydrocephalic brain[2, 5]. Evidence of an increase in extracellular space in the periventricular white matter[11, 13] is consistent with an increase in water content in the same area[12]. There are no reports on the changes in the components of biological water in hydrocephalic brain. The present study suggests that free water migrates into the extracellular space of the periventricular white matter at the acute and subactue stages, resulting in the development of interstitial oedema. There is a decrease in the bound water content in the same area, which was thought to be due to brain damage, such as demyelination of the white matter[14]. At the chronic stage, there was a decrease of the free water content as the development of the alternative pathway of cerebrospinal fluid absorption was enhanced[1, 4]. The bound water returned to the control level. The absorption of free water may be enhanced by its biological properties: free mobility and diffusability[6, 8].

Free water accumulation in the periventricular white matter most prominently seen in the acute and subacute stages of hydrocephalus may cause white matter damage[14] and alterations of energy metabolism of the hydrocephalic brain[7]. A mitochondrial dysfunction may be the cause of the alteration of energy metabolism of the hydrocephalic brain. The altered energy metabolism may recover in chronic stage, when the free water decreases in the periventricular white matter, leading to recovery of the same tissue region.

References

1. Bradbury MWB, Cserr HF, Westrop RJ (1981) Drainage of cerebral interstitial fluid into deep cervical lymph of the rabbit. Am J Physiol 240: F329–F336
2. Del Bigio MR, Bruni JE (1987) Cerebral water content in silicone oil-induced hydrocephalic rabbits. Pediatr Neurosci 13: 72–77
3. Drake JM, Potts DG, Lemaire C (1989) Magnetic resonance imaging of Silastic-induced canine hydrocephalus. Surg Neurol 31: 28–40
4. Erlich SS, McComb JG, Hyman S *et al* (1986) Ultrastructural morphology of the olfactory pathway for cerebrospinal fluid drainage in the rabbit. J Neurosurg 64: 466–473
5. Fishman RA, Greer M (1963) Experimental obstructive hydrocephalus. Changes in the cerebrum. Arch Neurol 8: 156–161
6. Furuse M, Gonda T, Inao S *et al* (1987) Thermal analysis on water components in brain tissue-quantitative determination of free and bound water fractions. Brain Nerve 39: 761–767
7. Higashi K, Asahisa H, Ueda N *et al* (1986) Cerebral blood flow and metabolism in experimental hydrocephalus. Neurol Res 8: 169–176
8. Kuntz ID, Zipp A (1977) Water in biological systems. N Engl J Med 297: 262–266
9. Lux WE Jr, Hochwald GM, Sahar A *et al* (1970) Periventricular water content. Effect of pressure in experimental chronic hydrocephalus. Arch Neurol 23: 475–479
10. McLone DG, Bondareff W, Raimondi AJ (1971) Brain edema in the hydrocephalic hy-3 mouse: submicroscopic morphology. J Neuropathol Exp Neurol 30: 627–637
11. Ogata J, Hochwald GM, Cravioto H *et al* (1972) Light and electron microscopic studies of experimental hydrocephalus. Ependymal and subependymal areas. Acta Neuropathol (Berl) 21: 213–223
12. Rosenberg CA, Saland L, Kyner WT (1983) Pathophysiology of periventricular tissue changes with raised CSF pressure in cats. J Neurosurg 59: 606–611
13. Rubin RC, Hochwald GM, Tiell M *et al* (1976) Hydrocephalus. I. Histological and ultrastructural changes in the preshunted cortical mantle. Surg Neurol 5: 109–114
14. Rubin RC, Hochwald GM, Tiell M *et al* (1976) Hydrocephalus. II. Cell number and size, and myelin content of the preshunted cerebral cortical mantle. Surg Neurol 5: 115–118
15. Weller RO, Mitchel J (1980) Cerebrospinal fluid edema and its sequelae in hydrocephaly. Adv Neurol 28: 111–123

Correspondence: N. Tamaki, M.D., Department of Neurosurgery, Kobe University School of Medicine, 5-1, Kusunoki-cho 7-chome, Chuo-ku, Kobe 650, Japan.

Acta Neurochirurgica, Suppl. 51, 357–361 (1990)

Venous and Cerebrospinal Fluid Outflow in Patients with Brain Swelling and Oedema

A. R. Shakhnovich, A. Y. Razumovsky, S. S. Gasparjan, and **V. I. Ozerova**

Burdenko Neurosurgical Institute, Moscow, USSR

Summary

67 patients with benign intracranial hypertension (BIH) and 44 with normal pressure hydrocephalus (NPH) were examined by employment of infusion tests. Brain swelling (decrease of ventricular size with normal or increased brain tissue density) was a characteristic feature of BIH. It may result from venous outflow disturbances leading to vascular engorgement. But later, the process appears to be independent from the increase of the dural sinus pressure. This was normal in patients with BIH and NPH. Despite absorption disturbances there was a strong positive correlation in NPH between cerebrospinal fluid- and dural sinus pressure, while in BIH such a correlation was absent. The data confirm a pathogenesis of brain swelling in BIH as an obstacle to venous outflow at the level of the bridging veins and venous lacunae, however, not at the level of the dural sinuses.

Introduction

Brain oedema (BE) is an increase of the water content of brain tissue, whereas brain swelling (BS) in an increase of brain volume. Amongst different causes of BE and BS it is important to consider the significance of disturbances of the venous and cerebrospinal fluid (CSF) outflow. Patients with Benign Intracranial Hypertension (BIH) and Normal Pressure Hydrocephalus (NPH) may serve as an adequate clinical model to study these disturbances. The development of BS in patients with BIH[2] may be attributed to disturbances of venous outflow[3]. In both clinical forms periventricular oedema is found, which according to some authors[5] may provide evidence for presence of accessory pathways of absorption of CSF. Employment of infusion tests may be the most adequate clinical method to study venous and CSF outflow. Aim of our current study was to analyze disturbances of venous and CSF outflow as a mechanism of BS in patients with BIH or NPH by infusion tests.

Material and Methods

A study was carried out in 67 patients with BIH (12 males and 55 females, age 14–63 years) and in 44 patients with NPH (36 males and 8 females, age 18–73 years). 12 patients with BIH and 20 with NPH were treated with ventriculo-atrial, or ventriculo-peritoneal shunt operations (Hakim medium). 11 patients developed hydrocephalus following subarachnoid haemorrhage, 19 after head trauma, 8 after inflammatory disease, while 6 had NPH of unknown etiology. The main characteristics of BIH were papilloedema, raised CSF pressure (P_{scf}), normal or low protein concentration in CSF, absence of neurological symptoms and signs except of those attributable to an increased intracranial pressure. Occurrence of BIH was most frequent in young women with hormonal disorders. Pressure-volume measurements were made by the constant pressure infusion test[7, 13]. The infusion test allows us to determine elasticity of the cerebrospinal fluid system (EL), of the CSF outflow resistance (R), of the formation rate of CSF (F), and intrasinus pressure (P_{is}). Changes of ventricle size on CT scans were quantitatively assessed as bicaudate index – the ratio of the bicaudate diameter and the distance between the inner skull surface along the same line. P_{is} and F were not measured in all patients, because these require CSF drainage.

Results and Discussion

BS was a characteristic feature of BIH. The main BS characteristics on the CT scan were a decrease of ventricle size in patients with normal or increased density in the white matter of the cerebral hemispheres. Changes of the ventricular size in patients with BIH and NPH on CT scans were quantitatively assessed as bicaudate index. The results were compared with 25 age related persons of a healthy group. The bicaudate index of the BIH group was significantly smaller ($12.3 \pm 1.6\%$) than in the healthy controls ($18.8 \pm 0.9\%$). The results are in accordance with data of Grant et al.[2], indicating brain swelling in BIH as reflected by a reduced ventricle size. Swelling associated with an increased density should be due to an increased blood volume. The two causes of swelling of white matter are oedema and/or an increased cerebral blood volume (CBV) raising tissue density in CT scan. However, an increased density may mask corresponding brain oedema (BE), which also must be considered as a mechanism of brain swelling (BS).

All patients with NPH had ventricular enlargement. The bicaudate index was significantly larger (33.3 ± 1.8%) as compared with the controls. Periventricular brain oedema was found in some patients with BIH and NPH. The characteristic feature of periventricular oedema was a decreased density of the periventricular brain tissue mainly in the anterior horn region. According to some authors[11] a main reason of the changes in ventricle size is the difference in elasticity of the cerebrospinal fluid system. Our previous[9] and present data, however, do not support this conclusion. EL was increased both, in BIH ($0.76 \pm 0.39\,\text{ml}^{-1}$) and in NPH ($0.60 \pm 0.37\,\text{ml}^{-1}$) with the difference not being significant. Another reason for a change of the ventricle size may be a difference in F. However, our data do not support this assumption. They demonstrate that the ventricle size does not depend on F. This parameter was normal in BIH ($0.28 \pm 0.13\,\text{ml/min}$) and in NPH ($0.45 \pm 0.24\,\text{ml/min}$).

Therefore, our well supported conclusion is that a decrease in ventricle size in patients with BIH is attributable to brain swelling secondary to an increase of CBV and/or development of BE. BS in BIH is combined either with a normal or increased brain tissue density (excluding periventricular regions in some patients). A simultaneous increase in CBV and brain water content cannot be distinguished from each other by employment of CT, since blood is denser than water obscuring in the CT scan a hypodensity inducing increased brain water content. Magnetic resonance imaging shows increased brain water self diffusion in patients with BIH[12]. Thus, BS and BIH may be always combined with an increase of CBV.

The question may be raised as to the mechanism of the increase in CBV in patients with BIH. Augmentation of cerebral blood volume may result from dilation of arterial, arteriolar, capillary, or venous blood vessels, although it is not known how much each of these segments contribute. The engorgement of the vascular system may not result from a paralysis of the resistance vessels. Patients with BIH had no CSF lactacidosis, which is considered a major mechanism of paralysis of resistance vessels in severe head injury[4]. Another mechanism explaining an increased CBV may

be associated with the decrease in cerebral perfusion pressure resulting from an increase in P_{csf} in patients with BIH. This may lead to a decrease in arterial resistance as a response of the cerebrovascular autoregulation[1].

An increase of P_{csf} may primarily result from disturbances of the CSF absorbtion. However, it was observed by ourselves that disturbances of CSF absorbtion were not present in all cases with BIH. Therefore, a major cause of the augmentation of the cerebral blood volume in patients with BIH may be an obstruction of the venous outflow. This in turn may increase pressure in the venous blood vessels impeding reabsorption of water from the brain parenchyma into the venous end of the capillaries, eventually resulting in accumulation of brain water, hence, in brain oedema.

These considerations are leading to the question as to the underlying cause of an increase in resistance of the venous outflow under these conditions. The venous outflow might be impeded in the cerebral dural sinuses leading there to a pressure rise. However, only in few patients with BIH, a minimal rise in P_{is} was seen, while the mean pressure level mostly was within the upper limit of the normal range. In spite of a certain variability between patients with BIH ($6.1 \pm 3.1\,\text{mm Hg}$) or with NPH ($4.2 \pm 1.5\,\text{mm Hg}$) significant differences were not found. Thus, there is no basis for considering presence of an outflow obstruction localized in the cerebral dural sinuses as a cause of venous engorgement and, thus, development of brain swelling.

An alternative mechanism of venous engorgement leading to brain swelling may be an increase in outflow resistance in the bridging veins and lateral lacunae before their entrance into the dural sinuses. The arachnoid villi which penetrate into the venous lacunae represent

Table 1. *Clinical Signs that Characterize the State of Venous and CSF Outflow Pathways in Patients with Benign Intracranial Hypertension*

Normal CSF outflow resistance	Primary disturbance of venous outflow pathways
Increased CSF outflow resistance Normal or increased subarachnoid space	Isolated primary disturbance of CSF outflow pathways or its combination with disturbance of venous outflow pathways
Increased CSF outflow resistance Decreased subarachnoid space	Different combinations of primary and secondary disturbances of venous and CSF outflow pathways

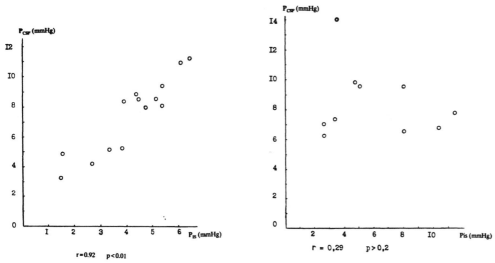

Fig. 1. Relationship between P_{is} and P_{csf} in patients with NPH (left) and BIH (right)

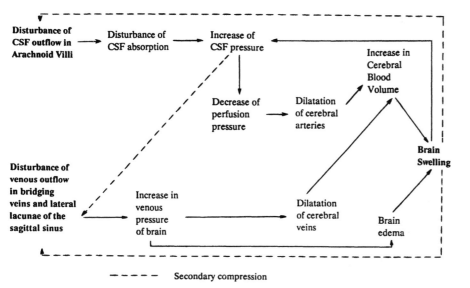

Fig. 2. Scheme of pathological processes that determine development of brain swelling in patients with Benign Intracranial Hypertension

the main outflow pathway of CSF into the cerebral dural sinuses. However, in patients either with NPH or BIH a correlation between P_{is} and R was not found. Thus, localization of a venous- or CSF outflow block in the dural sinuses can be ruled out. On the other hand, R was significantly increased in patients with NPH (15.8 ± 15.5) as well as in patients with BIH (25.5 ± 20.2) with the increase being more significant in the latter group.

A comparison of the data with findings obtained by CT investigations of the subarachnoid space may allow conclusions as to the character of the venous and CSF outflow problems in patients with BIH and brain swelling. Here, it is necessary to distinguish between the primary lesion in the bridging veins, lateral lacunae,

and arachnoid villi and a secondary compression of the bridging veins due to a rise in CSF pressure (cuff constriction[6]) and compression of these vessels, the venous lacunae and arachnoid villi by the swollen brain. By comparing the conditions of CSF outflow resistance and the morphology of the subarachnoid space it might be possible to identify both, the localization and character of the lesion (Table 1).

If R is normal, brain swelling my be explained only on the basis of a primary block of the venous outflow at the level of the bridging veins and venous lacunae. However, R was normal only in approximately one third of the patients with BIH and brain swelling. On the other hand, an increase of R with normal or an increased subarachnoid space may indicate presence of an isolated primary lesion of the arachnoid villi with or without obstruction of venous outflow through the bridging veins and venous lacunae. In case the rise in R correlates with a decrease or absence of the subarachnoid space one may assume a combination of a primary block of the venous- and CSF outflow of these pathways and a secondary compression by the swollen brain. Valuable information can be obtained on the state of CSF outflow in the dural sinuses by comparing P_{is} and P_{csf}. The left part of Fig. 1 shows a close coupling between P_{is} and P_{csf} in patients with NPH. Similar results were obtained by other investigators[8, 10]. Close coupling is characteristic of the normal state[14]. This may indicate a good draining function of the arachnoid villi. On the other hand, in BIH shown on the right side of Fig. 1 there is absolute uncoupling between P_{is} and P_{csf} indicating severe disturbances of the draining function of the arachnoid villi. Patients with NPH had tightly coupled P_{is} and P_{csf}, demonstrating a close relationship between the sinuses and the CSF system. In these patients the problem of reabsorption of CSF was related with an inhibiton of CSF outflow mainly across the arachnoid mater, however, not in the arachnoid villi suggesting the pathological process to be localized in the arachnoid mater.

The normal draining function of the arachnoid villi is sufficient to prevent an increase of P_{csf}, however, insufficient to prevent ventricular enlargement. In these patients a ventricle shunt operation is necessary to reduce ventricular size. In our material shunts were sucessful, if the CSF outflow resistance (R) was higher than normal (10 mm Hg/ml/min). Patients with BIH had uncoupling of P_{is} and P_{csf} and significant disturbances of CSF absorbtion as compared to patients with NPH. The current data indicating profound disturbances of the draining function in the arachnoid villi suggest localization of the pathological process in the lateral lacunae of the sagittal sinus. This may lead to disturbances of venous outflow from the brain resulting in brain swelling. Comparison of changes in venous- and CSF outflow in the patients with BIH makes possible an accurate localization of the pathological process in the venous lacunae of the dural sinuses and the intralacunar part of the arachnoid villi. Since patients with BIH mostly are young women with endocrine disorders, one may assume endocrine mechanisms underlying the stenosis of the venous lacunae. The results suggest, therefore, that the development of brain swelling in patients with BIH can be attributed to disturbances of CSF- and venous outflow across the lateral lacunae of the sagittal sinus and arachnoid villi. This

provides the basis for an involvement of different pathological mechanisms ultimately leading to an increase in brain volume, thus brain swelling. The most important interrelationships are schematically demonstrated in Fig. 2.

As seen in Fig. 2, expansion of CBV may result from arteriolar dilatation secondary to a decreased perfusion pressure due to an increase in CSF pressure on the basis of disturbances of CSF absorption. In addition, the increase in CBV may also be attributable to dilation of venules due to a rise in venous pressure from disturbances of venous outflow. An increase in venous pressure would also support development of brain oedema. Further, it is suggested in Fig. 2 that both, the rise in CSF pressure as well as the development of brain swelling causes secondary compression of the bridging veins and other cerebral veins, of the venous lacunae, and the arachnoid villi. Taken together, we conclude that brain swelling in patients with Benign Intracranial Hypertension results from primary and secondary disturbances of venous and CSF outflow in the bridging veins, lateral lacunae of the sagittal sinus, and the arachnoid villi.

References

1. Auer LM, Ishiyama N (1986) Cerebrovascular response to elevated intracranial pressure. In: Miller JD, Teasdale GM, Rowan JO *et al* (eds) Intracranial pressure VI. Springer, Berlin Heidelberg, New York, pp 399–403
2. Grant R, Condon B, Rowan J, Teasdale GM (1989) Benign intracranial hypertension: Brain swelling and cranial CSF volume. In: Hoff JT, Betz AL (eds) Intracranial pressure VII. Springer, Berlin Heidelberg New York, pp 344–345
3. Junk L (1986) Benign intracranial hypertension and normal pressure hydrocephalus: theoretical considerations. In: Miller JD, Teasdale GM, Rowan JO *et al* (eds) Intracranial pressure VI. Springer, Berlin Heidelberg New York, pp 447–450
4. Langfitt TW, Zimmerman RA (1985) Imaging and *in vivo* biochemistry of the brain in head injury. In: Becker DP, Povlishock JT (eds) Central nervous system trauma. Status report 1985. NIH, USA, pp 53–63
5. Milhorat TH, Clark RG *et al* (1970) Structural, ultrastructural and permeability changes in the ependyma and surrounding brain favouring equilibration in progressive hydrocephalus. Arch Neurol 22: 297–497
6. Nakagawa Y, Tsuru M, Furuse M *et al* (1974) Site and mechanisms for compression of the venous system during experimental intracranial hypertension. J Neurosurg 41: 427–434
7. Portnoy HD, Crossant PD (1976) A practical method for measuring hydrodynamics of cerebrospinal fluid. Surg Neurol 5: 273–277
8. Sahar A, Hochwald G, Ransohoff J (1970) Cerebrospinal fluid and cranial sinus pressures – Relationship in normal and hydrocephalic cats. Arch Neurol 23: 413–418

9. Shakhnovich AR, Razumovsky AY, Gasparian SS *et al* (1986) Elastic properties of the cerebrospinal system and hydrodynamic of the CSF in patients with intracranial hypertension. In: Miller JD, Teasdale GM, Rowan JO *et al* (eds) Intracranial pressure VI. Springer, Berlin Heidelberg New York, pp 89–95

10. Shulman K, Yarnell P, Ransohoff J (1964) Dural sinus pressure. In normal and hydrocephalic dogs. Arch Neurol 10: 575–580

11. Sklar FH, Beyer CW, Hagler JH *et al* (1980) The pressure volume function of brain elasticity and its relationship with ventricular size. In: Shulman K, Marmarou A, Miller JD *et al* (eds) Intracranial pressure IV. Springer, Berlin Heidelberg New York, pp 81–84

12. Soelberg-Sorensen P, Thomsen C, Gjerris F, Schmidt JF, Henriksen O (1989) Increased brain water self diffusion measured by magnetic resonance scanning in patients with pseudotumour cerebri. In: Hoff JT, Betz AL (eds) Intracranial pressure VII. Springer, Berlin Heidelberg New York, pp 419–421

13. Szewzykowsky J, Slivka ST, Kunicki A *et al* (1977) A fast method of estimating the elastance of the intracranial system. J Neurosurg 47: 19–26

14. Ueda Y, Nagai H, Kamiya K, Mase M (1989) Significance of cortical venous pulse pressure and superior sagittal sinus pulse pressure on ICP-pulse pressure. In: Hoff JT, Betz AL (eds) Intracranial pressure VII. Springer, Berlin Heidelberg New York, pp 166–168

Correspondence: A. R. Shakhnovich, Burdenko Neurosurgical Institute, Fadeev 5, 125047 Moscow, USSR.

Acta Neurochirurgica, Suppl. 51, 363–365 (1990)
© by Springer-Verlag 1990

*Various Disease Processes and Brain Oedema II
(Pseudotumour, Irradiation, Infection, Stress)*

Brain Water Accumulation in Pseudotumour Cerebri Demonstrated by MR-imaging of Brain Water Self-Diffusion

P. Soelberg Sørensen, C. Thomsen, F. Gjerris, and **O. Henriksen**

University Department of Neurology, Rigshospitalet, Copenhagen, Denmark

Summary

We studied brain water self diffusion in pseudotumour cerebri by MR-imaging using single spin echo pulse sequences with pulsed magnetic field gradients of different magnitude. The methods is based on the fact that the movement of water molecules is restricted in brain tissue and that accumulation of water in the brain tissue will enhance the self diffusion of water. In order to evaluate the brain water content in pseudotumour cerebri we compared the water self diffusion coefficient in various regions of the brain in pseudotumour patients with that of healthy controls. Ten patients with pseudotumour cerebri were studied. All had increased ICP and increased resistance to CSF outflow. All patients had normal conventional MR spin echo images without focal lesions and a normal sized ventricular system. All patients had abnormal diffusion images showing increased water diffusion. Some patients had in particular increased diffusion in the periventricular regions, others in the whole brain. The diffusion coefficients in all brain regions of interest were significantly higher in patients than in controls. The findings suggest that patients with pseudotumour cerebri have a convective transependymal flow of water causing an interstitial brain oedema and in addition an intracellular brain water accumulation.

Introduction

The pathophysiology of the intracranial hypertension in pseudotumour cerebri is still subject to debate. In the absence of an intracranial mass lesion or hydrocephalus, a number of other mechanisms have been suggested as possible explanations of the increased intracranial pressure: 1. a sustained increase in intracranial sinus pressure; 2. increased CSF formation rate; 3. increased resistance of CSF outflow at the level of the arachnoid villi; 4. increased brain volume owing to brain oedema and/or increased blood volume. Increased intracranial pressure caused by thrombosis of the major venous sinuses is usually not included under the term pseudotumour cerebri[7]. An increased brain volume in pseudotumour patients is indicated by the finding of small ventricles on CT[5], but it has not been clarified whether the brain swelling is caused by an interstitial brain water accumulation due to increased resistance to CSF outflow[1, 3] or by intracellular brain water accumulation[6].

By the use of a new magnetic resonance technique it has been possible to measure the molecular self diffusion of water in the brain. The method for measuring water self-diffusion is based on the fact that the movement of water molecules, the so-called Brownian motion, is restricted in brain tissue. The molecular diffusion of water is dependent on the content of free water in the tissue. Previous studies have shown that the self diffusion of water is higher in gray than in white matter[8, 9]. Accumulation of water in the brain, intra- or extracellularly would enhance the self diffusion of water in the brain tissue.

The purpose of the present study was to characterize the increased brain water volume in pseudotumour cerebri by *in vivo* quantitative measurements of water diffusion in the brain tissue by magnetic resonance scanning.

Materials and Methods

Ten consecutive patients with pseudotumour cerebri were studied. They were 8 women and 2 men with a median age of 35 years, range

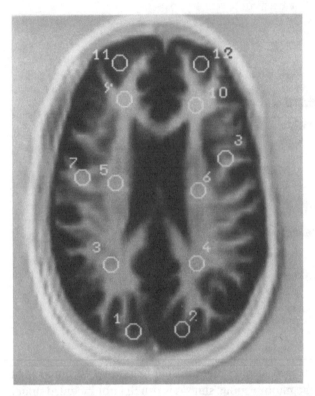

Fig. 1. Inversion recovery image showing regions of interest in which water self diffusion was measured: occipital cortex (*1* and *2*), occipital white matter (*3* and *4*), basal ganglia (*5* and *6*), central white matter (*7* and *8*), frontal white matter (*9* and *10*), frontal cortex (*11* and *12*)

23–60 years. All fulfilled the following diagnostic criteria: papilloedema; absence of focal neurological signs, except for those attributable to increased intracranial pressure; no intracranial mass or hydrocephalus on CT; and normal composition of CSF. All patients had increased intracranial pressure (ICP): mean steady state pressure 23–40 mm Hg, and all had increased resistance to CSF outflow measured by a lumbo-lumbar perfusion study[1]. The measurements of water diffusion in pseudotumour patients were compared with similar measurements in 7 healthy volunteers, 5 women and 2 men, with a median age of 31 years, range 17–50 years.

Magnetic resonance scanning was performed on a Siemens 1.5 Tesla wholebody scanner. All patients were imaged with a T_1-weighted and a T_2-weighted spin echo sequence. Quantitative diffusion measurements were obtained by using seven spin Hahn single echo sequences with pulsed gradients. The theoretical considerations and a detailed description of the method have been reported previously[8, 9]. Five slices were measured in each subject.

For statistical comparison of brain water self diffusion in patients and controls, the slice corresponding to the plane through the central part of the lateral ventricles was used. Corresponding areas in each hemisphere were evaluated: frontal cortex, frontal white matter, basal ganglia, central white matter, occipital white matter, and occipital cortex (Fig. 1).

Results

The diffusion images of the healthy subjects showed a regional difference with low diffusion coefficient in the

Table 1. *Water Self Diffusion Coefficients* $(10^{-9}\ m^2 \cdot s^{-1})$ *in Pseudotumour Patients and Control*

	Pseudotumour patients		Control subjects	
	Median	10%–90% percentiles	Median	10%–90% percentiles
Grey matter	3.54*	1.73–3.54	1.73	1.73–2.24
White matter	3.54*	1.73–3.54	1.22	0.71–2.24

* $p < 0.01$ compared with control subjects (Mann–Whitney).

Table 2. *Water Self Diffusion Coefficients* $(10^{-9}\ m^2 \cdot s^{-1})$ *in Pseudotumour Patients and Control*

Brain region	Pseudotumour patients		Control subjects	
	Median	Range	Median	Range
Occipital cortex	2.24	(1.22–7.07)*	1.73	(1.22–2.24)
Occipital white matter	2.24	(1.22–3.54)*	1.22	(0.25–1.73)
Basal ganglia	3.54	(2.24–7.07)*	2.24	(1.73–3.54)
Central white matter	3.54	(1.73–3.54)*	1.22	(0.25–1.73)
Frontal white matter	3.54	(2.24–3.54)*	1.73	(0.71–2.24)
Frontal cortex	3.54	(2.24–7.07)*	1.73	(1.22–2.24)

* $p < 0.01$ compared with control subjects (Mann–Whitney).

white matter (median $D = 1.22 \cdot 10^{-9}\ m^2 \cdot S^{-1}$) corresponding to 33% of water self diffusion coefficient at 37°C $(4.0 \cdot 10^{-9}\ m^2 \cdot s^{-1})$, and a higher diffusion coefficient in grey matter (median $D = 1.73 \cdot 10^{-9}\ m^2 \cdot s - 1$), corresponding to 55% of waters self diffusion in water (Table 1).

All pseudotumour patients had normal spin echo images without focal lesions and a normal sized ventricular system. In all patients with pseudotumour the diffusion images showed an increased diffusion. Five patients had in particular increased diffusion in the periventricular regions, where the diffusion coefficient in some regions was above the self diffusion coefficient of water in water, indicating a convective flow of water through the ependyma into the periventricular white matter. In 4 patients the water self diffusion was increased equally in the whole brain. In the patients with pseudotumour the water self diffusion coefficients both in gray and white matter were nearly equal to the diffusion coefficient of water in water (median

$D = 3.54 \cdot 10^{-9} \, m^2 \cdot s^{-1}$), in contrast to the normal finding in the controls of a higher water self diffusion in gray than in white matter (Table 1). The median diffusion coefficients in the various brain regions of interest were significantly higher in patients than in controls (Table 2).

Discussion

The findings of increased molecular self diffusion of water in the brain tissue of patients with pseudotumour cerebri indicate a decreased overall viscosity of the brain tissue caused by an increased volume of free water intra- and/or extracellularly. The very high water self diffusion coefficient in the periventricular regions of the brain, above the self diffusion of water in water, indicates a convective flow through the ependyma into the periventricular white matter. This would be in agreement with the hypothesis that an abnormal resistance to CSF outflow with a secondary interstitial oedema is of pathogenetic significance in pseudotumour cerebri[2]. Increased resistance of CSF outflow at the level of the arachnoid villi has been demonstrated in pseudotumour patients by a number of investigators[1, 3, 7].

An interstitial periventricular oedema alone, however, cannot explain the high water self diffusion seen more widely spread in the gray matter of the brain in pseudotumour patients. The extracellular volume constitutes only about 20% of the total water in the brain and even a substantial increase in extracellular volume cannot account for the difference in water self diffusion between pseudotumour patients and controls. Increased cerebral blood volume has been proposed as a pathogenetic factor in pseudotumour[4], but as the intravascular volume is less than 5% of the water space in the brain, the influence of cerebral blood volume and blood flow on the calculated diffusion coefficients is insignificant.

In conclusion, our findings suggest the presence of increased intra- and extracellular water content and support both the prevailing hypothesis of the pathophysiologic mechanism of the increased intracranial pressure in pseudotumour 1. an interstitial oedema due to increased resistance of CSF absorption and 2. a primary increase in brain water content due to intracellular accumulation of water. However, the mutual dependence of these two mechanisms is still unclear. In the future, measurements of brain water self diffusion may be of diagnostic value in patients with increased ICP of unknown origin.

References

1. Gjerris F, Sørensen PS, Vorstrup S, Paulson OB (1985) Intracranial pressure, conductance to cerebrospinal fluid outflow, and cerebral blood flow in patients with benign hypertension (pseudotumour cerebri). Ann Neurol 17: 158–162
2. Fishman RA (1980) Cereprospinal fluid in diseases of the nervous system. Saunders, Philadelphia, pp 128–139
3. Johnston I, Paterson A (1975) Benign intracranial hypertension. II Pressure and circulation. Brain 97: 301–312
4. Raichle ME, Grubb RLJ, Phelps ME *et al* (1978) Cerebral hemodynamics and metabolism in pseudotumour cerebri. Ann Neurol 4: 104–111
5. Reid AC, Teasdale GM, Matheson MS, Teasdale EM (1981) Serial ventricular volume measurements: further insights into the aetiology and pathogenesis of benign intracranial hypertension. J Neurol Neurosurg Psychiatry 44: 636–640
6. Sahs AL, Joynt RJ (1956) Brain swelling of unknown cause. Neurology 6: 791–802
7. Sørensen PS, Krogsaa B, Gjerris F (1988) Clinical course and prognosis of pseudotumour cerebri. A prospective study of 24 patients. Acta Neurol Scand 77: 164–172
8. Sørensen PS, Thomsen C, Gjerris F, Schmidt J, Kjær L, Henriksen O (1989) Increased brain water content in pseudotumour cerebri measured by magnetic resonance imaging of brain water self diffusion. Neurol Res 11: 160–164
9. Thomsen C, Henriksen O, Ring P (1987) *In vivo* measurement of water self diffusion in the human brain by magnetic resonance imaging. Acta Radiol 28: 353–361

Correspondence: P. Soelberg Sørensen, University Department of Neurology, Rigshospitalet, DK-2100 Copenhagen, Denmark.

Acta Neurochirurgica, Suppl. 51, 366–368 (1990)

Pseudotumour Cerebri-neurosurgical Considerations

T. Lundar and **H. Nornes**

Department of Neurosurgery, Rikshospitalet, Oslo, Norway

Summary

Pseudotumour was diagnosed in six patients aged 3–38 years during an 8 years period. The diagnosis was based on headache, papilloedema, normal CT scan and cerebrospinal fluid (CSF) composition. Additional clinical symptoms were nausea, VIth nerve palsy, ataxia, blurred vision and frank visual reduction over time. Sagittal sinus thrombosis was ruled out by angiography or magnetic resonance imaging.

In five of the six patients lumbar steady state infusion tests were performed to evaluate intracranial hydrodynamics and CSF resorbtion. All patients demonstrated a markedly increased opening pressure (range 13 to 48 mm Hg). CSF outflow resistance ranged from upper normal to pathologically increased levels (8–19 mm Hg/ml/min). Combined epidural intracranial pressure/middle cerebral artery blood velocity monitoring in 3 patients revealed a great number of B waves and a labile cerebral vasomotor state.

Pharmacological treatment was tried with digitoxin, acetazolamide, furosemide and/or corticosteroids. Two patients did well on long-term treatment with digitoxin and furosemide, respectively. In the other four patients the clinical development was unsatisfactory on medical treatment alone. They were subsequently operated with implantation of a lumboperitoneal, cisternoatrial or cisternoperitoneal shunt. Shunting rapidly reversed clinical signs and symptoms, except for a partial persistent visual loss in an 18 years old boy who had experienced symptoms for 3 years resistant to pharmacological treatment.

Introduction

Pseudotumour is characterized by choked disks, headache and often visual reduction. Many authors consider this condition innocent and self limiting; hence, the term benign intracranial hypertension. It is well known, however, that severe papilloedema may cause permanent visual reduction[3, 8]. Patients with severe or rapidly progressing symptoms and signs are more often referred to neurosurgical units. Under such circumstances, rapid diagnostic procedures must be undertaken and neurosurgical intervention may be pertinent.

Material and Methods

Pseudotumour was diagnosed in six patients during an 8 years period. Age, sex and clinical data are summarized in Table 1. All patients had normal cerebrospinal fluid composition. They demonstrated small or normal ventricular size on CT scans. Thrombosis of the sagittal or transverse sinuses were excluded by angiography or magnetic resonance imaging.

Epidural intracranial pressure monitoring was performed in 3 patients, combined with arterial blood pressure and continuous recording of blood velocity monitoring in the middle cerebral artery in two. Lumbar spinal infusion tests were performed through a 19 gauge needle in 5 of 6 patients. CSF outflow resistance (Ro) was calculated from the formula:

$$Ro = (Pp-Po)/\text{Infusion rate}^{7}$$

where Pp is the plateau pressure and Po the opening pressure.

Results

Patients 1 and 5 did well on pharmacological treatment (Table 1). In the others, however, the clinical response was not satisfactory. Patients 2 experienced permanent visual reduction after prolonged medical treatment. A lumbo-peritoneal shunt was implanted, and he has been without subjective complaints for 6 years. Patients 3, 4 and 6 demonstrated additional clinical symptoms (Table 1). Medical treatment failed to reverse their symptoms and signs, and a cisterno-atrial or cisterno-peritoneal shunt was implanted. This rapidly improved their condition with complete reversal of symptoms and signs within weeks.

All patients demonstrated increased intracranial pressure by lumbar puncture. The spinal infusion tests revealed pathological opening pressures and outflow resistance values ranging from upper normal to pathologically increased (Table 2). Epidural intracranial pressure (EDP) monitoring in 3 patients confirmed a labile, but increased EDP with a number of B waves.

Table 1. *Clinical Data*

Patient number	Sex/age	Symptoms	Signs	Medication	Shunt
1	k/15	headache	papilloedema	digitoxin	no
2	m/18	headache nausea vomiting	papilloedema visual reduction	digitoxin acetazolamide glycerol decadron	L-P* shunt
3	k/18	headache visual reduction	papilloedema (5 dioptries) left VIth nerve paresis	prednison furosemide	holter C-A** shunt
4	m/5	headache ataxia diplopia	papilloedema bilateral VIth nerve paresis	furosemide	Holter C-A** shunt
5	k/39	headache	(papilloedema)	furosemide	no
6	m/3	amblyopia ataxia	papilloedema	decadron	Holter C-P*** shunt

  * Lumbo-peritoneal shunt.
  ** Cisterno-atrial shunt.
*** Cisterno-peritoneal shunt.

Table 2. *Findings at lumbar steady state infusion tests*

Patient number	Opening pressure (Po) (mmHg)	Plateau pressure (Pp) (mmHg)	Infusion rate (ml/min)	Outflow resistance (Ro) (mmHg/ml/min)
2	48	65	1.5	10
3	34	>63	1.5	>19
4	12	36	1.5	16
	18–28	40	1.5	8–14
5	13	36	1.5	16
	9	21	1.5	8
6	15	35	1.5	14

Discussion

Pseudotumour cerebri is an entity of obscure etiology involving increased intracranial pressure. A number of conditions have been associated with pseudotumour: venous drainage obstruction, hormonal dysfunction, metabolic disorders, drug reactions and hematological disorders[6]. The cause of intracranial hypertension in pseudotumour patients has been attributed to an increased cerebral blood volume, cerebral oedema, or disordered CSF dynamics[5].

The spinal infusion tests revealed increased opening pressure in all patients. In the two small children, the test was performed during generalized anaesthesia including moderate hypocapnia (4 kPa), reducing the intracranial hypertension. CSF outflow resistance was also increased or in the upper normal range. Our findings confirm disturbed CSF dynamics in these patients[4]. Combined measurement of middle cerebral artery flow velocity and cerebral perfusion pressure monitoring performed in three patients disclosed an increased but labile intracranial pressure state with concomitant transient changes in blood velocity (B waves). This observation indicates fluctuations in the cerebral blood volume and a labile cerebral vasculature.

When pseudotumour patients do not respond favourably to pharmacological treatment, permanent visual reduction may ensue. This occurred in our patient No. 2, where the shunt implantation should have been performed at an earlier stage. In cases with additional symptoms, such as IIIrd and VIth nerve palsy, ataxia and slightly reduced consciousness, the situation appears more threatening and the time for expectation and evaluation of medical treatment may be limited. Indeed, our patient No. 6 presented with amblyopia for 36 hours, semidilated fixed pupils and severe ataxia. As lumbar puncture usually gives symptomatic relief in pseudotumour patients, a CSF diversion procedure is the natural choice in management from a neurosurgical point of view. Good results have been reported in small clinical series[1, 2]. In patients with dramatic symptoms and signs, the term benign intracranial hypertension appears inappropriate. Shunting may then be a neurosurgical emergency procedure. "Malignant pseudotumour" has been proposed as an appropriate term in such cases[6].

References

1. Ark GD van der, Kempe LG, Smith DR (1971) Pseudotumour cerebri treated with lumbar-peritoneal shunt. JAMA 217: 1832–1834
2. Beatty RA (1982) Cervical-peritoneal shunt in the treatment of pseudotumour cerebri. J Neurosurg 57: 853–855
3. Corbett JJ, Savino PJ, Thompson HS, Kansu T, Schatz NJ, Orr LS *et al* (1982) Visual loss in pseudotumour cerebri. Follow-up of 57 patients from five to 41 years and a profile of 14 patients with permanent severe visual loss. Arch Neurol 39: 461–474
4. Gerris F, Soelberg-Sørensen P, Vorstrup S, Paulson OB (1985) Intracranial pressure. Conductance to cerebrospinal fluid outflow, and cerebral blood flow in patients with benign intracranial hypertension. Ann Neurol 17: 158–162
5. Hoffman H (1986) How is pseudotumour cerebri diagnosed? Arch Neurol 43: 167–168
6. Kidron D, Pomeranz S (1989) Malignant pseudotumour cerebri. Report of two cases. J Neurosurg 71: 443–445
7. Marmarou A, Shulman K, Rosende RM (1978) A nonlinear analysis of the cerebrospinal fluid system and intracranial pressure dynamics. J Neurosurg 48: 332–344
8. Sklar F (1985) Pseudotumour cerebri. In: Wilkins RH, Rengachary SS (eds) Neurosurgery. McGraw-Hill, New York, pp 350–353

Correspondence: T. Lundar, M.D., Department of Neurosurgery, Rikshospitalet, Oslo, Norway.

Acta Neurochirurgica, Suppl. 51, 369–371 (1990)

Clinical Evaluation and Long-Term Follow-up in 16 Patients with Pseudotumour Cerebri

C. Arienta, M. Caroli, S. Balbi, A. Parma[1]**, E. Calappi,** and **R. Massei**[1]

Neurosurgical Institute and [1] Department of Anaesthesiology, University of Milan, Italy

Summary

Medical treatment must be promptly established in patients with pseudotumour cerebri. In fact, even though a spontaneous remission of the symptoms is recognized in the literature, we have not any predictive criteria of future visual impairment.

Introduction

Pseudotumour cerebri is a clinical syndrome of intracranial hypertension unassociated with a space-occupying lesion, hydrocephalus or frank brain oedema. The clinical findings are characterized by papilloedema, headache, occasional impairment of visual acuity or visual field defects. This condition is usually not accompanied by an alteration in the level of consciousness and focal neurological signs except for nonspecific signs of intracranial hypertension, such as oculomotor palsy. In the majority of cases pseudotumour cerebri is considered a benign clinical self-limiting condition. The authors present their experience on 16 selected patients with diagnosis of benign intracranial hypertension treated with pharmacological and/or surgical therapy and submitted to periodic follow-up with ophthalmological evaluation.

Material and Methods

Sixteen selected patients with diagnosis of benign intracranial hypertension (pseudotumour cerebri) were admitted to the Neurosurgical Institute of the University of Milan between 1976 and 1987. Only cases with a complete clinical and neurodiagnostic evaluation were considered in this study. Patients with clinical suspicion of pseudotumour cerebri who refused CSF pressure recording and patients only examined one time were excluded. Twelve patients were female and 4 were male. The age ranged from 12 to 51 years (average 28.6). The duration of symptomatology ranged from 2 weeks to 2 years (average 4 months).

The onset of the symptomatology was characterized by: headache (9 cases), blurring of vision (5 cases), diplopia (2 cases). Symptoms and signs at the time of admission are detailed in Table 1. Visual dysfunctions were represented by blurred vision and brief obscuration, but at the ophthalmological examination all patients had normal bilateral visual acuity. Bilateral papilloedema was found in all cases. In 4 cases a visual field defect (blind spot pathologically increased) was detected. In 5 patients an unilateral abducens nerve palsy was found.

Clinical conditions and associated factors were the following: remote history of otitis media (1 case), estrogenic therapy (1), oral contraceptives (1), obesity (4), menstrual irregularities (2), pregnancy (3). Endocrinologic evaluation was performed in all patients and in one case a moderate hypothyroidism was demonstrated. In 5 patients no concurrent conditions were demonstrated. Table 2 summarizes clinical features and associated factors.

In all cases cerebral CT scan with contrast medium infusion failed to demonstrate expanding lesions; ventricle size was normal in 9 cases, reduced in 7 cases. MRI was performed in 2 patients. All patients were submitted to angiographic study: in one case diagnosis of superior sagittal sinus thrombosis was made.

All patients underwent lumbar puncture and manometric reading of CSF pressure. In 7 patients the opening pressure was higher than 30 cm of water, in 9 patients the initial pressure was within normal limits but at 24–48 hours the recording discovered pathological findings (plateau-waves with a peak pressure of 60 cm of water and B-waves).

Table 1. *Symptoms and Signs at Time of Admission*

Visual dysfunction	7
Defect of visual field	4
Haemorrhagic papilloedema	6
Mild papilloedema	4
Moderate papilloedema	6
Abducens nerve palsy	5
Nausea and vomiting	7
Headache	12
Dizziness	3

Table 2. *Clinical Conditions and Associated Factors*

Remote history of otitis media	1
Estrogenic therapy	1
Oral contraceptives	1
Obesity	4
Pregnancy	3
Menstrual irregularities	2
Arterial hypertension	1
Recent head injury	1
Hypothyroidism	1
No concurrent conditions	5

Some patients had more than one factor.

Treatment. The medical and surgical treatment is summarized in Table 3. All patients but one were initially treated as following: dexamethasone (11 cases), dexamethasone and acetazolamide (2), dexamethasone, acetazolamide and ACTH (1), dexamethasone and furosemide (1). One patient (case n. 10), a pregnant woman, was submitted to external lumbar drainage for 10 days with prompt remission of her symptoms.

Two patients (n. 1 and n. 3) who failed to respond to medical therapy underwent surgical procedures with lumboperitoneal shunt. In some cases pharmacologic therapy was withdrawn when clinical remission of symptoms was achieved, but in most cases normalization of the optic disc occurred later.

We distinguished recovery from clinical symptomatology (ranging from 15 days to 6 months) and complete resolution of papilloe-

Table 3. *Treatment and Follow-up in 16 Patients with Pseudotumour cerebri*

N°	Sex/Age	Treatment	Recurrence	Treatment of the recurrence	Clin recovery Normal optic d.	Follow up
1	F/13	D+A (1 month)	2 years	D (1 month)	1 month 18months/2years	13 years Good
2	M/33	D (1 year)	/	/	6 months	13 years. Mild papilledema
3	M/49	D (1 month) Shunt	2 years	Shunt revision	2 months/15days 2 months/15days	2 year. Died Carcinoma)
4	F/50	D (1 month)	/	/	1 month 3 years	12 years Good
5	F/27	D (15 days)	/	/	1 month 1 month	11 years Good
6	F/23	D (15 days)	/	/	1 month 1 month	9 years Good
7	F/12	D (15 days)	/	/	1 month	9 years Mild Papilledema
8	F/39	D+F (15 days)	/	/	20 days 3 months	8 years Good
9	F/51	D (15 days) Shunt	/	/	1 month 1 month	9 years Good
10	F/26	Drainage (10 days)	/	/	15 days 15 days	9 years Good
11	M/14	D (1 month)	/	/	1 month 1 year	3 years Good
12	F/25	D (15 days)	/	/	1 month 6 months	2 year Good
13	F/14	D (1 month)	5 months	Shunt	1 month/15days 1 year/1month	3 years Good
14	F/25	D+A+ACTH (1 month)	/	/	1 month 1 year	8 years Good
15	F/46	D (20 days)	/	/	1 month 3 months	4 years Good
16	M/12	D+A (1 year)	/		4 months 1 year	3 years Good

D= dexamethasone
A= acetazolamide
F= furosemide

dema (ranging from 15 days to 2 years). Fourteen patients regained normal disc appearance, while in 2 cases (n. 2 and n. 7) persisting mild papilloedema was observed at follow-up examinations, respectively of 13 and 9 years. Three cases experienced recurrence of symptoms by 2 years after the initial diagnosis. In case n. 3, a patient affected by prostatic carcinoma, submitted to estrogenic therapy, recurrence of symptoms occurred because of malfunction of the lumboperitoneal shunt and increasing estrogenic therapy. Recovery of symptoms was obtained after shunt revision, but the patient died within 1 year with vertebral metastases. Case n. 1 experienced recurrence of papilloedema without clinical symptomatology 2 years after diagnosis, and a second cycle of steroid therapy was performed. Papilloedema disappeared after 2 years. Patient n. 13 received medical treatment at the first admission with success, but she had recurrence of symptoms after 5 months. On the second admission she underwent surgery because improvement of symptomatology was obtained with lumbar drainage.

Complete follow-up including serial ophthalmologic examination ranged from 2 to 13 years (average 7.4 years). During this period no patient developed permanent visual loss or defects of the visual field. No other recurrence of clinical symptoms was observed except for the three cases cited above. In case n. 12 with previous diagnosis of superior sagittal sinus thrombosis a MRI performed one year after discharge documented disappearance of venous obstruction.

Discussion

Although pseudotumour cerebri is usually considered a benign self-limiting entity, permanent visual loss has been reported in 2–24% of the cases[1, 2, 5, 7]. All the cases reported in the literature[3, 5, 7] confirm that both medical and surgical procedures do not restore visual acuity once deterioration of vision is present. Therefore the aim of the medical or surgical treatment should be to prevent permanent severe visual impairment.

Our series confirms the epidemiologic data reported in the literature: the preponderance of females, the average age of incidence, the association with hormonal factors[2, 6].

Many authors agree that several factors, such as the duration of symptoms, the degree of papilloedema, the number of recurrences, the disturbance of visual acuity are not related to permanent visual loss[5]. The favourable visual outcome in all our patients could be attributed to the fact that on admission no one had visual acuity complaints and to the prompt administration of medical therapy. In our experience dexamethasone (16 to 24 mg/day) is the drug of first choice, even though maintaining weight in obese patients is considered a limit of this pharmacological approach by some authors[2]. We did not observe this side effect in our four obese patients. In fact, three of them promptly recovered after 15 days of steroid therapy. The fourth patient who experienced recurrence of the symptomatology

after five months required a lumboperitoneal shunt. In four patients a combination of dexamethasone and diuretics was employed for tapering off the dose of steroids.

In our series, CSF pressure recording was only of diagnostic value. Some authors used this technique to determine the effectiveness of therapy[4, 7]. This was based in our study only on the clinical findings and the ophthalmological parameters.

Our experience suggests that pharmacological therapy is the treatment of choice in the management of pseudotumour cerebri. In most cases it succeeds in restoring normal clinical conditions and in preventing visual failure. A lumboperitoneal shunt is reserved for the few patients in whom medical therapy administered for at least 15 days has failed. In case of recurrence our strategy is based on these parameters: persistence of risk factors and the entity of the clinical symptomatology. For example, patient n. 13, an obese girl affected by hypothyroidism, with recurrence of haemorrhagic papilloedema and severe symptoms of increased intracranial pressure, was immediately submitted to surgical procedures. On the other hand, patient n. 1, without associated factors, with asymptomatic recurrence of mild papilloedema, received a second medical treatment.

References

1. Corbett JJ, Savino PJ, Thompson HS, Kansu T, Schatz NJ, Orr LS, Hopson D (1982) Visual loss in pseudotumour cerebri. Arch Neurol 39: 461–474
2. Greer M (1990) Pseudotumour cerebri. In: Youmans JR (ed) Neurological surgery, vol 5. Third ed. W. B. Saunders Company, Philadelphia, pp 3514–3530
3. Hoffman HJ (1987) Editorial. Pseudotumour cerebri. Surg Neurol 27: 405
4. Rivas JJ, Lobato RD, Muñoz JM, Esparza J, Portillo JM, Cordobés F, Castro S, Barcena A, Cabrera A (1983) Effect of acetazolamide and dexamethasone on the cerebrospinal fluid pressure in patients with pseudotumour cerebri. In: Intracranial pressure V. Springer, Berlin Heidelberg New York, pp 763–767
5. Rush JA (1980) Pseudotumour cerebri. Clinical profile and visual outcome in 63 patients. Mayo Clin Proc 55: 541–546
6. Sklar FH (1985) Pseudotumour cerebri. In: Wilkins RH, Rengachery SS (eds) Neurosurgery, vol 1. McGraw-Hill Book Company, pp 350–353
7. Soelberg Sørensen P, Krogsaa B, Gjerris F (1988) Clinical course and prognosis of pseudotumour cerebri. A prospective study of 24 patients. Acta Neurol Scand 77: 164–172

Correspondence: C. Arienta, M.D., Neurosurgical Institute, University of Milan, Milan, Italy.

Acta Neurochirurgica, Suppl. 51, 372–374 (1990)

Polyamine Accumulation and Vasogenic Oedema in the Genesis of Late Delayed Radiation Injury of the Central Nervous System (CNS)

P. H. Gutin, M. W. McDermott, G. Ross, P. H. Chan, S. F. Chen, K. J. Levin, O. Babuna, and L. J. Marton

Brain Tumour Research Center, University of California, San Francisco, California

Summary

Polyamine (PA) accumulation has been associated with blood-brain barrier (BBB) disruption and vasogenic oedema after cold injury. PAs and water content were measured in a rat spinal cord model of late-delayed radiation injury and were found to be elevated at paralysis. The elevated PA levels could be significantly reduced by treatment with difluoromethylornithine (DFMO). In unirradiated rats DFMO reduced putrescine to undetectable levels after 10–12 weeks. These data suggest that blockade of PA synthesis may be useful in treating the vasogenic oedema of radiation injury and may improve CNS radiation tolerance.

Introduction

Brain injury is the limiting factor in the radiation therapy of malignant gliomas. Even with radiation doses within the accepted therapeutic range some patients suffer from the early and late delayed side-effects of radiotherapy[7]. Breakdown of the BBB with vasogenic oedema, demyelination and white matter necrosis are the clinical and pathologic correlates.

Investigations into the cause of white matter radiation injury have focused on injury to the oligodendroglia and endothelial cells[1, 6, 8, 9]. Myers *et al.* have suggested that BBB disruption and vasogenic oedema contribute to increased tissue pressure, ischaemia and subsequent white matter infarction[6]. It follows that if chronic vasogenic oedema could be controlled, white matter necrosis might be prevented or reduced and the radiation tolerance of CNS tissues improved. Polyamines have been shown to be mediators of increased vascular permeability[3–5]. Koenig *et al.* demonstrated that polyamine levels were elevated after cold injury and hyperosmolar perfusion of the rat brain, both of which cause vasogenic oedema[3, 4]. In each instance the

oedema could be prevented by the administration of DFMO. Thus, there is experimental evidence to suggest that a class of drugs other than steroids may be effective in controlling vasogenic oedema after CNS injury.

Our work therefore sought to investigate: (1) the course of increased vascular permeability and vasogenic oedema in the irradiated rat spinal cord; (2) whether PA accumulation corresponded to any increase in vascular permeability; (3) the effect of DFMO on cord PA levels and increased permeability at paralysis and (4) the effect of long term administration of DFMO on PA levels in unirradiated spinal cord.

Materials and Methods

Model

A standardized model of late delayed radiation injury of the CNS was developed with the irradiation of the C2–T1 segment of the spinal cord in female Fisher 344 rats. Rats were irradiated under nembutal anaesthesia in single doses of 15–25 Gy, using 250 Kvp X-rays at a dose rate of 4.24 Gy/minute. Paralysis of one or more limbs was determined by 2 observers and defined as the end point for the experiment. Rats were observed over a period of 4–7 months

Polyamine Determination

The spinal cords of irradiated rats were harvested by injection of water under pressure into the lumbar spinal canal at L3–L4 after decapitation under ether anaesthesia. The irradiated spinal cord was submitted for tissue analysis. Spinal cord putrescine, spermidine and spermine levels were then determined by reversed-phase liquid chromatography after solid-phase extraction of dansyl derivatives of PAs as described by Kabra *et al.*[2]. PA levels were expressed as pmol/mg tissue.

Measurement of Spinal Cord Water Content and Na/K

The irradiated cord was divided into six 5 mm segments and each segment cut in half longitudinally, for water content and Na^+/K^+

measurements. Assays were performed on groups of 6 animals at 4 week intervals after a single dose of 23 Gy (ED_{100}) until paralysis or 22 weeks post treatment. Evans Blue dye (EB) was injected into 2 rats at the same time intervals to qualitatively assess the BBB. Water content was determined by taking the difference between wet and dry weights. Wet weights were taken at the time of sacrifice and dry weights determined after drying spinal cord tissue at 105 °C for 48 hours. Results were expressed as percentage weight of water.

Na$^+$/K$^+$ content was measured by first drying the spinal cord sections at 105 °C for 7 days, followed by suspension in 0.1 M HNO$_3$ for 7 days to extract Na$^+$/K$^+$. A 0.2 ml sample of this solution was then diluted to a volume of 10 ml with distilled water and placed in an atomic absorption spectrophotometer to measure Na$^+$/K$^+$ levels.

Effect of DFMO on Cord PA Levels After Irradiation

Rats were irradiated with a single dose of 23 Gy, followed to paralysis and then treated with 2 doses of DFMO 500 mg/kg I.P. 12 hours apart. Spinal cord PA levels in the irradiated segment of the rats treated with DFMO were then compared to irradiated controls.

Effect of DFMO on Cord PA Levels and Radiation Tolerance

Weanling rats were placed on diets of 0.5%, 1.0%, and 2.0% DFMO in drinking water. The 2.0% diet was not tolerated because of diarrhea and weight loss. PA levels in unirradiated spinal cords were assayed at 2–4 weeks intervals from the start of treatment to 24 weeks of DFMO administration.

Results

Model

A dose response relationship between radiation dose and paralysis was observed with an ED_{50} of 19.6 and ED_{100} of 23 Gy. Pathologic examination of the irradiated segment in paralyzed rats confirmed the prominent posterior and lateral location of white matter necrosis.

Polyamine Accumulation

Spinal cord putrescine (Pu) and spermine (Sp) levels were found to be significantly elevated at 20 weeks post-irradiation as compared to 1 week post-irradiation or unirradiated controls (P < 0.001; t-test) (Fig. 1). Spermidine levels were not significantly different. Studies are presently underway to determine the time course of tissue polyamine levels at 4 week intervals after irradiation until paralysis.

Spinal Cord Water Content and Na/K Levels

The lower 2 segments of the irradiated cervical cord, segments D and E, grossly appeared the most oedematous. As compared to the unirradiated internal controls the irradiated segments had a higher water content throughout the period after treatment (Fig. 2). After 16 weeks and up until paralysis, water content rose

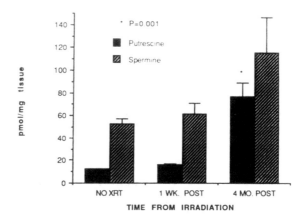

Fig. 1. Spinal cord putrescine and spermine levels significantly elevated at 4 months after irradiation (23 Gy) versus 1 week or in unirradiated rats. *XRT* Irradiation

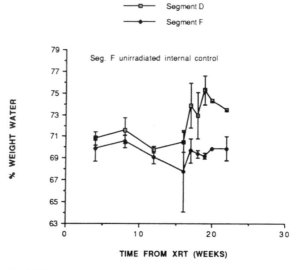

Fig. 2. The percent weight of water as a function of time in irradiated cord segment D versus nonirradiated lumbar cord segment F

significantly in the irradiated segments, particularly the lower segments D and E. From 4 weeks until paralysis there was no staining of the irradiated segment with EB. At paralysis there was a diffuse leakage of EB. Measurements of Na$^+$/K$^+$ revealed an early rise in Na$^+$ at 8 weeks, similar to the time course of increased water content, and a later but less abrupt rise near the onset of paralysis. Potassium values showed no significant fluctuations.

Effect of DFMO on Cord PA Levels After Irradiation

The administration of DFMO for 24 hours at the time of paralysis significantly reduced cord putrescine levels as compared to irradiated untreated controls (P = 0.005; t-test). Studies are presently underway to determine whether DFMO given at paralysis can reduce cord oedema and permeability.

Effect of Long Term Administration of DFMO on Cord PA Levels

Both 0.5% and 1.0% DFMO diets reduced cord putrescine to undetectable levels after 10–12 weeks. Spermidine and spermine levels were not significantly affected. The percentage paralysis for rats treated with long term DFMO after irradiation was lower than in untreated rats (13.3% vs 20.0% @ 19.5 Gy; 26.7% vs 53.0% @ 20.0 Gy; 66.7% vs 73.7% @ 21.0 Gy).

Discussion

That PAs may be involved in the disruption of the BBB after radiation injury is suggested by our findings of elevated Pu and Sp levels in the rat spinal cord at 4 months post irradiation as compared to levels at 1 week post treatment or in unirradiated spinal cords. This time point coincides with the time when spinal cord water content in irradiated segments is rising rapidly. DFMO given for a 24 hour period at the time of paralysis can also reduce Pu levels as compared to irradiated paralyzed controls. The long term administration of DFMO, or other PA inhibitors, may be an alternative to steroids in the treatment of vasogenic oedema. We have demonstrated that 0.5% and 1.0% DFMO are well tolerated by female F344 rats and that cord levels of putrescine are depleted significantly by this therapy. Presently, studies are underway to further define the time course of PA accumulation and its relationship to abnormal permeability and spinal cord blood flow. The effect of DFMO and other PA inhibitors on the oedema of radiation injury is being studied as is the effect of these agents on the dose-response (ED_{50}) after irradiation.

References

1. Delattre JY, Rosenblum MK, Thaler HT *et al* (1988) A model of radiation myelopathy in the rat. Brain 111: 1319–1336
2. Kabra PM, Lee HK, Lubich WP, Marton LJ (1986) Solid-phase extraction and determination of dansyl derivatives of unconjugated and acetylated polyamines by reverse-phase liquid chromatography: improved separation systems for polyamines in cerebrospinal fluid, urine and tissue. J Chromat 380: 19–32
3. Koenig H, Goldstone AD, Lu CY (1983) Blood-brain barrier breakdown in brain following cold injury is mediated by microvascular polyamines. Biochem Biophys Res Comm 116: 1039–1048
4. Koenig H, Goldstone AD, Lu CY (1989) Polyamines mediate the reversible opening of the blood-brain barrier by the intracarotid infusion of hyperosmolar mannitol. Brain Res 483: 110–116
5. Koenig H, Goldstone AD, Lu CY, Trout JJ (1989) Polyamines and Ca^{+2} mediate hyperosmolar opening of the blood-brain barrier. In vitro studies in isolated rat cerebral capillaries. J Neurochem 52: 1135–1142
6. Myers R, Rogers MA, Hornsey S (1986) A reappraisal of the roles of glial and vascular elements in the development of white matter necrosis in irradiated rat spinal cord. Br J Cancer [Suppl VII] 53: 221–223
7. Sheline GE, Wara WM, Smith V (1980) Therapeutic irradiation and brain injury. Int J Radiat Oncol Biol Phys 6: 1215–1228
8. Tamaki K, Sadoshima J, Baumbach GL *et al* (1984) Evidence that disruption of the blood-brain barrier precedes reduction in cerebral blood-flow in hypertensive encephalopathy. Hypertension [Suppl 1] 6: 175–181
9. Van Der Kogel AJ (1990) Central nervous system radiation injury in small animal models. In: Gutin PH, Leibel SA, Sheline GE (eds) Radiation injury to the nervous system (in press)

Correspondence: P. H. Gutin, M.D., Department of Neurosurgery, University of California, San Francisco, CA 94143-0112, U.S.A.

Acta Neurochirurgica, Suppl. 51, 375–377 (1990)
© by Springer-Verlag 1990

Dynamics of Tracer Distribution in Radiation Induced Brain Oedema in Rats

J. V. Lafuente[1], **E. Pouschman**[1], **J. Cervós-Navarro**[2], **H. S. Sharma**[2], **C. Schreiner**[2], and **M. Korves**[2]

[1] Department of Neurosciences, UPV, Lejona, Spain and [2] Institute of Neuropathology, Free University, Berlin

Summary

The dynamic behaviour of the distribution of Evans blue (EB), sodium fluorescein (SF), Lucifer yellow (LY) and horseradish peroxidase (HRP) was studied using standard light- and fluorescence microscopy following ultraviolet radiation induced brain oedema in the rat model. The cerebral cortex was irradiated after craniotomy (2 × 2 mm) under anaesthesia. The tracers were injected (iv) 30 min prior to radiation. Animals were perfused with glutaraldehyde through the heart at different survival periods ranging from 30 min to 24 h post irradiation. The results showed a remarkable difference in distribution and spread of these tracers in the oedematous brain following radiation. The extravasation of EB was evident in ipsilateral cortex 6 h after radiation which extended to the contralateral side at the end of the 24 h survival period. The HRP reaction product was seen in the necrotic area 3 h after radiation which further extended to the underlying white matter at 24 h survival. The LY stained the ipsilateral micronecrotic area 6 h after radiation, whereas a non-specific diffuse fluorescence of SF was noted at this time period. These results point out a specific selectivity of tracer distribution in oedematous brain following ultraviolet radiation.

Introduction

Many pathological conditions of the central nervous system are associated with a breakdown of the blood-brain barrier (BBB)[2, 4, 5, 7]. The extravasation of tracers across the BBB mainly depends on the protein binding capacity *in vivo* and the molecular diameter[4, 5, 6]. The BBB permeability is studied using tracers of different molecular size which can be visualized macroscopically or microscopically after administration in the systemic circulation[2]. Tracers with different molecular weights proved suitable to indicate the integrity of the BBB at different time points after the initial insult[4, 5].

Brain oedema is a serious consequence of any traumatic insult to brain in which the integrity of the BBB is altered[7]. The spread of oedema fluid, however, is mainly localised in the extracellular compartment. Electron microscopy demonstrated widening of the extracellular spaces without rupture of the cell membranes in various types of experimentally induced cerebral oedema. The swelling of astrocytes is regarded as a secondary phenomenon[3, 4, 7].

A study on the distribution of various sized tracers in ultraviolet irradiation induced oedema model is still lacking. Earlier studies demonstrated a size selective entry of the tracer following hyperosmotic opening of the BBB[6]. Therefore, in the present investigation we examined the magnitude and spread of four different sized tracers (EB and HRP as two markers of protein extravasation, and SF and LY as two small molecule tracers) in the oedematous brain following irradiation.

Materials and Methods

Ultraviolet radiation. Under Ketanest anaesthesia (15 mg/100 g) a left parietal craniotomy (2 × 2 mm) was done and the dura was removed. The left parietal cerebral cortex of 76 Wistar rats (200–250 gr) was irradiated with an ultraviolet lamp (Osram HBO-200, 15 cm above the cortex) for a period of 6 min (3).

Injection of tracers. Tracers were injected intravenously according to the survival times (30 min, 1 h, 3 h, 6 h, 12 h, 24 h). Evans blue was injected 30 min prior to the radiation whereas, the others (HRP, SF and LY) were administered 30 min before sacrifice.

Control group. Animals with craniotomy but without radiation served as control.

Light- and fluorescence microscopy. Animals were perfused with a glutaraldehyde based fixative *in situ* at 100 torr perfusion pressure preceded with a brief saline rinse. The brains were quickly removed and placed in the same fixative at 4 °C for 24 h. Routine paraffin and vibratome sections were made. The HRP reaction product was developed using a standard protocol. The intracerebral distribution of exudated plasma proteins at the cellular level was examined under the light microscope, whereas macroscopic distribution of the tracers was assessed using vibratome sections for fluorescence microscopy. Particular emphasis was placed on the intracerebral distribution of HRP and the corresponding fluorescence microscopical localization of the Evans blue tracer and the LY distribution.

Results

Ultraviolet radiation had specific effects on brain and was accompanied as a non-specific reaction with a focal tissue destruction. There was an intensive staining with EB and HRP of the ipsilateral irradiated cortex, whereas, the contralateral nonirradiated hemisphere was unstained (Fig. 1). The protein exudation across the cerebral vessels was prominent in the lesioned area and was diffusely distributed in the neuropil around the irradiated zone. In other regions, the penetration of EB and HRP was mainly confined to the walls of intracerebral vessels.

The EB staining of the damaged area was mild after 30 min irradiation. At this time period the occurrence of haemorrhage was very low. After 3 h postirradiation, the staining around the lesion was limited to the gray matter only. The spread of staining into adjacent white matter was evident after 6 h survival. With further increase in time, the spread and the intensity of the peripheral staining was increased. Thus, at the end of the 24 h period, the corpus callosum and the contralateral hemisphere were also stained. In the contralateral hemisphere the oedema was confined to the white matter alone without involving the overlying cortex.

The HRP reaction product was seen in the necrotic areas following 3 h after irradiation which extended to the deeper white matter after 24 h survival. The corpus callosum and the contralateral hemisphere were unstained at this time period (Fig. 2).

The LY stained the ipsilateral micronecrotic area 6 h after irradiation without spreading further into the

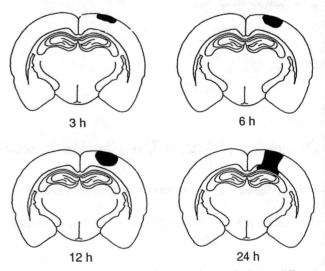

Fig. 2. HRP-reaction product after ultraviolet radiation at different survival periods. The spread of reaction product was limited to the underlying white matter within the ipsilateral hemisphere at 24 h after radiation

surrounding tissues. The LY fluorescence was limited to the area of cellular injury. At the same time period, a non specific diffuse fluorescence of SF was noted in the cortex.

Discussion

The most important new finding of this study is a specific selectivity of tracer distribution in the oedematous brain following ultraviolet irradiation. The cellular damage following ultraviolet irradiation was focal rather than diffuse. The border of the lesion was well marked. No spread of damage into the normal tissue was seen. A direct relationship between extent of oedema and the survival period after irradiation was observed indicating the validity of our model. It is quite likely that the peripheral staining which developed subsequently of the irradiation is due to the exposure itself and not due to the reaction of the tissue to injury.

There was a discrepancy in the vascular permeability to EB and HRP, which is quite likely due to their differences in protein binding capacity *in vivo*[4, 5, 6]. The specific staining of LY in this model indicates that the tracer could be used as a reliable indicator of cellular injury because it does not stain neuronal elements with an intact cell membrane[1]. This is further supported by the fact that LY injected intravenously does not penetrate into the brain when the BBB is intact[1, 3]. On the other hand, a diffuse fluorescence of SF is not a reliable indicator of cellular damage.

Our results further show that a minimal tissue damage produces a remarkable area of oedematization, the

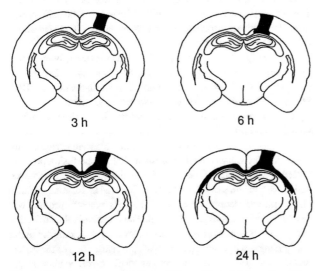

Fig. 1. Schematic representation of Evans blue staining after ultraviolet radiation at different survival periods. The spread of EB staining in the contralateral hemisphere is evident at 24 h after radiation

underlying mechanism(s) of such phenomena will require further investigation.

Acknowledgements

Authors are grateful to Accion Integrada 86 B (M.E.C.), Spain and DAAD, West Germany for financial support, Mrs. M. Schindler for art work and Mrs. Kristin Kemnitz for secretarial assistance.

References

1. Artigas J, Aruffo C, Sampaola S, Cruz-Sànchez F, Ferszt R, Cervòs-Navarro J (1987) Lucifer yellow as morphofunctional tracer of the blood brain barrier. In: Cervòs-Navarro J, Ferszt R (eds) Stroke and microcirculation. Raven Press, New York, pp 239–244

2. Houthoff HJ, Go KG, Huitema S (1981) The permeability of cerebral capillary endothelium in cold injury: comparison of an endogenous and exogenous protein tracer. In: Cervòs-Navarro J, Fritschka E (eds) Cerebral microcirculation and metabolism. Raven Press, New York, pp 331–336

3. Lafuente JV, Cervòs-Navarro J (1989) Variaciones del volumen tisular en un modelo de oedema cerebral experimental. Arch Neurobiol 52: 1–36

4. Olsson Y, Sharma HS, Petterson CAV (1990) Effects of p-chlorophenylalanine on microvascular permeability changes occurring in spinal cord trauma. An experimental study in the rat using ^{131}I-sodium and lanthanum tracers. Acta Neuropathol 79: 595–603

5. Sharma HS, Olsson Y, Dey PK (1990) Changes in blood-brain barrier and cerebral blood flow following elevation of circulating serotonin level in anaesthetized rats. Brain Res (in press)

6. Mayhan WG, Heistad DD (1985) Permeability of blood-brain barrier to various sized molecules. Am J Physiol 248: H712–H718

7. Vass K, Tomida S, Hossmann KA, Nowak TS, Klatzo I (1988) Microvascular disturbances and oedema formation after repetitive ischaemia of gerbil brain. Acta Neuropathol (Berl) 75: 288–294

Correspondence: J. V. Lafuente, M.D., Institute of Neuropathology, University Klinikum Steglitz, Free University Berlin, Hindenburgdamm 30, D-1000 Berlin 45.

Acta Neurochirurgica, Suppl. 51, 378–380 (1990)
© by Springer-Verlag 1990

Superoxide Dismutase Inhibits Brain Oedema Formation in Experimental Pneumococcal Meningitis

H. W. Pfister, U. Koedel, U. Dirnagl, R. L. Haberl, W. Feiden[1], and K. M. Einhäupl

Department of Neurology, University of Munich, and [1]Institute of Neuropathology, University of Munich, Federal Republic of Germany

Summary

The purpose of this study was to identify mediators of brain oedema formation in experimental pneumococcal meningitis. In a rat model of pneumococcal meningitis brain water content was significantly elevated 6 hours post infection (79.69% ± 0.24 compared to 78.94% ± 0.16 in the control group, mean ± SEM, $p < 0.05$). Brain oedema formation was completely blocked by superoxide dismutase (132,000 U/kg i.v. per 6 hours; n = 6), pretreatment with dexamethasone (3 mg/kg i.p., n = 3), or administration of dexamethasone at two hours after pneumococcal injection (n = 5). Pretreatment with indomethacin (10 mg/kg i.v., n = 5) attenuated the brain oedema formation. These findings suggest that oxygen derived free radicals act as mediators of brain oedema formation during the early phase of experimental bacterial meningitis. Cyclooxygenase metabolites may provide one possible source for the generation of oxygen derived free radicals in bacterial meningitis.

Introduction

Although great advances in antimicrobial chemotherapy have been achieved during the last decades, the mortality rate of bacterial meningits of the adult remains high. Meningitis due to pneumococci, the most frequent pathogen in adult bacterial meningits, still has a fatality rate of about 30%. Unfavourable clinical courses of bacterial meningitis are often associated with brain oedema formation, cerebral vascular complications and increased intracranial pressure (ICP). The precise pathophysiological mechanisms of the major intracranial complications are not completely understood. Therefore we established a rat model of experimental pneumococcal meningitis which allows the continuous monitoring of regional cerebral blood flow (rCBF) and the assessment of brain oedema formation[8]. The purpose of this study was to detect mediators of brain oedema formation in experimental pneumococcal meningitis.

Material and Methods

Adult male Wistar rats weighing 250 to 350 g were anaesthetized with 100 mg/kg i.p. thiobutabarbiturate (Inactin®). The rats were tracheostomized and mechanically ventilated. Mean arterial blood pressure (MABP) was measured with a Statham P23 pressure transducer connected to cannula in the femoral artery. Arterial blood samples were collected periodically from the arterial line and analyzed for paO_2, $paCO_2$, and pH. Body temperature was maintained at 38 °C by a rectal thermometer controlled heating pad. The animals were placed in a stereotactic frame. ICP was recorded with a Statham P23 pressure transducer via a catheter in the cisterna magna. rCBF was continuously measured by laser-Doppler flowmetry[3, 4] (LDF; Model BPM 403, TSI Inc., St. Paul, MN). For placement of the LDF-probe a craniotomy (diameter 5 mm) was made in the mediocaudal portion of the right parietal skull. The dura remained intact. MABP, ICP, endexspiratory pCO_2, and rCBF, were continuously recorded on a multichannel paper strip recorder and by a PC system for 6 hours after induction of meningitis. Meningitis was induced by intracisternal injection of 75 µl of 10^7 colony-forming units/ml pneumococci (n = 7). In control animals (n = 6), saline was injected.

Brain water content determination: Six hours post intracisternal inoculation, the animals were sacrificed by exsanguination. The brain was cut from the spinal cord at the cervicomedullary junction, removed and weighed in a glass dish. The brains were dried for 16 hours at 130 °C to stable weight. The brain water content was calculated by using the formula: brain water content = (wet weight-dry weight)/wet weight × 100.

Blocking substances: We investigated the effect of dexamethasone (3 mg/kg i.p.) which was given prior to the induction of meningitis (n = 3) or at two hours after pneumococcal injection (n = 5), indomethacin (10 mg/kg i.v., n = 5) and superoxide dismutase (free SOD, from bovine erythrocytes, given by continuous intravenous infusion in a dosage of 22,000 U/kg per hour).

The data of the different animal groups (rCBF, ICP and brain water content) were compared by using one-way analysis of variance and Student-Newman-Keuls multiple comparisons. A p value < 0.05 was considered significant.

Results

The physiological variables paO_2, $paCO_2$, pH, and MABP were within normal limits during the experiment in all groups. rCBF and ICP of controls did not change throughout the experiment. There was an increase in rCBF and ICP associated with brain oedema formation in untreated meningitis animals. rCBF increased to 211.1% ± 40.5 at 6 hours post infection (p.i., baseline 100%, mean ± S.D., p < 0.05, compared to controls). ICP increased from 2.9 ± 1.4 mm Hg to 10.4 ± 4.7 mm Hg at six hours p.i. (p < 0.05, compared to controls). Brain water content was significantly elevated 6 hours p.i. (79.69% ± 0.24 compared to 78.94% ± 0.16 in the control group, mean ± SEM, p < 0.05; Fig. 1). The increase in rCBF and ICP were prevented by pretreatment with dexamethasone and administration of SOD, delayed and attenuated by pretreatment with indomethacin and reversed by administration of dexamethasone 2 hours p.i. Brain oedema formation was blocked by pretreatment with dexamethasone (79.03% ± 0.10), administration of dexamethasone 2 hours p.i. (79.16% ± 0.19) and treatment with SOD (79.08% ± 0.09) and attenuated by pretreatment with indomethacin (79.25% ± 0.11) (Fig. 1).

Discussion

Bacterial meningitis is associated with increased intracranial pressure which is due to cerebral oedema, altered CSF absorption and increased cerebral blood

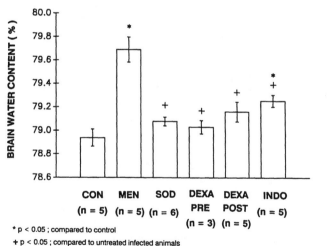

* p < 0.05 ; compared to control

+ p < 0.05 ; compared to untreated infected animals

Fig. 1. Brain water content determinations revealing brain oedema formation at six hours post pneumococcal injection in the untreated meningitis animals. The brain oedema formation was inhibited by treatment with superoxide dismutase (SOD), pretreatment with dexamethasone and administration of dexamethasone 2 hours after infection, and attenuated by pretreatment with indomethacin. *con*: controls; *men*: meningitis-no therapy; *dexa pre*: pretreatment with dexamethasone; *dexa post*: administration of dexamethasone 2 hours after pneumococcal infection; *indo*: indomethacin. *SOD*: superoxide dismutase. Data are expressed as means ± S.E.M. *p < 0.05, compared to controls; +p < 0.05, compared to untreated meningitis animals

volume due to inflammatory hyperemia or impaired autoregulation of the CBF. Brain oedema formation during bacterial meningitis was observed clinically, in pathological surveys, and in experimental bacterial meningitis[10, 11]. In our rat model, the intracisternal inoculation of living pneumococci produced an increase in rCBF and ICP, and brain oedema formation during the early phase of bacterial meningitis. Preliminary results in our rat model showed that the intracisternal injection of highly purified pneumococcal cell wall components (kindly provided by A. Tomasz from the Rockefeller university) were also capable of producing similar changes (unpublished data). The pathophysiological mechanisms of these alterations are not completely understood. A variety of factors are thought to be of relevance in the genesis of these changes: abnormalities of cerebral interstitial pH[1], endothelial adhesion of leucocytes, vasoactive cyclooxygenase metabolites[11, 12], free oxygen derived radicals[6], cytokines[7] or kinins[2, 13]. In this study we examined the effects of dexamethasone, indomethacin and superoxide dismutase. The major finding in our experiments was that the continuous infusion of free SOD prevents brain oedema formation and blocks the increase of rCBF and ICP observed in infected untreated rats. Oxygen derived free radicals could also be identified as mediators of vascular damage and brain oedema in other experimental models, such as acute severe arterial hypertension or fluid percussion brain injury[5]. The demonstrated effect of free SOD in our model shows that free SOD is able to reach the site where oxygen derived free radicals are generated. Since free SOD does not penetrate the blood-brain barrier, we assume that oxygen derived free radicals are produced intravascularly during the early phase of bacterial meningitis. The definite sources of generation of oxygen derived free radicals in bacterial meningitis are unknown. Oxygen derived free radicals may be generated during the metabolism of arachidonic acid via the transversion of prostaglandin G2 to Prostaglandin H2 (Fig. 2). The role of kinins, which may stimulate arachidonic acid metabolism in bacterial meningitis, is still unknown. Another source of oxygen derived free radicals in bacterial meningitis may be phagocytes. Histological examinations in our laboratory showed that SOD did not block the influx of polymorphonuclear leucocytes into the CSF (unpublished data). An important chemoattractant factor during bacterial meningitis seems to be the complement system. In addition, cytokines which are known to be increased in the CSF during bacterial meningitis have shown *in vitro* chemoattractant prop-

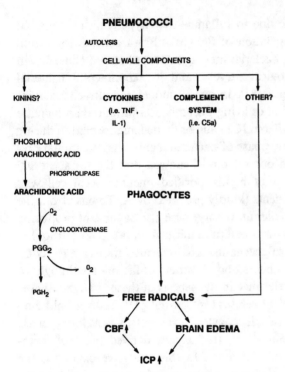

Fig. 2. Proposed mechanisms for increased intracranial pressure in experimental pneumococcal meningitis

erties and stimulate phagocytes to release oxygen derived free radicals.

Quagliarello *et al.*[9] investigated the nature of the brain oedema observed in their rat meningitis model. These authors identified morphologic alterations of the blood-brain barrier during experimental bacterial meningitis in the rat. They demonstrated an early (within 4 hours) increase of pinocytotic vesicle formation and separation of intercellular tight junctions. In addition, these morphologic changes were associated with increases in J^{125} albumin in the CSF. Ongoing immunohistochemic investigations in our laboratory with biotin labeled albumin (i.v.) focus attention on the temporal pattern of vasogenic brain oedema formation. The brain oedema observed during bacterial meningitis may not only be of vasogenic origin, but also may have cytotoxic (from inflammatory mediators in the exudate) and interstitial (impaired CSF absorption) components. Regardless of the cause of brain oedema, it may lead to increased intracranial pressure with the danger of coning or ischaemic lesions due to decreased cerebral perfusion pressure.

In conclusion, our data suggest that oxygen derived free radicals are involved as mediators in the increases of rCBF and ICP and brain oedema formation during the early phase of experimental bacterial meningitis. Cyclooxygenase metabolites may be one possible source for the generation of oxygen derived free radicals in bacterial meningitis.

References

1. Andersen NEO, Gyring J, Hansen AJ, Laursen H, Siesjö BK (1989) Brain acidosis in experimental pneumococcal meningitis. J Cereb Blood Flow Metab 9: 381–387
2. Baethmann A, Maier-Hauff K, Kempski O, Unterberg A, Wahl M, Schürer L (1988) Mediators of brain oedema and secondary brain damage. Crit Care Med 16: 972–978
3. Dirnagl U, Kaplan B, Jacewicz M, Pulsinelli W (1989) Laser-Doppler-Flowmetry for the estimation of CBF changes: a validation using autoradiography in a rat stroke model. J Cereb Blood Flow Metab 9: 589–596
4. Haberl RL, Heizer ML, Marmarou A, Ellis EF (1989 a) Laser-Doppler assessment of brain microcirculation: effect of systemic alterations. Am J Physiol 256: H1247–H1254
5. Kontos HA (1989) Oxygen radicals in experimental brain injury. In: Hoff JT, Betz AL (eds) Intracranial pressure, vol VII. Springer, Berlin, Heidelberg, New York, pp 787–798
6. McCord JM (1974) Free radicals and inflammation: Protection of synovial fluid by superoxide dismutase. Science 185: 529–531
7. Mustafa MM, Lebel MH, Ramilo O, Olsen KD, Reisch JS, Beutler B, McCracken GH (1989) Correlation of interleukin-1β and cachectin concentrations in cerebrospinal fluid and outcome from bacterial meningitis. J Pediatr 115: 208–213
8. Pfister HW, Koedel U, Haberl RL, Dirnagl U, Feiden W, Ruckdeschel G, Einhäupl KM (1990) Microvascular changes during the early phase of experimental bacterial meningitis. J Cereb Blood Flow Metab (in press)
9. Quagliarello VJ, Long WJ, Scheld WM (1986) Morphological alterations of the blood brain barrier with experimental meningitis in the rat. J Clin Invest 77: 1084–1095
10. Täuber MG, Khayam-Bashi H, Sande MA (1985) Effects of ampicillin and corticosteroids on brain water content, cerebrospinal fluid pressure, and cerebrospinal fluid lactate levels in experimental pneumococcal meningitis. J Infect Dis 151: 528–534
11. Tuomanen E (1987) Molecular mechanisms of inflammation in experimental pneumococcal meningitis. Ped Infect Dis 6: 1146–1149
12. Tureen JH, Stella FB, Clyman RI, Mauray F, Sande MA (1987) Effect of indomethacin on brain water content, cerebrospinal fluid white blood cell response and prostaglandin E_2 levels in cerebrospinal fluid in experimental pneumococcal meningitis in rabbits. Ped Infect Dis 6: 1151–1153
13. Wahl M, Young AR, Edvinsson L, Wagner F (1983) Effects of bradykinin on pial arteries and arterioles *in vitro* and *in situ*. J Cereb Blood Flow Metab 3: 231–237

Correspondence: H. W. Pfister, M.D., Department of Neurology, Ludwig-Maximilians-University of Munich, Klinikum Großhadern, Marchioninistrasse 15, D-8000 München 70, Federal Republic of Germany.

Acta Neurochirurgica, Suppl. 51, 381–382 (1990)

Perifocal Brain Oedema in Experimental Brain Abscess in Rats

Y. Nakagawa, K. Shinno, K. Okajima, and **K. Matsumoto**[1]

Department of Neurosurgery, Kagawa National Children's Hospital, Zentsuji city, Kagawa, Japan,
[1] Department of Neurosurgery, University Tokushima, Japan

Summary

An experimental cerebral abscess model in which staphylococcus aureus was inoculated into the brain parenchyma of rats was evaluated for perifocal brain oedema. Blood-brain barrier permeability was studied using various kinds of tracers; sodium fluorescein as a small molecule and Evans blue and HRP as macromolecular tracers. Brain abscess with clear delineation of fibrous capsule formation were found in all animals. Extravasation of tracers was demonstrated in the capsule (Evans blue), extracellular space (HRP) and white matter (sodium fluorescein). There were two types of oedema formation, so called vasogenic oedema and cytotoxic brain oedema.

Introduction

The introduction of computed tomography (CT) and new antibodies has reduced the mortality from brain abscess. However, patients still have mild to severe morbidity, such as epileptic seizures or neurological deficits. Despite surgical intervention for the brain abscess, perifocal brain oedema is demonstrated by CT for a long time afterwards. We believe that the postoperative clinical manifestations mostly correlate with perifocal brain oedema. In order to investigate the pathological factors, we studied blood-brain barrier permeability and ultrastructural changes of the microvasculature in both of the capsule of the brain abscess and the cortex around the abscess using an experimental model with rats.

Materials and Methods

Inoculation of abscess: Wistar rats unselected with regard to sex and weighing from 300 to 400 gm were used in this study. Anaesthetized with an intraperitoneal injection of pentobarbital (50 mg/kg), the spontaneously breathing animals were placed into a stereotactic frame. The scalp was prepared using providone paint, and a 1 cm paramedian incision was made over the right calvarium. A single 0.8 mm burr hole was drilled 4 mm to the right of the mid line just posterior to the coronal suture. A No. 27 G needle affixed to a 5 μl syringe was manoeuvered with the aid of a micromanipulator. The needle was inserted stereotactically into the white matter 3 mm in depth. An inoculation of fluid of Staphylococcus aureus (ATCC25923) bacterium was made over 30 minutes. Ten minutes elapsed after the termination of the injection and the needle was withdrawn over an additional 15 minutes[5]. Wounds were irrigated with provindone and saline solution and closed. To investigate blood-brain barrier permeability we used three kinds of tracers; sodium fluorescein as a small molecule and Evans blue and horseradish peroxidase (HRP; Sigma type II) as macromulecular tracers. The animals were perfused and fixed through the ascending aorta with a mixture of 4% paraformaldehyde and 0.5% glutaraldehyde in 0.1 M phosphate buffer at pH 7.4 on days 8 or 14. The brain was removed after transcardiac perfusion and studied ultrastructurally as described previously[2].

Results

Brain abscesses with clear delineation of fibrous capsule formation were found in all animals. Evans blue was observed mainly in the capsule of the brain abscess. Fluorescein staining was diffusely noted around the capsule and significantly in the white matter. Microscopically, a thin capsule consisting of small blood vessels, proliferating fibroblast and numerous macrophages was demonstrated. There was marked oedema in the cortex around the abscess. Ultrastructurally, mild to severe perifocal oedema around the abscess was found. There was fenestration between the endothelial cells of the microvasculature. The basal lamina of the neovasculature was poorly differentiated and deposition of lamina was revealed in some areas. In the cortex around the abscess, a marked angionecrosis of the microvasculature was observed. Extravasation of HRP was demonstrated in the extracellular space, basement membrane and interendothelial space of the microvasculature of the cortex (Fig. 1). In the white matter, there was remarkable swelling of dendrites, however, reactive

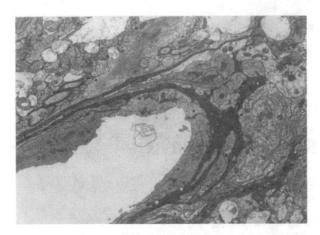

Fig. 1. Remarkable extravasation and reaction product of HRP in the extracellular space, basal lamina and interendothelial space

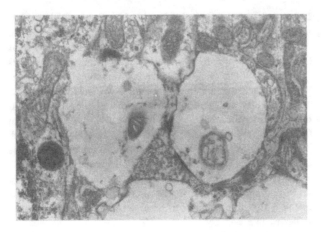

Fig. 2. Abscess on 16th day with marked swelling of dendrites

products of HRP were not found in the cytoplasm (Fig. 2). These findings indicate that perifocal oedema around the abscess may be a manifestation of two pathological situations; one, dilatation of extracellular space induced by dysfunction of BBB, and the other, swelling of the dendrites caused by cytotoxicity.

Discussion

Although significant progress has been made in imaging technology, microbiology and antibiotic treatment of brain abscess, much controversy still surrounds the treatment of the abscess-associated oedema. It is well known that brain oedema around the brain abscess is remarkable as one of the typical findings demonstrated on CT and that it exists for a long time after the surgical treatment of brain abscess. Therefore, supplemental use of steroids to control cerebral oedema is important as well as surgical treatment and appropriate antibiotic selection. However, there were few reports about blood-brain permeability in brain oedema associated with brain abscess[5]. In this study, extravasation of Evans blue and sodium fluorescein indicate that marked cerebral oedema may be caused by neovasculature in the membrane of the brain abscess. However, reactive product of HRP was demonstrated in the space between endothelial cells, basal lamina and in the extracellular space. This study showed leakage of large molecules through the tight junctions of the microvasculature of cortex indicating abnormal blood-brain barrier permeability of cerebral vessels[1, 3]. On the other hand, swelling of dendrites was also remarkable in the white matter which indicates cytotoxic brain oedema. It may correlate with the occurrence of epileptic seizures in the patients with brain abscess.

References

1. Cervos-Navarro J, Nakagawa Y, Sampaola (1985) Early changes in blood-brain barrier permeability after MCA occlusion in rats. In: Inaba Y, Klatzo I, Spatz M (eds) Brain oedema, Springer, Berlin Heidelberg New York, pp 210–214
2. Nakagawa Y, Cervos Navarro J, Artigas J (1985) Tracer study on a paracellular route in experimental hydrocephalus. Acta Neuropathol (Berl) 65: 247–254
3. Nakagawa Y, Fujimoto N, Matsumoto K, Cervos-Navarro J (1987) Blood-brain barrier permeability in acute cerebral ischaemia after MCA occlusion and reperfusion in the rat. Stroke and microcirculation. Raven Press, New York, pp 271–276
4. Schroeder KA, Mckeever PE, Schaberg DR, Hoff JT (1987) Effect of dexamethasone on experimental brain abscess. J Neurosurg 66: 264–269
5. Wallenfang T, Bohl J, Kretzschmar K (1980) Evolution of brain abscess in cats; Formation of capsule and resolution of brain oedema. Neurosurg Rev 3: 101–111

Correspondence: Y. Nakagawa, Department of Neurosurgery, Kagawa National Children's Hospital, Zentsuji-cho, Zentsuji city, Kagawa 765, Japan.

Acta Neurochirurgica, Suppl. 51, 383–386 (1990)
© by Springer-Verlag 1990

Brain Oedema and Cellular Changes Induced by Acute Heat Stress in Young Rats

H. S. Sharma and J. Cervós-Navarro

Institute of Neuropathology, University Klinikum Steglitz, Free University Berlin

Summary

Exposure of young animals (70–80 g, Age 6–7 weeks) to heat stress (HS) at 38 °C for 4 h in a B.O.D. incubator (rel humid 50–55%, wind vel 28.6 cm/sec) resulted in a 4.41% increased brain water content from the control value. Morphological studies in parietal cerebral cortex at light microscopical level revealed chromatolysis and appearance of dark neurons. Electron microscopy of similar regions showed perivascular oedema, vacuolation and collapsed microvessels. The swelling of astrocytes and of postsynaptic membranes was frequent. A diffuse infiltration of lanthanum (La[NO$_3$]$_3$) in endothelial cell cytoplasm and in the vesicles was very common. Occasionally, the lanthanum was seen in the basement membrane but the tight junctions were mainly intact. At this time period, a significant increase in blood-brain barrier (BBB) permeability as well as 5-HT levels in brain and plasma were observed. Pretreatment with p-CPA (a 5-HT synthesis inhibitor) prevented the increase of brain water content, BBB permeability and 5-HT levels in brain and plasma. Cyproheptadine (a 5-HT$_2$ receptor antagonist) treatment significantly reduced the occurrence of increased brain water content and the BBB permeability. The 5-HT level continued to remain high. These results point out a probable role of 5-HT in pathophysiology of HS via 5-HT$_2$ receptors.

Introduction

Heat stress (HS) causes severe neurological symptoms and/or death in children during summer months in many parts of India. The aetiological factors are still unclear. Earlier we reported a profound increase of blood-brain barrier (BBB) permeability to proteins in young rats subjected to a 4 h HS at 38 °C in a BOD incubator which was closely related with elevated 5-HT levels in brain and plasma[6]. It is quite likely that extravasation of serum proteins into the brain extracellular space causing vasogenic oedema may be instrumental in precipitating the neurological symptoms. A marked increase in brain volume in a closed cranial compartment is likely to induce cellular damage and even death.

In the present investigation, the brain water content, BBB permeability and 5-HT levels were measured following HS in young rats. The extent of cellular damage was examined using routine light and electron microscopical techniques. We took the advantage of pharmacological agents which modify the serotonin content and/or its action on the cerebral vessels in order to explore suitable therapeutic measures.

Materials and Methods

Animals. Experiments were carried out on 40 inbred male Wistar rats (body wt 80–90 g, age 6–7 weeks) housed at controlled ambient temperature (22 ± 1 °C) with 12 h light and dark schedule. The rat food and tap water were supplied *ad libitum*.

Heat exposure. Animals were exposed to 4 h HS in a B.O.D. incubator at 38 °C (rel humid 50–55%, wind vel 28.6 cm/sec)[6].

Stress symptoms and physiological variables. The body temperature, behavioural salivation and prostration, occurrence of haemorrhagic spots in stomach were recorded in each animal. In addition, the mean arterial blood pressure (MABP), arterial pH, PaO$_2$, and PaCO$_2$ were recorded using a standard protocol.

Morphology. Animals were perfused *in situ* with a fixative containing 2% paraformaldehyde and 2.5% glutaraldehyde in a 0.1 M sodium-potassium phosphate buffer (pH 7.4) containing 2.5% lanathanum nitrate (La[NO$_3$]$_3$) preceded by a brief saline rinse[4]. The routine paraffin embedded sections were stained with Nissl or haematoxylin and eosin. For electron microscopy, the selected brain regions were post fixed in 1% OsO$_4$, dehydrated in graded alcohol and embedded in Epon 812. Sections were cut using a diamond knife on LKB ultramicrotome, contrasted with lead citrate and uranyl acetate and examined in a Philips transmission electron microscope[7].

Blood-brain barrier permeability, brain oedema and 5-HT level. The BBB permeability to Evans blue (3 ml/kg of a 2% solution) and ^{131}-I-sodium (100 µCi/kg) was measured[6]. The brain water content and volume of swelling were determined[2]. The 5-HT in plasma and brain was measured fluorometrically according to Snyder *et al.*[9].

Drug treatments. The following drugs were used.

(a) p-chlorophenylalanine (p-CPA, 5-HT synthesis inhibitor, Sigma Chemical Co., USA). Animals were treated with p-CPA

Table 1. *Stress Symptoms and Physiological Variables Following 4 h HS (38° C) in Young Rats and Their Modification by Drugs.* The compounds were injected prior to HS (for details see text)

Parameters	Control	Heat stress	p-CPA + heat stress	Cyroheptadine + heat stress
A. Stress symptoms	n = 5	n = 5	n = 5	n = 5
1. △ °C Rectal temp.	0.23 ± 0.02	3.88 ± 0.23*	3.38 ± 0.22*	3.83 ± 0.33*
2. Salivation	Nil	+ + + +	+ + + +	+ + + +
3. Prostration	Nil	+ + + +	+ + +	+ + + +
4. Number of gastric ulcers per group	Nil	34 ± 8	28 ± 12	30 ± 10
B. Physiological variables	n = 5	n = 5	n = 5	n = 5
1. MABP torr	100 ± 4	78 ± 3*	82 ± 2	80 ± 2
2. Arterial pH	7.38 ± 0.04	7.36 ± 0.06	7.37 ± 0.05	7.36 ± 0.03
3. PaO_2 torr	79.34 ± 0.23	81.64 ± 0.36*	80.78 ± 0.24	80.08 ± 0.33
4. $PaCO_2$ torr	33.64 ± 0.14	32.87 ± 0.22*	33.17 ± 0.33	33.28 ± 0.21

MABP = mean arterial blood pressure. Values are Mean ± SD. * $p < 0.05$ Dunnett test for multiple group comparison.

(100 mg/kg, ip) daily for 3 consecutive days. On the fourth day these animals were subjected to a 4 h HS.

(b) Cyproheptadine (Periactin, Merck Sharp and Dohme, India). Cyproheptadine was injected intraperitoneally (15 mg/kg) 30 min prior to HS.

Control group. The intact animals served as control.

Statistical analysis. Dunnett's test for multiple group comparison was used to evaluate statistical significance[1].

Results

Stress symptoms and physiological variables. Exposure of animals to 4 h HS resulted in profound stress symp-

toms. A mild hypotension, increased PaO_2 and decreased $PaCO_2$ was observed at this time period. Pretreatment with p-CPA and cyproheptadine did not alter these parameters significantly (Table 1).

Morphology. At light microscopical level, chromatolysis and appearance of dark neurons in parietal cerebral cortex was frequent (Fig. 1). The cortex was spongy in appearance and contained many neurons with distorted cell bodies; some were swollen and others were shrunken. Many neurons were devoid of a prominent nucleus and in some neurons the nucleolus was

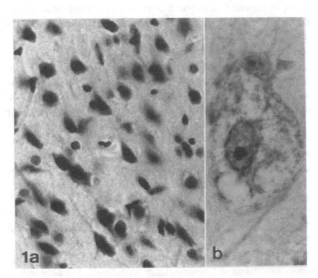

Fig. 1. Dark neurons in parietal cerebral cortex after 4 h heat stress (a, × 70). In similar region, chromatolysis is evident (b, × 170). Nissl staining

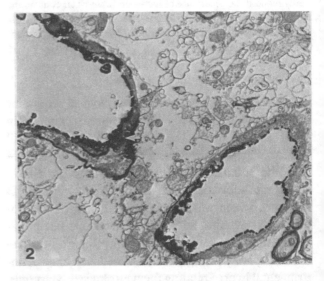

Fig. 2. Heat stress induced perivascular oedema and vacuolation in the parietal cerebral cortex. A diffuse uptake of lanthanum in the endothelial cell cytoplasm and in the basement membrane is evident (× 6,000)

Table 2. *Brain Edema, 5-HT Level and BBB Permeability Following 4 h HS (38° C) in Young Rats and Their Modification by Drugs.* The compounds were injected prior to HS (for details see text)

Parameters	Control	Heat stress	p-CPA + heat stress	Cyproheptadine + heat stress
A. Brain oedema	n = 5	n = 5	n = 5	n = 5
1. Water content %	77.86 ± 0.21	81.34 ± 0.32*	78.14 ± 0.23	78.83 ± 0.18
2. Volume swelling %	–	15.71	1.26	4.38
B. 5-HT	n = 5	n = 5	n = 5	n = 5
1. Brain µg/g	0.68 ± 0.14	2.87 ± 0.34*	0.81 ± 0.21	3.12 ± 0.32*
2. Plasma µg/ml	0.20 ± 0.04	1.79 ± 0.12*	0.32 ± 0.14	2.09 ± 0.18*
C. BBB permeability	n = 5	n = 5	n = 5	n = 5
1. Evans blue mg %	0.24 ± 0.06	1.56 ± 0.23*	0.34 ± 0.14	0.38 ± 0.25
2. ^{131}I-sodium %	0.30 ± 0.05	2.27 ± 0.14*	0.46 ± 0.12	0.50 ± 0.23

Values are Mean ± SD. * $p < 0.05$ Dunnett test for multiple group comparison. Volume swelling was calculated from the differences in water content according to the formula of Elliott and Jasper[2].

shifted towards the periphery. The extracellular compartment was expanded. At the ultrastructural level, swelling of astrocytes, postsynaptic membranes and perivascular region was quite prominent. The occurrence of collapsed vessels with profound uptake of lanthanum into the endothelial cell cytoplasm and in the vesicles was observed. Occasionally the lanthanum was seen in the basement membrane (Fig. 2), however the tight junctions were mainly intact.

BBB permeability, brain oedema and 5-HT level. The BBB permeability to EBA and ^{131}I-sodium was increased by 550% and 656% from the control group, respectively. At this time period, the brain water content was increased by 4.41% from the control group. Pretreatment with p-CPA and cyproheptadine significantly reduced the occurrence of increased permeability and brain oedema. The 5-HT level in untreated animals was significantly elevated in brain and plasma. This increase in 5-HT level was prevented by prior treatment with p-CPA, whereas the 5-HT level continued to remain high in animals with cyproheptadine (Table 2).

Discussion

The salient new findings of the present study is profound brain oedema and marked cellular changes following 4 h HS in young rats. It is quite likely that increased permeability of the BBB could play an important role in manifesting these changes.

The probable mechanism(s) underlying the in-

creased BBB permeability, brain oedema and cellular changes following HS is not known in all details. However, circumstantial evidence speaks in favour of the involvement of 5-HT as one of the neurochemical mediators[10]. This is evident from the results obtained with drug treatments. Thus the increase in brain water content and BBB permeability were almost absent in animals in which the 5-HT level did not increase from the control value as a result of p-CPA pretreatment. 5-HT is known to cause a decrease in electrical resistance of cerebral endothelium and thus allowing transport of micromolecular substances[4, 8]. This action of 5-HT is mediated through 5-HT$_2$ receptors. That 5-HT$_2$ receptors are involved in mediating BBB permeability and brain oedema in HS is evident from the results obtained with cyproheptadine treatment. A profound increase of 5-HT without altered BBB permeability and brain water content with cyproheptadine treatment indicates that the binding of serotonin to 5-HT$_2$ receptors is important in initiating its action.

A direct action of 5-HT on membrane permeability leading to cellular damage and cell death is quite likely[3, 7]. If serotonin is involved in the mechanisms of cellular damage in HS, one would expect a morphological preservation in drug treated animals, a feature which requires further investigation.

Acknowledgements

This study was supported by grants from Alexander von Humboldt foundation, FRG to HSS. The skillful technical assistance of Mrs. Elisabeth Scherer and Franziska Drum is highly appreciated. The secretarial assistance of Mrs. Aruna Misra is acknowledged with thanks.

References

1. Dunnett CW (1955) A multiple comparison procedure for comparing several treatments with a control. J Am Statis Assoc 50: 1096–1121
2. Elliott KAC, Jasper H (1949) Measurement of experimentally induced brain swelling and shrinkage. Am J Physiol 157: 122–129
3. Essman W (1978) Serotonin in health and disease, vol 3, The central nervous system. Spectrum, New York
4. Olesen SP (1989) An electrophysiological study of microvascular permeability and its modulation by chemical mediators. Acta Physiol Scand 136 [Suppl] 579: 1–28
5. Olsson Y, Sharma HS, Petterson CAV (1990) Effects of p-chlorophenylalanine on microvascular permeability changes in spinal cord trauma. An experimental study in the rat using ^{131}I-sodium and lanthanum tracers. Acta Neuropathol (Berl) 79: 595–603
6. Sharma HS, Dey PK (1987) Effect of acute longterm heat exposure on regional blood-brain barrier permeability, cerebral blood flow and 5-HT level in conscious normotensive young rats. Brain Res 424: 153–162
7. Sharma HS, Olsson Y (1990) Oedema formation and cellular alterations following spinal cord injury in rat and their modification with p-chlorophenylalanine. Acta Neuropathol (Berl) 79: 604–610
8. Sharma HS, Olsson Y, Dey PK (1990) Changes in blood-brain barrier and cerebral blood flow following elevation of circulating serotonin level in anaesthetized rats. Brain Res (in press)
9. Snyder SH, Axelrod J, Zweig M (1965) A sensitive and specific fluorescence assay for tissue serotonin. Biochem Pharmacol 14: 831–835
10. Wahl M, Unterberg A, Baethmann A, Schilling L (1988) Mediators of blood-brain barrier dysfunction and formation of vasogenic brain oedema. J Cereb Blood Flow Metab 8: 621–634

Correspondence: Dr. H. S. Sharma, Ph.D., Institute of Neuropathology, University Klinikum Steglitz, Free University Berlin, Hindenburgdamm 30, D-1000 Berlin 45.

Acta Neurochirurgica, Suppl. 51, 387–390 (1990)

Brain Oedema, Mass Effects, CBF

Automated Time-averaged Analysis of Craniospinal Compliance

(Short Pulse Response)

I. R. Piper, J. D. Miller, I. R. Whittle, and **A. Lawson**

Department of Clinical Neurosciences, Edinburgh, Scotland

Summary

We have developed an automated method [Short Pulse Response (SPR)] of measuring craniospinal compliance using an electronic square wave pressure generator to produce a small (0.05 ml) and reproducible transient volume increase in the CSF space (pulse duration 100 msec). In experimental models of intracranial hypertension, arterial hypertension, arterial hypotension and arterial hypercarbia in cats, the new method accurately followed pyhsiological changes in compliance when compared to the manual volume-pressure injection method. The VPR overestimated compliance compared to the new SPR method (by 20% to 162%, mean = 77%). The SPR method was less variable between sequential measurements with a coefficient of variation (CV) ranging from 0.6% to 9.6% (mean CV = 2.6%), compared with a CV ranging from 5.6% to 48% (mean CV = 17%) for the VPR method. Repeated compliance measurements by the new method over a 12 hour period, produced no neuropathological evidence of either blood brain barrier breakdown or tissue damage resulting from the repeated volume injections.

Introduction

Raised intracranial pressure (ICP) is common in patients with head injury and other forms of brain injury and leads to further secondary brain damage[5]. Measuring craniospinal compliance offers the potential for early detection and prediction of raised ICP[4]. The most commonly used methods of measuring craniospinal compliance depend upon manual injection of known volumes of fluid into the CSF space with measurement of the resultant increase in CSF pressure. In our hands, these methods are too variable for routine clinical use. Present manual injection methods of measuring com-

pliance such as the volume pressure response (VPR) have other drawbacks, being time-consuming and causing infection and secondary rises in ICP[1]. Measurement of craniospinal compliance has consequently been little used clinically despite the value of the information provided. The objective of this project was to modify the VPR technique from a manual to an automatic injection sequence as a first step towards improving the reproducibility of measurement.

Materials and Methods

The method is illustrated in Fig. 1. A pressure generator allows the triggered production of rectangular wave pressure pulses of 150 mm Hg amplitude (P1) and 100 msec duration (T1). When these pressure pulses are applied to the end of the injection tubing resistance (R1) a small volume injection results (approx. 0.05 ml) into the CSF space. Provided the injection tubing resistance (R1) is considerably less than the CSF outflow resistance (R2) the injected volume bolus will leave the system through R1 rather than through the much higher resistance R2.

Compliance is calculated from the peak amplitude of the intracranial pressure response to this volume increment where the ICP response was time-averaged using a biological signal averager which facilitates separation of the pressure response from background noise. The mechanical term craniospinal compliance is equivalent in electrical terms to the capacitance (C1) and can be calculated from the standard capacitance charge equation (see below) provided the following are known: injection tubing resistance (R1), peak input pulse pressure (P1), input pulse duration (T1), and the intracranial pressure (Vc) after T1 seconds.

$$Vc_{(T1)} = P1(1-e^{-T1/R1C1}).$$

Twelve cats weighing between 2.5 and 5.5 Kg, were studied in three groups: an ICP group (n = 4), a BP group (n = 4) and a CO_2 group (n = 4). All animals were anaesthetised by alpha-chloralose given as a 1% solution (75 mg/kg I.V.). Both femoral veins were cannulated for infusion of fluids and drugs. Both femoral arteries were cannulated for arterial blood gas sampling and to allow monitoring of arterial pressure through placement of a Camino catheter-tip pressure transducer into the descending thoracic aorta. The animals were intubated, paralysed (pancuronium 0.7 mg I.V. repeated hourly) then ventilated with intermittent positive pressure ventilation. Ventilatory volume was adjusted as required to keep the animals normocarbic ($PaCO_2$ 35–40 mm Hg). Body temperature was maintained constant at 38 degrees Celsius. The animal was placed in a stereotactic frame in the sphinix position, the scalp and temporal muscles reflected, and bilateral burr holes were drilled in the skull 7 mm posterior and 4 mm lateral to bregma for cannulation of the lateral ventricles with 2 inch and 9 gauge needles. A midline incision was made in the posterior fossa, and a 19 gauge 2 inch metal needle was placed into the cisterna magna. The needle insertion holes were sealed by cyanoacrylate adhesive to prevent CSF leaks. ICP was monitored through the left ventricular cannula using an optimally damped fluid-filled catheter transducer system. The right ventricular cannula was used for infusion of saline or Hartman's solution into the CSF space for altering the baseline ICP. The cisterna magna cannula was connected to the SPR method pressure pulse generator for applying input pressure pulses to the CSF system for measurement of craniospinal compliance. Through the positioning of a 3-way tap, the cisterna magna cannula could be redirected to the VPR method test apparatus which consisted of a 30 cm low compliance tube, a 3 way tap and a 1 ml syringe. The ICP response to the SPR and VPR pressure and volume pulses were recorded through the left lateral ventricular cannula. All pressure transducers were calibrated against a water column and zeroed with reference to the plane of the sterotactic frame ear bars. Figure 1 also shows sample input and

output SPR method pressure pulses recorded from an animal at normal ICP (13 mm Hg).

At each level of ICP, BP of CO_2, four sequential measurements of craniospinal compliance were performed by both the SPR and VPR methods with five minutes separating each measurement. For the SPR method, the compliance for a given level of ICP was the average of the four sequential compliance measurements, where each measurement was the signal average of 50 stimulus presentations. For the VPR method, the compliance for a given level of ICP was calculated from the mean pressure response after rapid injection of 0.05, 0.1, 0.15, and 0.2 ml of saline. All comparisons between two groups of data were tested statistically with a students paired t-test. Curve fits were performed on some data to characterize the trend of the data. A power function was found to be the approximating function that best fit the shape or general trend of the data. The curve fits were performed using a least squares regression.

Results and Discussion

In the ICP group, ICP was raised in five stages from 10 to 100 mm Hg by infusion of saline (3–60 ml/hour) into the right lateral ventricle. So that all animals could start from the same ICP baseline, ICP was raised, by slow intraventricular infusion (3 ml/hour) of saline, to a starting baseline level of between 10 and 15 mm Hg. In two animals the SPR preceded the VPR method, while with the remaining two the VPR preceded the SPR method. Both methods show a decrease in compliance with increasing ICP (Fig. 2A). All VPR compliance values from all four animals were greater than the corresponding SPR compliance values, indicating that the overestimation is independent of whether com-

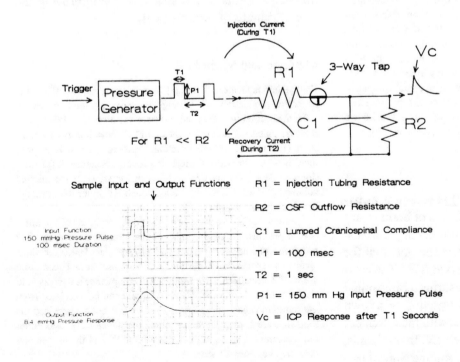

Fig. 1. Summary of method with sample input and output functions

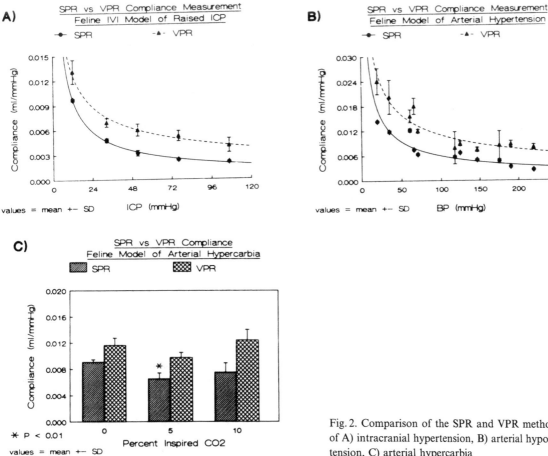

Fig. 2. Comparison of the SPR and VPR method in feline models of A) intracranial hypertension, B) arterial hypotension and hypertension, C) arterial hypercarbia

pliance was first measured by the SPR or VPR method. The VPR method overestimated compliance compared to the SPR method (range = 20% to 162%, mean = 77%) and also demonstrated greater variability between measurements. The amount of variability was expressed by the coefficient of variation (CV). The CV for the SPR method ranged from 0.6% to 9.3% with a mean variation between measurement of 2.6%. The VPR method showed a CV ranging from 5.6% to 48% with a mean variation between measurements of 17%.

In the BP group, the BP was first lowered and then raised over a range of BP from 20 mm Hg to 220 mm Hg (Fig. 2B). BP was lowered with intravenous infusion (10–40 ml/hour) of trimetaphan camsylate (2.5 mg/ml). BP was raised with intravenous infusion (6–40 ml/hour) of Angiotensin II (100 μg/ml) and supplemented as required with adrenalin (0.2 mg/ml at 10–40 ml/hour). Both methods show a decrease in compliance with increasing BP. Similar to the results with raised ICP the VPR method overestimates compliance compared to the SPR method and also demonstrates greater vari-

ability between measurements. The amount of overestimation ranged from 28% to 214% with a mean of 89%. The coefficient of variation (CV) between the four sequential measurements at each level of BP for the SPR method ranged from 0.4% to 4.6% with a mean variation between measurements of 2.56%. The VPR method showed a CV ranging from 3.8% to 49.3% with a mean variation between measurements of 16%.

In the CO_2 group, arterial CO_2 tension was raised in two stages by the administration of 5% and 10% CO_2 in the inspired gas mixture (Fig. 2C). With 5% inspired CO_2 the $PaCO_2$ increased from a baseline of 37.6 mm Hg to 59.8 mm Hg (P < 0.01) increasing further still with 10% CO_2 to 117 mm HG (P < 0.01). ICP also increased significantly rising from a baseline value of 13 mm Hg to 25 mm Hg (P < 0.01) and increasing to 34 mm Hg (P < 0.01) with 10% CO_2. Associated with these changes, both methods showed a decrease in compliance with 5% CO_2 although only SPR method showed a statistically significant decrease (P < 0.01). With 10% inspired CO_2, compliance increased towards control levels, although this change did not reach statistical significance. There was a large variation in compliance across animals. However, within individual animals, similar to the ICP and BP groups, the SPR method showed less variation (CV = 1.8% to 10.9% with a mean of 4.9%) compared with the VPR method (CV = 3.4% to 41% with a mean of 14.2%). The amount of overestimation of the VPR method compared with the SPR method ranged from 21% to 171% with a mean of 52%. Both methods showed a decrease in compliance with 5% CO_2 although only the SPR method showed a statistically significant decrease. With 10% inspired CO_2, both methods showed an increase in compliance towards control levels, although this change did not reach statistical significance.

To determine whether the repeated application of pressure pulses into the CSF space induced by the SPR method produced damage to the blood brain barrier or to the neuronal parenchyma, animals were prepared as previously described with the exception that intraventricular cannulae were not placed and cranial surgery, other than exposure of the foramen magnum, was not performed. This was to eliminate surgery and ICP monitoring as a source of trauma to the brain. Pressure pulses were applied through the cisterna magna cannula as normal. A sequence of fifty pressure pulses was applied, once every 15 minutes for 8 hours. Forty minutes before the end of the protocol, 8 ml (2 ml/Kg body weight) of a 2% Evans Blue Dye was given intravenously to assess major blood brain barrier disruption[3]. At the end of the protocol, trans-cardiac perfusion-fixation using a formaldehyde, acetate, methanol buffer mixture (40:1:1) was performed as described by the method of Brierley and Brown[2]. After the brain was perfusion fixed, it was removed, sectioned and stained with Haematoxylin, Eosin, and Luxol Fast blue stains. Histological sections of brains were examined by a neuropathologist and showed normal brain parenchyma, no evidence of blood brain barrier disruption and no evidence of cortical, deep nuclear or white matter damage.

In conclusion, craniospinal compliance measured by the new SPR method appears to be less variable than the VPR method. Work is in progress to develop a clinical version of this experimental method.

Acknowledgements

The authors would like to thank Dr. Alex Gordon for preparing brain specimens for light microscopical analysis. This work was supported by a project grant from the Medical Research Council (G8804953N).

References

1. Avezaat CJJ, Van Eijndhoven JHM, Wyper DJ (1979) Cerebrospinal fluid pulse pressure and intracranial volume-pressure relationships. J Neurol Neurosurg Psychiatry 42: 687–700
2. Brown AW, Brierley JB (1968) The nature, distribution and earliest stages of anoxic-ischaemic nerve cell damage in the rat brain as defined by the optical microscope. Br J Exp Path 49: 87–106
3. Mackenzie ET, Strandgaard S, Graham DI, Jones JV, Harper AM, Farrar JK (1976) Effects of acutely induced hypertension in cats on pial arteriolar calibre, local cerebral blood flow, and the blood-brain barrier. Circ Res 39: 33–41
4. Maset AL, Marmarou A, Ward JD *et al* (1987) Pressure-volume index in head injury. J Neurosurg 67: 832–840
5. Miller JD, Sweet RC, Narayan R, Becker DP (1978) Early insults in the injured brain. JAMA 240: 439–442

Correspondence: Professor J. D. Miller, Department of Clinical Neurosciences, Western General Hospital, Crewe Road South, Edinburgh, Scotland EH4 2XU.

Acta Neurochirurgica, Suppl. 51, 391–393 (1990)

Changes in Epidural Pulse Pressure in Brain Oedema Following Experimental Focal Ischaemia

M. Mase, H. Nagai, H. Mabe, K. Kamiya, T. Matsumoto, and **Y. Ueda**

Department of Neurosurgery, Nagoya City University Medical School, Nagoya, Japan

Summary

This study was carried out to clarify the changes of pulse pressure of the intracranial pressure pulse wave in ischaemic brain oedema. Intracranial pressure and PP were measured in two groups of anaesthetized dogs; 1) increased volume of cerebrospinal fluid by cisternal saline injection (control group), 2) brain oedema caused by focal ischaemia (oedema group). Ischaemia was induced by 2 hours of occlusion of the anterior, middle cerebral and internal carotid arteries. The canine focal ischaemic model showed consistent ischaemic damage in the caudate nucleus and produced brain oedema successfully. PP increased linearly with rising ICP to 35 mm Hg, and PP in the oedema group was significantly smaller than that in the control group at the same ICP value. The slopes of the regression equation of ICP and PP were significantly different between the oedema and control group (oedema: 0.057 ± 0.029, control: 0.106 ± 0.009), mean $\pm$ SD, $P < 0.005$).

These results suggest that PP is easily affected by ischaemic brain oedema, which indicates increase of the brain tissue in the cranium. We conclude that PP is affected even at the same ICP value when intracranial components have altered.

Introduction

It is well known that epidural pulse pressure (PP) increases with rising intracranial pressure (ICP). However, PP at the same ICP is not always identical in various intracranial pathologies[3,6]. Many authors have investigated PP at increased states of ICP, but few studies related to brain oedema have been done. This study was carried out in order to clarify the changes of PP in brain oedema following focal ischaemia.

Materials and Methods

16 adult mongrel dogs of both sexes, weighting 8 to 12 kg, were anaesthetized with intravenous thiopental sodium (10 mg/kg) for endotracheal intubation and catheterization of the femoral artery and vein. Anaesthesia was maintained with intermittent intramuscular ketamine hydrochloride (10 mg/kg) and continuous infusion

of alcuronium chloride (0.3 mg/kg/h). The arterial blood pressure (ABP) was measured with a Statham P50 pressure transducer and arterial blood was sampled intermittently from the arterial catheter. $PaCO_2$ (35 ± 3 mm Hg) and PaO_2 (>100 mm Hg) were maintained within physiological limits by ventilator adjustment. Body temperature was maintained at approximately 37 °C using a heat pad. The ICP pulse wave was measured with a Toyota pressure transducer which was placed epidurally through a burr hole in the parietal region. The ICP and ABP were recorded by a data recorder (Sony PC-108) and analyzed with a minicomputer (DEC PDP 11/23). The mean ICP and mean ABP were calculated by adding one third of the pulse pressure to the diastolic pressure. The PP of the ICP pulse wave was measured from the bottom to the highest peak of the pulse wave. The mean PP was calculated by averaging 20 PP readings. ICP and PP were evaluated statistically with a two-tailed unpaired t-test in the following groups.

Control group. Increased volume of cerebrospinal fluid (CSF) by cisternal saline injection (n = 5).

Oedema group. Brain oedema caused by focal ischaemia (n = 11). Ischaemia was induced by electro-coagulation of the right anterior cerebral artery and by clipping the right middle cerebral artery and right internal carotid artery transorbitaly. The brain was reperfused for 6 hours after 2 hours of ischaemia. The orbit was sealed with bone cement to prevent CSF leakage. The ischaemic areas were identified by Evans blue, triphenyl tetrazolium chloride (TTC) or histological examination. Water content of the brain was measured by the wet-dry weight method.

Results

Ischaemic areas and water content. TTC staining showed ischaemic areas on the right caudate nucleus and part of the frontotemporal cortex. Histological examination showed ischaemic changes of the neuronal cells in the right caudate nucleus. Leakage of Evans blue, which was only slight, was seen in the right caudate nucleus in 4 dogs. Water content of the ischaemic hemisphere ($78.31 \pm 1.49\%$, mean $\pm$ SD) was higher

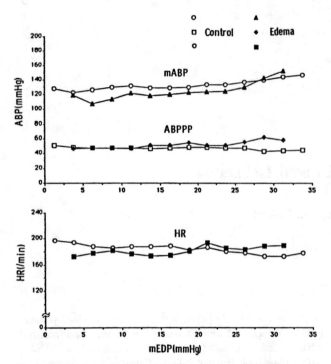

Fig. 1. The changes of mean arterial blood pressure (mABP), arterial pulse pressure (ABPPP) and heart rate (HR) in rising mean epidural pressure (mEDP). There was no significant difference between the oedema and control group. [Reprinted with permission from Mase M: Brain and Nerve (Tokyo) 42: 643–649, 1990]

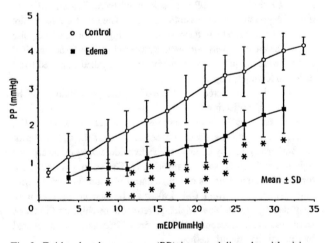

Fig. 2. Epidural pulse pressure (PP) increased linearly with rising mean epidural pressure (mEDP), and PP in the oedema group was smaller than that in the control group (*p<0.02, **p<0.005, ***p<0.0005). [Reprinted with permission from Mase M: Brain and Nerve (Tokyo) 42: 643–649, 1990]

than that of the intact hemisphere (77.30 ± 1.11%), but there was no significant difference (p<0.09).

Physiological parameters. There was no significant difference between the oedema and control group in ABP, arterial pulse pressure and heart rate (Fig. 1). $PaCO_2$ of the oedema and control group [35.2 ± 1.1

and 34.7 ± 0.7 mm Hg (mean ± SD)] were not significantly different.

ICP and PP. ICP increased gradually after reperfusion, and reached a maximum mean of 26.7 mm Hg in the oedema group. PP increased linearly with rising ICP in both groups, when ICP was less than 35 mm Hg. PP in the oedema group was smaller than that in the control group at the same ICP value (Fig. 2). The slopes of regression equation of ICP and PP were significantly different between the oedema and control group (oedema: 0.057 ± 0.029, control: 0.106 ± 0.009. mean ± SD, p< 0.005)

Discussion

The canine focal ischaemic model which we used in this study was described by Ohno[8]. He mentioned that the model showed significant reduction of cerebral blood flow (CBF) around the sylvian fissure and kept the skull closed. It was demonstrated that the model led to consistent ischaemic damage in the caudate nucleus and produced brain oedema resulting in ICP elevation. We think that this model is suitable to evaluate the changes of ICP and PP in brain oedema following focal ischaemia.

It became apparent that PP in the oedema group was smaller than that in the control group at the same ICP value. We consider two possibilities to account for the reduced PP in brain oedema. The first is a decreased volume of the pulsatile vascular bed because of brain oedema. Several papers suggest that the ICP pulse almost always originates from the pulse of the arterioles[1, 4]. Brain oedema increases tissue pressure which compresses the capillary bed mechanically resulting in reduced CBF[5]. During recirculation after focal ischaemia, there is a marked diversity of regional CBF and CBF in the ischaemic areas does not always recover[11]. We assume that CSF in this model also is not fully recovered after reperfusion and the development of brain oedema compresses the arterioles resulting in a smaller PP. Because we did not measure CBF during recirculation, the possibility has not yet been proven adversely.

The second possibility is the damping of ICP pulse transmission by brain oedema. The brain has a viscoelastic property. Aoyagi measured the Young modulus of the brain in experimental cold injury *in vivo* and mentioned that the Young modulus of slightly oedematous brain and central necrotic tissue was smaller than that of normal brain tissue[2]. This parameter has not yet been measured in ischaemic brain oedema. However, if the young modulus of the oedematous brain in ischaemia is as small as that in cold injury, the energy of ICP pulsation may be absorbed by distortion, resulting in smaller PP.

PP increases with rising ICP, and PP at the same ICP is not always identical in various intracranial pathologies. The distribution of intracranial components [brain, CSF and cerebral blood volume (CBV)] also changes in various intracranial conditions. Matsumoto reported that PP in hypercapnia was larger than following cisternal saline injection to the same ICP increase and that this was due to the fact that hypercapnia produced a significant increase in $\triangle$ V (pulsatile change of CBV)[6]. In the present study, we demonstrated that PP in the oedema group was smaller than that in the control group at the same ICP value. Brain oedema indicates an increase of brain tissue volume in the cranium. In summary PP in the state of increased brain tissue volume is smaller, and PP in the state of increased CBV larger, than PP in the state of increased CSF volume. We would like to conclude that PP is easily affected even at the same ICP value when intracranial components have altered.

It may be impossible to speculate on the underlying intracranial pathology only on the basis of the absolute PP value, since the ICP pulse wave is affected by many factors, such as changes of intracranial components, cerebrovascular tonicity[9], intracranial elasticity[10], the compensatory capacity of the cranio-spinal cavity[7], etc. However, by measuring ICP continuously, we may to some extent derive conclusions from the changes of PP on the changes of the intracranial condition.

References

1. Adolph RJ, Fukusumi H. Fowler NO (1967) Origin of cerebrospinal fluid pulsations. Am J Physiol 222: 840–846
2. Aoyagi N, Masuzawa H, Sano K *et al* (1980) Compliance of the brain. Brain and Nerve 32: 47–56
3. Avezaat CJJ, van Eijndhoven JHM, Wyper DJ (1980) Effects of hypercapnia and arterial hypotension and hypertension on cerebrospinal fluid pulse pressure and intracranial volume-pressure relationships. J Neurol Neurosurg Psychiatry 42: 222–234
4. Hirai O, Handa H, Ishikawa M (1982) Intracranial pressure pulse waveform: Considerations about its origin and methods of estimating intracranial pressure dynamics. Brain and Nerve 34: 1059–1065
5. Iannotti F, Hoff JT, Schielke GP (1985) Brain tissue pressure in focal cerebral ischaemia. J Neurosurg 62: 83–89
6. Matsumoto T (1985) Influence of intracranial components on intracranial pressure pulse wave. Neurol Med Chir (Tokyo) 25: 411–417
7. Matsumoto T, Nagai H, Kasuga Y *et al* (1986) Changes in intracranial pressure (ICP) pulse wave following hydrocephalus. Acta Neurochir (Wien) 82: 50–56
8. Ohno M (1986) Correlation between EEG, CBF, and ICP on experimental focal ischaemic brain produced by transorbital approach in dogs. Jpn J Stroke 8: 69–79
9. Portnoy HD, Chopp M (1981) Cerebrospinal fluid pulse wave form analysis during hypercapnia and hypoxia. Neurosurgery 9: 14–27
10. Szewczykowski J, Sliwka S, Kunicki A *et al* (1977) A fast method of estimating the elastance of the intracranial system. J Neurosurg 47: 19–26
11. Tamura A, Asano T, Sano K (1980) Correlation between rCBF and histological changes following temporary middle cerebral artery occlusion. Stroke 11: 487–493

Correspondence: M. Mase, M.D., Department of Neurosurgery, Nagoya City University Medical School, 1 Kawasumi, Mizuho-cho, Mizuho-ku, Nagoya, 467, Japan.

Acta Neurochirurgica, Suppl. 51, 394–396 (1990)

Hemodynamic Effects of Decompressive Craniectomy in Cold Induced Brain Oedema

A. Rinaldi, A. Mangiola, C. Anile, G. Maira, P. Amante, and **A. Ferraresi**[1]

Institutes of Neurosurgery and [1] Physiology, Catholic University, Rome, Italy

Summary

An experimental model of cold induced brain oedema was carried out in 7 albino rabbits by topic application of fluid N_2 on the skull to examine the favourable effects of an appropriate tailored cranio-dural opening both on the whole brain perfusion and on the local dynamics of the oedematous fluid. After positioning the animals in a stereotactic frame the following parameters were recorded: ICP, carotid blood pressure and flow velocity, EEG, EKG. Histological staining with Evan's blue was utilized to check fluid extravasation. In all the animals the lesion was followed both by an ICP increase and a reduction in the carotid blood flow velocity so indicating an increase in the cerebrovascular resistances. In 4 animals the removal of the craniodural flap over the lesion dramatically reduced the ICP values and normalized the blood flow velocity. No difference in the Evan's blue distribution was noticed between the two groups of animals.

Introduction

Different viewpoints have been expressed about the value of decompressive craniectomy as a last resource to control untreatable intracranial pressure (ICP) elevations[2,5]. We have introduced this technique in our therapeutic protocol, conscious of contrary opinions, *i.e.* that craniectomy favours "disastrous" brain swelling. As a consequence an experimental study was carried out to examine the effects of an appropriate tailored cranio-dural opening both on the whole brain perfusion and on the local fluid dynamics in cold induced brain injury. Following cold induced brain oedema there is evidence for a progressive increase in the local tissue pressure associated with a reduction in the regional cerebral blood flow (Bruce *et al.* 1972). As the lesion reaches a critical mass effect able to compromise the volume buffering capacity of the whole intracranial system (ICS), an increase in the cerebrovascular resistances (CVR) could induce a vicious circle worsening the initial oedema formation by a retroactive action on the vascular-tissue pressure gradients.

The cranio-dural removal, opening the ICS to the atmosphere and reducing ICP, could favour the cerebral venous outflow from the skull, so decreasing the CVR and disrupting the linkage between primary and secondary pathological events.

Methods

The experiments were carried out in 10 albino rabbits (3–4 kg b.w.) under general anaesthesia (Ketamine, Tubocurarine) and assisted ventilation. This animal was selected because both the anatomy and the functional significance of the carotid-vertebral system and the anastomotic circle at the base of the skull (mimicking the Willis poligone) show strict analogy with that of the human being (Meyer *et al.* 1986). Blood gases and body temperature were continuously monitored and kept within normal limits. The following parameters were simultaneously recorded: 1. ICP by a 18 G needle inserted stereotactically in the left lateral ventricle; 2. carotid blood pressure (CBP) by catheter into the left common carotid artery (CCA); 3. internal carotid artery blood flow velocity (ICABFV) by means of a continuous wave 8 MHz Doppler probe firmly placed on the right common carotid artery after ligature of the homolateral external branch in order to estimate CVR variations; 4. electrocortical activity (ECoA) through screw electrodes applied on the cranial vault; 5. electrocardiac activity by needle electrodes inserted at the extremities.

After positioning the animal on the stereotactic frame the experimental protocol started with surgical removal of a right parietal rectangular shaped (1.5 × 1.0 cm) cranio-dural flap and dural plan reconstruction by a thin silastic layer sealed to the external boundaries of the residual skull to prevent cerebrospinal fluid (CSF) leakage and to allow brain displacement without damage. In 3 control animals this initial step was followed by the recording procedure lasting for 3 hours, without any other manipulation. In the remaining 7 rabbits the bone flap was then rigidly replaced and sealed by acrylic cement. Thereafter, a cold lesion was induced through the replaced bone by 1 min topical application of fluid N_2 (−180 °C) utilizing an appropriate sized tube (1 cm of diameter) sealed to the skull.

The lesion was followed by a period of observation up to 4 hours.

In 4 of the 7 animals, as high values of ICP were touched (> 50 mm Hg), usually in 2 hours from the lesion, the bone flap was again and definitively removed and the recording procedure continued to reach the 4 hours of observation as the non-craniectomized experimental model.

In all the animals, half an hour before sacrifice, 1 cc/kg body weight 2% Evan's blue (EB) solution was administered intravenously. At the end of the experiment the brain was stored in buffered formaline solution and 3 days later cut in slices for macroscopic and microscopic examination.

Results

Basal conditions and craniectomized control animals

The diastolic values of ICP, CBP and ICABFV ranged in basal conditions (the pressure transducers were referenced to the external acoustic meatus) between 7–15, 120–150 mm Hg and 3–6 cm/sec respectively. The ECoA was represented by 4–8 Hz activity of about 200 microvolts amplitude symmetrically in both hemi-

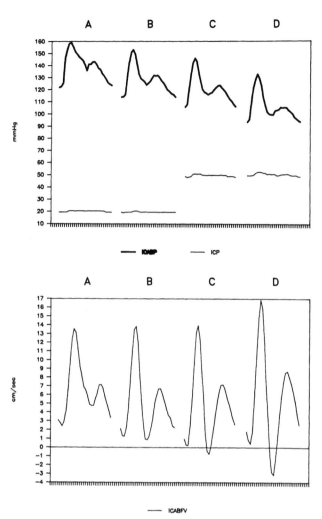

Fig. 2. Intracranial pressure (ICP), carotid blood pressure (CBP) and internal carotid artery blood flow velocity (ICABFV) synchronous recordings after cold lesion in a craniectomized animal. Craniectomy was carried out immediately after C and before D

spheres. Heart frequency (HF) was between 120 and 150 Hz. No significant variation in these parameters was noticed in the 3 control animals during the observation period. The post-mortem histologic staining with EB showed no extravasation of the dye in the brain tissue.

Non-craniectomized cold lesioned animals

After lesion, while CBP and HF remained relatively stable, ICP showed a progressive increase up to 40–60 mm Hg within 2–3 hours. This rise was concomitant with a reduction in the diastolic ICABFV and an increase in the ratio between systolic and diastolic value. The latter reached and even passed zero flow velocity, thus indicating a remarkable increment in the CVR (Fig. 1).

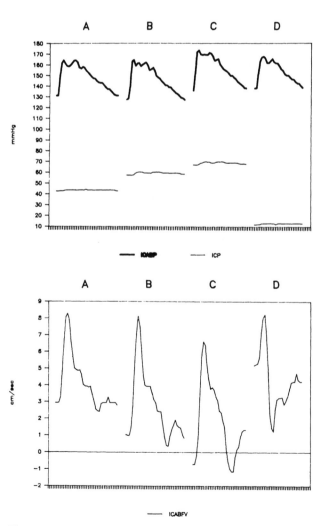

Fig. 1. Intracranial pressure (ICP), carotid blood pressure (CBP) and internal carotid artery blood flow velocity (ICABFV) synchronous recordings after cold lesion in a non-craniectomized animal

ECoA showed a triphasic trend with a generalized hypovoltage immediately after the cold lesion, followed by a temporary restoration and evolving afterwards in electrocortical silence.

At the cross examination the lesion appeared as a cortical area of hyperemia well dermarcated from the surrounding brain tissue. There was a remarkable staining of the brain parenchyma with EB, which spread from the boundaries of the lesion to the midline.

Craniectomized cold lesioned animals

In this group of 4 animals the first 2 hours of experiment showed the same findings as in the previous group. The bone removal was immediately followed by a dramatic ICP decrease to basal levels with simultaneous ICABFV restoration towards normal values, thus indicating a decrease in the CVR (Fig. 2). A partial ECoA recovery was later observed followed by signs of spontaneous breathing.

Pathological gross examination and Evans blue staining did not differ significantly from non-craniectomized cold lesioned animals. In particular no additional brain damage due to mass displacement was noticed.

Discussion and Conclusions

These preliminary results do not seem to provide arguments against the use of decompressive craniectomy in the treatment of severe intracranial hypertension due to severe brain oedema.

In contrast, they support the suggestion, that this surgical procedure, if carried out in a timely and proper way, could represent an efficient therapeutic tool able to mechanically disrupt the events' cascade (similar to the vasodilatatory cascade) which links the primary to the secondary brain damage following a cold induced lesion.

The opening of the ICS to atmospheric pressure, by-passing the action of the Starling resistor (Portnoy *et al.* 1983), relieves the effects of increased ICP on the venous outflow resistance and rebalances the cerebral inflow-outflow regulation. As a consequence an arterio-venous gradient is restored; at the same time the transmural pressure at the capillary bed is reduced following the decrease in CSF and/or venous pressure. These phenomena on one hand favour the flow through the vessels, on the other hand improve the fluid dynamics by reducing the formation and increasing the absorption of the brain oedema. The short observation period probably could explain the lack of a direct evidence of this mechanism in our experimental model.

Acknowledgements

We are grateful to Mrs. Anna Capuano and Mr. Pietro Santini for their precious technical support.

References

1. Bruce DA, Vapalahti M, Schutz H, Langfitt TW (1972) rCBF, CMRO$_2$ and intracranial pressure following a local cold injury of the cortex. In: Brock M, Dietz H (eds) Intracranial pressure I. Springer, Berlin Heidelberg New York
2. Clark K, Nash TM, Hutchison GC (1968) The failure of circumferential craniotomy in acute traumatic cerebral swelling. J Neurosurg 29: 367–371
3. Meyer FB, Anderson RE, Sundt TM, Yaksh TL (1986) Intracellular brain pH, indicator tissue perfusion, EEG and histology in severe and moderate focal cortical ischaemia in the rabbit. J Cereb Blood Flow Metab 6: 71–78
4. Portnoy HD, Chopp M, Branch C (1983) Hydraulic model of myogenic autoregulation and the cerebrovascular bed: the effects of altering systemic arterial pressure. Neurosurgery 13: 482–498
5. Ransohoff J, Benjamin MV, Gage EL, Epstein F (1971) Emicraniectomy in the management of acute subdural haematoma. J Neurosurg 34: 70–76

Correspondence: C. Anile, M.D., Istituto di Neurochirurgia, Policlinico Gemelli, Largo Gemelli 8, I-00168 Roma, Italy.

Acta Neurochirurgica, Suppl. 51, 397–400 (1990)

Blood Flow and Metabolism in Vasogenic Oedema

L. N. Sutton[1, 2], **J. Greenberg**[2], and **F. Welsh**[2]

[1] Children's Hospital of Philadelphia and [2] University of Pennsylvania, School of Medicine, Philadelphia, Pennsylvania, U.S.A.

Summary

The relationship between white matter cerebral blood flow (CBF) and glucose metabolism (LCMRgl) was studied in a plasma infusion model of vasogenic oedema in cats. LCBF as determined by iodoantipyrine was found to be significantly decreased in oedematous white matter (17.3 ± 1.5 ml/100 gm/min) when compared with contralateral control white matter (24.8 ± 1.8 ml/100 gm/min). If the values for oedematous brain were corrected for dilution, however, the LCBF averaged 25.3 ± 1.7 ml/100 gm/min, which was the same as control.

LCMRgl was found to be significantly increased in plasma-infused white matter (16.3 ± 2.2 µmol/100 gm/min), compared with control white matter (10.7 ± 1.3). This difference remained despite correction for dilution and recalculation of LCMRgl values based on altered kinetic constants found in oedematous brain. A similar increase in LCMRgl was noted with saline infusion oedema.

It is concluded that increased tissue water does not alter CBF, but does induce an increase in anaerobic metabolism.

Introduction

Cerebral blood flow: Blood flow has been reported to be decreased[1, 3, 4, 7, 8, 12, 13], increased[2] and unchanged[10, 11, 23] in oedematous brain. Thus, despite numerous studies, it remains unclear whether or not CBF is altered by brain oedema. Those reports which have demonstrated alterations in CBF have proposed a number of possible mechanisms: (1) a generalized increase in ICP with a consequent decrease in cerebral perfusion pressure, (2) a local increase in tissue pressure due to a focal mass effect and microvascular compression, (3) a possible vasoactive factor within plasma, and (4) a direct effect of oedema on metabolism resulting in a decrease in CBF due to altered metabolic autoregulation.

Our group has also examined the effects of vasogenic oedema on CBF[23]. In doing so, we have attempted to account for dilution, and also to remove the confounding influence of an oedema producing lesion by using an infusion model, as described by Marmarou et al.[11].

Brain energy metabolism and glucose utilization: The problems in studying brain metabolism are in many respects similar to those encountered in studying CBF. Reulen et al.[15, 16] reported diminished energy reserves in the form of decreased tissue concentrations of ATP, and phosphocreatine (Pcr) and increased lactate in oedematous white matter, and interpreted these changes as indicating local tissue hypoxia, perhaps as a result of decreased rCBF. Our group has also measured ATP, PCr and lactate in experimental oedema, and found that if a correction was made for dilution using the NAD^+-NADH sum, ATP and PCr were unchanged in both cold lesion and infusion oedema when compared with control white matter[25]. The dilution corrected lactate, however, was increased. It was concluded that oedema resulted in an increase in anaerobic metabolism, but that the energy requirements of oedematous tissue were maintained.

In order to further evaluate this hypothesis, we sought to measure the local cerebral metabolic rate for glucose (LCMRgl) in plasma and saline infusion oedema using the 2-deoxyglucose method[20, 21, 22]. We hypothesized that an increase in flux through the glycolytic pathway would be reflected in an increase in LCMRgl, accounting for the increased lactate found in oedematous brain.

Methods

Five cats were anaesthetized with ketamine and artificially ventilated while blood pressure, ICP, total CO_2 and arterial blood gases were monitored. Large decompressive craniectomies were performed to prevent the development of intracranial hypertension, and infusion oedema was produced by infusing stained autologous plasma or saline into the subcortical white matter by a No. 25 needle over four hours to a total volume of 0.7 ml. When the infusion was completed, 200 µCi of ^{14}C-labeled iodoantipyrine was infused via a femoral venous catheter over a 45 second period while 20 ml arterial samples were obtained every 2 to 3 seconds and placed in miniscintillation vials. The animals were sacrificed by decapitation. The brains were sectioned throughout the area of oedema in the coronal plane, and tissue samples of oedema and contralateral control white matter were obtained for scintillation counting and water content determination by gravimetry[9].

Local CBF was determined for oedematous regions and paired white matter control samples using the method described by Reivich *et al.*[14] and Sakurada[17] using ^{14}C-iodoantipyrine.

CBF is usually expressed as ml/100 gm wet tissue/min, but oedematous brain is diluted with extracellular water and some correction for dilution must be applied[23, 24, 25, 2]. It is not sufficient merely to express CBF in terms of dry weight of tissue, since oedema may alter the dry weight of the tissue due to a transudation of solids[5]. CBF values were therefore corrected to "normal" water content by multiplying by an oedema dilution correction factor C, defined as:

$$C = \frac{\% \, H_2O \, (\text{infusate}) - \% \, H_2O \, (\text{control white matter})}{\% \, H_2O \, (\text{infusate}) - \% \, H_2O \, (\text{oedematous white matter})}$$

which is calculated for each oedematous sample by measuring the water content of the oedematous sample and its paired contralateral hemisphere control sample, and the water content of the infusate (0.92 for plasma, 0.991 for saline)[24, 25, 23].

For the examination of the brain energy metabolism and glucose utilisation the same animals were used as for the CBF studies. Each animal received a bolus injection of 350 µCi of ^{3}H-labeled 2-deoxy-D-glucose (2-DG) through a femoral venous catheter 30 minutes after completion of the oedema infusion. Arterial blood samples (0.1 ml) were obtained immediately prior to administration of the ^{3}H-2-DG and at intervals initially of 15 to 30 seconds and later at intervals of 2 to 10 minutes for determination of plasma ^{3}H-2-DG concentration. Blood samples were also taken every 10 to 15 minutes for determination of the plasma glucose level. Samples were immediately centrifuged, and 30 µl plasma was pipetted into miniscintillation vials along with 0.4 ml water and 4 ml scintillation cocktail and placed in the scintillation counter.

Local cerebral metabolic rate for glucose (LCMRgl) was calculated using the operational equation of Sokoloff *et al.*[20, 21, 22] as modified by Savaki *et al.*[18] for a changing arterial glucose concentration. The numerator of the equation, (C_T-C_E) was determined by measuring the total counts in the brain sample (C_T) and subtracting the amount of unphosphorylated [3]H 2-deoxyglucose remaining in the tissue (C_E). C_E was calculated from the arterial curve and the average kinetic constants as determined for white matter according to Sokoloff *et al.*[20]. The denominator of the equation, the product of the "lumped constant" (LC) and the integrated precursor specific activity in the tissue, was estimated using the plasma [3]H 2 DG time course and average kinetic contants using the equation of Savaki *et al.*[18]. It was assumed that the lumped constant was not different from that of the normal brain[6]. Therefore, an "uncorrected" LCMRgl was calculated using the LC determined for the normal anaesthetized cat (0.411 for white matter)[20].

Results and Discussion

Cerebral blood flow was compared for oedematous white matter and control hemisphere white matter for both plasma-infusion oedema and saline-infusion oedema. LCBF in plasma infused white matter averaged 17.3 ± 1.5 ml/100 gm/min which was significantly less than control values which averaged 24.8 ± 1.8 ml/

100 gm/min (p < .01 paired T-test). CBF was lower in the oedematous sample in all but one of 15 samples obtained from 5 cats when compared with the paired control sample from the same animal.

Plasma infusion resulted in a significant increase in water content, however (74.2% versus 67.0% for control hemisphere white matter), resulting in a mean dilution correction factor of 1.6. If the LCBF values for the oedematous samples are corrected for dilution, the flow averaged 25.3 ± 1.7 ml/100 gm/min, which was virtually the same as control.

Saline infusion resulted in more variability in flow in the oedematous white matter. No significant difference was found, however, between flow in oedematous and control samples, whether or not a correction for dilution was applied.

Glucose utilisation: The 2 DG method depends on accurate estimation of the rate constants to determine the numerator of the operational equation. These rate constants, K_1^*, K_2^*, K_3^* represent the rate constants for carrier-mediated transport of deoxyglucose from plasma to tissue (K_1^*), for carrier-mediated transport back from tissue to plasma (K_2^*) and for phosphorylation by hexokinase (K_3^*). The rate constants used in these calculations are those determined for normal white matter, however, and it is possible that the presence of extracellular brain oedema might cause them to change and introduce an error in the calculation of C_E. For example, K_1^* and K_2^* are influenced by both blood flow and transport of [3]H 2 DG across the blood brain barrier[21], and it is possible that increased diffusion distances for glucose transport might alter the rate constants and therefore the C_E term of the operational equation.

In order to test this possibility, a series of 3 cats was studied to measure directly the amount of unphosphorylated [3]H 2 DG (C_E) in both oedematous and control white matter in the same animal. In all three animals, the amount of unphosphorylated [3]H 2 DG was higher in the oedematous white matter by an overall average of 25%. Corrected values for LCMRgl were then calculated for the oedematous samples by multiplying the calculated C_E values by 1.25, and then recalculating LCMRgl according to the equation of Savaki *et al.*[18]. These values were then multiplied by the dilution factor as determined by gravimetry. The values obtained for plasma infused white matter [LCMRgl (corr)] averaged 19.6 ± 3.0 µmol/100 gm/min, which was significantly greater than control (p < 0.01, paired T-test).

Thus, although the calculated C_E values underestimated the amount of unphosphorylated 2 DG, the magnitude of this error did not appear to be of sufficient magnitude to explain the increased rate of glucose metabolism in the samples of oedematous brain.

Saline-infused cats showed a similar difference between LCMRgl in control (10.8 ± 0.56) and oedematous (18.2 ± 1.1) brain. C_E was not directly measured for saline infusion oedema, and hence no corrected values were calculated.

Previous studies have demonstrated variable[4] and decreased[13] CMRgl in cold lesion oedema, but these reports did not attempt to compensate for dilution or possible alterations in the rate constants. In addition, it is possible that there was depression in the functional state of the subcortical white matter brought about by the cortical cold injury. In the present study, it was found that CMRgl was increased in plasma and saline infused white matter, and, at least in plasma oedema, this effect was not explained on the basis of miscalculation of C_E. Since C_E was not directly measured in saline oedema, we cannot conclude that the increase in CMRgl in oedema was not due to an alteration in the rate constants leading to a miscalculation in C_E. However, it is very likely that the correction factor would be in the same direction as was the case for plasma-induced oedema, leading to an even larger CMRgl. The increase in CMRgl thus appears to be a function of increased extracellular water, and not a function of any metabolically active substance within the plasma itself.

A unifying hypothesis — Our work suggests that vasogenic oedema does not alter rCBF, and that oedematous brain maintains normal levels of ATP and PCr. Lactic acid, however, is increased in oedema, as is LCMRgl. These results may be explained by postulating an increase in anaerobic metabolism, resulting in the rather inefficient use of glucose to produce high-energy phosphates and a mild lactic acidosis. Why this would occur remains unknown. Since CBF is normal, it is assumed that blood flow is not limiting. It is possible, however, that the increased extracellular space seen in oedematous brain results in a relative impediment to oxygen diffusion, and a relative hypoxia. Whatever the mechanism, the increase in anaerobic metabolism, however inefficient, appears to provide sufficient high-energy phosphate stores so that (except for a mild lactic acidosis) the cellular energy machinery remains largely unaffected.

References

1. Blasberg RG, Gazendam J, Patlak CS, Fenstermacher JD (1980) Quantitative autoradiographic studies of brain oedema and a comparison of multo-isotope autoradiographic techniques. In: Cervos-navarro J, Ferszt R (eds) Brain oedema. Raven Press, New York, pp 255–270
2. Bothe HW, van der Kerckhoff W, Paschen W *et al* (1982) Dissociation between blood flow and metabolic disturbances in oedema associated with experimental abscess in cats. In: Go KG, Baethmann A (eds) Recent progress in study and therapy of oedema. Plenum Press, New York, pp 355–363
3. Bruce DA, Vapalahti M, Schutz H, Langfitt TW (1972) rCBF, $CMRO_2$ and intracranial pressure following a local cold injury of the cortex. In: Brock M, Dietz H (eds) Intracranial pressure. Springer, Berlin Heidelberg, New York, pp 85–89
4. Dick AR, Nelson SR, Turner PL (1980) Quantified regional cerebral glucose consumption, rCBF and oedema and the effects of papaverine in cats with cortical cold injury. In: Shulman K, Marmarou A, Miller JD *et al* (eds) Intracranial pressure IV. Springer, Berlin Heidelberg, New York, pp 261–267
5. Fenske A (1973) Extracellular space and electrolyte distribution in cortex and white matter of dog brain in cold induced oedema. Acta Neurochir (Wien) 28: 81–94
6. Ginsberg MD, Reivich M (1979) Use of the 2-deoxyglucose method of local cerebral glucose utilization in the abnormal brain: Evaluation of the lumped constant during ischaemia. Acta Neurol Scand 60 (72): 226–227
7. Lammertsma AA, Wise RJS, Jones T (1984) Regional cerebral blood flow and oxygen utilization in oedema associated with cerebral tumours. In: Go GK, Baethamnn A (eds) Recent progress in the study and therapy of brain oedema. Plenum Press, New York, pp 331–334
8. Lammertsma AA, Wise RJS, Cox TCS *et al* (1985) Measurement of blood flow, oxygen extraction ratio and fractional blood volume in human brain tumours and surrounding oedematous brain. Br J Radiol 58: 725–734
9. Marmarou A, Poll W, Shulman K, Bhagauaw H (1978) A simple gravimetric technique for measurement of cerebral oedema. J Neurosurg 49: 530–537
10. Marmarou A, Takagi H, Walstra G, Shulman K (1980) The role of brain tissue pressure in autoregulation of CBF in areas of brain oedema. In: Shulman K, Marmarou A, Miller JD *et al* (eds) Intracranial pressure IV. Springer, Berlin Heidelberg New York, pp 257–260
11. Marmarou A, Takagi H, Shulman K (1980) Biomechanics of brain oedema and effects on local blood flow. In: Cervos-Navarro S, Ferszt R (eds) Advances in Neurology, vol 28: Brain oedema. Raven Press, New York, pp 345–357
12. Marshall LF, Bruce DA, Graham DI, Langfitt TW (1976) Alterations in behavior brain electrical activity, cerebral blood flow, and intracranial pressure produced by triethlyl tin sulfate induced cerebral oedema. Stroke 7: 21–25
13. Pappius HM, Savaki HE, Fieschi C *et al* (1979) Osmotic opening of the blood-brain barrier and local cerebral glucose utilization. Ann Neurol 5: 211–219
14. Reivich M, Jehle J, Solokoff L, Kety SJ (1969) Measurement of regional cerebral blood flow with antipyrine-C^{14} in awake cats. J Appl Physiol 27: 296
15. Reulen HJ, Medzihradsky F, Enzenbach R *et al* (1969) Electrolytes, fluids, and energy metabolism in human cerebral oedema. Arch Neurol 21: 517–525

16. Reulen HJ, Samii M, Fenske A *et al* (1971) Energy metabolism and electrolyte distribution in cold injury oedema. In: Head injuries. Churchill-Livingston, London, pp 232–239

17. Sakurada O, Kennedy C, Jehle J *et al* (1978) Measurement of local cerebral blood flow with iodo[^{14}C] antipyrine. Am J Physiol 234: H59–H66

18. Savaki HE, Davidsen L, Smith C, Sokoloff L (1980) Measurement of free glucose turnover in brain. J Neurochem 32: 495–502

19. Schmiedek P *et al* (1974) Energy state and glycolysis in human cerebral oedema. J Neurosurg 40: 351–364

20. Sokoloff L, Reivich M, Kennedy C, Des rosiers MH, Patlak CS, Pettigrew KD, Sakurada O, Shinohara M (1977) The [^{14}C] deoxyglucose method for the measurement of local cerebral glucose utilization: Theory, procedure, and normal values in the conscious and anaesthetized albino rat. J Neurochem 28: 897

21. Sokoloff L (1980) The [^{14}C] deoxyglucose method for the quantative determination of local cerebral glucose utilization: theoretical and practical considerations. In: Passoneau JV *et al* (eds) Cerebral metabolis and neural function. Williams & Wilkins, Baltimore, pp 319–330

22. Sokoloff L (1981) Localization of functional activity in the central nervous sytem by measurement of glucose utilization with radioactive deoxyglucose. J Cereb Blood Flow Metab 1: 7–36

23. Sutton LN, Barranco D, Greenberg J *et al* (1989) Cerebral blood flow and glucose metabolism in experimental brain oedema. J Neurosurg 71: 868–874

24. Sutton LN, Bruce DA, Welsh FA, Jaggi J (1980) Metabolic and electrophysiologic consequences of vasogenic oedema. In: Cervos-Navarro J, Ferszt R (eds) Advances in neurology, vol 28: Brain oedema. Raven Press, New York, pp 241–254

25. Sutton LN, Welsh FA, Bruce DA (1980) Bioenergetics of acute vasogenic oedema. J Neurosurg 53 (4): 470–476

Correspondence: L. N. Sutton, M.D., Children's Hospital of Philadelphia, 34th Street & Civic Center Boulevard, Philadelphia, PA 19104, U.S.A.

Acta Neurochirurgica, Suppl. 51, 401–403 (1990)
© by Springer-Verlag 1990

Effect of Mannitol on Intracranial Pressure-Volume Status and Cerebral Haemodynamics in Brain Oedema

S. Inao, H. Kuchiwaki, A. Wachi[1], K. Andoh, M. Nagasaka, K. Sugita, and **M. Furuse[2]**

Department of Neurosurgery, Nagoya University School of Medicine, [1]Department of Neurosurgery, Juntendoh University, [2]Department of Neurosurgery, Nakatsugawa Municipal General Hospital, Nagoya, Japan

Summary

Continuous measurement of CBV and CBF by means of laser-Doppler flowmetry was performed to analyze the temporal profile of cerebrovascular response to water accumulation in an infusion model of brain oedema. Using this method, the effect of mannitol on CBV was also examined to determine how mannitol improves the intracranial pressure-volume relationship. The results show that:

(1) The presence of an oedema generator elicits a continuous reduction of CBV and a transient reduction of CBF. With cessation of this mechanical force of oedema production, CBV and CBF return to control levels.

(2) Mannitol increases CBV and PVI. The improvement of PVI might be due to an increase in CBV as well as ICP reduction.

Introduction

Cerebral blood volume (CBV) is an important factor in the relationship between intracranial pressure (ICP) and cerebral blood flow (CBF). In this study, laser-Doppler flowmetry was introduced to continuously measure CBV as well as CBF in brain infusion oedema. We first describe the temporal profile of CBV and CBF as well as the tissue pressure gradient in oedematous tissue to determine whether or not the cerebrovascular system is affected by the production of oedema. Then, using this oedema model, we measured ICP, pressure volume index (PVI) and CBV following mannitol administration to examine the relationship between PVI and CBV, and to determine the effect of mannitol on the improvement of intracranial pressure-volume status.

Materials and Methods

Thirty five adult cats were used for this study

Experiment I (n = 21): The animals were anaesthetized with ketamine hydrochloride, and maintained with chloralose-urethane.

They were tracheostomized and ventilated. Catheters were placed in the femoral artery and vein. A No. 25 spinal needle was inserted in the white matter of the right hemisphere via a burr hole using stereotactic technique (coordinates 19 AP, 9.0 LAT, −9 V). Brain odema was produced by infusion of physiological saline (0.25 ml/120 minutes) through this needle, as described by Marmarou[3]. The platinum electrodes for regional CBF measurement (hydrogen clearance method) were inserted in both the oedematous and normal white matter. ICP was monitored in the cisterna magna and PVI was measured. In 10 cats, tissue pressure in oedematous and control brain was measured with a catheter tip transducer (Tokai-Rika Co., Nagoya). In another 11 cats CBF and CBV in the oedematous tissue were continously measured by means of a laser-Doppler flowmeter[1,6].

Experiment II (n = 14): After producing brain oedema and observing the parameters for 30 minutes post-infusion, mannitol was administered intravenously 2 g/Kg/3 minutes. MABP, ICP, PVI and haematocrit were measured. CBF (H_2 clearance method), and CBF and CBV (laser-Doppler method) were also measured in the normal and oedematous tissue for 30 minutes following administration of mannitol. Brain tissue water content was measured by dry-weighing method in each experimental animal.

Laser-Doppler flow analysis: Laser-Doppler method to measure CBF and CBV is based on the principle that light scattered by moving red blood cells experiences a frequently shift that is proportional to the average velocity of red cells. Electrical signals generated by the photodetector contain information about the frequency, which is related to the blood cell velocity, and the power, which is related to the blood volume[1]. Blood flow is computed as follows:

$$CBF = CBV \times Velocity$$

The laser-Doppler flowmeter (ALF 2100, Advance Co, Tokyo) used in this study had the He-Ne laser beam with an optical output power of 2 mW at the probe and a wavelength of 632.8 nm. The fiber-optic probe diameter of 0.8 mm (which conducts laser light to the tissue and carry backscattered light to a photodetector), is placed in the brain tissue for continuous monitoring the local blood flow and blood volume in the sampling area. The laser flowmeter processes all the frequency and power information from the voltage signals, and the flow values (volts) are recorded on a digital printer every minute.

Results

Experiment I: The tissue pressure gradient between normal and oedematous tissue gradually increased to 1.4 ± 0.5 mm Hg at the end of oedema formation and decreased slightly (1.2 ± 0.5 mm Hg) after completion of the infusion. Comparative CBF data obtained both by the hydrogen clearance technique and by laser flowmeter were nearly correlated ($r = 0.70$). Following oedema production, CBF decreased significantly ($100.6 \pm 1.5\%$ to $93.0 \pm 3.8\%$, $p < 0.05$) at 15 minutes after the beginning of tissue infusion, returned to control value within 60 minutes and remained close to the control value for the remainder of the oedema formation period. After finishing oedema production CBF rapidly increased to $118.1 \pm 9.7\%$ ($p < 0.05$) and then returned to control within 30 minutes. CBV decreased significantly at 15 minutes following the beginning of oedema production ($99.0 \pm 1.8\%$ to $90.8 \pm 3.8\%$, $p < 0.05$), and then steadily decreased reaching $84.0 \pm 3.2\%$ ($p < .0.01$), just before completion of oedema production (120 minutes post infusion). CBF gradually returned to control value within 15 minutes after completion of oedema production.

Experiment II: MABP significantly decreased from control value (125.5 ± 5.6 mm Hg to 89.1 ± 7.7 mm Hg, $p < 0.01$), immediately after mannitol administration, returned to control value 10 minutes post mannitol injection, and remained at control values for the remainder of the experiment. Hematocrit decreased dramatically ($35.1 \pm 1.7\%$ to $22.1 \pm 1.4\%$, $p < 0.01$) 10 minutes post bolus injection of mannitol and remained low ($27.0 \pm 1.8\%$, $p < 0.01$) at 30 minutes post mannitol. After mannitol administration ICP increased significantly (10.1 ± 1.1 mm Hg to 13.4 ± 0.7 mm Hg), $p < 0.05$), then returned to control value by 10 minutes post mannitol, and gradually decreased to 6.8 ± 0.9 mm Hg ($p < 0.05$) 30 minutes post mannitol. PVI was found to be markedly elevated as early as completion of mannitol administration (1.07 ± 0.11 ml to 2.86 ± 0.32 ml, $p < 0.01$). PVI peaked (3.20 ± 0.28 ml) at 10 minutes post mannitol, gradually decreased to 1.98 ± 0.16 ml by 30 minutes post mannitol, but always remained higher than control ($p < 0.01$). CBF obtained by laser method in the normal tissue increased significantly ($100.4 \pm 5.4\%$ to $125.8 \pm 5.7\%$, $p < 0.01$) 5 minutes post mannitol, then gradually decreased to $121.0 \pm 9.8\%$ at 30 minutes post mannitol. In the oedematous brain CBF increased following mannitol, but there was not a significant difference from the control value at any experimental time.

CBV in the normal tissue increased ($93.6 \pm 2.8\%$ to $147.8 \pm 12.1\%$, $p < 0.01$) 5 minutes post mannitol, then gradually decreased to $119.2 \pm 5.6\%$ at 30 minutes post mannitol, but always remained higher than control. In the oedematous tissue CBV increased ($91.4 \pm 7.8\%$ to $136.7 \pm 9.6\%$, $p < 0.01$) 5 minutes post mannitol, then gradually returned to control by 30 minutes post mannitol. Water content did not decrease significantly in either oedematous or normal white matter at 30 minutes post mannitol.

Discussion

This is the first report describing a continuous measurement of CBV and CBF by means of laser-Doppler flowmetry in oedematous tissue. Comparative CBF data obtained utilizing both H_2 clearance and laser-Doppler methods were linearly correlated[1, 6].

The presence of an oedema generator affects the cerebrovascular system. CBF decreased in the initial stage of brain infusion. CBV steadily decreased throughout the oedema production period. With cessation of the mechanical force of fluid infusion, CBF transiently increased, then returned to control values. CBV also returned to control values after stopping brain infusion. These results show that the excess accumulation of water in brain tissue itself does not decrease CBF[4], and that the driving pressure of oedema fluid as well as the chemical composition of oedema fluid reported by others are important factors affecting the cerebrovascular system.

Immediately after mannitol administration ICP temporarily increased. This was associated with elevation of both CBV[5] and PVI. Subsequently, ICP gradually decreased while CBV and PVI remained elevated without evidence of tissue water reduction 30 minutes post mannitol administration[2]. These results suggest that the effect of mannitol on ICP reduction might be due to decreased CSF volume[7] and/or undetectable tissue water reduction. The increase in PVI which indicates intracranial pressure-volume improvement may be due to both an increase in CBV as well as ICP reduction.

References

1. Haberl R, Heizer M, Marmarou A et al (1989) Laser-Doppler assessment of brain microcirculation: effect of systemic alterations. Am J Physiol 256: H1247–H1254
2. Inao S, Fatouros PP, Marmarou A (1989) Effect of mannitol on experimental infusion oedema. In: Hoff JT, Betz AL (eds) Intracranial pressure VII. Springer, Berlin Heidelberg New York, pp 447–450

3. Marmarou A, Tanaka K, Shulman K (1982) The brain response to infusion oedema; dynamics of fluid resolution. In: Hartmann A, Brock M (eds) Treatment of cerebral oedema. Springer, Berlin Heidelberg, pp 11–18

4. Marmarou A, Takagi H, Shulman K (1980) Biomechanics of brain oedema and effets on local cerebral blood flow. In: Cervos-Navaro J, Ferszt R (eds) Advances in neurology, vol 28: Brain oedema. Raven Press, New York, pp 345–358

5. Ravussin P, Archer DP, Tyler JL *et al* (1986) Effects of rapid mannitol infusion on cerebral blood volume. A positoron emission tomographic study in dogs and man. J Neurosurg 64: 104–113

6. Saeki Y, Sato A, Sato Y *et al* (1990) Effects of stimulation of cervical smypathetic trunks with various frequencies of the local cortical cerebral blood flow measured by laser Doppler flowmetry in the rat. Jap J Physiol 40: 15–32

7. Takagi H, Saitoh T, Kitahara T *et al* (1983) The mechanism of ICP reducing effect of mannitol. In: Ishi S, Nagai H, Brock M (eds) Intracranial pressure V. Springer, Berlin Heidelberg New York, pp 729–733

Correspondence: S. Inao, M.D., Department of Neurosurgery, Nagoya University School of Medicine, 65 Tsurumai-cho, Showa-ku, Nagoya 466, Japan.

Acta Neurochirurgica, Suppl. 51, 404–406 (1990)
© by Springer-Verlag 1990

Effect of Gammahydroxybutyrate on Intracranial Pressure, Mean Systemic Arterial Pressure and Cerebral Perfusion Pressure in Experimentally Induced Brain Oedema of the Rat

C. A. M. Plangger

Department of Neurosurgery, University of Innsbruck, Austria

Summary

The effect of gammahydroxybutyrate (GHB) on ICP, systemic arterial pressure and cerebral perfusion pressure in the experimentally induced brain oedema of the rat was examined. 400 mg/kg GHB reduced significantly ICP (11.74 ± 1.20 mm Hg; control: 16.20 ± 8.89 mm Hg; $p < 0.01$) while increasing mean systemic arterial pressure (109.89 ± 6.35 mm Hg; control: $89,65 \pm 4.22$ mm Hg; $p < 0.05$) and cerebral perfusion pressure (98.11 ± 6.79 mm Hg; control: 73.84 ± 5.25 mm Hg; $p < 0.02$). In the dose-effect curve 200 mg/kg GHB show an increase in mean systemic arterial pressure from 89.60 ± 9.35 mm Hg to 97.60 ± 3.48 mm Hg ($p < 0.02$) and 400 mg/kg GHB to 108.00 ± 5.20 mm Hg ($p < 0.001$) mean systemic arterial pressure. Thus, the decrease in intracranial pressure is not due to a reduction in the mean systemic arterial pressure, but GHB does reduce the ICP while increasing mean systemic arterial pressure and cerebral perfusion pressure. GHB may be a useful adjunct to neurosurgical therapy in controlling elevated ICP.

Introduction

In the treatment of raised intracranial pressure (ICP) an agent is needed which can similarly reduce ICP and protect the brain without reducing systemic arterial pressure and cerebral perfusion pressure. Gammahydroxybutyrate (GHB), an analogue of the neuroinhibitory transmitter gamma-amino-butyric acid (GABA), is an anaesthetic induction agent which may be useful in the control of elevated ICP. It is a naturally occuring substance in mammalian brain[7] and is present in highest concentrations in the hippocampus, midbrain, diencephalon and cerebellum. Because it rapidly produces a reversible trance-like state with a wide margin of safety, it has been widely used as anaesthetic agent or anaesthetic adjuvant[10]. Use of GHB in non-operated comatose patients showed an improvement in outcome compared to a barbiturate treated group

and its use in controlling elevated ICP in head injury has been advocated[8]. But there are no experimental data of the effects of GHB on ICP and systemic arterial pressure.

Methods

Male wistar rats (220–300 g) were anaesthesized by intraperitoneal injection of Inactin (120 mg/kg body weight, Byk-Gulden, Konstanz), placed on a heatable operating-table and maintained at body temperature of 37 ± 1 °C. The animals were tracheotomized and a catheter inserted in the right jugular vein. Systemic arterial blood pressure was continuously recorded by a catheter in the right femoral artery, through which also determination of the acid-base-balance and P_{CO_2} was possible. The brain tissue pressure was continuously measured using an electromagnetic Statham transducer P 23 Db and recorded on a two-channel-recorder (Linseis L 650). The rats were functionally nephrectomized by ligating the veins and arteries of the kidneys. A small burr hole was made on the left frontal part of the cranium and a Wick-catheter introduced for determination of brain tissue pressure[4,5,6]. Within 50 minutes 25 ml aqua bidest. were infused at a rate of 0.5 ml/min through the vena jugularis catheter and thus brain oedema of the cytotoxic type induced[4, 5, 11]. After 50 minutes the infusion was discontinued and 484 mg/kg of the sodium salt of gammahydroxybutyric acid corresponding to 400 mg/kg GHB (Na salt; Sigma Chemical Co., St. Louis, Mo, USA) was injected i.v. in the treatment group followed by continuous recording of ICP and systemic arterial pressure for at least three hours. Animals without administration of GHB served as controls. Additionally the relationship between the dose of gammahydroxybutyric acid and the response on the sytemic arterial pressure was studied so as to get a dose-effect curve.

Results

The functioning of the Wick-catheter was reflected by the typical wave form of the ICP recording with respiratory and pulse components. Infusion of 0.5 ml/min aqua bidest. led within 50 minutes to a marked increase of brain tissue pressure from 4.43 ± 0.25 mm Hg to 15.11 ± 0.76 mm Hg (n = 21). Systemic arterial pressure increased during this period from 75.48 ± 5.23 mm Hg to 83.66 ± 3.88 mm Hg (n = 11) while cerebral perfusion pressure (CPP) decreased slightly from 70.82 ± 5.46 mm Hg to 68.70 ± 4.70 mm Hg (n = 11).

Intravenous bolus injection of 400 mg/kg GBH caused a decrease in the ICP from 14.48 ± 1.57 mm Hg to 11.74 ± 1.20 mm Hg (n = 8) (in controls from 15.55 ± 0.72 mm Hg to 16.20 ± 0.89 mm Hg (n = 13); p < 0.01) after 10 minutes (Fig. 1). The systemic arterial pressure of the animals treated with GHB rose from 81.22 ± 4.33 mm Hg to 109.85 ± 6.35 mm Hg (n = 8) (controls from 91.19 ± 6.75 to 89.65 ± 4.22 mm Hg (n = 3); p < 0.05) and remained for 20 minutes significantly (p < 0.05) higher than the control levels. The cerebral perfusion pressure, defined as the difference between mean arterial pressure and ICP, rose from 66.73 ± 5.70 mm Hg to 98.11 ± 6.79 mm Hg (n = 8) (controls from 74.72 ± 7.29 to 73.84 ± 5.25 mm Hg (n = 3); p < 0.02) after the injection of GHB. The cerebral perfusion pressure remained for 20 minutes significantly (p < 0.05) higher than the control values (Fig. 2).

In the dose-effect curve 200 mg/kg GHB increased systemic arterial pressure from 89.60 ± 4.35 mm Hg (n = 5) to 97.60 ± 3.48 mm Hg; (p < 0.02 in Student's two tailed t-test for paired data) and 400 mg/lg to 108.00 ± 5.20 mm Hg (n = 5); (p < 0.001).

The decrease in the ICP is not due to a reduction of the mean systemic arterial pressure, but GHB does

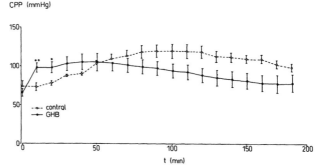

Fig. 2. Cerebral perfusion pressure (CPP) following infusion of 25 ml aqua bidest. and subsequent application of either 400 mg/kg gammahydroxybutyrate (GHB) in 2 ml isotonic saline (closed symbols, n = 8) or 2 ml saline alone (open symbols, n = 3). Arithmetic means ± SEM. **p < 0.02, compared with controls in Student's two tailed t-test for unpaired data. *p < 0.05

lower the ICP, while increasing systemic arterial pressure and cerebral perfusion pressure.

Discussion

Barbiturates have been used in the treatment of raised ICP causing a reduction in cerebral blood flow and in brain metabolism. This has been the basis of their application in cases of intracranial hypertension in an endeavour to reduce the ICP. Unfortunately, barbiturates also act to reduce arterial blood pressure and the arterial hypertension may be proportionally greater than the reduction in ICP, resulting in a net fall in cerebral perfusion pressure[3]. Thus one is still searching for agents which can similarly reduce ICP and protect the brain without reducing arterial pressure and cerebral perfusion pressure.

Wolfson et al.[12] demonstrated that the administration of gammabutyrolactone (GBL), an analogue of GHB, results in striking dose-dependent reduction in cerebral glucose utilization throughout the brain. The most likely explanation appears to be diminished energy demand resulting from widespread inhibition of neuronal activity like that caused by GABA or hypothermia. Kuschinsky et al.[2] examined the relationship between local cerebral glucose utilization (LCGU) and local CBF (LCBF) during the action of GHB (900 mg/kg i.v.) in conscious rats. LCGU was markedly depressed in all structures examined, whereas LCBF was differently affected in that not related changes were observed. Global glucose utilization was markedly depressed (− 51%), whereas global blood flow was not significantly altered. The observed depression in glucose utilization represents a decrease in brain energy

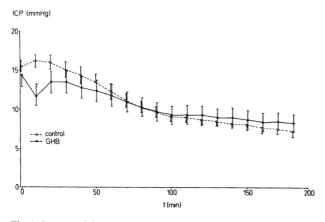

Fig. 1. Intracranial pressure (ICP) following infusion of 25 ml aqua bidest. and subsequent application of either 400 mg/kg gammahydroxybutyrate (GHB) in 2 ml isotonic saline (closed symbols, n = 8) or 2 ml saline alone (open symbols, n = 13). Arithmetic means ± SEM. *p < 0.01 compared with controls in Student's two tailed t-test for unpaired data

metabolism, because the utilization of fuels other than glucose can be excluded.

Various reports[1, 8, 9] have shown a reduction of raised intracranial pressure by gammahydroxybutyric acid in head injured patients, indicating an improvement in outcome compared to a barbiturate treated group[1], but there have been no systematic experimental investigations into the effect of GHB on ICP, systemic arterial pressure and cerebral perfusion pressure. The relatively short lasting effect (10–30 minutes) may necessitate continuous infusion of the substance in a clinical situation. In the experimentally induced oedema of the rat GHB reduces the ICP and protects the brain without reducing mean systemic arterial pressure and cerebral perfusion pressure. GHB may be a useful adjunct to neurosurgical therapy in controlling elevated ICP, decreasing cerebral metabolism and glucose utilization was well.

References

1. Escuret E, Baldy-Moulinier M, Roquefeuil B, Frerebeau P (1979) Gammyhydroxybutyrate as a substitute for barbiturate therapy in comatose patients with head injuries. Acta Neurol Scand 60 [Suppl 72]: 38–39
2. Kuschinsky W, Suda S, Sokoloff L (1985) Influence of gammahydroxybutyrate on the relationship between local cerebral glucose utilisation and local cerebral blood flow in the rat brain. J Cerebr Blood Flow 5: 58–64
3. Miller JD (1985) Head injury and brain ischaemia-implications for therapy. Br J Anaesth 57: 120–130
4. Plangger C (1990) Effect of gammahydroxybutyrate on intracranial pressure, mean systemic arterial pressure and cerebral perfusion pressure in experimentall induced brain oedema of the rat. Zbl Neurochir (in press)
5. Plangger C, Völkl H (1990) Effect of furosemide, bumetanide and mannitol on intracranial pressure in experimental brain oedema of the rat. Zbl Neurochir (in press)
6. Pöll W, Brock M, Markakis E, Winkelmüller W, Dietz H (1972) Brain tissue pressure. In: Brock M, Dietz H (eds) Intracranial pressure. Springer, Berlin Heidelberg New York, pp 188–194
7. Roth RH, Giarman NJ (1970) Natural occurrence of gammahydroxybutyrate in mammalian brain. Biochem Pharmacol 19: 1087–1093
8. Strong AJ (1984) Gammahydroxybutyric acid and intracranial pressure. Lancet 1: 1304
9. Strong AJ, Howd A, Hunt JM (1983) Reduction of raised intracranial pressure (ICP) by gamma-hydroxybutyric acid following severe head injury. Br J Surg 70: 303
10. Vickers MD (1969) Gammahydroxybutyric acid. Int Anesth Clin 7: 75–89
11. Weed LH, Mc Kibben PS (1919) Pressure changes in the cerebrospinal fluid following intravenous injection of solutions of various concentrations. Am J Physiol 48: 512–530
12. Wolfson LI, Sakurada O, Sokoloff L (1977) Effects of gammabutyrolactone on local cerebral glucose utilisation in the rat. J Neurochem 29: 777–783

Correspondence: Dr. C. A. M. Plangger, Department of Neurosurgery, University of Innsbruck, Anichstrasse 35, A-6020 Innsbruck, Austria.

Acta Neurochirurgica, Suppl. 51, 407–408 (1990)

Evaluation of Oedema Spreading and Intracranial Volume Reserve Using CT Image Numerical Analysis

Z. Czernicki and **J. Walecki**

Department of Neurosurgery, Medical Research Centre, Polish Academy of Sciences, Warsaw, Poland

Summary

Using the cold lesion model vasogenic oedema was produced in 8 cats. The intracranial volume reserve was evaluated applying CT image numerical analysis (CTINA). Summarizing all voxels corresponding to CSF and brain tissue in 8 CT slices, the brain/CSF ratio was computed. The animals were studied at different time intervals after the lesion (2 h, 4 h, 6 h) then sacrificed and the spreading of oedema was estimated based on the BBB disturbances. The brain/CSF ratios varied from mean value of 0.043 in control animals to mean values of 0.022 in animals 6 hours after the lesion.

The CTINA method was found very useful as a noninvasive screening method to study brain oedema development with a promising application in clinical practice.

Introduction

The spreading of oedema leads to brain volume enlargement and intracranial hypertension. The noninvasive determination of the intracranial volume reserve is till now a very difficult problem. It is possible to study the intracranial volume-pressure parameters using, for instance infusion tests, however the method is invasive and therefore not applicable in cases with intracranial hypertension. The other possibility is the CT image numerical analysis (CTINA) used before in trauma patients[3] and in experiments with epidural balloon compression[2]. The method was found in those studies to be practical and informative.

In the present study the small and slow changes in the intracranial volume reserve during oedema development were examined. The very well known model of vasogenic oedema produced by cold lesion was chosen[4].

Materials and Methods

The studies were performed on eight cats weighing from 2,800 to 3,600 g. All animals were anaesthetized and tracheostomized. As a blood-brain barrier (BBB) marker a 2% Evans blue solution was used. The marker was injected 2 hours before the animals were sacrificed. The cold lesion was applied in the left parieto-occipital region according to a previously described procedure[1]. A temperature of $-70\,°C$ was administered to the cortex for one minute. The study of oedema development lasted six hours, then the animals were sacrificed, the brains were taken out, cut in slices and examined for BBB changes.

During brain oedema development all animals were examined by a CT-scan [Somatom DR2 Siemens] before the lesion and then in 2-hour intervals. All eight slices were taken from brain tissue above the tentorium. A high resolution programme was used. The method principles were presented and discussed previously[3]. In the present study the results were calculated as CSF/brain tissue ratio. All voxels corresponding to the CSF density and brain tissue were summarized separately in each slice. Then the CSF/brain tissue ratio for the whole intracranial space was computed. The CT data were correlated with the development of the vasogenic oedema.

Results and Discussion

The results obtained are presented in Table 1. The control value of the CSF/brain tissue ratio varied in a relative wide range and probably is age dependent.

However, the first decrease in CTINA values was seen already after 2 hours of oedema development but the change was found statistically not significant. In this phase, the oedema, according to the BBB studies, was limited to one gyrus. The lateral ventricle was

Table 1. *CTINA Values After Cold Lesion*

Control	2 h after lesion	4 h after lesion	6 h after lesion
0.043 ± 0.007	0.038 ± 0.003	$0.030 \pm 0.003\,^*$	$0.022 \pm 0.002\,^*$

* Statistically significant.

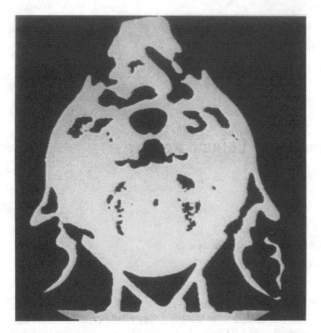

Fig. 1. CT image numerical analysis picture obtained in control cat 2/88

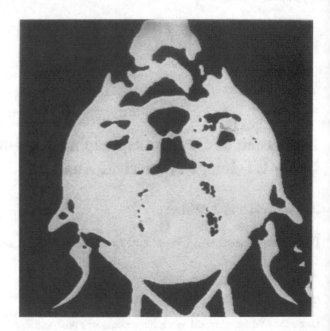

Fig. 2. CT image numerical analysis picture obtained in cat 4/90 6 hours after cold lesion

slightly compressed and no shift was observed. Later, vasogenic oedema was spreading to other brain regions. Figs. 1 and 2 show the CTINA pictures of control brain and 6 hours after cold lesion.

One can observe a marked reduction of black dots area – corresponding to the CSF voxels. The values given in table 1 are decreased statistically significant for studies at 4 hours and 6 hours after the lesion.

In the last examination 6 hours after the lesion, the intracranial volume reserve was reduced to about 2%; by two times smaller than the control value.

Information on the intracranial volume reserve is very important in clinical practice. In a previous experimental study using the epidural balloon compression, the first changes of the CTINA numbers were found also very early at a ballon volume of $0.5 \, \text{ml}^2$. Thus, the CTINA was confirmed to be very informative, and useful as a noninvasive method to study the intracranial volume relation.

In further investigations the method will be applied in clinical conditions.

Conclusions

1. The CT image numerical analysis enables to detect early changes of the intracranial volume relation.

2. The methods is noninvasive and has a possible clinical application in cases with developing brain oedema.

References

1. Czernicki Z (1979) Treatment of experimental brain oedema following sudden decompression, surgical wound and cold lesion with vasoprotective drugs and the proteinase inhibitor "Trasylol". Acta Neurochir (Wien) 50: 311–326
2. Czernicki Z, Walecki J, Jurkiewicz J, Grochowski W, Tychmanowicz K (1990) Intracranial volume reserve studies using CT images numerical analysis and lumbal infusion tests in cats. Przegl Radiol (in press)
3. Czernicki Z, Walecki J, Połacin A (1990) Intracranial volume reserve determination using CT images numerical analysis in trauma and tumour patients. Przegl Radiol (in press)
4. Reulen HJ, Graham R, Spatz M, Klatzo I (1977) Role of pressure gradients and bulk flow in dynamics of vasogenic brain oedema. J Neurosurg 46: 24–35

Correspondence: Z. Czernicki, M.D., Department of Neurosurgery, Medical Research Centre, Polish Academy of Sciences, Barska 16, 02-325 Warsaw, Poland.

Acta Neurochirurgica, Suppl. 51, 409–410 (1990)

Quantitative Analysis of CT Visualized Periventricular and Perifocal Brain Oedema by Means of CSF Dynamics Infusion Test

J. Eelmäe, J. Tjuvajev, and **A. Tikk**

Tartu University Hospital, Estonia, USSR

Summary

Brain oedema of different pathogenesis cannot be differentiated with CSF infusion tests. In cases with periventricular oedema the CSS viscoelastic parameters are changed more towards the decompensated state than in cases with perifocal oedema.

Introduction

The aim of this study was to determine the possibility of quantitative description of CT visualized periventricular (PV) and perifocal (PF) oedema by means of repeated bolus injection method.

Materials and Methods

This prospective study was carried out on 81 patients, who were treated in the intensive care unit or the neurosurgical and neurological departments of Tartu University Hospital. Computed tomography studies (CT) were performed using "Delta-Scan-190" (Ohio Nuclear, USA) and a slice thickness of 10 mm. To achieve maximum objectiveness CT Evans' index (EI)[2] was used for the evaluation of ventricular dimensions. Four groups according to the enlargement of ventricular size were formed: $EI > 0.3$ (hydrocephalus), $EI = 0.2–0.3$ (encephalopathy); $EI = 0.1–0.2$ (norm) and $EI < 0,1$ (brain oedema). The severity of PV and PF oedema was divided into four grades visually (no oedema, small, medium and large oedema). The investigation of CSF dynamics by repeated bolus injection technique[6] was followed. After inserting a spinal needle into the lumbar subarachnoid space the CSF pressure was measured continuously (transducer MP-4, amplifier RP-3, Nihon Kohden-Japan; paperprinter PS-1-02, USSR). 20 ml of saline were injected by 2 ml repeated boluses (mean injection rate 1.7 ± 0.2 ml/min) and after every bolus CSF pressure was measured (Fig. 1). CSF pathway obstruction was excluded by Queckenstedt's maneuvre. The mean time of test was 24.0 ± 5.9 min. From the obtained curves (Fig. 2) the complex parameters of CSF dynamics were calculated including baseline of intracranial pressure (ICP), pulsatile amplitude (A), elastance (E), compliance (C), pressure-volume index (PVI), resistance to CSF absorbtion (R) and Ayala index (AI)[1].

Results

In 184 patients with mixed diagnosis, normal values of the above parameters were calculated: $ICP < 15$ mm Hg, $A < 10$ mm Hg, $E = 2–3$ mm Hg/ml, $C = 0.2–0.6$ ml/mm Hg, $PVI > 12$ ml, $R < 10$ mm Hg/ml/min and $AI > 2.3$ ml/mm Hg. The values are in line with those of other authors[3, 4, 5]. The correlation matrix for this CSF dynamic parameters and ventricular size was composed in Table 1.

Table 1 demonstrates the strong correlation between ICP, C, and E, which is understandable mathematically, as well as the real positive correlation with ventricular size and C. The results of this study also show that the PF oedema grade was in correlation with Ao ($r = 0.24$, $p < 0.05$); A 10 ($r = 0.29$, $p < 0.01$) and PVI 20 ($r = 0.27$, $p < 0.05$).

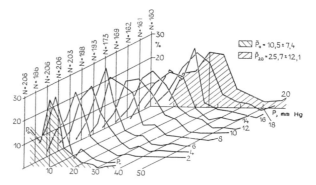

Fig. 1. The number of patients according to the CSF pressure on several stages of the infusion test. P_0 opening pressure of CSF, P_{20} pressure of CSF after 20 ml saline injection, P_- pressure after withdrawal of 2–3 ml CSF, 2.4.6 ... 20 pressure of CSF after corresponding injection of the saline, N number of patients on corresponding stages

Table 1

	EI	ICP	A	E	PVI	C
ICP	−0.160					
A	−0.065	0.990*				
E	−0.232	0.874*	0.869*			
PVI	0.094	−0.388	−0.370	−0.667		
C	0.300*	−0.908*	−0.873*	−0.896*	0.645	
R	−0.074	−0.434	−0.420	0.481	−0.413	0.228

* −p < 0.05

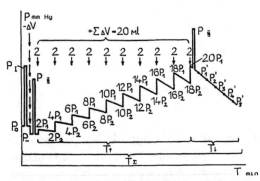

Fig. 2. Schematic representation of resultant curve during infusion test

The PV oedema grade correlates stronger with viscoelastic parameters of the craniospinal system (CSS): Po (r = 0.27, p < 0.025); P10 (r = 0.42, p < 0.001); Ao (r = 0.32, p < 0.005); A1 (r = 0.31, p < 0.25) and E2 (r = 0.28, p < 0.025).

Discussion

Numerous investigations have shown significant positive correlation between ventricular size (EI) and compliance of CSS. It means, that oedematous brain is characterized by a decrease of C and increase of ICP, E, and A. The negative correlation between EI and R is mainly due to the nature of the investigated group. Many patients were included with large ventricles ("ex vacuo") who did not need any shunting procedures.

It was also found, that CSF dynamic parameters, obtained by means of the described infusion test, are better suited for quantitative estimation of periventricular oedema than for the estimation of perifocal oedema.

References

1. Ayala G (1935) Physiopathologie der Mechanik der Liquor cerebrospinalis und der Rachidealquotient. Mschr Psychiat Neurol Stuttgart, Bd 3, 1866–1873: 184–206
2. Evans W (1942) An encephalometric ratio for estimating ventricular enlargement and cerebral atrophy. Arch Neurol Psychiat 47: 931–937
3. Kosteljanetz M (1985) Pressure-volume conditions in patients with subarachnoid and/or intraventricular haemorrhage. J Neurosurg 63: 398–403
4. Pollay M (1985) Pathology of the cerebrospinal fluid circulation. In: Crockard A *et al* (eds) Neurosurgery. The scientific basis of clinical practice. Oxford, pp 279–298
5. Tans J, Poortvliet D (1982) Intracranial volume-pressure relationship in man. Part 1: Calculation of pressure-volume index. J Neurosurg 56: 524–528
6. Tikk AA, Eelmae JM (1985) CSF hydrodynamics studied by means of a simple constant volume injection technique.: Abstracts of the 6th International Symposium on Intracranial Pressure. ICP and Mechanism of Brain Damage. 9th June-13th, 1985, Glasgow. Scotland/Glasgow, pp II: 10

Correspondence: J. Eelmäe, Tartu University Hospital, Estonia, USSR.

Acta Neurochirurgica, Suppl. 51, 411–413 (1990)

VPF and Interstitial Fluid Pressure in Brain Oedema

E. Sirovskiy, V. Kornienko, A. Moshkin, V. Amcheslavskiy, G. Ingorokva, and L. Glazman

N.N. Burdenko Research Institute of Neurosurgery, Moskow, USSR

Summary

Monitoring of VFP and local brain interstitial fluid pressure was performed in 169 patients after removal of hemispheric gliomas, basal and subtentorial tumours. On the basis of CT-data 97% of the patients had postoperative oedema of various severity and spreading. The location of the tumour determined both the degree and severity of oedema as well as VFP and ISFP.

Different mechanisms of oedema formation may be involved depending on localization of the tumour. Thus, we can propose a hypothesis of the mechanisms of oedema development in neurosurgical pathology.

Introduction

The aim of the study was to evaluate ventricular fluid pressure (VFP), brain interstitial fluid pressure (ISFP), degree and severity of oedema spreading depending on the location of brain tumour.

Materials and Methods

We have analyzed the results of VFP and ISFP monitoring, performed in 169 neurosurgical patients during the early postoperative period. The first group (Group I) included 57 patients after removal of basal extracerebral tumours, a second group (Group II) 67 patients after removal of hemispheric gliomas of various location and a third group 45 patients after removal of subtentorial tumours.

VFP was measured in one of the lateral ventricles. Local ISFP was measured with a specially designed catheter[5]. The method of the ISFP measurement was identical to methods described in literature[1, 4, 6].

Brain oedema was confirmed by means of CT. According to the CT-studies 3 degrees of oedema spreading could be delineated: local (up to 30% of tissue area), regional (30–60%) and diffuse (more than 60%).

Results and Discussion

1. *CT-data.* Practically all patients (97%) had postoperative oedema of various severity and spreading, depending on the location of the tumour. Table 1 shows CT-density data in the zones of maximal oedema. Thus

Table 1. *VFP, ISFP, and Density of Oedematous Tissue in the Patients After the Removal of Basal (I), Intracerebral (II), and Subtentorial (III) Tumours*

	I	II	III
VFP (mm Hg)	12.1 ± 2.9	27.1 ± 3.6	21.4 ± 3.5
ISFP (mm Hg)	−0.8 ± 1.5	7.8 ± 5.6	1.1 ± 1.6
Density (H unit)	26.8 ± 7.2	20.8 ± 6.3	22.4 ± 6.4

the lowest average density (maximal accumulation of oedema fluid) was observed after removal of intracerebral tumours (20.8 ± 6.3 H units). In these cases the oedema had a limited (local) extension and the severity of its manifestation was definitely greater than in the two other groups.

After the removal of basal or subtentorial tumours the oedema fluid accumulation in the cerebral hemispheres was less pronounced (average density 24.7 ± 6.8 and 22.4 ± 6.4 H units) but showed a tendency to diffuse spreading.

We can speculate: the farther from the brain stem and diencephalic region the tumour was, the more often oedema had local character.

2. *VFP and ISFP.* Fig. 1 shows the distribution of VFP and ISFP in the three groups of patients. VFP always exceeds ISFP, which in most cases is ranging from −5 to 0 mm Hg (e.g. "negative"").

In the group of patients operated for basal tumours the mean VFP was 12.1 ± 2.9 mm Hg (ranging from 0 to 35 mm). The mean ISFP was −0.8 ± 1.5 mm Hg (ranging from −20 to +16 mm Hg).

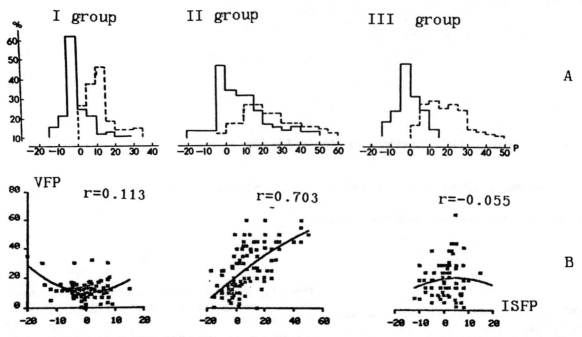

Fig. 1. The histograms of VFP (dotted line) and ISFP (solid line) in patients after removal of basal (I), hemispheric (II) and subtentorial (III) tumours (A). Correlation between VFP and ISFP in these groups of patients (B)

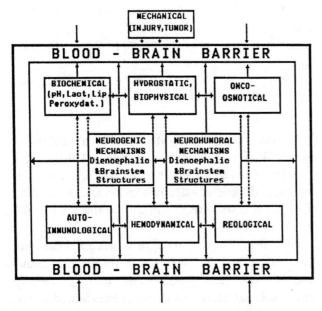

Fig. 2. Schematic model of the mechanisms and factors of cerebral oedema development

In the group of patient with removal of hemispheric gliomas the mean VFP was 27.1 ± 0.4 mm Hg (ranging from 1 to 70 mm Hg). The mean ISFP in "normal" brain tissue or in the non-damaged hemisphere was −0.7 ± 2.1 mm Hg (ranging from −18 to 45 mm Hg). The mean ISFP in the perifocal zone of the removed tumour (region of maximal oedema) was 7.8 ± 5.6 mm Hg (ranging from 1 to 60 mm Hg).

In the group of patients with removal of subtentorial tumours the mean VFP was 21.4 ± 3.6 mm Hg (ranging from 0 to 60 mm Hg). The mean ISFP was 1.1 ± 1.6 mm Hg (ranging from −15 + 15 mm Hg),

The differences between groups in VFP and ISFP in the zones of maximal oedema were statistically significant ($p < 0.05$).

In patients after removal of basal tumours VFP usually was "normal" or had a tendency to hypotension. Also ISFP in general was not increased, but rather "negative". There was no evident dependency between VFP and ISFP. We suppose that oedema fluid is mainly accumulated in cells rather that in the interstitial space.

In patients after removal of hemispheric gliomas VFP in general was increased and a correlation existed between VFP and ISPF ($r = 0.7$; $p < 0.05$), (Fig. 1). Maximal accumulation of oedema fluid was seen in the perifocal zone of the removed tumour. The same area was characterized by a marked increase of ISFP, probably due to a local increase in interstitial fluid volume. It is supposed that VFP increase is a consequence of the raised interstitial fluid pressure.

In patients after removal of subtentorial tumours VFP was moderately increased (Table 1). According to the CT-densities the diffuse fluid accumulation in brain tissue was more pronounced than in the group with basal tumours. Mean ISFP was positive and slightly higher than in the patients of group I (Table 1). We suppose that oedema fluid accumulates in the interstitial space. Nevertheless the raised VFP is not correlated with brain oedema as there was no dependency between VFP and ISFP (Fig. 1).

3. *Mechanisms of oedema development.* Based on the results of our investigations and on modern concepts of mechanisms of brain oedema development[1, 2, 4, 7] we propose a hypothesis which allows to unify the mechanisms of cerebral oedema development in neurosurgical pathology. We assume that both specific and non-specific mechanisms participate in cerebral oedema development (Fig. 2). Specific mechanisms may be of neurogenic and neurohumoral origin and may cause typical reactions of diffuse character. Non-specific mechanisms may be biochemical, onco-osmotic, hemodynamic and other factors and may be responsible for local oedema (Fig. 2). The operative trauma starts a number of biochemical reactions of cascade type. Different specific and non-specific oedema mechanisms begin to act depending on the location of the trauma. The longer is a distance of the local injury from the brain stem and diencephalic structures the lesser will be diffuse oedema spreading. The local character of oedema in superficial hemispheric tumours shows that it is mainly triggered by non-specific factors; *i.e.* this process is caused by a local disturbance of BBB permeability. The closer the pathological focus is located to the brain stem and diencephalic structures the more likely neurogenic and neurohumoral mechanisms are involved in the oedema development.

At the same time there are principal differences between the mechanisms of development of generalized cerebral oedema in diencephalic and subtentorial lesions.

In subtentorial lesions the involvement of brain stem vasomotor centers is responsible for a generalized change of the vascular tone and an increase of vascular permeability for water and serum proteins, resulting in vasogenic brain oedema. In contrast, in diencephalic lesions the hypothalamic centers responsible for the metabolism and trophic regulation of tissues are involved. Subsequent generalized disorders of cell trophic regulation and of cellular membrane permeability finally leads to hyperhydration. This process is typical for cellular, so-called cytotoxic oedema. It is evident that this concept is mostly schematic and oedema development is much more complicated.

However, the suggested unification of the mechanisms of oedema development allows to understand the principle conditions in which vasogenic, cytotoxic or hydrostatic oedema mainly occurs.

References

1. Brock M, Furuse M, Weber R, Dietz H (1975) Brain tissue pressure gradients. In: Lundberg N, Ponten U, Brocks M (eds) Intracranial pressure II. Springer, Berlin Heidelberg New York
2. Klatzo I (1976) Neuropathological aspects of brain oedema. J Neuropathology 26: 1–13
3. Konovalov A, Sirovskiy E, Kornienko V, Ingorokva G, Lobanov S (1989) Postoperative oedema of the brain. J Voprosi Neurochirurg 11, 1: 3–7
4. Marmarou A, Shapiro K, Poll W, Shulman K (1976) Studies of kinetics of fluid movements with brain tissue. In: Beks IWF, Bosch DA, Brock M (eds) Intracranial pressure III. Springer, Berlin Heidelberg New York
5. Paltsev E, Sirovskiy E (1982) In: Biomechanika-79. Riga, pp 68–77
6. Pöll W, Brock M, Markakis E, Winkelmüller W, Dietz H (1972) Brain tissue pressure in: Brock M, Dietz H (eds) Intracranial pressure. Spinger, Berlin Heidelberg New York
7. Reulen HJ, Kreysch HG (1973) Measurement of brain tissue pressure in cold induced cerebral oedema. Acta Neurochir (Wien) 29: 29–40
8. Reulen HJ, Graham R, Klatzo I (1975) Development of pressure gradients within brain tissue during the formation of vasogenic brain oedema. In: Intracranial pressure 11. Springer, Berlin Heidelberg New York

Correspondence: E. Sirovskiy, M.D., Chief of the ICU, N.N. Burdenko Research Institute of Neurosurgery, Fadeev-Str. 5, Moskow, 125047, USSR.

Author Index

Jörn Bo Madsen, Georg Emil Cold

The Effects of Anaesthetics upon Cerebral Circulation and Metabolism

1990. 14 figures. Approx. 150 pages.
Cloth DM 89,– öS 623,–
ISBN 3-211-82198-8

During the last decade, the effects of anaesthetics on cerebral blood flow, cerebral metabolic rate of oxygen and intracranial pressure, have been studied experimentally and clinically. In this review studies of CBF and CMRO2 during craniotomy have been performed with the classical technique described by Kety and Schmidt. In chapter 1 general considerations concerning the effects of anaesthetics on cerebral blood flow and metabolism are reviewed. In chapters 2 and 3, the effects of inhalation agents and hypnotics on flow and metabolism are considered. Chapters 4 and 5 cover the effects of central analgetics and neuromuscular blocking agents. In chapter 6, the effects of other drugs in common use in neuroanaesthetic practice are summarized. Chapter 7 considers the effects of drugs used for controlled hypotension. In chapter 8, the application of Kety's method in studies of CBF and metabolism is reviewed, the studies of cerebral circulation and metabolism during nine different techniques of anaesthesia for craniotomy are presented, and other studies of cerebral circulation during neuroanaesthesia are reviewed. In chapter 9, considerations concerning central and cerebral hemodynamics during anaesthesia in the sitting position are considered.

This review, is primarily addressed to anaesthetists, but it will also be of interest to those working within neurosurgery, neuroradiology and clinical neurophysiology.

Springer-Verlag Wien New York
Moelkerbastei 5, P.O. Box 367, A-1011 Wien
Heidelberger Platz 3, D-1000 Berlin 33
175 Fifth Avenue, New York, NY 10010, USA
37-3, Hongo 3-chome, Bunkyo-ku, Tokyo 113, Japan

Printed in the United States
by Baker & Taylor Publisher Services